Nonlinear Dynamical Control Systems

Henk Nijmeijer Arjan van der Schaft

Nonlinear Dynamical Control Systems

With 32 Illustrations

Springer-Verlag
New York Berlin Heidelberg
London Paris Tokyo Hong Kong

Henk Nijmeijer
Arjan van der Schaft
Department of Applied Mathematics
University of Twente
P.O. Box 217
7500 AE Enschede
The Netherlands

Library of Congress Cataloging-in-Publication Data
Nijmeijer, H. (Henk), 1955-
Nonlinear dynamical control systems / Henk Nijmeijer, Arjan van der Schaft.
p. cm.
ISBN 0-387-97234-X
1. Control theory. 2. Nonlinear theories. 3. Geometry, Differential. I. Schaft. A. J. van der. II. Title.
QA402.3.N55 1990
629.8'312—dc20 89-26360

Printed on acid-free paper

Camera-ready copy supplied by the authors using ChiWriter.
Printed and bound by R.R. Donnelley & Sons, Harrisonburg, Virginia.
Printed in the United States of America.

9 8 7 6 5 4 3 2 1

ISBN 0-387-97234-X Springer-Verlag New York Berlin Heidelberg
ISBN 3-540-97234-X Springer-Verlag Berlin Heidelberg New York

Preface

This textbook on the differential geometric approach to nonlinear control grew out of a set of lecture notes, which were prepared for a course on nonlinear system theory, given by us for the first time during the fall semester of 1988. The audience consisted mostly of graduate students, taking part in the Dutch national Graduate Program on Systems and Control. The aim of this course is to give a general introduction to modern nonlinear control theory (with an emphasis on the differential geometric approach), as well as to provide students specializing in nonlinear control theory with a firm starting point for doing research in this area.

One of our primary objectives was to give a self-contained treatment of all the topics to be included. Since the literature on nonlinear geometric control theory is rapidly expanding this forced us to limit ourselves in the choice of topics. The task of selecting topics was further aggravated by the continual shift in emphasis in the nonlinear control literature over the last years. Therefore, we decided to concentrate on some rather solid and clear-cut achievements of modern nonlinear control, which can be expected to be of remaining interest in the near future. Needless to say, there is also a personal bias in the topics we have finally selected. Furthermore, it was impossible not to be influenced by the trendsetting book “Nonlinear Control Systems: an Introduction”, written by A. Isidori in 1985 (Lecture Notes in Control and Information Sciences, 72, Springer).

A second main goal was to illustrate the theory presented with examples stemming from various fields of application. As a result, Chapter 1 starts with a discussion of some characteristic examples of nonlinear control systems, which will serve as illustration throughout the subsequent chapters, besides several other examples.

Thirdly, we decided to include a rather extensive and self-contained treatment of the necessary mathematical background on differential geometry. Especially the required theory on Lie brackets, (co-) distributions and Frobenius' Theorem is covered in detail. However, some rudimentary knowledge about the fundamentals of differential geometry (manifolds, tangent space, vectorfields) will greatly facilitate the reading of the book. Furthermore, the reader is supposed to be familiar with the basic concepts of linear system theory; especially some acquaintance with linear geometric control theory will be very helpful.

Modern nonlinear control theory, in particular the differential geometric approach, has emerged during the seventies in a rather successful attempt to deal with basic questions in the state space formulation of nonlinear control systems, including the problems of controllability and observability, and (minimal) realization theory. It was also motivated by optimal control theory, in particular the Maximum Principle and its relation with controllability issues. The theory gained strong impetus at the end of the seventies and beginning of the eighties by the introduction of several new concepts, most of them having as their crucial part nonlinear *feedback*. Let us illustrate this with two papers, which can be seen as benchmarks in this development. First, there is the paper by Brockett on "Feedback invariants for nonlinear systems" (Proc. VIIth IFAC World Congress, Helsinki, pp. 1115-1120, 1978), which deals with the basic question to what extent the structure of a nonlinear control system can be changed by (static state) feedback. A direct outgrowth of this paper has been the theory on feedback linearization of nonlinear control systems. Secondly, in the paper "Nonlinear decoupling via feedback: a differential geometric approach" by Isidori, Krener, Gori-Giorgi & Monaco (IEEE Trans. Automat. Control, AC-26, pp. 341-345, 1981) the concept of a controlled invariant distribution is used for various sorts of decoupling problems (independently, a similar approach was taken by Hirschorn ("(A,B)-invariant distributions and disturbance decoupling of nonlinear systems", SIAM J. Contr. Optimiz. 19, pp. 1-19, 1981)). It is worth mentioning that the concept of a controlled invariant distribution is a nonlinear generalization of the concept of a controlled invariant subspace, which is the cornerstone in what is usually called linear geometric control theory (see the trendsetting book of Wonham, "Linear Multivariable Control", Springer, first edition 1974, third edition 1985). In fact, a substantial part of the research on nonlinear control theory in the eighties has been involved with the "translation" to the nonlinear domain of solutions of various feedback synthesis problems obtained in linear geometric control theory. Connected with the concept of (controlled) invariant distributions, the above mentioned IEEE paper also stressed the usefulness of special choices of state space coordinates, in which the system structure becomes more transparant. The search for various kinds of nonlinear normal forms, usually connected to some algorithm such as the nonlinear D^*-algorithm, the Hirschorn algorithm or the dynamic extension algorithm, has been another major trend in the eighties.

At this moment it is difficult to say what will be the prevailing trends in nonlinear control theory in the near future. Without doubt the feedback stabilization problem, which has recently obtained a strong renewed interest, will be a fruitful area. Also adaptive control of nonlinear systems, or, more modestly, the search for adaptive versions of current nonlinear control schemes is likely going to be very important, as well as digital implementation (discretization) of (continuous-time based) control strategies. Moreover, it seems that nonlinear control theory is at a point in its development where more attention should be paid to the special (physical) structure of some classes of nonlinear control systems, notably in connection with classical notions of passivity, stability and symmetry, and notions stemming from bifurcation theory and dynamical systems.

The contents of the book are organized as follows:
Chapter 1 starts with an exposition of four examples of nonlinear control systems, which will be used as illustration for the theory through the rest of the book. A few generalities concerning the definition of nonlinear control systems in state space form are briefly discussed, and some typical phenomena occurring in nonlinear differential (or difference) equations are touched upon, in order to put the study of nonlinear control systems also into the perspective of nonlinear dynamics. **Chapter 2** provides the necessary differential geometric background for the rest of the book. Section 2.1 deals with some fundamentals of differential geometry, while in Section 2.2 vectorfields, Lie brackets, (co-)distributions and Frobenius' Theorem are treated in some detail. For the reader's convenience we have included a quick survey of Section 2.1, as well as a short summary of Section 2.2 containing a list of useful properties and identities. In **Chapter 3** some aspects of controllability and observability are treated with an emphasis on nonlinear rank conditions that generalize the well-known Kalman rank conditions for controllability and observability of linear systems, and on the role of invariant distributions in obtaining local decompositions similar to the linear Kalman decompositions. **Chapter 4** is concerned with various input-output representations of nonlinear control systems, and thus provides a link with a more input-output oriented approach to nonlinear control systems, without actually going into this. Conditions for invariance of an output under a particular input, which will be crucial for the theory of decoupling in later chapters, are derived in the analytic as well as in the smooth case. In **Chapter 5** we discuss some problems concerning the transformation of nonlinear systems into simpler forms, using state-space and feedback transformations, while **Chapter 6** contains the full solution of the local feedback linearization problem

(using static state feedback). In **Chapter 7** the fundamental notion of a controlled invariant distribution is introduced, and applied to the local disturbance decoupling problem. **Chapters 8** and **9** are concerned with the input-output decoupling problem; using an analytic, respectively a geometric approach. In **Chapter 10** some aspects of the local feedback stabilization problem are treated. **Chapter 11** deals with the notion of a controlled invariant submanifold and its applications to stabilization, interconnected systems and inverse systems. In **Chapter 12** a specific class of nonlinear control systems, roughly speaking mechanical control systems, is treated in some detail. Finally, in **Chapters 13** and **14** a part of the theory developed in the preceding chapters is generalized to general continuous-time systems $\dot{x} = f(x,u)$, $y = h(x,u)$, respectively to discrete-time systems.

At the end of every chapter we have added bibliographical notes about the main sources we have used, as well as some (very partial) historical information. Furthermore we have occasionally added some references to related work and further developments. We like to stress that the references are by no means meant to be complete, or are even carefully selected, and we sincerely apologize to those authors whose important contributions were inadvertently not included in the references.

As already mentioned before, many topics of interest could not be included in the present book. Notable omissions are in particular realization theory, conditions for local controllability, observer design, left- and right-invertibility, global issues in decoupling and linearization by feedback, global stabilization, singular perturbation methods and high-gain feedback, sliding mode techniques, differential algebraic methods, and, last but not least, nonlinear optimal control theory. (We also like to refer to the very recent second edition of Isidori's "Nonlinear Control Systems" (Springer, 1989) for a coverage of some additional topics.)

Acknowledgements

The present book forms an account of some of our views on nonlinear control theory, which have been formed in contacts with many people from the nonlinear control community, and we like to thank them all for stimulating conversations and creating an enjoyable atmosphere at various meetings. In particular we like to express our gratitude to Peter Crouch, Jessy Grizzle, Riccardo Marino, Witold Respondek and Hans Schumacher for the very pleasant and fruitful cooperation we have had on some joint research endeavors. We thank the graduate students attending the course on nonlinear system theory of the Graduate Program on Systems and Control in the fall semester of 1988, for serving as an excellent and responsive audience for a first "try-out" for parts of this book. Special thanks go to our Ph.D. students Harry Berghuis, Antonio Campos Ruiz, Henri Huijberts and Leo van der Wegen for their assistance in correcting and proof reading the present manuscript. Of course, the responsibility for all remaining errors and omissions in the book remains ours. We like to thank Dirk Aeyels and Hans Schumacher for very helpful comments on parts of the text. We are very much indebted to our former supervisor Jan C. Willems for the many inspiring discussions we have had throughout the past decade.

Over the years the Systems and Control Group of the Department of Applied Mathematics of the University of Twente has offered us excellent surroundings for our research and teaching activities. It is a pleasure to thank all our colleagues for creating this pleasant working atmosphere. Special thanks go to our secretary Marja Langkamp for her invaluable assistance throughout the years. We are most grateful to Anja Broeksma, Marjo Quekel, Jeane Slag-Vije and Marja Langkamp for their skilful typing of the manuscript. We thank them for remaining cheerful and patient, despite the length of the manuscript. Also we thank Mr. M.W. van der Mey for his contribution in preparing the figures. Finally we thank Eduardo Sontag for his publishing recommendation, and the staff at the Springer Verlag office in New York for the pleasant cooperation during the preparation of this book.

Enschede, October 1989,

Henk Nijmeijer
Arjan van der Schaft

Contents

1 Introduction

This book is concerned with nonlinear control systems described by either (ordinary) differential equations or difference equations with an emphasis on the first class of systems. That is, the systems under consideration are of the following type

$$\begin{cases} \dot{x}(t) = f(x(t),u(t)), \\ y(t) = h(x(t),u(t)), \end{cases} \tag{1.1}$$

or

$$\begin{cases} x(k+1) = f(x(k),u(k)), \\ y(k) = h(x(k),u(k)), \end{cases} \tag{1.2}$$

where x denotes the state of the system, u the control and y the output of the system. Before we will discuss in some depth the general definitions and assumptions on the systems (1.1) or (1.2) we focus on four examples of control systems which fit into (1.1) or (1.2), and which serve as a motivation for considering nonlinear control systems. Particular control questions will not yet be addressed, but are deferred to the later chapters. As one will see, the examples are taken from rather different scientific disciplines such as robotics, aeronautics, economics and biology.

Example 1.1 (Robot-Arm Control) Consider a frictionless, rigid two-link robot manipulator (or double pendulum) with control torques u_1 and u_2 applied at the joints.

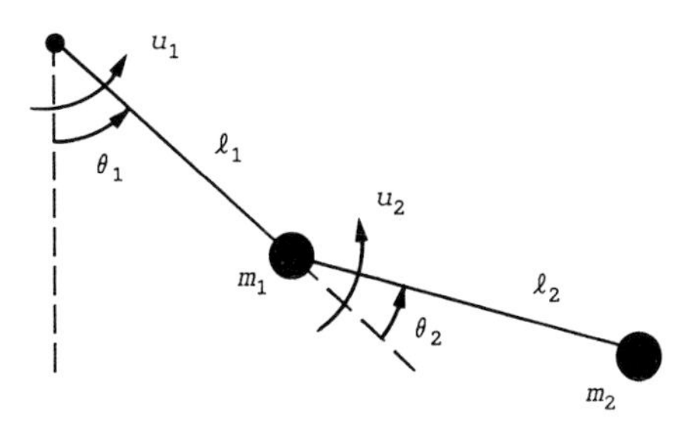

Fig. 1.1. Two-link robot manipulator.

The dynamics of such a robot-arm may be obtained via the *Euler-Lagrange* formalism. Let $\theta = (\theta_1, \theta_2)$ and $\dot{\theta} = (\dot{\theta}_1, \dot{\theta}_2)$ and define the *Lagrangian function*

$$L(\theta, \dot{\theta}) = T(\theta, \dot{\theta}) - V(\theta) , \tag{1.3}$$

where $T(\theta, \dot{\theta})$ is the *kinetic energy* and $V(\theta)$ the *potential energy*. For the above configuration with rigid massless links the kinetic energy is computed as the sum of the kinetic energies T_1 and T_2 of the masses m_1, respectively m_2. This yields

$$T_1(\dot{\theta}) = \tfrac{1}{2} m_1 \ell_1^2 \dot{\theta}_1^2,$$

$$T_2(\theta, \dot{\theta}) = \tfrac{1}{2} m_2 \left(\ell_1^2 \dot{\theta}_1^2 + \ell_2^2 (\dot{\theta}_1 + \dot{\theta}_2)^2 + 2\ell_1 \ell_2 (\cos \theta_2) \dot{\theta}_1 (\dot{\theta}_1 + \dot{\theta}_2) \right),$$

and similarly the potential energy V is the sum of the potential energies V_1 and V_2 of the two masses;

$$V_1(\theta) = - m_1 g \ell_1 \cos \theta_1 ,$$

$$V_2(\theta) = - m_2 g \ell_1 \cos \theta_1 - m_2 g \ell_2 \cos (\theta_1 + \theta_2).$$

Therefore,

$$\begin{aligned} L(\theta, \dot{\theta}) = {} & \tfrac{1}{2} m_1 \ell_1^2 \dot{\theta}_1^2 + \tfrac{1}{2} m_2 \left(\ell_1^2 \dot{\theta}_1^2 + \ell_2^2 (\dot{\theta}_1 + \dot{\theta}_2)^2 + 2\ell_1 \ell_2 (\cos \theta_2) \dot{\theta}_1 (\dot{\theta}_1 + \dot{\theta}_2) \right) \\ & + m_1 g \ell_1 \cos \theta_1 + m_2 g \ell_1 \cos \theta_1 + m_2 g \ell_2 \cos (\theta_1 + \theta_2). \end{aligned} \tag{1.4}$$

Now the celebrated *Euler-Lagrange* equations are

$$\frac{d}{dt} \left(\frac{\partial L}{\partial \dot{\theta}_i} \right) - \frac{\partial L}{\partial \theta_i} = u_i , \qquad i = 1,2, \tag{1.5}$$

which yields in this case the vector equation

$$M(\theta)\ddot{\theta} + C(\theta, \dot{\theta}) + k(\theta) = u \tag{1.6}$$

where $\ddot{\theta} = (\ddot{\theta}_1, \ddot{\theta}_2)$, $u = (u_1, u_2)$, and

$$M(\theta) = \begin{pmatrix} m_1 \ell_1^2 + m_2 \ell_1^2 + m_2 \ell_2^2 + 2m_2 \ell_1 \ell_2 \cos \theta_2 & m_2 \ell_2^2 + m_2 \ell_1 \ell_2 \cos \theta_2 \\ m_2 \ell_2^2 + m_2 \ell_1 \ell_2 \cos \theta_2 & m_2 \ell_2^2 \end{pmatrix}, \tag{1.7.a}$$

$$C(\theta, \dot{\theta}) = \begin{pmatrix} -m_2 \ell_1 \ell_2 (\sin \theta_2) \dot{\theta}_2 (2\dot{\theta}_1 + \dot{\theta}_2) \\ m_2 \ell_1 \ell_2 (\sin \theta_2) \dot{\theta}_1^2 \end{pmatrix}, \tag{1.7.b}$$

$$k(\theta) = - \begin{pmatrix} m_1 g \ell_1 \sin \theta_1 + m_2 g \ell_1 \sin \theta_1 + m_2 g \ell_2 \sin (\theta_1 + \theta_2) \\ m_2 g \ell_2 \sin (\theta_1 + \theta_2) \end{pmatrix}. \tag{1.7.c}$$

In (1.6) the term $k(\theta)$ represents the *gravitational force* and the term $C(\theta,\dot{\theta})$ reflects the *centripetal* and *Coriolis forces*. Note that the matrix $M(\theta)$ in (1.6) has as determinant $m_1 m_2 \ell_1^2 \ell_2^2 + m_2^2 \ell_1^2 \ell_2^2 - m_2^2 \ell_1^2 \ell_2^2 \cos^2\theta_2$, which is positive for all θ. Therefore (1.6) is equivalent to the vector equation

$$\ddot{\theta} = - M(\theta)^{-1} C(\theta,\dot{\theta}) - M(\theta)^{-1} k(\theta) + M(\theta)^{-1} u. \tag{1.8}$$

Equation (1.8) describes the dynamical behavior of a two-link robot manipulator. It clearly constitutes a nonlinear control system with as state space $(\theta_1,\dot{\theta}_1,\theta_2,\dot{\theta}_2) \in S^1 \times \mathbb{R} \times S^1 \times \mathbb{R} \cong TS^1 \times TS^1$. Often the purpose of controlling a robot arm is that of using the end effector for doing some prescribed task. Though we did not incorporate the robot hand — which is more difficult to model — in the model, it is clear that the interesting outputs of the model would be the Cartesian coordinates of the end point rather than the angles θ_1 and θ_2 between the separate links.

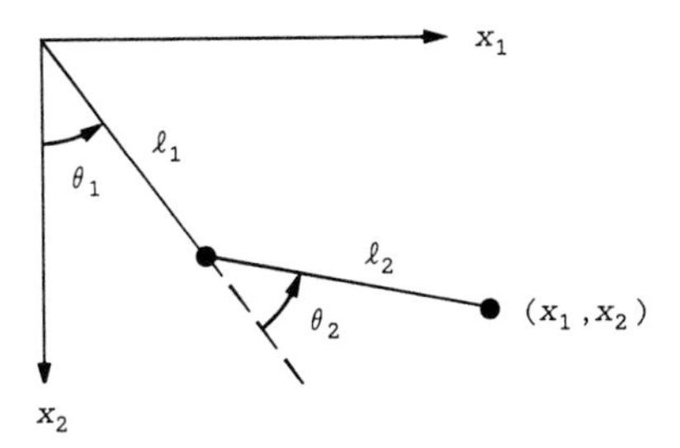

Fig. 1.2. End point of two-link robot arm.

Denoting the Cartesian coordinates of the endpoint as y_1 and y_2 we obtain the output functions

$$\begin{cases} y_1 = \ell_1 \sin \theta_1 + \ell_2 \sin (\theta_1 + \theta_2), \\ y_2 = \ell_1 \cos \theta_1 + \ell_2 \cos (\theta_1 + \theta_2). \end{cases} \tag{1.9}$$

This is what is called the *direct kinematics* for the robot arm. Of course, in practice the more important question is how to determine the angles θ_1 and θ_2 when the end position (y_1,y_2) is given (possibly as a function of time). This is the so called *inverse kinematics problem* for the robot arm.

Computing the Jacobian of the right-hand side of (1.9) we obtain

$$J(\theta_1,\theta_2) = \begin{bmatrix} \ell_1 \cos\theta_1 + \ell_2 \cos(\theta_1+\theta_2) & \ell_2 \cos(\theta_1+\theta_2) \\ -\ell_1 \sin\theta_1 - \ell_2 \sin(\theta_1+\theta_2) & -\ell_2 \sin(\theta_1+\theta_2) \end{bmatrix}, \tag{1.10}$$

and thus

$$\begin{aligned} \det J(\theta_1,\theta_2) &= \ell_1\ell_2 \sin\theta_1 \cos(\theta_1+\theta_2) - \ell_1\ell_2 \cos\theta_1 \sin(\theta_1+\theta_2) \\ &= -\ell_1\ell_2 \sin\theta_2 . \end{aligned} \tag{1.11}$$

Hence for any point (θ_1,θ_2) with $\theta_2 \neq k\pi$, $k \in \mathbb{Z}$, we see that rank $J(\theta_1,\theta_2) = 2$ and so we may apply at these points the inverse function theorem, yielding θ_1 and θ_2 as a nonlinear function of (y_1,y_2).

We conclude this discussion on robot arm control with the remark that the approach given here may be extended to various more complicated configurations. For example one can equally well handle an m-link robot manipulator with control torques applied at each joint by using the Euler-Lagrange formalism. Of course the analysis in obtaining the dynamical equations as well as in the direct and inverse kinematics becomes much more involved. The study of this kind of nonlinear control systems needs further investigation. □

Example 1.2 (Spacecraft Attitude Control) In this example we study the dynamics describing the spacecraft attitude with gas jet or momentum exchange actuators. The equations describing the attitude control of a spacecraft are basically those of a rotating rigid body with extra terms giving the effect of the control torques. Therefore one may separate the equations into *kinematic* equations relating the angular position with the angular velocity and *dynamic* equations describing the evolution of angular velocity (or, equivalently, angular momentum). The kinematic equations can be represented as follows. The angular position is described by a rotation matrix R. R transforms an inertially fixed set of orthonormal axes, say e_1, e_2, e_3 into a set of orthonormal axes r_1, r_2, r_3 (with the same orientation as e_1, e_2, e_3), which are fixed in the spacecraft and have as origin the center of mass of the spacecraft, thus

$$R\, e_i = r_i \quad \text{for } i = 1,2,3.$$

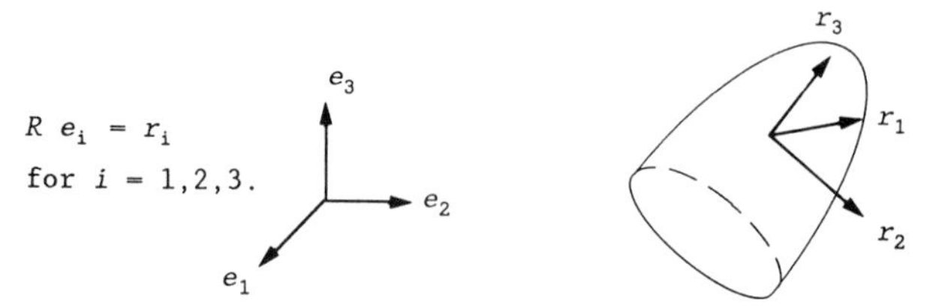

Fig. 1.3. Angular position.

The evolution of R may now be expressed as

$$\dot{R}(t) = - R(t)\ S(\omega(t)) \tag{1.12}$$

were $\omega(t)$ is the angular velocity of the spacecraft at time t (with respect to the axes in the spacecraft) and $S(\omega)$ is a 3×3-matrix defined by

$$S(\omega) = \begin{bmatrix} 0 & \omega_3 & -\omega_2 \\ -\omega_3 & 0 & \omega_1 \\ \omega_2 & -\omega_1 & 0 \end{bmatrix},$$

with $\omega = (\omega_1, \omega_2, \omega_3)$. An alternative (local) description of (1.12) is obtained as follows. The angular position may be described locally by three angles ϕ, θ, ψ, which represent consecutive clockwise rotations about the axes r_1, r_2 and r_3, respectively. Setting r_i to be the standard i-th basis vector in $\mathbb{R}^3$ we obtain the kinematic equations as follows.

$$\begin{bmatrix} \omega_1 \\ \omega_2 \\ \omega_3 \end{bmatrix} = \begin{bmatrix} \dot{\varphi} \\ 0 \\ 0 \end{bmatrix} + \begin{bmatrix} 1 & 0 & 0 \\ 0 & \cos\phi & \sin\phi \\ 0 & -\sin\phi & \cos\phi \end{bmatrix} \begin{bmatrix} 0 \\ \dot{\theta} \\ 0 \end{bmatrix} +$$

$$\begin{bmatrix} 1 & 0 & 0 \\ 0 & \cos\phi & \sin\phi \\ 0 & -\sin\phi & \cos\phi \end{bmatrix} \begin{bmatrix} \cos\theta & 0 & -\sin\theta \\ 0 & 1 & 0 \\ \sin\theta & 0 & \cos\theta \end{bmatrix} \begin{bmatrix} 0 \\ 0 \\ \dot{\psi} \end{bmatrix}$$

Therefore,

$$\begin{bmatrix} \dot{\phi} \\ \dot{\theta} \\ \dot{\psi} \end{bmatrix} = \begin{bmatrix} 1 & \sin\phi \cdot \tan\theta & \cos\phi \cdot \tan\theta \\ 0 & \cos\phi & -\sin\phi \\ 0 & \sin\phi(\cos\theta)^{-1} & \cos\phi(\cos\theta)^{-1} \end{bmatrix} \begin{bmatrix} \omega_1 \\ \omega_2 \\ \omega_3 \end{bmatrix} \tag{1.13}$$

Clearly, this description is only locally valid in the region $-\pi/2 < \theta < \pi/2$, but it serves to show that the equations (1.12) evolve on a three dimensional space (which in fact is the Lie group SO(3) of 3×3 real orthogonal matrices with determinant 1). The dynamic equations obviously depend on how the spacecraft is controlled. We consider two typical situations.

I. Gas Jet Actuators

Let J be the inertia matrix of the spacecraft, h the angular momentum of the spacecraft with respect to the inertial axes e_1, e_2, e_3, and $b_1, b_2, \ldots, b_m$ the axes about which the corresponding control torque of magnitude $\|b_i\| u_i$ is applied by means of opposing pairs of gas jets. Here $\|\cdot\|$ denotes the standard Euclidean norm on $\mathbb{R}^3$. Using a momentum balance about the center of mass one obtains the dynamic equations for the controlled spacecraft as

$$\begin{cases} \dot{R} = -RS(\omega), \\ J\dot{\omega} = S(\omega)J\omega + \sum_{i=1}^{m} b_i u_i . \end{cases} \tag{1.14}$$

II. Momentum Wheel Actuators

We assume that we have m wheels with the i-th wheel spinning about an axis b_i, which is fixed in the spacecraft, such that the center of mass of the i-th wheel lies on the axis b_i and a torque $-\|b_i\|u_i$ is applied to the i-th wheel about the axis b_i by a motor fixed in the spacecraft. Consequently an equal and opposite torque $\|b_i\|u_i$ is exerted by the wheel on the spacecraft. Then, a more complicated momentum balance yields

$$\sum_{i=1}^{m} \bar{J}_i(\omega+\nu_i) + J^*\omega = Rh, \quad \dot{h} = 0, \tag{1.15}$$

$$\bar{J}_i(\omega+\nu_i) = Rh_i, \quad b_i^T R\dot{h}_i = -\|b_i\|^2 u_i, \tag{1.16}$$

where J^* is the inertia matrix of the spacecraft without wheels, $\bar{J}_i$ is the inertia matrix of the i-th wheel, h is the total constant momentum of the system, h_i is the angular momentum of the i-th wheel both measured with respect to the inertial frame e_1, e_2, e_3, and ν_i is the angular velocity of the i-th wheel relative to the axes r_1, r_2, r_3. Assume that b_i is a principal axis for wheel i and assume the i-th wheel is symmetric about b_i. Then $\bar{J}_i = J_i + (\bar{J}_i - J_i)$, where $J_i = b_i b_i^T \bar{j}_i / \|b_i\|^2$ and $\bar{j}_i$ is the inertia-moment of the i-th wheel about the axis b_i. Clearly $\bar{J}_i - J_i$ is a positive semi-definite matrix so we may define a positive definite matrix J via $J = J^* + \sum_{i=1}^{m} (\bar{J}_i - J_i)$. Let $v = \sum_{i=1}^{m} J_i(\omega+\nu_i)$, then (1.15) reduces to,

$$J\omega + v = Rh, \tag{1.17}$$

and from (1.16) we obtain

$$\dot{v} = -\sum_{i=1}^{m} b_i u_i . \tag{1.18}$$

Differentiating (1.17) and substitution of (1.12) and (1.18) yield the following closed set of equations describing the control system

$$\begin{cases} \dot{R} = -RS(\omega), \\ J\dot{\omega} = -RS(\omega)h + \sum_{i=1}^{m} b_i u_i , \\ \dot{h} = 0. \end{cases} \tag{1.19}$$

Both spacecraft attitude control models (1.14), respectively (1.19), show that the dynamics are typically nonlinear for two reasons, namely the state space of (1.14) resp. (1.19) equals the Cartesian product $SO(3)\times \mathbb{R}^3$ — here $SO(3)$ denotes the Lie group of 3×3 real orthogonal matrices with determinant 1 — and in both models nonlinear terms $\omega_i r_{jk}$ appear (where $R = (r_{jk})$ with $j,k = 1, 2, 3$). Both phenomena are essential in a further analysis of the controlled spacecraft.

Next consider again the model with gas jet actuators. It is easily seen that $(R,\omega,u) = (I_3,0,0)$ forms an equilibrium for the system (1.14). Linearizing the dynamics (1.14) around $(I_3,0,0)$ yields

$$\begin{aligned} \dot{\bar{R}} &= 0 , \\ J\dot{\bar{\omega}} &= \sum_{i=1}^{m} b_i \bar{u}_i \ . \end{aligned} \tag{1.20}$$

Obviously, this linearized model (1.20) does not reveal any of the essential features of the original model (1.14), like for instance stability, controllability, etc. This shows, that for a better understanding of the controlled spacecraft, one has to develop a nonlinear analysis rather than just studying the linearization of such a model. □

Example 1.3 (Control of a Closed Economy) The following equations describe the evolution of a closed economy in discrete time.

$$Y(k+1) = Y(k) + \alpha\big(C(Y(k))+I(Y(k),\ R(k),\ K(k))+P(k)^{-1}G(k)-Y(k)\big) \tag{1.21}$$

$$R(k+1) = R(k) + \beta\big(L(Y(k),\ R(k))-P(k)^{-1}M(k)\big) \tag{1.22}$$

$$K(k+1) = K(k) + I\big(Y(k),\ R(k),\ K(k)\big) \tag{1.23}$$

$$Y(k) = F(N(k),\ K(k)) \tag{1.24}$$

$$N(k) = H(W(k),\ P(k)) \tag{1.25}$$

In this model the quantities have the following interpretation:

Y : real output

C : real private consumption

I : real private net investment

R : nominal interest rate

K : real capital stock

P : price level

G : nominal government spending

L : real money demand

M : nominal money stock

N : labour demand

W : nominal wage rate

α and β are positive constants.

Equation (1.21) is a dynamic *IS* (Investments-Savings) equation and (1.22) is a dynamic *LM* (Loan-Money) equation. The capital accumulation is described via the dynamic Keynesian equation (1.23). Equation (1.24) represents a macro-economic production function and (1.25) defines the labour demand as a function of the real wage rate.

The equations (1.21-25) typically describe a dynamic economic system. To bring it into the form of a control system we have to distinguish control variables and to-be-controlled variables ("outputs"). One way to do so is as follows. Interprete G and M as the "controls" of the system (which in an economic context are labeled as *instruments* or *instrument variables*), W as a known *exogenous variable* (so a prescribed known control function) and the real output Y and the price level P as the *target variables* (the to-be-controlled variables). To bring the model (1.21-25) into a state space form, one rewrites the equations (1.24) and (1.25). Suppose $(\bar{Y},\bar{R},\bar{K},\bar{W},\bar{G},\bar{M},\bar{N})$ is a particular steady-state solution of (1.21-25). Then the relation

$$N = H(W,P) \tag{1.26}$$

holds at the steady state $(\bar{N},\bar{W},\bar{P})$ and provided

$$\frac{\partial H}{\partial P}(\bar{W},\bar{P}) \neq 0 \ , \tag{1.27}$$

we may locally apply the Implicit Function Theorem yielding locally P as a function of N and W, say

$$P = \tilde{H}(W,N), \tag{1.28}$$

which satisfies $\bar{P} = \tilde{H}(\bar{W},\bar{N})$. Similarly, the relation

$$Y = F(N,K) \ , \tag{1.29}$$

which holds at $(\bar{Y},\bar{N},\bar{K})$ may locally be transformed into

$$N = \tilde{F}(Y,K), \tag{1.30}$$

with $\bar{N} = \tilde{F}(\bar{Y},\bar{K})$, provided that

$$\frac{\partial F}{\partial N}(\bar{N},\bar{K}) \neq 0. \tag{1.31}$$

Assuming that (1.27) and (1.31) hold, we find for the second target variable

$$P(k) = \tilde{H}(W(k),N(k)) = \tilde{H}(W(k),\tilde{F}(Y(k),K(k))). \tag{1.32}$$

Altogether we have obtained — locally — a model of the following form

$$\begin{cases} Y(k+1) = f_1(Y(k),R(k),K(k),W(k),G(k)), \\ R(k+1) = f_2(Y(k),R(k),K(k),W(k),M(k)), \\ K(k+1) = f_3(Y(k),R(k),K(k)) \end{cases} \tag{1.33}$$

$$\begin{cases} Q_1(k) = Y(k), \\ Q_2(k) = P(k) = \tilde{H}(W(k),\tilde{F}(Y(k),K(k))), \end{cases} \tag{1.34}$$

where Q_1 and Q_2 denote the target variables and the functions f_1, f_2 and f_3 follow from (1.21-23) and (1.32). Therefore the model of the closed economy as described here is a set of difference equations on the state space (Y,R,K) together with output equations given by (1.34). Note that in this example the state space may not be $\mathbb{R}^3$ but rather some nontrivial region in $\mathbb{R}^3$. As is clear from the definition of the functions f_1, f_2 and f_3, (see (1.21-23)) the dynamics (1.33) are typically nonlinear, which can not be avoided even by assuming a simple structure on the functions C, L and I. Although almost always in the economic literature, when dealing with a model of this type, one directly starts with the linearized version of the model described by (1.33) and (1.34), it seems that a further analysis incorporating the nonlinearities is necessary for a closer study. □

Example 1.4 (A Model of a Mixed-Culture Bioreactor) Let us study a model of the dynamics of a culture of two cellstrains that are differentiated by their sensitivity to an external growth-inhibiting agent. The model is based on a description of the unstable fermentations that occur with micro-organisms altered by recombinant-DNA techniques. In such a mixed-culture bioreactor we distinguish two celltypes, namely the *inhibitor resistant cells* and the *inhibitor sensitive cells*. Their cell-densities will be denoted as x_1, respectively x_2. In addition, let S and I represent the concentration of rate-limiting substrate and inhibitor in the fermentation medium. The interactions of the two cell populations are illustrated in the following diagram.

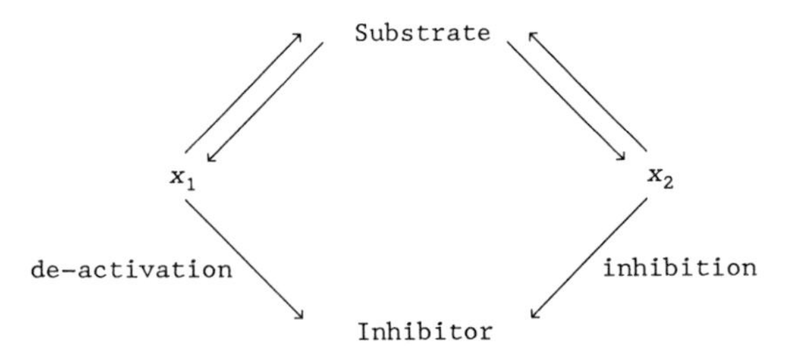

Fig. 1.4. Diagram of two cell populations.

We consider a continuous mixed-culture chemostat of fixed volume with a constant inlet substrate concentration S_f. There are two control parameters in the model, namely the dilution rate D and the inlet concentration of the inhibitor I_f. After a certain residence time the model takes the following form (using material balances of the chemostat)

$$\frac{d}{dt}\begin{pmatrix} x_1 \\ x_2 \\ I \end{pmatrix} = \begin{pmatrix} \mu_1(S)x_1 \\ \mu_2(S,I)x_2 \\ -px_1 I \end{pmatrix} + \begin{pmatrix} -x_1 \\ -x_2 \\ -I \end{pmatrix} u_1 + \begin{pmatrix} 0 \\ 0 \\ 1 \end{pmatrix} u_2 , \tag{1.35}$$

where

$\mu_1(S) = \frac{\mu^1 S}{K+S}$, the growth rate of species 1 ,

$\mu_2(S,I) = \frac{\mu^2 S}{K+S} \cdot \frac{K_I}{K_I+I}$, the growth rate of species 2 ,

μ^1, μ^2, K, K_I are specific constants describing the growth rates and p a constant reflecting the rate proportional to $x_1 I$ with which inhibitor-resistant species deactivate the inhibitor.

$u_1 = D$, the dilution rate ,

$u_2 = DI_f$, the total inhibitor addition rate ,

$S = S_f - x_1/Y_1 - x_2/Y_2$, the substrate rate ,

I_f : the inlet inhibitor concentration ,

S_f : the (constant) inlet substrate concentration ,

Y_1 : the yield of species 1 ,

Y_2 : the yield of species 2 .

For the above model (1.35) one can work out a complete steady-state analysis (which is beyond the scope of this book). A few interesting things about the model can be immediately stated. It seems reasonable to impose the condition that (1.35) has an equilibrium point (x_1^0, x_2^0, I^0) in the positive orthant $x_1 > 0$, $x_2 > 0$, $I > 0$. This implies some additional constraints on the parameters in (1.35). In particular it follows from the existence of such an equilibrium point that the right hand side of (1.35) vanishes for suitably selected controls u_1^0, u_2^0. Therefore it follows that in (x_1^0, x_2^0, I^0) one has $\mu_1(S) = \mu_2(S,I)$ and so it is necessary that $\mu^2 \geq \mu^1$, and I^0 is determined as

$$I^0 = \left(\frac{\mu^2}{\mu^1} - 1\right) K_I > 0 . \tag{1.36}$$

It is reasonable to assume that the inhibitor feed concentration I_f exceeds this value I^0, so

$$I_f = \frac{u_2}{u_1} \geq I^0,$$

or equivalently

$$u_2 - u_1 K_I \left(\frac{\mu^2}{\mu^1} - 1\right) \geq 0, \tag{1.37}$$

which puts an extra constraint on the inputs of the system (1.35). Often one imposes an additional constraint on the inputs u_1 and u_2 in order to prevent that the species 2 will wash out, but it is not necessary to assume that in a first analysis of the model (1.35). Altogether we conclude that the model description of the mixed-culture bioreactor leads to a complex nonlinear model with state space $\mathbb{R}^+ \times \mathbb{R}^+ \times \mathbb{R}^+$ and controls u_1 and u_2 satisfying the constraint equation (1.37). □

The above examples clearly exhibit the structure of a nonlinear control system, which in continuous time is of the form (1.1) or in discrete-time of the form (1.2). Clearly, control systems as described by either (1.1) or (1.2) are much more general than their standard linear counterparts, i.e. in continous time

$$\begin{aligned} \dot{x} &= Ax + Bu \\ y &= Cx + Du \end{aligned} \tag{1.38}$$

or, in discrete time

$$\begin{aligned} x(k+1) &= Ax(k) + Bu(k) \\ y(k) &= Cx(k) + Du(k) \end{aligned} \tag{1.39}$$

where the matrices A, B, C and D are properly dimensioned. A large part of the control literature is devoted to such linear systems and many structural properties and problems have been satisfactorily dealt with in the literature. Our emphasis will be on the study of similar aspects for the nonlinear systems (1.1) respectively (1.2). We next discuss some basic assumptions for (especially continuous time) nonlinear systems. A continuous time nonlinear control system is usually given by equations

$$\begin{aligned} \dot{x}(t) &= f(x(t),u(t)), \\ y(t) &= h(x(t),u(t)), \end{aligned} \tag{1.1}$$

where $x \in \mathbb{R}^n$, $u \in U \subset \mathbb{R}^m$ and $y \in \mathbb{R}^p$ denote respectively the *state*, the *input* (*control*) and the *output* of the system. The "system map" $f : \mathbb{R}^n \times \mathbb{R}^m \to \mathbb{R}^n$ is assumed to be a smooth mapping. In this context *smooth* means C^∞, though many results which will be given in the next chapters hold under weaker conditions (in many circumstances f only needs to be sufficiently many times continuously differentiable with respect to x and u). Sometimes it will be useful to strengthen the smoothness condition and to require that f is (real) *analytic*. Similarly we assume the output map $h : \mathbb{R}^n \times \mathbb{R}^m \to \mathbb{R}^p$ to be smooth or analytic. So (1.1) is a shorthand notation for

$$\left\{ \begin{aligned} \dot{x}_1(t) &= f_1(x_1(t),\ldots,x_n(t),u_1(t),\ldots,u_m(t)), \\ &\vdots \\ \dot{x}_n(t) &= f_n(x_1(t),\ldots,x_n(t),u_1(t),\ldots,u_m(t)), \end{aligned} \right. \tag{1.40a}$$

$$\left\{ \begin{aligned} y_1(t) &= h_1(x_1(t),\ldots,x_n(t),u_1(t),\ldots,u_m(t)), \\ &\vdots \\ y_p(t) &= h_p(x_1(t),\ldots,x_n(t),u_1(t),\ldots,u_m(t)), \end{aligned} \right. \tag{1.40b}$$

Together with (1.40a,b) we have to specify a class of *admissible controls* $\mathcal{U}$ for the system. Of course the input functions we consider are functions $u : \mathbb{R}^+ = [0,\infty) \to U$. Here $\mathbb{R}^+$ (or $\mathbb{R}$) denotes the time axis. A main requirement for u is that $\mathcal{U}$ is closed under *concatenation*, i.e. when $u_1(\cdot)$ and $u_2(\cdot)$ both belong to $\mathcal{U}$ then for any $\bar{t}$ also $\bar{u}(\cdot) \in \mathcal{U}$, where $\bar{u}(\cdot)$ is defined as

$$\bar{u}(t) = \left\{ \begin{array}{ll} u_1(t), & t < \bar{t}, \\ u_2(t), & t \geq \bar{t}. \end{array} \right. \tag{1.41}$$

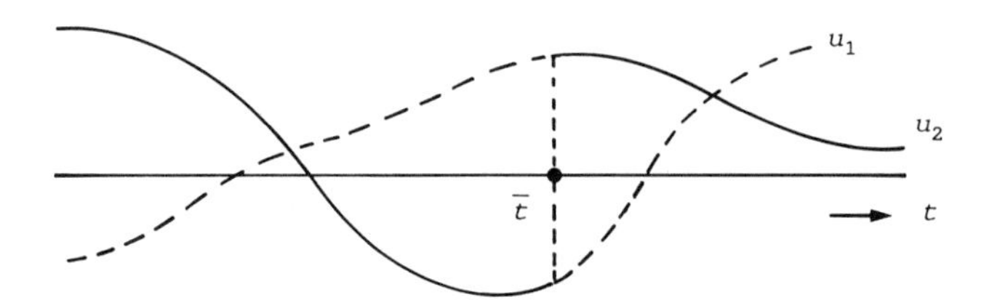

Fig. 1.5. Concatenation of $u_1(\cdot)$ and $u_2(\cdot)$.

One possible and in many cases acceptable choice for $\mathcal{U}$ is the set of piecewise continuous from the right functions on $\mathbb{R}^m$, which is obviously closed under concatenation. *Throughout* we will assume that $\mathcal{U}$ at least *contains* this set of piecewise continuous from the right functions.

Next we have to make sure that solutions for (1.40) exist, at least locally. That is, consider for a given admissible control $\bar{u}(\cdot) \in \mathcal{U}$ and an arbitrary initial state $x_0 \in \mathbb{R}^n$, the differential equation

$$\begin{aligned} \dot{x}(t) &= f(x(t),\bar{u}(t)), \\ x(t_0) &= x_0, \end{aligned} \tag{1.42}$$

If $\bar{u}(\cdot)$ is a piecewise constant input function then for t sufficiently small, there exists a unique solution $x(t)$ of (1.42). To guarantee that unique solutions exist for more general inputs (for instance piecewise continuous controls) we impose what is called a *local Lipschitz condition* on f. That is, there is a neighborhood N of x_0 in $\mathbb{R}^n$ such that for each input $\bar{u}(\cdot) \in \mathcal{U}$ we have

$$\|f(x,\bar{u}(t))-f(z,\bar{u}(t))\| \le K \|x-z\|, \tag{1.43}$$

for all x, $z \in N$ and all $t \in (t_0-\epsilon, t_0+\epsilon)$, where $K > 0$ and $\epsilon > 0$ are constants and $\|.\|$ denotes the usual Euclidean norm. The unique local solution of (1.42), will be denoted as $x(t,t_0,x_0,\bar{u})$. In the same manner the corresponding output function given by (1.40b) will be written as $y(t,t_0,x_0,\bar{u})$. Note that once $x(t,t_0,x_0,\bar{u})$ is determined $y(t,t_0,x_0,\bar{u})$ follows directly from (1.40b).

The above conditions only guarantee the existence of $x(t,t_0,x_0,\bar{u})$ for $|t-t_0|$ sufficiently small. For the linear system (1.38) each piecewise constant input function $\bar{u}(\cdot)$ yields a globally defined solution $x(t,t_0,x_0,\bar{u})$ and thus the piecewise constant inputs form a well defined class of admissible controls. We will not enter here the difficult problem under which extra conditions the solutions of the nonlinear equation (1.42) are defined for all t. Even when (1.42) has global solutions for all constant input functions $\bar{u}$, it may happen that no global solution exists when allowing for piecewise constant controls. This is illustrated in the following example.

Example 1.5 Consider on $\mathbb{R}^2$ the system

$$\begin{aligned} \dot{x}_1 &= (1+x_2^2)u, \\ \dot{x}_2 &= (1+x_1^2)(1-u) \end{aligned} \tag{1.44}$$

Take $(x_1(0),\ x_2(0)) = (0,0)$. For constant inputs solutions of (1.44) are defined for all t. Now we construct a piecewise constant control $\bar{u}(\cdot)$ for which the solution of (1.42) blows up in finite time. Let $b_{-1} = 0$, $a_0 = 1$ and

$$b_n = a_n + \frac{1}{1+n^2},\quad a_{n+1} = b_n + \frac{1}{1+(1+n)^2}\ .$$

Let $\lim_{n\to\infty} a_n = T < \infty$ and define $\bar{u}(\cdot)$ on the interval $[0,T]$ by

$$\bar{u}(t) = \begin{cases} 0\ , & a_n \le t < b_n\ , \\ 1\ , & b_n \le t < a_{n+1}\ . \end{cases}$$

Then the solution $x(t,0,0,\bar{u})$ is well defined for all $t \in [0,T)$ but $x(T,0,0,\bar{u})$ does not exist. □

Solutions of (1.42) which are defined for all t are called *complete*. From the above example we may conclude that further restrictions on the admissible controls have to be imposed in order to guarantee completeness or one has to be content at first instance with local small time solutions of (1.42).

Another interesting phenomenon is that the setting as presented so far does not directly cover the Examples 1.1, 1.2 and 1.4. The essential observation is that the state space and/or the input space and output space in these examples are not necessarily Euclidean spaces but rather *manifolds* (see Chapter 2). For instance the state space of Example 1.1 consists of $(\theta_1,\dot{\theta}_1,\theta_2,\dot{\theta}_2)$, with θ_1 and θ_2 the angles defined in figure 1.1 and $\dot{\theta}_1$ and $\dot{\theta}_2$ the corresponding angular velocities. Clearly θ_1 and θ_2 belong to $(-\pi,\pi]$ rather than $\mathbb{R}$, and a point $\theta + k\cdot 2\pi$, $k \in \mathbb{Z}$, will be identified with θ. However in understanding the solutions of differential equations on such a manifold no difficulties arise because one can equally well consider the controlled differential equation in (1.1) on an open neighborhood of $\mathbb{R}^n$ and thus interprete the solutions of such differential equations as a solution defined on a neighborhood in $\mathbb{R}^n$. This is in fact the process of using *coordinate charts* for a manifold, as will be extensively dealt with in Chapter 2. When a solution of the differential equation tends to leave the neighborhood under consideration, another neighborhood may be taken on which again (1.42) is considered. A very simple example may illustrate this.

Example 1.6 (A system on S^1) Consider the 1-dimensional sphere $S^1 = \{(x_1,x_2)\,|\,x_1^2+x_2^2 = 1\}$ with unit tangent vector at a point $(x_1,x_2) \in S^1$

equal to $(x_2, -x_1)^T$

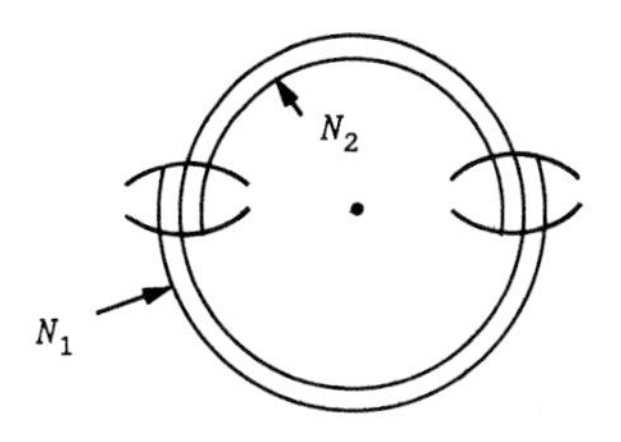

Fig. 1.6. The sphere S^1.

Consider on S^1 the control system

$$\frac{d}{dt}\begin{pmatrix} x_1 \\ x_2 \end{pmatrix} = \begin{pmatrix} x_2 \\ -x_1 \end{pmatrix} u. \tag{1.45}$$

Because S^1 is a 1-dimensional manifold this control system can also be described in a local fashion as in (1.40). As neighborhoods we take N_1 and N_2, see figure 1.6. and the control system reads as

$$\dot{\theta} = u, \tag{1.46}$$

with the constraint that $\theta(t)$ belongs to N_1 or N_2. When a solution leaves N_1 one continues to consider the differential equation on N_2 and so forth. □

There is a particular class of continuous time nonlinear systems we will often consider in this book. That are the *input-linear* or *affine systems* which are described as follows

$$\dot{x}(t) = f(x(t)) + \sum_{i=1}^{m} g_i(x(t))u_i(t), \tag{1.47}$$

together with some output equation only depending on the state. In (1.47) we assume f, $g_1, \ldots, g_m$ to be smooth mappings from $\mathbb{R}^n$ into $\mathbb{R}^n$. The distinctive feature of these systems is that the control $u = (u_1, \ldots, u_m)$ appears *linearly* (or better, affine) in the differential equation (1.47). This type of control system is often encountered in applications, see for instance the examples at the beginning of this chapter.

We remark that everything which has been stated so far for *time-invariant* systems of the form (1.1) in principle can directly be extended to *time-varying* nonlinear systems

$$\begin{aligned}\dot{x}(t) &= f(x(t),u(t),t),\\ y(t) &= h(x(t),u(t),t).\end{aligned} \tag{1.48}$$

The trick is to extend the state space of (1.48) with the time-variable t, namely to (1.48) we add the equation

$$\dot{t} = 1. \tag{1.49}$$

Then (1.48) together with (1.49) forms a system of the form (1.1).

Let us end the discussion of defining continuous time nonlinear systems with some comments. Considering the controlled differential equation in (1.1) we basically deal with a system described by the following commutative diagram

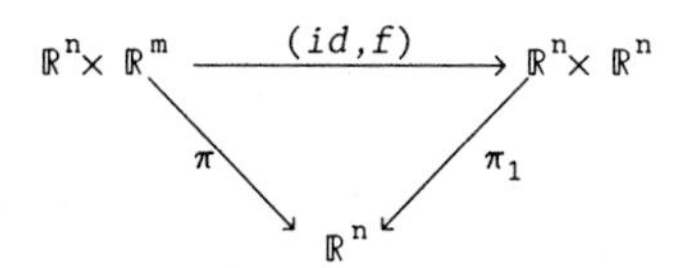

Fig. 1.7. The control system $\dot{x} = f(x,u)$ on $\mathbb{R}^n$.

where $(id,f)(x,u) = (x,f(x,u))$, $\pi(x,u) = x$ and $\pi_1(x,z) = x$. Mathematically this can be seen as the *local* description of a control system on a manifold, while a *global* description is as follows:

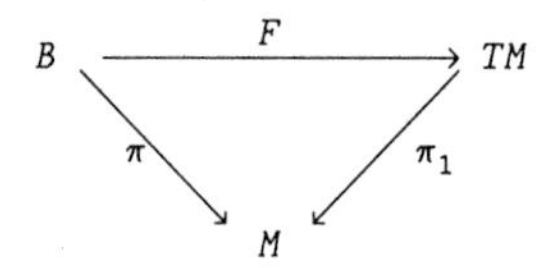

Fig. 1.8. The control system $\dot{x} = f(x,u)$ on M.

where M denotes the state space manifold, $\pi : B \to M$ a *fiber bundle* whose fibers $\pi^{-1}(x)$, $x \in M$, denote the *state dependent* input spaces, TM the tangent space of M ($TM = \cup\, T_xM$, where T_xM is the tangent space at x in M consisting of all velocity vectors at x and $\cup$ stands for union over all x in M), and $\pi_1 : TM \to M$ the canonical projection of TM on M, and $F : B \to TM$ represents the dynamics of the systems, i.e. for any point (x,u) in B, $f(x,u)$, where $F(x,u) = (x,f(x,u))$, is the velocity vector at the point $x \in M$. Note that locally (i.e. using local coordinates for the manifolds) this representation is precisely as given above. The mathematical description given by the commutative diagram in figure 1.8 has some interesting advantages; in particular when studying global questions for a nonlinear control system. Moreover there are examples which can be described correctly in a global manner only by using this framework.

Example 1.7 (A system on TS^2) Consider a spherical pendulum with a gas jet control which is always directed in the tangent space. We suppose that the magnitude and direction of the jet is completely adjustable within the tangent plane. In this situation the state space is TS^2, the tangent space of the 2-sphere S^2, i.e. $TS^2 = \cup T_pS^2$, the union of all T_pS^2, the tangent plane at p to the sphere S^2. Let $\pi : TS^2 \longrightarrow S^2$ be the canonical projection, then B is a fiber bundle over TS^2 where the fibers are defined as follows. In each point $x \in TS^2$ the fiber above x equals $\pi^{-1}(\pi(x))$. Notice that in this way the manifold B locally is diffeomorphic to $TS^2 \times \mathbb{R}^2$ but B itself is not diffeomorphic to $TS^2 \times \mathbb{R}^2$. Observe that $B = TS^2 \times \mathbb{R}^2$ would imply that the control system could be written as a smooth system $\dot{x} = f(x) + g_1(x)u_1 + g_2(x)u_2$, however g_1 (as well as g_2) has to vanish at some point x ("you cannot comb the hairs on a sphere"). This illustrates that the state-manifold and input-manifold not appear as the usual Cartesian product. □

In many cases, however, the fiber bundle $\pi : B \longrightarrow M$ is a *trivial* bundle, i.e. equals a *product* $M \times U$ for some input space U. In this case an alternative but equivalent global description of the continuous time nonlinear control system (1.1) is provided by a *family* of (globally defined) vectorfields on the state space manifold M, *parametrized* by the inputs $u \in U$. In fact, this will be the setting that will be used throughout the subsequent chapters. Only in Chapter 13 we will give a further discussion on the global setting as depicted in figure 1.8 for a general bundle $\pi : B \to M$.

So far we have discussed various aspects of nonlinear systems described by (1.1). Let us next briefly concentrate on the dynamical behavior of the dynamics (1.1) in case the input u is identically zero (or equals some interesting constant reference value). The dynamics then reads as

$$\dot{x} = f(x,0) =: \tilde{f}(x) , \qquad (1.50)$$

which in case of a linear system (1.38) yields the linear dynamics

$$\dot{x} = Ax . \qquad (1.51)$$

There are several features in which the nonlinear dynamics (1.50) and the linear ones (1.51) may differ. A first distinction occurs in the equilibrium points of (1.50) and (1.51). A point x_0 is called an *equilibrium point* of (1.50) if $\tilde{f}(x_0) = 0$, which is equivalent to the fact that $x(t) \equiv x_0$ is a solution of the differential equation (1.50). Obviously, the set of equilibrium points of the linear system (1.51) form a linear subspace of the state space, whereas the system (1.50) may possess several *isolated* equilibrium points. As an example, one could take the 1-dimensional system (1.50) with $\tilde{f}(x) = x(1-x)$, having equilibrium points at $x = 0$ and $x = 1$. Besides the difference in structure of the set of equilibrium points of (1.50) and (1.51) a similar difference appears in the *periodic orbits* of the systems. The system (1.50) is said to have a *periodic solution of a period $T > 0$ if* there exists a solution $x(t)$ of (1.50) with $x(t) = x(t+T)$ for all t, and T is the smallest real number for which this holds true. The linear differential equation (1.51) possesses a periodic solution if and only if the matrix A has a pair of (conjugate) purely imaginary eigenvalues. If this is the case the system (1.51) has an infinite number of periodic orbits of the same period, all lying in a linear subspace of the state space. In contrast with the situation for the linear dynamics (1.51) the nonlinear system (1.50) may possess a unique or a finite number of periodic orbits with possibly different periods. The following example forms a simple illustration of this.

Example 1.8 Consider on $\mathbb{R}^2$ the dynamics

$$\frac{d}{dt}\begin{bmatrix} x_1 \\ x_2 \end{bmatrix} = \begin{bmatrix} x_2 + x_1(1 - x_1^2 - x_2^2) \\ -x_1 + x_2(1 - x_1^2 - x_2^2) \end{bmatrix} . \tag{1.52}$$

The system (1.52) has an equilibrium point at the origin. Moreover an easy computation shows that the circle $x_1^2 + x_2^2 = 1$ forms a periodic solution of the system (1.52). In fact, $(x_1(t), x_2(t)) = (\cos t, -\sin t)$ is the solution of (1.52) with initial condition $(x_1(0), x_2(0)) = (1,0)$ and which has period $T = 2\pi$. □

Partly as a consequence of the forementioned differences the *qualitative behavior* of the systems (1.50) and (1.51) can differ substantially. Assuming the system (1.50) to be complete (the linear system (1.51) automatically is complete) the study of the qualitative behavior of (1.50) refers to the "behavior in the large" of (1.50), i.e. what happens with solutions $x(t)$ of (1.50) when t goes to infinity? The next example shows that, contrary to a linear system, a periodic orbit of (1.50) may exhibit attracting properties.

Example 1.9 (See Example 1.8.) Consider again the dynamics (1.52). The *phase portrait* of (1.52) is such that any nontrivial solution of (1.52) starting inside the circle $x_1^2 + x_2^2 = 1$, spirals towards the periodic orbit $x_1^2 + x_2^2 = 1$, while solutions starting outside $x_1^2 + x_2^2 = 1$ also spiral towards this circle. So we may conclude that, except for the equilibrium (0,0), all solutions of (1.52) tend towards the set $x_1^2 + x_2^2 = 1$.

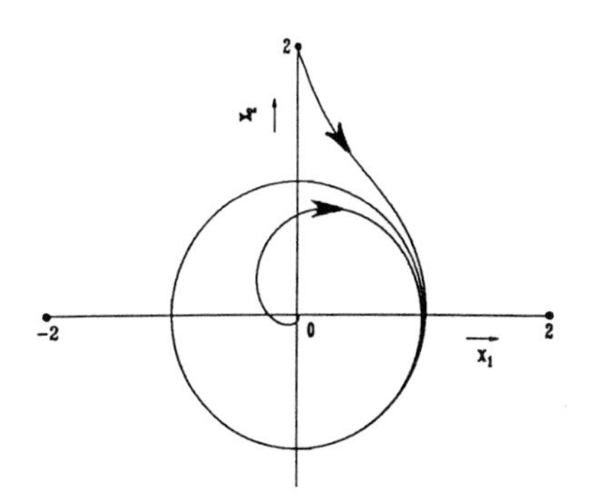

Fig. 1.9. Phase portrait of (1.52). □

The situation as described in Example 1.9 is quite common for planar nonlinear differential equations (1.50). Furthermore for higher dimensional systems a lot more complications can arise. In particular, the positive limit set of (1.50), i.e. the set of limit points of solutions $x(t)$ of (1.50) when t tends to infinity, may have a very wild or even chaotic structure. Although the study of the qualitative behavior of nonlinear systems is beyond the scope of this text, we will come back to some aspects of this, in particular those concerning stability and stabilization, later on in Chapter 10.

Finally we will briefly discuss discrete-time nonlinear systems given as

$$\begin{aligned} x(k+1) &= f(x(k),u(k)), \\ y(k) &= h(x(k),u(k)), \end{aligned} \tag{1.2}$$

where as before x, u and y denote respectively the state, the input and the output. Assuming $x \in \mathbb{R}^n$, $u \in \mathbb{R}^m$ and $y \in \mathbb{R}^p$, (1.2) is a shorthand writing for

$$\begin{cases} x_1(k+1) = f_1(x_1(k),\dots,x_n(k),\ u_1(k),\dots,u_m(k)), \\ \quad\vdots \\ x_n(k+1) = f_n(x_1(k),\dots,x_n(k),\ u_1(k),\dots,u_m(k)), \end{cases} \tag{1.53a}$$

$$\begin{cases} y_1(k) = h_1(x_1(k),\ldots,x_n(k),\ u_1(k),\ldots,u_m(k)), \\ \vdots \\ y_p(k) = h_p(x_1(k),\ldots,x_n(k),\ u_1(k),\ldots,u_m(k)), \end{cases} \tag{1.53b}$$

Again we assume $f : \mathbb{R}^n \times \mathbb{R}^m \to \mathbb{R}^n$ and $h : \mathbb{R}^n \times \mathbb{R}^m \to \mathbb{R}^p$ to be smooth, though this is in what follows not very essential. Together with (1.53a,b) we have to specify a class of admissible controls $\mathcal{U}$, which for a discrete time system may be any set of time functions $u : \mathbb{Z} \to \mathbb{R}^m$ or $u : \mathbb{Z}^+ \to \mathbb{R}^m$ that is closed under concatenation. The important observation is that (in contrast with continuous time systems) the difference equation

$$\begin{aligned} x(k+1) &= f(x(k),u(k)), \\ x(k_0) &= x_0, \end{aligned} \tag{1.54}$$

admits a well defined (forward) solution $x(k)$, $k \geq k_0$, for any admissible control $u(\cdot) \in \mathcal{U}$ and any arbitrary initial state $x_0 \in \mathbb{R}^n$. As before such a solution will usually be denoted as $x(k,k_0,x_0,u)$ (and similarly the output will be written as $y(k,k_0,x_0,u)$). No Lipschitz condition is needed to discuss solutions of (1.54). Of course the setting for discrete time systems straightforwardly extends to more general state-, input- and output spaces (e.g. manifolds). We refer to Example 1.3 for an illustration.

We conclude this chapter with the remark that discrete-time nonlinear dynamics exhibit similar phenomena as their continuous-time counterparts such as, for instance, isolated periodic orbits and "strange" positive limit sets.

Notes and References

We have discussed some examples of typically nonlinear control systems arising from different sources. These examples have been chosen in order to provide motivation for the development of a structural analysis of nonlinear systems in the next chapters. Example 1.1 is one of the simplest robot-arm configurations. For a further discussion on this example, as well as an account of more advanced manipulators we refer to [AS],[Cra],[Pa]. A detailed exposition of Example 1.2 has been given in [Ar2],[CB],[Cr1],[Cr2],[NvdS],[SG]. The economic Example 1.3 has been taken from [WK], see also [Nij]. Example 1.4 on a mixed-culture bioreactor is extensively discussed in [Ol],[HK]. Example 1.5 may be found in [Su].

For general information on existence and uniqueness of differential equations we refer to textbooks as [Ar1],[CL],[HS]. A further discussion about the possibly complicated behavior of dynamical systems may be found in for instance [Ar2],[GH],[HS]. For more background on the formulation of nonlinear control systems on manifolds (including systems modelled on fiber bundles) we refer to e.g. [Su],[Br],[Lo],[Wi],[vdS].

[Ar1] V.I. Arnold, **Ordinary differential equations**, MIT Press, Cambridge (MA), 1980.

[Ar2] V.I. Arnold, **Méthodes mathématiques de la mécanique classique**, Editions MIR, Moscou, 1976.

[AS] H. Asada, J.J.E. Slotine, **Robot analysis and control**, John Wiley & Sons, New York, 1986.

[Br] R.W. Brockett, "Global descriptions of nonlinear control problems; vector bundles and nonlinear control theory", Notes for a CBMS conference, manuscript, 1980.

[CL] E.A. Coddington, H. Levinson, **Theory of ordinary differential equations**, Mc Graw-Hill, New York, 1955.

[Cra] J.J. Craig, **Introduction to robotics, mechanics and control**, Addison Wesley, Reading, 1986.

[CB] P.E. Crouch, B. Bonnard, "An appraisal of linear analytic system theory with applications to attitude control", ESA ESTEC Contract report 1980.

[Cr1] P.E. Crouch, "Application of linear analytic systems theory to attitude control", ESA ESTEC report 1981.

[Cr2] P.E. Crouch, "Spacecraft attitude control and stabilization: applications of geometric control theory to rigid body models", IEEE Trans. Aut. Contr. AC-29, pp. 321-331, 1984.

[GH] J. Guckenheimer, P. Holmes, **Nonlinear oscillations, dynamical systems and bifurcations of vectorfields**, Springer Verlag, Berlin, 1983.

[HK] K.A. Hoo, J.C. Kantor, "Global linearization and control of a mixed-culture bioreactor with competition and external inhibition", Math. Biosci. 82, pp. 43-62, 1986.

[HS] M.W. Hirsch, S. Smale, **Differential equations, dynamical systems, and linear algebra**, Academic Press, New York, 1974.

[Lo] C. Lobry, "Controlabilité des systèmes non linéaires", SIAM J. Contr. 8, p.p. 573-605, 1970.

[Nij] H. Nijmeijer, "On dynamic decoupling and dynamic path controllability in economic systems", Journ. of Economic Dynamics and Control, 13, pp. 21-39, 1989.

[NvdS] H. Nijmeijer, A.J. van der Schaft, "Controlled invariance for nonlinear systems: two worked examples", IEEE Trans. Aut. Contr., AC-29, pp. 361-364, 1984.

[Ol] D.F. Ollis, "Competition between two species when only one has antibiotic resistance: Chemostat analysis", paper presented at AIChE meeting, San Francisco, 1984.

[Pa] R.P. Paul, **Robot manipulators: mathematics, programming and control**, MIT Press, Cambridge (MA), 1981.

[vdS] A.J. van der Schaft, **System theoretic descriptions of physical systems**, CWI Tract 3, Centrum voor Wiskunde en Informatica, Amsterdam, 1984.

[Su] H.J. Sussmann, "Existence and uniqueness of minimal realizations of nonlinear systems", Math. Systems Theory 10, pp. 263-284, 1977.

[SG] J.L. Synge, B.A. Griffiths, **Principles of mechanics**, McGraw-Hill, New York, 1959.

[Wi] J.C. Willems, "System theoretic models for the analysis of physical systems", Ricerche di Automatica 10, pp. 71-106, 1979.

[WK] H.W. Wohltmann, W. Krömer, "Sufficient conditions for dynamic path controllability of economic systems", Journ. of Economic Dynamics and Control 7, pp. 315-330, 1984.

2
Manifolds, Vectorfields, Lie Brackets, Distributions

In the previous chapter we have seen that many nonlinear control systems

$$\begin{aligned} \dot{x} &= f(x,u) \\ y &= h(x,u) \end{aligned} \tag{2.1}$$

are properly defined on a state space which is not equal to Euclidean space $\mathbb{R}^n$, but instead is a curved n-dimensional subset of $\mathbb{R}^m$ for some m, called a *manifold*. As a consequence, the equations (2.1) usually do not describe the system everywhere, but only on a *part* of the state space, and for a different part of the state space we generally need another representation of the system in equations like (2.1). In geometric language we say that (2.1) is a local coordinate expression of the system we wish to describe, and that in order to cover the *whole* system more than one coordinate expression is needed. For patching together these different expressions the notion of *coordinate transformation* is instrumental.

In the first part of this chapter (Section 2.1) we will develop the mathematical machinery to do this in a proper way. The approach taken here enables us to define a nonlinear control system on a curved state space independently of any choice of local coordinates. This so-called coordinate-free viewpoint is often illuminating, and can provide shortcuts in calculations which may be very tedious in local coordinates. The material covered in Section 2.1 is not easy to grasp at first reading. On the other hand, in order to be able to read Section 2.2 and the next chapters it is not really necessary to understand this material in full detail. In fact for most of the chapters to follow a rough understanding of the main ideas will be sufficient. Therefore we give *before* Section 2.1 a rough and intuitive *survey* of the material covered. One is advised first to read this survey and to pass swiftly over Section 2.1, and then to re-read Section 2.1 at later occasions.

For the second part of the chapter (Section 2.2) we will follow a different path. Since this material is much more at the heart of the contents of this book we will first go through it in reasonable detail, and give a short summary of the material afterwards (Section 2.3).

In Section 2.2 we will dwell more specifically upon the properties of sets of ordinary differential equations (without inputs) defined on a manifold, which in a coordinate-free approach are described as *vectorfields*. We show how under certain conditions there exist proper choices of local coordinates in which the vectorfield or the collection of vectorfields take an easy form, which is very much amenable for further analysis. These tools will prove to be instrumental for much of the theory developed in the chapters that will follow.

2.0 Survey of Section 2.1

Manifold

Consider a topological space M (i.e., we know what subsets $U \subset M$ are called open). Suppose that for any $p \in M$ there exists an open set U containing p, and a bijection φ mapping U onto some open subset of $\mathbb{R}^n$ for some fixed n. On $\mathbb{R}^n$ we have the natural coordinate functions $r_i(a_1, \ldots, a_n) = a_i$, $i \in \underline{n} := \{1, \ldots, n\}$. By composition with φ we obtain coordinate functions x_i, $i \in \underline{n}$, on U by letting

$$x_i = r_i \circ \varphi, \qquad i \in \underline{n} . \tag{2.2}$$

In this way the grid defined on $\varphi(U) \subset \mathbb{R}^n$ by the coordinate functions r_i, $i \in \underline{n}$, transforms into a grid on $U \subset M$

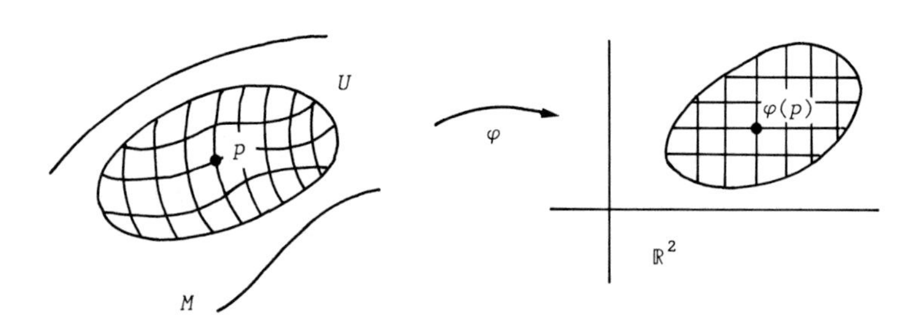

Fig. 2.1. A coordinate chart.

The open set U together with the map φ is called a *coordinate chart* and is also denoted as $(U, x_1, \ldots, x_n)$. On $\mathbb{R}^n$ there is a natural notion of differentiability, and we would like to transfer this notion to M, so that we can talk about differentiable functions on M. In order to do this we have to impose extra conditions on the coordinate charts (U,φ) as above. First of all we require that for any chart (U,φ) the map φ is a *homeomorphism* (φ and φ^{-1} are continuous). Secondly we require that all charts are (C^∞-)compatible. This means that for any two charts (U,φ), (V,ψ) with $U \cap V \neq \emptyset$ the map

$$S = \psi \circ \varphi^{-1}: \varphi(U \cap V) \subset \mathbb{R}^n \to \psi(U \cap V) \subset \mathbb{R}^n \tag{2.3}$$

is *smooth* (C^∞), i.e. derivatives up to arbitrary order exist.

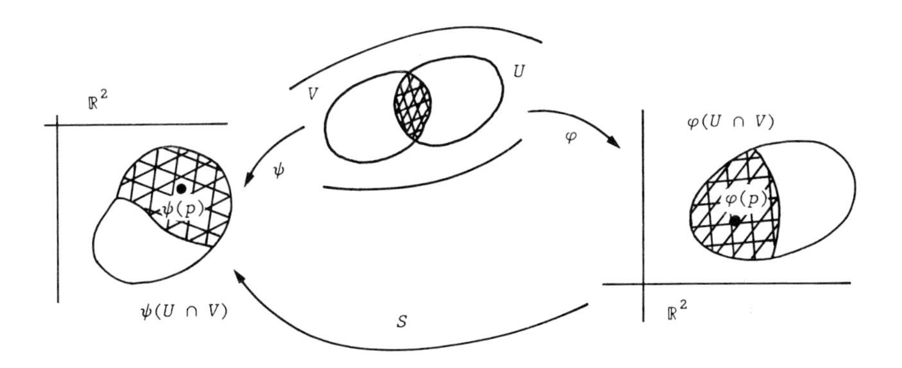

Fig. 2.2. Coordinate transformation.

Let $z_1, \ldots, z_n$ be the coordinate functions on V corresponding to ψ, i.e. $z_i = r_i \circ \psi$, $i \in \underline{n}$. The map S is called a *coordinate transformation* (from coordinates $x = (x_1, \ldots, x_n)$ to $z = (z_1, \ldots, z_n)$), in fact we can write

$$z_i = S_i(x_1, \ldots, x_n), \qquad i \in \underline{n} . \tag{2.4}$$

Finally, M together with its (compatible) coordinate charts is called a smooth *manifold*.

Maps and functions

Let M_1 and M_2 be smooth manifolds of dimension n_1, respectively n_2. Then $F: M_1 \to M_2$ is called a smooth map, briefly map, if for any $p \in M_1$ there exist coordinate charts (U,φ) about p for M_2, and (V,ψ) about $F(p)$ for M_2, such that the map

$$\hat{F} = \psi \circ F \circ \varphi^{-1}: \varphi(U) \subset \mathbb{R}^{n_1} \to \psi(V) \subset \mathbb{R}^{n_2} \tag{2.5}$$

is a smooth map. We call $\hat{F}$ the expression of F in local coordinates, or *local representative* of F. Usually we omit the caret, and write $\hat{F}$ in the local coordinates $(x_1, \ldots, x_{n_1}), (z_1, \ldots, z_{n_2})$ corresponding to φ, resp. ψ, as

$$F(x) = \left(F_1(x_1, \ldots, x_{n_1}), \ldots, F_{n_2}(x_1, \ldots, x_{n_1})\right)^T \tag{2.6}$$

Note that $(x_1, \ldots, x_{n_1})$ here is interpreted as a *point* in $\mathbb{R}^{n_1}$, i.e. the local coordinates of some (not specified) point $p \in M_1$.

The rank of the map F at a point p is defined as the rank of the Jacobian matrix of the map (2.6) from $\mathbb{R}^{n_1}$ to $\mathbb{R}^{n_2}$ calculated in $\varphi(p)$. If F is bijective, and F and F^{-1} are smooth then F is called a *diffeomorphism*. Accordingly, a map $f\colon M \to \mathbb{R}$ is called a smooth function, briefly function, if for any $p \in M$ there exists a coordinate chart (U,φ) about p such that

$$\hat{f} = f\circ\varphi^{-1}\colon \varphi(U) \subset \mathbb{R}^n \to \mathbb{R} \tag{2.7}$$

is a smooth function. Usually we omit the caret in the local coordinate expression $\hat{f}$ and denote it as $f(x_1,\ldots,x_n)$, where $x_1,\ldots,x_n$ are local coordinates.

Submanifolds

Let M be a smooth manifold of dimension n. A non-empty open set $V \subset M$ is itself a smooth manifold of dimension n with coordinate charts obtained by *restricting* the coordinate charts for M to V. V is called an *open* submanifold of M. A subset $P \subset M$ is called a submanifold of dimension $m < n$ if for each $p \in P$ there exists an *adapted* coordinate chart $(U,x_1,\ldots,x_n)$ *for* M about p such that

$$P \cap U = \{q \in U \mid x_i(q) = x_i(p),\ i = m+1,\ldots,n\} \tag{2.8}$$

If we take the subset topology on P (i.e. the open sets on P are of the form $P \cap U$, with U open in M), and coordinate charts $(P \cap U,\ \varphi|P)$, with (U,φ) coordinate chart for M satisfying (2.8) and $\varphi|P$ the restriction of φ to P, then P is itself a manifold. Notice that if $x_1,\ldots,x_n$ are adapted local coordinates for M then $x_1,\ldots,x_m$ are local coordinates for P.

An important theorem concerning submanifolds is the following. Let $F\colon M_1 \to M_2$ be a smooth map with $\dim M_i = n_i$, $i = 1,2$. Let $p_2 \in M_2$ and suppose that $F^{-1}(p_2)$ is nonempty, and suppose furthermore that the rank of F in every point of $F^{-1}(p_2)$ equals the dimension of M_2. Then

$$F^{-1}(p_2) \text{ is a submanifold of } M_1, \text{ of dimension } n_1 - n_2. \tag{2.9}$$

Tangent space

Let M be a smooth manifold of dimension n. The *tangent space* T_pM in $p \in M$ is geometrically clear:

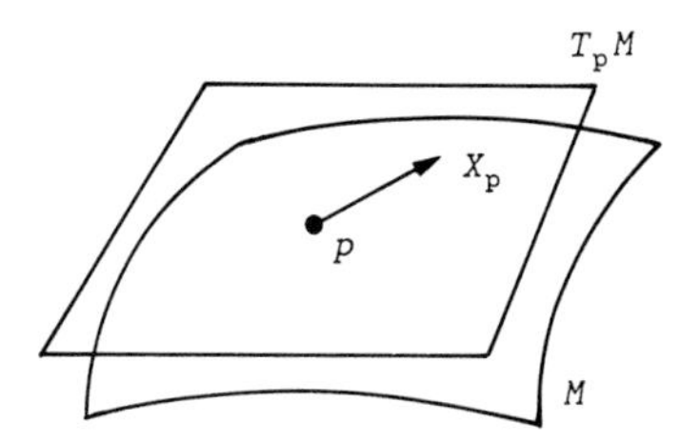

Fig. 2.3. Tangent space T_pM at $p \in M$.

Formally we define T_pM as the linear space of *derivations*; an element $X_p \in T_pM$ acts on functions f defined in a neighbourhood of p, i.e.

$$X_p(f) \in \mathbb{R} \quad . \tag{2.10}$$

Geometrically $X_p(f)$ is the directional derivative of f in the tangent direction X_p. Said otherwise, let c: $(-\epsilon,\epsilon) \to M$ be a curve on M with $c(0) = p$ and $c'(0) = X_p$, then $X_p(f)$ equals $\frac{df(c(t))}{dt}\Big|_{t=0}$.

The *tangent space* $T_a\mathbb{R}^n$ to the smooth manifold $\mathbb{R}^n$ in any point $a \in \mathbb{R}^n$ can be identified with $\mathbb{R}^n$ itself. The natural basis for $T_a\mathbb{R}^n$ is $\{\frac{\partial}{\partial r_1}\Big|_a,\ldots,\frac{\partial}{\partial r_n}\Big|_a\}$, where as before $r_i : \mathbb{R}^n \to \mathbb{R}$ are the natural coordinate functions for $\mathbb{R}^n$. Let now $(U,\varphi) = (U,x_1,\ldots,x_n)$ be a coordinate chart about p, then we obtain a basis for T_pM in the following way. Let F: $M_1 \to M_2$ be any smooth map. Then the *tangent map* of F at a point $p \in M_1$ is the linear map

$$F_{*p}: T_pM_1 \to T_{F(p)}M_2 \quad , \tag{2.11}$$

defined as follows. Let $X_p \in T_pM_1$, and f a smooth function on M_2 about $F(p)$. Then

$$F_{*p}X_p(f) = X_p(f \circ F) \quad . \tag{2.12}$$

It follows that if F is a diffeomorphism, then F_{*p} is a non-singular linear map, for all $p \in M_1$. In particular, any coordinate map φ: $U \subset M \to \varphi(U) \subset \mathbb{R}^n$ is a diffeomorphism, and so $\varphi_{*p}: T_pM \to T_{\varphi(p)}\mathbb{R}^n$ has an inverse φ_{*p}^{-1}. Define

$$\frac{\partial}{\partial x_i}\Big|_p = \varphi_{*p}^{-1}\frac{\partial}{\partial r_i}\Big|_{\varphi(p)}, \qquad i \in \underline{n}, \tag{2.13}$$

then $\{\frac{\partial}{\partial x_1}\Big|_p,\ldots,\frac{\partial}{\partial x_n}\Big|_p\}$ is a basis for T_pM. By definition we have for any smooth f around p

$$\frac{\partial}{\partial x_i}\Big|_p(f) = \varphi_{*p}^{-1}\frac{\partial}{\partial r_i}\Big|_{\varphi(p)}(f) = \frac{\partial}{\partial r_i}(f\circ\varphi^{-1})\Big|_{\varphi(p)}$$
$$= \frac{\partial\hat{f}}{\partial r_i}\big(x_1(p),\ldots,x_n(p)\big) \tag{2.14}$$

with $\hat{f}$ the local representative of f. Instead of $\frac{\partial}{\partial x_i}\Big|_p(f)$ we will usually simply write $\frac{\partial f}{\partial x_i}(p)$, and we conclude that $\frac{\partial f}{\partial x_i}(p)$ is obtained by differentiating the local representative $\hat{f}$ of f with respect to its i-th argument.

Let now (U,φ) and (V,ψ) be overlapping coordinate charts yielding a coordinate transformation $z = S(x)$ with $S = \psi\circ\varphi^{-1}$. Let $X_p \in T_pM$, with $p \in U \cap V$, be expressed in the basis corresponding to (U,φ) as

$$X_p = \sum_{i=1}^{n}\alpha_i\frac{\partial}{\partial x_i}\Big|_p \tag{2.15}$$

and in the basis for T_pM corresponding to (V,ψ) as

$$X_p = \sum_{i=1}^{n}\beta_i\frac{\partial}{\partial z_i}\Big|_p \tag{2.16}$$

then the coefficients $\alpha = (\alpha_1,\ldots,\alpha_n)^T$ and $\beta = (\beta_1,\ldots,\beta_n)^T$ are contravariantly related as (with $\frac{\partial S}{\partial x}$ the Jacobian matrix of S)

$$\beta = \frac{\partial S}{\partial x}\big(x(p)\big)\alpha \tag{2.17}$$

In general let $F\colon M_1 \to M_2$ be a smooth map, and let $(U,x_1,\ldots,x_n)$, $(V,z_1,\ldots,z_n)$ be coordinate charts about p, resp. $F(p)$, then

$$F_{*p}\left(\frac{\partial}{\partial x_i}\Big|_p\right)(z_j) = \frac{\partial}{\partial x_i}(z_j\circ F)\Big|_p = \frac{\partial(z_j\circ F)}{\partial x_i}(p). \tag{2.18}$$

So in coordinate bases for T_pM_1 and $T_{F(p)}M_2$ the tangent map F_{*p} equals the Jacobian matrix of F expressed in these local coordinates.

Tangent bundle

The tangent bundle TM of M is defined as $\bigcup_{p\in M} T_pM$. It is itself a smooth manifold with dimension $2n$ with natural local coordinates $(x_1,\ldots,x_n,v_1,\ldots,v_n)$ defined as follows. Let $(x_1,\ldots,x_n)$ be local coordinates about p. Then the coordinate values of $X_p = \sum_{i=1}^{n}\alpha_i\frac{\partial}{\partial x_i}\Big|_p$ are $\big(x_1(p),\ldots,x_n(p),\alpha_1,\ldots,\alpha_n\big)$, i.e. $x_i(X_p) = x_i(p)$ and $v_i(X_p) = \alpha_i$, $i \in \underline{n}$.

2.1 Manifolds, Coordinate Transformations, Tangent Space

2.1.1 Differentiability, Manifolds, Submanifolds

Let f be a function from an open set $A \subset \mathbb{R}^n$ into $\mathbb{R}$, and let $k > 0$ be a positive integer. The function f is called C^k (k times continuously differentiable) if it possesses continuous partial derivatives of all orders $\leq k$ on A. If f is C^k for all k then f is C^∞ or *smooth*. If f is *real analytic* (expandable in a power series in its arguments about each point of A) then f is called C^ω. (Of course, f being C^ω implies that f is C^∞.) For $i \in \underline{n}$ let r_i be the natural slot or coordinate function on $\mathbb{R}^n$

$$r_i(a_1,\ldots,a_n) = a_i \quad . \tag{2.19}$$

A map f from an open set $A \subset \mathbb{R}^n$ into $\mathbb{R}^n$ is C^∞ (resp. C^ω, C^k) if each of its slot functions

$$f_i = r_i \circ f\colon A \to \mathbb{R} \tag{2.20}$$

is C^∞ (resp. C^ω, C^k). Henceforth we will mainly work in the smooth (C^∞) setting, although everything can be adapted to the C^k-setting, with k large enough. Sometimes we will make some additional remarks about the C^ω-case.

Let us now define the notion of a smooth *manifold*. The basic idea is that a smooth manifold is a set which locally can be identified with $\mathbb{R}^n$ together with the intrinsic notion of differentiability defined on $\mathbb{R}^n$. First we need the notion of a topological space, in particular the technical notion of a Hausdorff topological space with countable basis.

Digression Let M be a set. A topological structure or topology on M is a collection T of subsets of M satisfying

(i) the union of any number of subsets in T belongs again to T.

(ii) the intersection of any finite number of subsets in T belongs to T.

(iii) M and the empty set belong to T.

The elements of T are called *open sets* of M, and complements of these *closed sets*. The set M together with the topology T is called a *topological space*. A basis for a topology T on M is a collection $B_T \subset T$ of open sets such that every open set can be written as a union of elements of B_T (Example: a basis for the usual topology on R are the open finite intervals (a,b), $a,b \in \mathbb{R}$.) A neighborhood of a point $p \in M$ is any open set which contains p. A topological space M is *Hausdorff* if any two different points p_1 and p_2 have disjoint neighborhoods. A mapping $F: M_1 \to M_2$ between two topological spaces is called continuous if $F^{-1}(O_2)$ is an open set of M_1 for any open set O_2 of M_2. F is called a homeomorphism if F is bijective and F as well as F^{-1} are continuous. Then M_1 and M_2 are called *homeomorphic*.

Definition 2.1 *A Hausdorff topological space M with countable basis is called a topological manifold of dimension n if for any $p \in M$ there exists a homeomorphism φ from some neighborhood U of p onto an open set of $\mathbb{R}^n$.*

Let M be a topological manifold of dimension n. A pair (U,φ) with U an open set of M, and φ a homeomorphism from U onto an open set of $\mathbb{R}^n$, is called a *coordinate chart* or *coordinate neighborhood*. The functions

$$x_i = r_i \circ \varphi, \qquad i \in \underline{n} \ , \tag{2.21}$$

are called *local coordinate functions* and the values $x_1(p),\ldots,x_n(p)$ of a point $p \in U$ are called the local coordinates of p. The coordinate neighborhood (U,φ) will be also denoted by $(U,x_1,\ldots,x_n)$. Let $f: M \to \mathbb{R}$ be a map. Then f yields a function $\hat{f}: \varphi(U) \subset \mathbb{R}^n \to \mathbb{R}$ defined as $\hat{f} = f \circ \varphi^{-1}$, that is for $p \in U$

$$f(p) = \hat{f}\big(x_1(p),\ldots,x_n(p)\big) \ . \tag{2.22}$$

The function $\hat{f}$ is called the *local representative* of f.

It is customary to omit the caret, and to use the same letter "f" for f as defined on M and for $\hat{f}$, its expression in local coordinates. Hence instead of $\hat{f}\big(x_1(p),\ldots,x_n(p)\big)$ we write $f\big(x_1(p),\ldots,x_n(p)\big)$. Furthermore usually we delete the dependence on p, and write $f(x_1,\ldots,x_n)$ where $(x_1,\ldots,x_n)$ are the local coordinates of some (unspecified) point $p \in U$.

Let now (U,φ) and (V,ψ) be two coordinate charts on M which overlap, i.e. $U \cap V \neq \emptyset$. Let $x_i = r_i \circ \varphi$ and $z_i = r_i \circ \psi$ be the corresponding coordinate functions. We consider the map (actually homeomorphism), cf. Fig. 2.2,

$$S := \psi \circ \varphi^{-1}: \underset{\mathbb{R}^n}{\underset{\cap}{\varphi(U \cap V)}} \to \underset{\mathbb{R}^n}{\underset{\cap}{\psi(U \cap V)}} . \tag{2.23}$$

This map is called the *coordinate transformation* from local coordinates $(x_1,\ldots,x_n)$ to local coordinates $(z_1,\ldots,z_n)$ on $U \cap V$. In the natural coordinates for $\mathbb{R}^n$ we write

$$z_i = S_i(x_1, \ldots, x_n) \quad , \quad i \in \underline{n} \ , \tag{2.24}$$

and conversely, denoting the i-th component of $S^{-1} = \varphi \circ \psi^{-1}$ as S_i^{-1}

$$x_i = S_i^{-1}(z_1, \ldots, z_n) \qquad i \in \underline{n} \ . \tag{2.25}$$

Any map $f\colon M \to \mathbb{R}$ can be expressed on $U \cap V$ as $f(x_1, \ldots, x_n)$ $(= f \circ \varphi^{-1})$, or alternatively as $f(z_1, \ldots, z_n)$ $(= f \circ \psi^{-1})$.

Example 2.2 Let $M = \mathbb{R}^2 \backslash \{(x_1, x_2) \mid x_1 \geq 0, x_2 = 0\}$, and let $U = V = M$. Let $\varphi = id$ with coordinate functions x_1, x_2 and define ψ by

$$\psi^{-1}(a_1, a_2) = (a_1 \cos a_2, \ a_1 \sin a_2) \ . \tag{2.26}$$

The coordinate functions z_1, z_2 corresponding to ψ are related with x_1, x_2 by the map $S^{-1} = \varphi \circ \psi^{-1}$ given as

$$\begin{aligned} x_1 &= z_1 \cos z_2 , \\ x_2 &= z_1 \sin z_2 . \end{aligned} \tag{2.27}$$

Thus (x_1, x_2) are Cartesian coordinates, and (z_1, z_2) are polar coordinates. The function $f\colon M \to \mathbb{R}$ expressed in coordinates (x_1, x_2) as $x_1^2 + x_2^2$ is expressed in coordinates (z_1, z_2) as z_1^2. □

Definition 2.3 *Two coordinate charts (U, φ), (V, ψ) are called C^∞-compatible if $U \cap V = \emptyset$, or in case $U \cap V \neq \emptyset$, if the coordinate transformation $S = \varphi \circ \psi^{-1}$ and the inverse coordinate transformation $S^{-1} = \psi \circ \varphi^{-1}$ are both C^∞.*

Definition 2.4 *A C^∞-atlas on a topological manifold M is a collection $\mathcal{D} = \{U_\alpha, \varphi_\alpha\}_{\alpha \in A}$ of pairwise C^∞-compatible coordinate charts with the property that $\bigcup_{\alpha \in A} U_\alpha = M$. $\mathcal{D}$ is called a maximal C^∞-atlas or smooth differentiable structure if any coordinate chart (V, ψ) which is C^∞-compatible with every $(U_\alpha, \varphi_\alpha) \in \mathcal{D}$ is also in $\mathcal{D}$.*

Definition 2.5 *A smooth manifold is a topological manifold endowed with a smooth differentiable structure.*

Remark We can also define C^ω- or C^k-compatibility by requiring that S and S^{-1} in Definition 2.3 are C^ω, respectively C^k. Replacing C^∞-compatibility by C^ω-, respectively C^k-compatibility in Definitions 2.4 and 2.5 then results in the notion of a C^ω- (real analytic), respectively C^k-, manifold.

Example 2.6 Take $M = \mathbb{R}^n$, with atlas consisting of the single chart $(U = \mathbb{R}^n,\ \varphi = id)$. □

Example 2.7 Let M be any open set of $\mathbb{R}^n$ with atlas consisting of the single chart $(U = M,\ \varphi = id$ restricted to $U)$. □

Example 2.8 Let $M = Gl(n)$ be the set of all $n{\times}n$ invertible real matrices. Identify in the obvious way an $n{\times}n$ matrix with a point in $\mathbb{R}^{n^2}$. The determinant then becomes a continuous function on $\mathbb{R}^{n^2}$, and $Gl(n)$ receives by Example 2.7 a smooth manifold structure as the open set of $\mathbb{R}^{n^2}$ where the determinant function does not vanish. □

Example 2.9 An open set P of a smooth manifold M is itself a smooth manifold. P endowed with the subset topology is a topological manifold. (A set $V \subset P$ is open in the subset topology on P if $V = P \cap U$ for some open set U on M.) For any coordinate chart (U,φ) on M let now (U',φ') with $U' = N \cap U$, $\varphi' = \varphi$ restricted to N, be a coordinate chart for N. This defines a C^∞-atlas for N. □

Example 2.10 The product $M_1 \times M_2$ of two smooth manifolds is itself a smooth manifold by taking coordinate charts $\left(U_1 \times U_2, (\varphi_1, \varphi_2)\right)$ with (U_1, φ_1), (U_2, φ_2) coordinate charts for M_1, resp. M_2. (Obviously, (φ_1, φ_2) is defined as $(\varphi_1, \varphi_2)(u_1, u_2) = \left(\varphi_1(u_1), \varphi_2(u_2)\right)$, $u_1 \in U_1$, $u_2 \in U_2$.) □

The most important tools in the sequel will be the *inverse function theorem* and the *implicit function theorem*, known from calculus, which we state next for completeness (without proof).

Theorem 2.11 (Inverse Function Theorem) *Let $U \subset \mathbb{R}^n$ be open, and let $f : U \to \mathbb{R}^n$ be a smooth map. If the Jacobian matrix $\frac{\partial f}{\partial x}$ is non-singular at some point $p \in U$, then there exists a neighborhood $V \subset U$ of p such that $f : V \to f(V)$ is a diffeomorphism (i.e., $f : V \to f(V)$ is bijective and $f^{-1} : f(V) \to V$ is smooth).*

Theorem 2.12 (Implicit Function Theorem) *Let $W \subset \mathbb{R}^n \times \mathbb{R}^m$ be open, and let $f : W \to \mathbb{R}^m$ be a smooth map, satisfying $f(p,q) = 0$ for some point $(p,q) \in W$. If the Jacobian matrix of $f(x,y)$, $x \in \mathbb{R}^n$, $y \in \mathbb{R}^m$, at the point (q,p) is such that the submatrix composed of its last m columns, i.e. $\frac{\partial f}{\partial y}(p,q)$ is non-singular, then there exists a neighborhood $U \subset W$ of (p,q)*

and a neighborhood $V \subset \mathbb{R}^n$ of p, and a unique smooth map $g : V \to \mathbb{R}^m$ such that $g(p) = q$ and

$$\{(x,y) \in U \mid f(x,y) = 0\} = \{(x,g(x)) \mid x \in V\} .$$

(In particular $f(x,g(x)) = 0$ for all $x \in V$.)

Many smooth manifolds are generated in the following way.

Theorem 2.13 *Let $f_1, \ldots, f_m$, $m \le n$, be smooth functions on $\mathbb{R}^n$ (or on an open part of $\mathbb{R}^n$). Define*

$$N = \{x \in \mathbb{R}^n \mid f_1(x) = \cdots = f_m(x) = 0\} \tag{2.28}$$

and assume that $N \neq \emptyset$. Suppose that the rank of the Jacobian matrix of $f = (f_1, \ldots, f_m)^T$

$$\frac{\partial f}{\partial x}(x) = \begin{pmatrix} \frac{\partial f_1}{\partial x_1}(x) & \cdots & \frac{\partial f_1}{\partial x_n}(x) \\ \vdots & & \vdots \\ \frac{\partial f_m}{\partial x_1}(x) & \cdots & \frac{\partial f_m}{\partial x_n}(x) \end{pmatrix} \tag{2.29}$$

is m at each $x \in N$. Then N is a manifold of dimension $n - m$.

Proof Let $x^0 \in N$. By permuting the coordinates $x_1, \ldots, x_n$ we may assume that the matrix

$$\begin{pmatrix} \frac{\partial f_1}{\partial x_1} & \cdots & \frac{\partial f_1}{\partial x_m} \\ \vdots & & \vdots \\ \frac{\partial f_m}{\partial x_1} & \cdots & \frac{\partial f_m}{\partial x_m} \end{pmatrix} \tag{2.30}$$

is non-singular at x^0. By the Implicit Function Theorem there exists a neighborhood $V \subset \mathbb{R}^n$ of x^0, a neighborhood $W \subset \mathbb{R}^{n-m}$ of $(x^0_{m+1}, \ldots, x^0_n)$ and a smooth map g: $W \to \mathbb{R}^m$ such that $N \cap V$ equals

$$\{(g_1(x_{m+1}, \ldots, x_n), \ldots, g_m(x_{m+1}, \ldots, x_n), x_{m+1}, \ldots, x_n) \mid (x_{m+1}, \ldots, x_n) \in W\}.$$

Now define a coordinate chart (U,φ) on N about x^0 by setting

$$U = N \cap V, \tag{2.31}$$

$$\varphi(g_1(x_{m+1}, \ldots, x_n), \ldots, g_m(x_{m+1}, \ldots, x_n), x_{m+1}, \ldots, x_n) = (x_{m+1}, \ldots, x_n).$$

Doing this for any $x^0 \in N$, N becomes a smooth manifold of dimension $n - m$. □

Example 2.14 The circle S^1 is a smooth manifold of dimension 1, since $S^1 = \{(x_1,x_2) \mid x_1^2 + x_2^2 - 1 = 0\}$. □

Example 2.15 Consider the group $O(n)$ of orthogonal $n \times n$-matrices (i.e. for $A \in O(n)$ we have $A^T A = I_n$). This group can be given the structure of a smooth manifold in the following way. Define the map f from $Gl(n)$ (which by Example 2.8 can be identified with an open set of $\mathbb{R}^{n^2}$) to the space of *symmetric* $n \times n$-matrices (which in the obvious way can be identified with $\mathbb{R}^{\frac{1}{2}n(n+1)}$)

$$f(A) = A^T A. \tag{2.32}$$

Then $O(n) = \{A \in Gl(n) \mid f(A) = I_n\}$. It can be checked (see Exercise 2.3) that the rank of the Jacobian of f (seen as a mapping from $\mathbb{R}^{n^2}$ to $\mathbb{R}^{\frac{1}{2}n(n+1)}$) in $A \in O(n)$ is $\frac{1}{2}n(n+1)$. Therefore $O(n)$ is a smooth manifold of dimension $n^2 - \frac{1}{2}n(n+1) = \frac{1}{2}n(n-1)$. Actually $O(n)$ consists of two disconnected parts: the elements in $O(n)$ with determinant +1, denoted by $SO(n)$, and the elements with determinant -1. □

Consider now two smooth manifolds M_1 and M_2 of dimension n_1, respectively n_2. A map $F: M_1 \to M_2$ is *smooth* if for each $p \in M_1$ there exist coordinate neighborhoods (U,φ) of M_1 about p, respectively (V,ψ) of M_2 about $F(p)$, such that the expression of F in the corresponding local coordinates $x_1,\ldots,x_{n_1}$, respectively $z_1,\ldots,z_{n_2}$, i.e. the local representative $\hat{F} = \psi \circ F \circ \varphi^{-1}$, or

$$z_i = \hat{F}_i(x_1,\ldots,x_{n_1}) \quad , \quad i \in \underline{n_2} \ , \tag{2.33}$$

is a smooth map from $\varphi(U) \subset \mathbb{R}^{n_1}$ to $\psi(V) \subset \mathbb{R}^{n_2}$.

F is called a *diffeomorphism* if F is bijective and both F and F^{-1} are smooth. Then M_1 and M_2 are called *diffeomorphic*, in particular $\dim M_1 = \dim M_2$. Note that if (U,φ) is a coordinate chart on a smooth manifold M then according to this definition $\varphi: U \to \mathbb{R}^n$ is a smooth map, and $\varphi: U \to \varphi(U)$ is a diffeomorphism.

The *rank* of a smooth map $F: M_1 \to M_2$ at a point $p \in M_1$ is defined as the rank of the Jacobian matrix $\frac{\partial F}{\partial x}(p)$ of F expressed in local coordinates (it is easily seen that this rank is independent of the choice of local coordinates), and is denoted as $\operatorname{rank}_p(F)$. If $\operatorname{rank}_p(F) = \dim M_1$ for all $p \in M_1$, then F is called an *immersion*, and if $\operatorname{rank}_p(F) = \dim M_2$ for all $p \in M_1$, then F is called a submersion or a *regular map*. A set of smooth functions $f_1, \ldots, f_k$ defined on some neighborhood U of p in M is called *independent* if the map $f = (f_1, \ldots, f_k)^T : U \to \mathbb{R}^k$ is such that $\operatorname{rank}_p(f) = k$ for all $p \in M$.

Proposition 2.16 *Let N and M be smooth manifolds, both of dimension n. Let $F : N \to M$ satisfy $\operatorname{rank}_p(F) = n$ for some $p \in N$. Then there is a neighborhood U of p such that $F : U \to F(U)$ is a diffeomorphism. Moreover if $\operatorname{rank}_p F = n$ for every $p \in N$ and F is bijective, then $F : N \to M$ is a diffeomorphism.*

Proof Choose coordinate charts (U,φ) about p and (V,ψ) about $F(p)$. By definition of $\operatorname{rank}_p(F)$ the rank of the Jacobian matrix of

$$\hat{F} = \psi \circ F \circ \varphi^{-1} : \varphi(U) \to \psi(V) \tag{2.34}$$

is n in $\varphi(p)$. Hence by the Inverse Function Theorem (Theorem 2.11) there exists a neighborhood $W \subset \varphi(U)$ such that $\hat{F} : W \to \hat{F}(W)$ is a diffeomorphism. Let now $V = \varphi^{-1}(W)$, then $F : V \to F(V)$ is a diffeomorphism. □

The following two consequences of the Inverse Function Theorem will be often used in the sequel.

Proposition 2.17 *Suppose that* $\dim M = n$, *and that $f_1, \ldots, f_n$ are independent functions about $p \in M$. Then there exists a neighborhood U about p such that $(U, f_1, \ldots, f_n)$ is a coordinate chart.*

Proof By the Inverse Function Theorem there exists a neighborhood U about p such that $f = (f_1, \ldots, f_n)^T : U \to f(U) \subset \mathbb{R}^n$ is a diffeomorphism. □

Proposition 2.18 *Suppose that* $\dim M = n$, *and that $f_1, \ldots, f_k$, $k \leq n$, are independent functions about $p \in M$. Then there exists a neighborhood U about p, and functions $x_{k+1}, \ldots, x_n$ such that $(U, f_1, \ldots, f_k, x_{k+1}, \ldots, x_n)$ is a coordinate chart.*

Proof Take an arbitrary coordinate chart $(\tilde{U}, x_1, \ldots, x_n)$ about p. By permuting $x_1, \ldots, x_n$ we can ensure that $f_1, \ldots, f_k, x_{k+1}, \ldots, x_n$ are independent functions about p. Then use Proposition 2.17. □

The next important notion is that of a submanifold. As remarked before (Example 2.9), an open set $V \subset M$ is itself a smooth manifold of dimension n (= dim M), with coordinate charts obtained by *restricting* the coordinate charts for M to V. V is called an *open submanifold* of M.

A subset $P \subset M$ is called a *submanifold* of dimension $m < n$ if for each $p \in P$ there exists a coordinate chart $(U,\varphi) = (U,x_1,\ldots,x_n)$ *for* M about p such that

$$P \cap U = \{q \in U \mid x_i(q) = x_i(p),\ i = m+1,\ldots,n\}. \tag{2.35}$$

$(U,x_1,\ldots,x_n)$ is called an *adapted* coordinate chart (adapted to P). If we take the subset topology on P (i.e., the open sets on P are of the form $P \cap U$, with U open in M), then P becomes a manifold as follows. Let (U,φ) be an adapted coordinate chart, then we let $(P \cap U,\varphi|P)$ to be a coordinate chart for P. Notice that by (2.32) the coordinates for P are $x_1,\ldots,x_m$.

An important theorem concerning submanifolds, which generalizes Theorem 2.13, is the following.

Theorem 2.19 *Let* $F : M_1 \to M_2$ *be a smooth map with* dim $M_i = n_i$, $i = 1,2$. *Let* $p_2 \in M_2$ *and suppose that* $F^{-1}(p_2)$ *is non-empty. Assume that the rank of* F *in every point of* $F^{-1}(p_2)$ *equals the dimension of* M_2. *Then*

$$F^{-1}(p_2) \text{ is a submanifold of } M_1, \text{ of dimension } n_1 - n_2. \tag{2.36}$$

Proof Let $(U,z_1,\ldots,z_{n_2})$ be a coordinate neighborhood of $p_2 \in M_2$. Let $p_1 \in F^{-1}(p_2)$. Since $F: F^{-1}(p_2) \to M_2$ is regular it follows from Proposition 2.18 that the collection of functions $x_i = z_i \circ F$, $i = 1,\ldots,n_2$ forms part of a coordinate system for M_1 about p_1, which can be completed to a coordinate system $(x_1,\ldots,x_{n_2},x_{n_2+1},\ldots,x_{n_1})$ on a neighborhood U of P_1. Then

$$F^{-1}(p_2) \cap U = \{q \in F^{-1}(p_2) \mid x_i(q) = x_i(p_1),\ i = 1,\ldots,n_1\}, \tag{2.37}$$

and so F^{-1} is a submanifold. □

It follows from Theorem 2.19 that Theorem 2.13 can be sharpened in the following way: N given by (2.28) is not only itself a smooth manifold, but also a submanifold of $\mathbb{R}^n$. Actually this situation is very general: it can be proved (but is outside the scope of these notes) that for an *arbitrary* smooth manifold M, there exists some Euclidian space $\mathbb{R}^m$ and a submanifold $P \subset \mathbb{R}^m$, such that M and P are diffeomorphic. Hence every abstract smooth manifold can be identified with a submanifold of some $\mathbb{R}^m$.

Finally, let F: $M_1 \to M_2$ be an injective immersion. One may expect that the image of M_1 under F is a submanifold of M_2. This is *not* exactly true. If fact only the following can be proved.

Proposition 2.20 *Let $F : M_1 \to M_2$ be an immersion and let $p_1 \in M_1$. Then there exist coordinate charts $(U, x_1, \ldots, x_{n_1})$ about p_1 and $(V, z_1, \ldots, z_{n_2})$ about $p_2 = F(p_1)$ such that F in these local coordinates equals*

$$(a_1, \ldots, a_{n_1}) \to (a_1, \ldots, a_{n_1}, 0, \ldots, 0) \tag{2.38}$$

For the proof of this proposition (which is a special case of the Rank Theorem for smooth maps) we refer to Exercise 2.5.

It follows from Proposition 2.20 that if F is an immersion then for any $p_1 \in M_1$ there exists a neighborhood U of p_1 such that $F(M_1 \cap U)$ is a submanifold of M_2. However in general $F(M_1)$ will *not* be a submanifold. Instead $F(M_1)$ is called an *immersed submanifold*. Things that can go wrong are of the following nature.

Fig. 2.4. $F(M_1)$ is an immersed submanifold.

The problems are caused by the fact that the topology of an immersed submanifold $F(M_1)$ may properly contain the topology on $F(M_1)$ as a subset of M_2, i.e. there may exist sets $F(U)$, with U open in M_1, which *cannot* be written as $V \cap F(M_1)$ for some open set V in M_2. If these topologies are equal, then we are in the situation of a normal submanifold.

2.1.2 Tangent Vectors, Tangent Space, Tangent Mappings

Let M be a smooth manifold of dimension n. If M is a submanifold of some $\mathbb{R}^m$, then geometrically the idea of tangent vector and tangent space at a point $p \in M$ is clear.

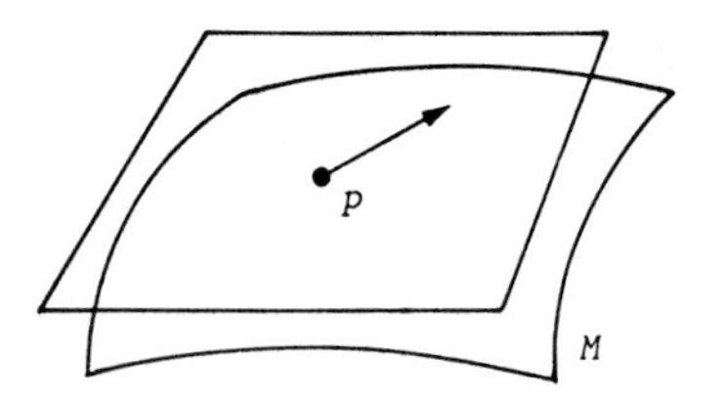

Fig. 2.5. Tangent space at p ∈ M.

We first proceed however in a much more formal way. Later on we will see how everything fits together. Denote by $C^\infty(p)$ the set of smooth functions defined on a neighborhood of p.

Definition 2.21 *We define the tangent space T_pM to M at p to be the linear space of mappings X_p: $C^\infty(p) \to \mathbb{R}$ satisfying for all $f,g \in C^\infty(p)$*

$$\begin{aligned} &(i) \quad X_p(\alpha f+\beta g) = \alpha X_p(f) + \beta X_p(g), \quad \alpha,\beta \in \mathbb{R}, \ (linearity) \\ &(ii) \quad X_p(fg) = X_p(f)\, g(p) + f(p) X_p(g) , \quad (product\ rule) \end{aligned} \tag{2.39}$$

with the vector space operations in T_pM defined by

$$\begin{aligned} (X_p+Y_p)(f) &= X_p(f) + Y_p(f) , \\ (\alpha X_p)(f) &= \alpha\, X_p(f) , \qquad \alpha \in \mathbb{R} . \end{aligned} \tag{2.40}$$

A tangent vector M at p is any $X_p \in T_pM$.

Remark The geometric meaning of $X_p(f)$ is that the function f is differentiated in $p \in M$ along a curve c: $(-\epsilon,\epsilon) \to M$ with $c(0) = p$ and $c'(0) = X_p$, i.e. $X_p(f) = (f \circ c)'(0)$. This will be made clear later on.

Let now F: $M \to N$ be smooth. Then we define a map (called the *tangent map* of F at p)

$$F_{*p}: T_pM \to T_{F(p)}N$$

as follows. Let $X_p \in T_pM$. For any $f \in C^\infty\big(F(p)\big)$ set

$$F_{*p}X_p(f) = X_p(f \circ F) . \tag{2.41}$$

It is easily checked that $F_{*p}X_p$ is indeed an element of $T_{F(p)}N$: part (i) of Definition 2.21 is trivally satisfied while for (ii) we compute

$$F_{*p}X_p(fg) = X_p\big((f\circ F).(g\circ F)\big) = X_p(f\circ F).g\big(F(p)\big) + f\big(F(p)\big).X_p(g\circ F)$$

$$= F_{*p}X_p(f).g\big(F(p)\big) + f\big(F(p)\big).F_{*p}X_p(g) \tag{2.42}$$

Furthermore F_{*p} is easily seen to be a *linear* map.

Proposition 2.22 The following properties of tangent maps are immediate.

(a) *If $H = G\circ F$ is a composition of smooth maps, then $H_{*p} = G_{*F(p)}\circ F_{*p}$.*

(b) *Let $id: M \to M$ be the identity mapping, then $id_{*p}: T_pM \to T_pM$ is the identity matrix I.*

(c) *For a diffeomorphism $F : M \to N$ we have*

$$I = F^{-1}_{*F(p)}\circ F_{*p} = F_{*p}\circ F^{-1}_{*F(p)}. \tag{2.43}$$

Hence $\dim T_pM = \dim T_{F(p)}N = \operatorname{rank}(F_{*p}) = \operatorname{rank}(F^{-1}_{*F(p)})$.

Let M be a smooth manifold, and let (U,φ) be a coordinate chart about $p \in M$. Recalling that any open set of a smooth manifold is itself a smooth manifold of the same dimension, we have that $\varphi: U \to \varphi(U) \subset \mathbb{R}^n$ is a diffeomorphism. Therefore by Proposition 2.22 (c)

$$\varphi_{*p}: T_pM \to T_{\varphi(p)}\mathbb{R}^n \tag{2.44}$$

is a linear bijection. Hence the study of tangent spaces to an arbitrary manifold can be reduced to tangent spaces of $\mathbb{R}^n$. So let us first consider the tangent space $T_a\mathbb{R}^n$ for $a \in \mathbb{R}^n$. Geometrically it is clear that $T_a\mathbb{R}^n$ for any $a \in \mathbb{R}^n$ can be identified with $\mathbb{R}^n$. Formally we proceed as follows. Define elements $E_{1a},\ldots,E_{na}$ in $T_a\mathbb{R}^n$ by letting

$$E_{ia}(f) = \frac{\partial f}{\partial r_i}(a) \;, \qquad i \in \underline{n}, \qquad f \in C^\infty(a) \;, \tag{2.45}$$

i.e., E_{ia} equals the directional derivatives $\frac{\partial}{\partial r_i}\Big|_a$, $i \in \underline{n}$. It is easily checked that E_{ia}, $i \in \underline{n}$ are in $T_a\mathbb{R}^n$. In particular we have

$$E_{ia}(r_j) = \delta_{ij} \;, \qquad i,j \in \underline{n} \;, \tag{2.46}$$

for $r_1,\ldots,r_n$ the natural coordinate functions on $\mathbb{R}^n$. Since $r_1,\ldots,r_n$ are independent it follows that $E_{1a},\ldots,E_{na}$ are independent vectors in $T_a\mathbb{R}^n$. We have to prove that $\{E_{1a},\ldots,E_{na}\}$ is a *basis* for $T_a\mathbb{R}^n$. For this we need the following simple lemma.

Lemma 2.23 *Let $f \in C^\infty(a)$, with $f(a) = 0$. Then for $(z_1,\ldots,z_n)$ in a sphere about the origin*

$$f(a_1+z_1,\ldots,a_n+z_n) = \sum_{i=1}^{n} z_i g_i(a_1+z_1,\ldots,a_n+z_n) , \tag{2.47}$$

for certain functions $g_1,\ldots,g_n$ *satisfying* $g_i(a) = \frac{\partial f}{\partial r_i}(a)$.

Proof Define $h(t,z) = f(a+tz)$, $0 \le t \le 1$. Then

$$\begin{aligned} f(a+z) &= \int_0^1 \frac{\partial h}{\partial t}(t,z)dt = \int_0^1 \sum_{i=1}^{n} \frac{\partial f}{\partial r_i}(a+tz).z_i dt \\ &= \sum_{i=1}^{n} z_i . \int_0^1 \frac{\partial f}{\partial r_i}(a+tz)dt =: \sum_{i=1}^{n} z_i g_i(a+z) . \end{aligned} \tag{2.48}$$

□

By this lemma we can write for any $f \in C^\infty(a)$

$$f = f(a) + \sum_{i=1}^{n} (r_i - a_i)g_i , \text{ with } g_i(a) = \frac{\partial f}{\partial r_i}(a) . \tag{2.49}$$

Therefore for an arbitrary element $E_a \in T_a\mathbb{R}^n$ we have by the product rule

$$\begin{aligned} E_a(f) &= \sum_{i=1}^{n} E_a(r_i)g_i(a) + \sum_{i=1}^{n} (r_i(a)-a_i)E_a(g_i) \\ &= \sum_{i=1}^{n} E_a(r_i)E_{ia}(f) =: \sum_{i=1}^{n} \alpha_i E_{ia}(f) . \end{aligned} \tag{2.50}$$

Hence indeed $\{E_{1a},\ldots,E_{na}\}$ is a basis of $T_a\mathbb{R}^n$. Since $\varphi_{*p}: T_pM \to T_{\varphi(p)}\mathbb{R}^n$ is a bijection it follows that we can define a natural basis for T_pM (natural with respect to the coordinate chart $(U,\varphi) = (U,x_1,\ldots,x_n)$)

$$\frac{\partial}{\partial x_1}\Big|_p , \ldots , \frac{\partial}{\partial x_n}\Big|_p ,$$

by letting

$$\frac{\partial}{\partial x_i}\Big|_p := \varphi_{*p}^{-1}(E_{i\varphi(p)}) \quad , \quad i \in \underline{n} . \tag{2.51}$$

With this definition we have for any $f \in C^\infty(p)$, $p \in M$,

$$\begin{aligned} \frac{\partial}{\partial x_i}\Big|_p(f) &= \varphi_{*p}^{-1}(E_{i\varphi(p)})(f) = E_{i\varphi(p)}(f\circ\varphi^{-1}) \\ &= E_{i\varphi(p)}(\hat{f}) = \frac{\partial \hat{f}}{\partial r_i}\big(x_1(p),\ldots,x_n(p)\big) , \end{aligned} \tag{2.52}$$

where $\hat{f}$ is the local representative of f. Hence $\frac{\partial}{\partial x_i}\Big|_p(f)$ is just the i-th partial derivative of f expressed in local coordinates $x_1,\ldots,x_n$. Instead of the cumbersome notation $\frac{\partial}{\partial x_i}\Big|_p(f)$ we will usually simply write

$$\frac{\partial}{\partial x_i}\Big|_p (f) = \frac{\partial f}{\partial x_i}(p), \qquad f \in C^\infty(p) \quad . \tag{2.53}$$

Let now X_p be an arbitrary element of T_pM. It follows that we can write

$$X_p = \sum_{i=1}^{n} \alpha_i \frac{\partial}{\partial x_i}\Big|_p \tag{2.54}$$

for certain constants $\alpha_1, \ldots, \alpha_n \in \mathbb{R}$. Then for $f \in C^\infty(p)$

$$X_p(f) = \sum_{i=1}^{n} \alpha_i \frac{\partial f}{\partial x_i}(p), \tag{2.55}$$

and by (2.52)

$$X_p(f) = \sum_{i=1}^{n} \alpha_i \frac{\partial \hat{f}}{\partial r_i} \left(x_1(p), \ldots, x_n(p)\right). \tag{2.56}$$

Hence $X_p(f)$ is just the directional derivative of the function f (or its local representative $\hat{f}$) in the direction of the vector $(\alpha_1, \ldots, \alpha_n)^T$. Notice that $(\alpha_1, \ldots, \alpha_n)^T$ is the vector representation of X_p in the basis $\frac{\partial}{\partial x_1}\Big|_p, \ldots, \frac{\partial}{\partial x_n}\Big|_p$ for T_pM, and equivalently of $\varphi_{*p}X_p$ in the basis $E_{i\varphi(p)} = \frac{\partial}{\partial r_i}\Big|_{\varphi(p)}$, $i \in \underline{n}$, for $T_{\varphi(p)}\mathbb{R}^n$.

Let now (U,φ) and (V,ψ) be coordinate charts about p, with local coordinates $x_1, \ldots, x_n$, respectively $z_1, \ldots, z_n$. Let $\frac{\partial}{\partial x_i}\Big|_p = \varphi_{*p}^{-1}E_{i\varphi(p)}$ and $\frac{\partial}{\partial z_i}\Big|_p = \psi_{*p}^{-1}E_{i\psi(p)}$ be the corresponding bases of T_pM. Denote the coordinate transformation as $z = \psi \circ \varphi^{-1}(x) = S(x)$. Let $X_p \in T_pM$ be represented in these bases as $\sum_{i=1}^{n} \alpha_i \frac{\partial}{\partial x_i}\Big|_p$, respectively $\sum_{i=1}^{n} \beta_i \frac{\partial}{\partial z_i}\Big|_p$. The coefficients α_i and β_i are related as

$$\beta_i = \sum_{j=1}^{n} \frac{\partial S_i}{\partial x_j} \left(x(p)\right)\alpha_j \quad , \tag{2.57}$$

or, with $\alpha = (\alpha_1, \ldots, \alpha_n)^T$ and $\beta = (\beta_1, \ldots, \beta_n)^T$,

$$\beta = \frac{\partial S}{\partial x} \left(x(p)\right)\alpha \quad . \tag{2.58}$$

(Classically it is said that under a coordinate transformation S the coefficients of a tangent vector transform in a *contravariant* fashion).

The above can be conveniently used to derive a standard formula which gives the matrix representation of the tangent map F_{*p} relative to local coordinates. Namely let $F\colon M_1 \to M_2$, and let $(U, x_1, \ldots, x_{n_1})$ and $(V, z_1, \ldots, z_{n_2})$ be coordinate charts about $p \in M_1$, respectively about $F(p) \in M_2$, resulting in the basis $\frac{\partial}{\partial x_i}\Big|_p$, $i \in \underline{n_1}$ respectively $\frac{\partial}{\partial z_j}\Big|_{F(p)}$,

$j \in \underline{n}_2$. Then

$$F_{*p}\left(\frac{\partial}{\partial x_i}\bigg|_p\right)(z_j) = \frac{\partial}{\partial x_i}(z_j \circ F)\bigg|_p = \frac{\partial F_j}{\partial x_i}(p) \tag{2.59}$$

where $F_j = z_j \circ F$ is the j-th component of F. Hence F_{*p} in bases for T_pM_1 and $T_{F(p)}M_2$ is given by the Jacobian matrix of the local representative $\hat{F}$ of F.

Corollary 2.24 $\operatorname{rank}(F_{*p}) = \operatorname{rank}\left(\frac{\partial F}{\partial x}(x(p))\right)$.

The *tangent bundle* of a smooth manifold M is simply defined as

$$TM = \bigcup_{p \in M} T_pM \tag{2.60}$$

It follows that there is a natural projection π: $TM \to M$ taking a tangent vector $X_p \in T_pM \subset TM$ to $p \in M$. TM can be given a smooth manifold structure as follows. First notice that $T\mathbb{R}^n = \bigcup_{a \in \mathbb{R}^n} T_a\mathbb{R}^n \cong \bigcup_{a \in \mathbb{R}^n} \mathbb{R}^n \cong \mathbb{R}^n \times \mathbb{R}^n = \mathbb{R}^{2n}$, and in this way $T\mathbb{R}^n$ evidently is a smooth manifold. Then for a coordinate chart $(U,\varphi) = (U,x_1,\ldots,x_n)$ on M, we define a coordinate chart $(\tilde{U},\tilde{\varphi})$ on TM by setting

$$\begin{aligned} \tilde{U} &= \pi^{-1}(U) \\ \tilde{\varphi}(X_p) &= \left(x_1(p),\ldots,x_n(p),x_{1*p}X_p,\ldots,x_{n*p}X_p\right) \end{aligned} \tag{2.61}$$

with $x_{i*p}X_p \in T_{x_i(p)}\mathbb{R} \cong \mathbb{R}$ $(x_i : U \subset M \to \mathbb{R})$, $i \in \underline{n}$. This defines a C^∞-atlas for TM. The resulting local coordinates for TM are also denoted as $x_1,\ldots,x_n,v_1,\ldots,v_n$ (with $v_i(X_p) = x_{i*p}X_p$). They are called *natural* because of the following. Let r be the natural coordinate function on $\mathbb{R}$. For an $X_p \in T_pM$ expressed in the basis $\frac{\partial}{\partial x_1}\big|_p,\ldots,\frac{\partial}{\partial x_n}\big|_p$ as $X_p = \sum_{i=1}^n \alpha_i \frac{\partial}{\partial x_i}\big|_p$ we have

$$x_{i*p}X_p(r) = X_p(r \circ x_i) = X_p(x_i) = \alpha_i \tag{2.62}$$

and so $x_{i*p}X_p = \alpha_i \frac{\partial}{\partial r}$. By the identification $T_{x_i(p)}\mathbb{R} \cong \mathbb{R}$ (i.e. $\alpha_i \frac{\partial}{\partial r}$ with α_i) we therefore obtain $v_i(X_p) = x_{i*p}X_p = \alpha_i$.

Finally, let F: $M_1 \to M_2$ be a smooth map, then we define the *tangent mapping*

$$F_* : TM_1 \to TM_2$$

as the union of all tangent mappings $F_{*p}: T_pM_1 \to T_{F(p)}M_2$ for $p \in M_1$. Let $x = (x_1, \ldots, x_{n_1})$ be local coordinates for M_1, and $z = (z_1, \ldots, z_{n_2})$ for M_2, then in the natural coordinates $(x,v) = (x_1, \ldots, x_{n_1}, v_1, \ldots, v_{n_1})$ for M_1 respectively $(z,w) = (z_1, \ldots, z_{n_2}, w_1, \ldots, w_{n_2})$ for M_2, the tangent mapping F_* is given as

$$F_*(x,v) = \left(F(x), \frac{\partial F}{\partial x}(x)v\right) \tag{2.63}$$

2.2 Vectorfields, Lie Brackets, Distributions, Frobenius' Theorem, Differential One-Forms

2.2.1 Vectorfields, Lie Brackets, Lie Algebras

Definition 2.25 *A smooth vectorfield X on a smooth manifold M is defined as a map*

$$X: M \to TM \tag{2.64}$$

satisfying

$$\pi \circ X = \text{identity on } M \tag{2.65}$$

(with π the natural projection from TM on M).

Replacing "smooth" throughout by C^ω, resp. C^k, in the above definition, the vectorfield is called C^ω, resp. C^k. In the sequel we will drop the adjective "smooth". *Thus throughout we will assume that all maps, functions, manifolds, vectorfields are smooth* (C^∞), unless stated explicity otherwise.

It follows that a vectorfield is a map which assigns to every $p \in M$ a tangent vector $X_p \in T_pM$ in a smooth way, as illustrated in the next figure.

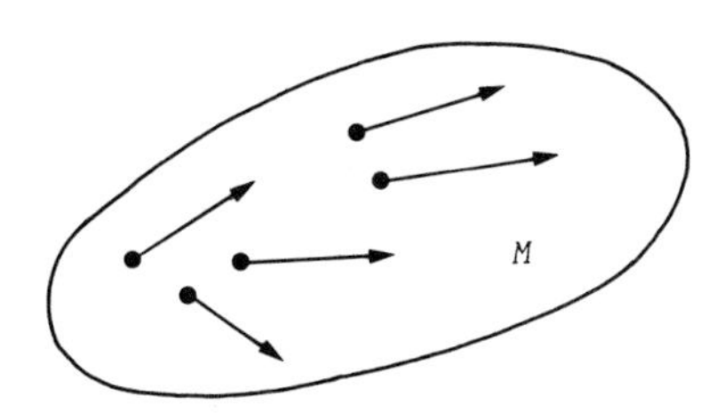

Fig. 2.6. Geometrical picture of a vectorfield.

Let $(U,\varphi) = (U,x_1,\ldots,x_n)$ be a local coordinate chart for M, inducing the natural local coordinate chart $(\pi^{-1}(U),\varphi_*) = (U,x_1,\ldots,x_n,v_1,\ldots,v_n)$ for TM. Then the local representative of $X\colon M \to TM$ is the map

$$\hat{X} = \varphi_* \circ X \circ \varphi^{-1}\colon \varphi(U) \subset \mathbb{R}^n \to T\varphi(U) \cong \varphi(U) \times \mathbb{R}^n \tag{2.66}$$

which, because of (2.65), can be written as

$$\hat{X}(x_1,\ldots,x_n) = \big(x_1,\ldots,x_n,\hat{X}_1(x),\ldots,\hat{X}_n(x)\big) , \tag{2.67}$$

for some functions $\hat{X}_i(x_1,\ldots,x_n)$, $i \in n$. In fact as follows from the preceding section these functions $\hat{X}_i(x)$ are given by the formula

$$X(p) = \sum_{i=1}^{n} \hat{X}_i\big(x(p)\big)\frac{\partial}{\partial x_i}\Big|_p . \tag{2.68}$$

Equivalently, the local representative of X is given by the vectorfield $\hat{X}$ on $\varphi(U) \subset \mathbb{R}^n$ given as

$$\hat{X}\big(x(p)\big) = \varphi_{*p}X(p) = \sum_{i=1}^{n} \hat{X}_i\big(x(p)\big)\frac{\partial}{\partial r_i}\Big|_{\varphi(p)} . \tag{2.69}$$

Notice also that if we write

$$X(p) = \sum_{i=1}^{n} X_i(p)\frac{\partial}{\partial x_i}\Big|_p , \quad p \in M , \tag{2.70}$$

for functions $X_1,\ldots,X_n\colon U \subset M \to \mathbb{R}$, then it follows that $\hat{X}_i = X_i \circ \varphi^{-1}$, i.e. $\hat{X}_i$ is the local representative of these functions X_i.

It is customary (but at first reading a little confusing!) to omit all the carets. Furthermore often $\frac{\partial}{\partial r_i}$ in (2.69) will be simply replaced by $\frac{\partial}{\partial x_i}$. Hence we usually write X in local coordinates $x_1,\ldots,x_n$ as $\sum_{i=1}^{n} X_i(x)\frac{\partial}{\partial x_i}$, or as the vector

$$X(x_1,\ldots,x_n) = \begin{bmatrix} X_1(x_1,\ldots,x_n) \\ \vdots \\ X_n(x_1,\ldots,x_n) \end{bmatrix} . \tag{2.71}$$

where X_i are of course the functions $\hat{X}_i$ from (2.67)-(2.69).

Let now $\sigma\colon (a,b) \to M$ be a smooth curve in M. For $t \in (a,b)$ we define

$$\dot{\sigma}(t) := \sigma_{*t}\left(\frac{\partial}{\partial t}\Big|_t\right) \in T_{\sigma(t)}M \tag{2.72}$$

(with t the natural coordinate on $(a,b) \subset \mathbb{R}$). We say that σ is an *integral curve* of a given vectorfield X on M if

$$\dot{\sigma}(t) = X\big(\sigma(t)\big), \qquad \forall t \in (a,b). \tag{2.73}$$

In local coordinates $x_1,\dots,x_n$ this just means that $\sigma(t) = \big(\sigma_1(t),\dots,\sigma_n(t)\big)$ is a solution of the set of differential equations

$$\begin{cases} \dot{\sigma}_1(t) = X_1\big(\sigma_1(t),\dots,\sigma_n(t)\big) \\ \quad\vdots \\ \dot{\sigma}_n(t) = X_n\big(\sigma_1(t),\dots,\sigma_n(t)\big) \end{cases} \qquad t \in (a,b) , \tag{2.74}$$

with X_i as in (2.71). So, to a vectorfield X given in local coordinates as in (2.71) we associate in a one-to-one way the set of differential equations

$$\begin{cases} \dot{x}_1 = X_1(x_1,\dots,x_n) \\ \quad\vdots \\ \dot{x}_n = X_n(x_1,\dots,x_n) \end{cases} \tag{2.75}$$

also abbreviated as

$$\dot{x} = X(x) , \tag{2.76}$$

where $x = (x_1,\dots,x_n)$ is the vector of local coordinates for M. (Note the slight abuse of notation, since x on the left-hand side is actually a *column*-vector in $\mathbb{R}^n$.)

By the existence and uniqueness theorem for smooth differential equations it follows that for any $p \in M$ there exists an interval (a,b) of maximal length containing 0 and a unique integral curve $\sigma(t)$, $t \in (a,b)$ with $\sigma(0) = p$. If for every p we have $(a,b) = (-\infty,\infty)$, and so solutions are defined for all time t the vectorfield X is called *complete*. Note that vectorfields on compact manifolds are always complete, the only thing that can go wrong in general is that in finite time solutions tend to infinity (or to the boundary of the manifold, which itself does not belong to the manifold). In any case, for every bounded set $U \subset M$, there exists an interval (a,b) containing 0 such that for any $p \in U$ the integral curves $\sigma_p(t)$ with $\sigma_p(0) = p$ are defined for all $t \in (a,b)$. This allows us to define on U a set of maps (time t-integral or *flow*)

$$X^t\colon U \to M \quad , \quad t \in (a,b) , \tag{2.77}$$

by letting $X^t(p)$ be the solution of the differential equation (2.75) for time t with initial condition at time 0 the point p, i.e. $X^t(p) = \sigma_p(t)$.

It follows from the theory of differential equations that the maps X^t are *smooth*.

By definition a vectorfield X defines in any $p \in M$ a tangent vector $X(p)$. For $f\colon M \to \mathbb{R}$ this yields in any $p \in M$ the directional derivative $X(p)(f)$. Hence by varying p we obtain a smooth function $X(f)$ defined as

$$X(f)(p) := X(p)(f) \ . \tag{2.78}$$

The function $X(f)\colon M \to \mathbb{R}$ will be called the total derivative of f along the vectorfield X, or the *Lie derivative* of f along X and is also denoted as $L_X f$. Notice that if X is expressed in local coordinates as the vector $(X_1(x),\ldots,X_n(x))^T$ then we have

$$L_X f(p) = X(f)(p) = \sum_{i=1}^{n} \frac{\partial f}{\partial x_i}\big(x_1(p),\ldots,x_n(p)\big) X_i\big(x_1(p),\ldots,x_n(p)\big). \tag{2.79}$$

Furthermore we have

$$X(f)(p) = \lim_{h\to 0} \frac{f\big(X^h(p)\big) - f(p)}{h}. \tag{2.80}$$

It is now clear how we can give a global, coordinate-free definition of a smooth nonlinear control system, given in local coordinates as

$$\dot{x} = f(x,u). \tag{2.81}$$

Indeed let M be the state space of the control system and let U be the input space, then the system is given by a smooth map (the system map)

$$f\colon M \times U \to TM \tag{2.82}$$

with the requirement that $\pi \circ f$ equals the natural projection of $M \times U$ onto M. With the same abuse of notation as above, f is represented in local coordinates x for M, natural local coordinates (x,v) for TM, and local coordinates u for U as

$$f(x,u) = \big(x, f(x,u)\big) \tag{2.83}$$

and so we recover the local coordinate expression $\dot{x} = f(x,u)$.

Remark In Chapter 1 it was indicated that in some cases the above definition is still not general enough. The problem is in writing the *product* $M \times U$; this implies that the input space is globally *independent* of the state of the system. In order to deal with situations where this is not the case we have to replace $M \times U$ by a *fiber bundle* above the base space M, with fibers diffeomorphic to U. As a result we have only *locally* a product $M \times U$. This is discussed in Section 13.5.

In case the system map $f: M \times U \to TM$ is *affine* in the u-variables we write (with the addition and multiplication defined in the linear space T_xM)

$$f(x,u) = f(x) + \sum_{j=1}^{m} g_j(x)u_j \tag{2.84}$$

for some functions $f, g_1, \ldots, g_m: M \to TM$ satisfying $\pi \circ f = \pi \circ g_j$ = identity on M, which hence are *vectorfields* on M.

Now let us return to the study of vectorfields. Since tangent vectors transform under a coordinate transformation in a contravariant fashion (see (2.58)) also vectorfields do. In fact let X be given in local coordinates $(U,\varphi) = (U, x_1, \ldots, x_n)$ as $X = \sum_{i=1}^{n} X_i(x)\frac{\partial}{\partial x_i}$ and in local coordinates $(V,\psi) = (V, z_1, \ldots, z_n)$ as $X = \sum_{i=1}^{n} Z_i(z)\frac{\partial}{\partial z_i}$, then with $z = S(x)$ we have

$$\begin{pmatrix} Z_1(S(x)) \\ \vdots \\ Z_n(S(x)) \end{pmatrix} = \frac{\partial S}{\partial x}(x) \begin{pmatrix} X_1(x) \\ \vdots \\ X_n(x) \end{pmatrix} \tag{2.85}$$

For convenience we will introduce a new notation. Let $F: M \to N$ be a diffeomorphism, and let X be a vectorfield on M. Then we can define a vectorfield Y on N by letting

$$Y_{F(p)} = F_{*p}X_p \quad , \quad \text{for any } p \in M \; . \tag{2.86}$$

If F is *not* a diffeomorphism then y is a well-defined vectorfield on N if and only if F and X are such that $F_{*p_1}X_{p_1} = F_{*p_2}X_{p_2}$, whenever $F(p_1) = F(p_2)$. We will abbreviate (2.86) as

$$Y = F_* X \; . \tag{2.87}$$

If $F_*X = Y$ for vectorfields X on M and Y on N (F not necessarily being a diffeomorphism) then we say that X and Y are *F-related*. Note that by (2.86) we have for any function $g: N \to \mathbb{R}$

$$F_*X(g) = \big(X(g \circ F)\big) \circ F^{-1} \; . \tag{2.88}$$

Hence the Lie derivative of g along F_*X in a point $p \in N$ is computed by taking the Lie derivative of the function $g \circ F: M \to \mathbb{R}$ in the point $F^{-1}(p) \in M$.

We warn the reader that we are now using the notation F_* in two, slightly different, ways: (a) as a map $F_*: TM \to TN$, and (b) as a map from vectorfields on M to vectorfields on N.

The following theorem shows that outside equilibria vectorfields can be given a very simple form.

Theorem 2.26 (Flow-box Theorem) *Let X be a vectorfield on M with $X(p) \neq 0$. Then there exists a coordinate chart $(U, x_1, \ldots, x_n)$ around p such that*

$$X = \frac{\partial}{\partial x_1} \qquad \text{on } U \ . \tag{2.89}$$

Geometrically this means that around p the integral curves of X are of the form $x_i(q) = constant$, $i = 2, \ldots, n$.

Proof Let $(V, \psi) = (V, z_1, \ldots, z_n)$ be a coordinate chart with $\psi(p) = 0$ and $\psi(V)$ bounded such that $\psi_{*p}X = \frac{\partial}{\partial r_1}\Big|_0$. Define locally around 0 the map $T: \mathbb{R}^n \to \mathbb{R}^n$

$$T(a_1, \ldots, a_n) = (\psi_* X)^{a^1}(0, a^2, \ldots, a^n) \tag{2.90}$$

(i.e. the time-a^1 integral of the vectorfield $\psi_* X$ in local coordinates $(z_1, \ldots, z_n)$. It is easily checked that

$$T_*\left(\frac{\partial}{\partial r_1}\right) = \psi_* X \tag{2.91}$$

and that the T_{*0} equals the identity matrix, implying that T is a coordinate transformation. Hence $S := T^{-1}$ is the desired coordinate transformation. □

For X and Y any two (smooth) vectorfields on M, we define a new vectorfield, denoted as $[X,Y]$ and called the *Lie bracket* of X and Y by setting

$$[X,Y]_p(f) = X_p\big(Y(f)\big) - Y_p\big(X(f)\big). \tag{2.92}$$

In order that $[X,Y]_p \in T_pM$ we have to check conditions (i), (ii) of Definition 2.21. Condition (i) is trivial, while (ii) follows from

$$\begin{aligned}
[X,Y]_p(fg) &= X_p\big(Y(fg)\big) - Y_p\big(X(fg)\big) = \\
&= X_p\{Y(f)\cdot g + f\cdot Y(g)\} - Y_p\{X(f)\cdot g + f\cdot X(g)\} = \\
&= X_p\big(Y(f)\big)g(p) + Y_p(f)X_p(g) + X_p(f)Y_p(g) + f(p)X_p\big(Y(g)\big) + \\
&\quad - Y_p\big(X(f)\big)g(p) - X_p(f)Y_p(g) - Y_p(f)X_p(g) - f(p)Y_p\big(X(g)\big) \\
&= [X,Y]_p(f)\cdot g(p) + f(p)\cdot[X,Y]_p(g).
\end{aligned} \tag{2.93}$$

If X and Y are given in local coordinates $(x_1,\dots,x_n)$ as the vectors $X(x) = \big(X_1(x),\dots,X_n(x)\big)^T$, respectively $Y(x) = \big(Y_1(x),\dots,Y_n(x)\big)^T$, then $[X,Y]\big(x(p)\big)$ is given as the vector

$$\left[\frac{\partial Y}{\partial x}\big(x(p)\big)\right]X\big(x(p)\big) - \left[\frac{\partial X}{\partial x}\big(x(p)\big)\right]Y\big(x(p)\big), \tag{2.94}$$

as follows from computing (2.92) for the coordinate functions $x_1,\dots,x_n$. Indeed let $X_q = \sum_{i=1}^{n} X_i\big(x(q)\big)\frac{\partial}{\partial x_i}\Big|_q$ and $Y_q = \sum_{i=1}^{n} Y_i\big(x(q)\big)\frac{\partial}{\partial x_i}\Big|_q$ for q in the coordinate chart, then for $j \in \underline{n}$

$$\begin{aligned}
&[X,Y]_p(x_j) = X_p\big(Y(x_j)\big) - Y_p\big(X(x_j)\big) \\
&= X_p(Y_j) - Y_p(X_j) = \sum_{i=1}^{n}\left[\frac{\partial Y_j}{\partial x_i}X_i - \frac{\partial X_j}{\partial x_i}Y_i\right]\big(x(p)\big)
\end{aligned} \tag{2.95}$$

and therefore

$$[X,Y] = \sum_{j=1}^{n}\left(\sum_{i=1}^{n}\frac{\partial X_j}{\partial x_i}X_i - \frac{\partial X_j}{\partial x_i}Y_i\right)\frac{\partial}{\partial x_j}. \tag{2.96}$$

It immediately follows from (2.94) that $[X,Y]_p$ depends in a smooth way on p, so that indeed $[X,Y]$ is a smooth vectorfield.

The following properties of the Lie bracket follow immediately from the definition.

Proposition 2.27 *For any vectorfields X, Y, Z and functions f, g on a manifold M*

(a) $[fX,gY] = fg[X,Y] + f\cdot X(g)\cdot Y - g\cdot Y(f)\cdot X$,

(b) $[X,Y] = -[Y,X]$, (2.97)

(c) $[[X,Y],Z] + [[Y,Z],X] + [[Z,X],Y] = 0$ (Jacobi-identity).

Before going on we give the general definition of a *Lie algebra*.

Definition 2.28 *A vector space V (over $\mathbb{R}$) is a Lie algebra if in addition to the linear structure there is a binary operation $V \times V \to V$, denoted by $[\ ,\ ]$, satisfying*

(i) $[\alpha_1 v_1 + \alpha_2 v_2, w] = \alpha_1[v_1,w] + \alpha_2[v_2,w]$, $\forall v_1,v_2,w \in V$, $\forall \alpha_1,\alpha_2 \in \mathbb{R}$, *(bilinearity)*

(ii) $[v,w] = -[w,v]$, $\forall v,w \in V$ *(anti-symmetry)*, (2.98)

(iii) $[v,[w,z]] + [w,[z,v]] + [z,[v,w]] = 0$, $\forall v,w,z \in V$ *(Jacobi-identity)*

A subalgebra of a Lie algebra $(V,[\ ,\ ])$ *is a linear subspace* $V' \subset V$ *such that* $[v',w'] \in V'$ *for all* $v',w' \in V'$.

Remark 2.29 The most well-known example of a Lie algebra is the linear space of $n\times n$ matrices with bracket operation

$$[A,B] = AB - BA \quad , \quad A,B \ n\times n \text{ matrices} . \tag{2.99}$$

An example of a subalgebra of this Lie algebra is the space of skew-symmetric matrices.

Now let us consider the linear space of C^∞ vectorfields on M, denoted by $V^\infty(M)$, and take as bracket operation the Lie bracket of two vectorfields defined above. Properties (ii) and (iii) in Definition 2.28 follow from Proposition 2.27 (b), respectively (c), while property (i) is trivially satisfied. Hence $V^\infty(M)$ together with the Lie bracket is a Lie algebra, in fact an infinite-dimensional Lie algebra.

A crucial property of the Lie bracket is the following

Proposition 2.30 *Let* $F: M \to N$, *and suppose that* $F_* X_i = Y_i$, $i = 1,2$, *for vectorfields* X_1,X_2 *and* Y_1,Y_2 *on* M *respectively* N. *Then*

$$F_*[X_1,X_2] = [Y_1,Y_2] . \tag{2.100}$$

Proof By (2.88) we have for any function $g: N \to \mathbb{R}$

$$Y_i(g)\circ F = X_i(g\circ F) \quad , \quad i = 1,2 . \tag{2.101}$$

Therefore

$$\begin{aligned} \{[Y_1,Y_2](g)\}\circ F &= \{Y_1(Y_2(g))\}\circ F - \{Y_2(Y_1(g))\}\circ F \\ &= X_1(Y_2(g)\circ F) - X_2(Y_1(g)\circ F) \end{aligned} \tag{2.102}$$

by (2.101) with g replaced by $Y_2(g)$, respectively $Y_1(g)$. By another application of (2.101) this equals

$$X_1(X_2(g\circ F)) - X_2(X_1(g\circ F)) = [X_1,X_2](g\circ F) , \tag{2.103}$$

and hence by (2.101) with X_i and Y_i replaced by $[X_1,X_2]$, respectively $[Y_1,Y_2]$, we have $F_*[X_1,X_2] = [Y_1,Y_2]$. □

In order to develop an *interpretation* of $[X,Y]$ we first prove some lemmas. The first is immediate.

Lemma 2.31 *Let $F: M \to N$ be a diffeomorphism, and X a vectorfield on M with flow X^t. Then the flow Y^t of the vectorfield $Y = F_*X$ on N equals $F \circ X^t \circ F^{-1}$.*

Proof $(F_*X)(g)(q) = \big(X(g \circ F)\big) \cdot \big(F^{-1}(q)\big) =$

$$\lim_{h \to 0} \frac{1}{h}\Big[(g \circ F)\Big(X^h\big(F^{-1}(q)\big)\Big) - (g \circ F)\big(F^{-1}(q)\big)\Big] = \tag{2.104}$$

$$\lim_{h \to 0} \frac{1}{h}\Big[g\big(F \circ X^h \circ F^{-1}(q)\big) - g(q)\Big] .$$

□

This lemma just expresses that if $F_*X = Y$, then F maps the integral curves of X onto the integral curves of Y.

Corollary 2.32 *$F_*X = X$ for $F: M \to M$, if and only if $X^t \circ F = F \circ X^t$ for all t.*

Secondly we need the following derivational interpretation of the Lie bracket.

Theorem 2.33 *For any vectorfields X and Y on M,*

$$[X,Y](p) = \lim_{h \to 0} \frac{1}{h}\big[(X_*^{-h}Y)(p) - Y(p)\big] \quad , \; p \in M. \tag{2.105}$$

Proof Write out the right-hand side of (2.105) in local coordinates, and check equality with (2.94). □

We see that $[X,Y]$ can be interpreted in some sense as the "derivative" of the vectorfield Y along X. It is therefore also denoted as L_XY, the Lie derivative of Y along X. The following lemma is crucial in the interpretation of $[X,Y]$.

Lemma 2.34 *Let X and Y be vectorfields, with flows X^t, Y^t. Then $[X,Y] = 0$ if and only if $X^t \circ Y^s = Y^s \circ X^t$, for all s,t for which X^t and Y^s are defined.*

Proof $Y^s \circ X^t = X^t \circ Y^s$ for all s if and only if $X_*^tY = Y$ by Corollary 2.32. If this is true for all t then by Theorem 2.33 $[X,Y] = 0$. Conversely assume that $[X,Y] = 0$, so that

$$0 = \lim_{h \to 0} \frac{1}{h}\big[(X_*^{-1}Y)(q) - Y(q)\big] \quad , \quad \text{for all } q . \tag{2.106}$$

Given $p \in M$, consider the curve $c: (-\epsilon, \epsilon) \to T_pM$ given by $c(t) = (X_*^tY)(p)$.

Then

$$\begin{aligned} c'(t) &= \lim_{h\to 0} \frac{1}{h}\left[c(t-h) - c(t)\right] \\ &= \lim_{h\to 0} \frac{1}{h}\left[(X_*^{t-h}Y)_p - (X_*^{t}Y)_p\right] \\ &= \lim_{h\to 0} \frac{1}{h}\left[X_*^{t}(X_*^{-h}Y)_{X^{-t}(p)} - X_*^{t}(Y)_{X^{-t}(p)}\right] \\ &= X_*^{t} \lim_{h\to 0} \frac{1}{h}\left[(X_*^{-h}Y)_{X^{-t}(p)} - Y_{X^{-t}(p)}\right] \\ &= 0, \text{ by (2.106) for } q = X^{-t}(p) \end{aligned} \tag{2.107}$$

Consequently $c(t) = c(0)$, so that $X_*^t Y = Y$ for all t. As remarked before this implies that $Y^s \circ X^t = X^t \circ Y^s$ for all s,t. □

It follows that $[X,Y] = 0$ if and only if the flows of X and Y commute, i.e.

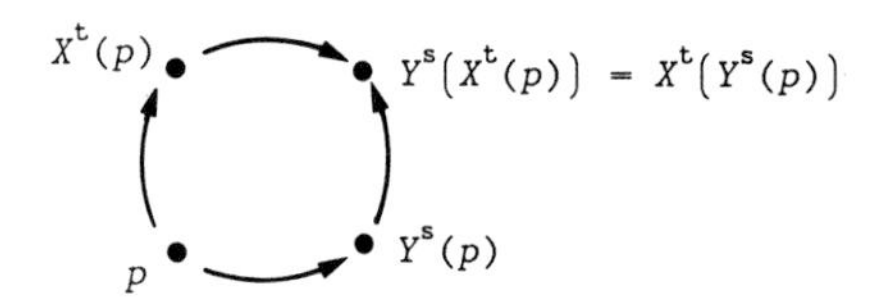

Fig. 2.7. Commuting vectorfields X and Y.

Hence if $[X,Y](p) \neq 0$ then $Y^s(X^t(p)) \neq X^t(Y^s(p))$ for some t and s, and so the Lie bracket is a measure for this difference. This will be instrumental in understanding the *controllability* properties of a nonlinear control system, as dealt with in Chapter 3.

Example 2.35 Consider the following simplified model of maneuvering an automobile,

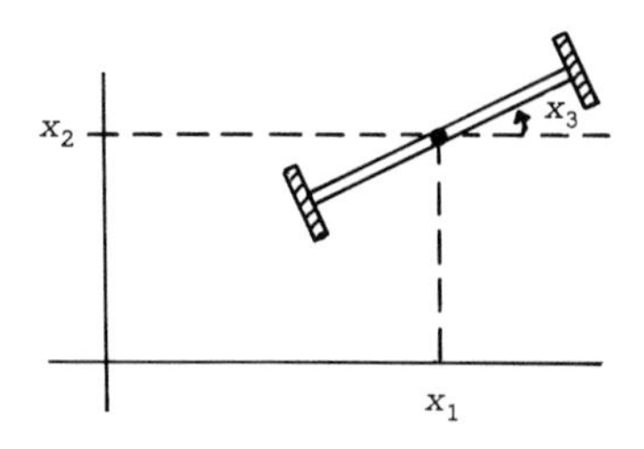

Fig. 2.8. The front axis of a car.

i.e. the middle of the axis linking the front wheels has position $(x_1, x_2) \in \mathbb{R}^2$, while the rotation of this axis is given by the angle x_3. The configuration manifold is thus $\mathbb{R}^2 \times S^1$ with local coordinates x_1, x_2, x_3. Consider the two vectorfields

$$X = \sin x_3 \frac{\partial}{\partial x_1} + \cos x_3 \frac{\partial}{\partial x_2} \quad \text{(rolling)} \ , \qquad Y = \frac{\partial}{\partial x_3} \quad \text{(rotation)} \ . \tag{2.108}$$

The Lie bracket $[X,Y]$ is computed as

$$\left[\begin{pmatrix} \sin x_3 \\ \cos x_3 \\ 0 \end{pmatrix}, \begin{pmatrix} 0 \\ 0 \\ 1 \end{pmatrix} \right] = - \begin{pmatrix} 0 & 0 & \cos x_3 \\ 0 & 0 & \sin x_3 \\ 0 & 0 & 0 \end{pmatrix} \begin{pmatrix} 0 \\ 0 \\ 1 \end{pmatrix} = \begin{pmatrix} -\cos x_3 \\ \sin x_3 \\ 0 \end{pmatrix} \tag{2.109}$$

and thus the vectorfields X,Y do not commute. This also follows from the following computation. Start in $x(0) = x_0 = (x_{10}, x_{20}, x_{30})^T$. Rolling during time h yields the position $x(h) = (x_{10} + h \sin x_{30}, x_{20} + h \cos x_{30}, x_{30})^T$. Then rotation during time h yields $x(2h) = (x_{10} + h \sin x_{30}, x_{20} + h \cos x_{30}, x_{30} + h)^T$. Rolling back during time h results in $x(3h) = (x_{10} + h \sin x_{30} - h \sin(x_{30} + h), x_{20} + h \cos x_{30} - h \cos(x_{30} + h), x_{30} + h)^T$. Finally rotating back during time h gives the end position

$$x(4h) = \begin{pmatrix} x_{10} + h \sin x_{30} - h \sin(x_{30} + h) \\ x_{20} + h \cos x_{30} - h \cos(x_{30} + h) \\ x_{30} \end{pmatrix} = Y^{-h} \circ X^{-h} \circ Y^{h} \circ X^{h}(x_0). \tag{2.110}$$

Noting that

$$\sin(x_{30} + h) = \sin x_{30} + h \cos x_{30} + \text{h.o.t. (higher order terms)},$$
$$\cos(x_{30} + h) = \cos x_{30} + h \sin x_{30} + \text{h.o.t.},$$

we obtain

$$x(4h) = x_0 + h^2 \begin{pmatrix} -\cos x_{30} \\ \sin x_{30} \\ 0 \end{pmatrix} + \text{h.o.t.} = x_0 + h^2 [X,Y](x_0) + \text{h.o.t.} \tag{2.111}$$

□

In Theorem 2.26 we have already shown that if $X(p) \neq 0$ then there is a coordinate chart $(U, x_1, \ldots, x_n)$ such that on U we have $X = \frac{\partial}{\partial x_1}$. If Y is another vectorfield, linearly independent from X in a neighborhood of p, then we may expect to find a coordinate chart such that

$$X = \frac{\partial}{\partial x_1}, \qquad Y = \frac{\partial}{\partial x_2} \ . \tag{2.112}$$

However it is immediate that $[\frac{\partial}{\partial x_1}, \frac{\partial}{\partial x_2}] = 0$ and hence by Proposition 2.30 a necessary condition for the existence of such a coordinate chart is that $[X,Y] = 0$. The next theorem shows that this condition is also *sufficient*.

Theorem 2.36 *Let $X_1,\dots,X_k$ be linearly independent vectorfields in a neighborhood of p, satisfying $[X_i,X_j] = 0$, $i,j \in \underline{k}$, then there is a coordinate chart $(U,x_1,\dots,x_n)$ around p such that on U*

$$X_i = \frac{\partial}{\partial x_i} \quad , \quad i \in \underline{k} \ . \tag{2.113}$$

Proof Denote $Z_i = \psi_* X_i$, $i \in \underline{k}$, Z_i vectorfields on $\psi(V)$. As in the proof of Theorem 2.26 we can take a coordinate chart $(V,\psi) = (V,z_1,\dots,z_n)$ around p such that $\psi(p) = 0$, and

$$Z_i(0) = \frac{\partial}{\partial r_i}\Big|_0 \quad , \quad i \in \underline{k} \ .$$

Define the map T: $\mathbb{R}^n \to \mathbb{R}^n$, defined in a neighborhood of 0, by

$$T(a_1,\dots,a_n) = Z_1^{a_1}\Big(Z_2^{a_2}(\cdots\cdots(Z_k^{a_k}(0,\dots,0,a_{k+1},\dots,a_n))\cdots)\Big) \ . \tag{2.114}$$

We compute that

$$T_{*0}\left(\frac{\partial}{\partial r_i}\Big|_0\right) = \begin{cases} Z_i(0) = \dfrac{\partial}{\partial r_i}\Big|_0 & , \quad i = 1,\dots,k \ , \\ \dfrac{\partial}{\partial r_i}\Big|_0 & , \quad i = k+1,\dots,n \ , \end{cases} \tag{2.115}$$

so that by the inverse function theorem T is a diffeomorphism around 0, which can be used as a coordinate transformation. Moreover precisely as in Theorem 2.26 we have that

$$T_* \frac{\partial}{\partial r_1} = Z_1 \ . \tag{2.116}$$

On the other hand, since $[X_i,X_j] = 0$, $i,j \in \underline{k}$, we can by Lemma 2.34 change the order of integration in any way, i.e., for any $i \in \underline{k}$ we can first integrate in (2.114) along Z_i

$$T(a_1,\dots,a_n) = Z_i^{a_i}\Big(Z_1^{a_1}(\cdots\cdots(Z_k^{a_k}(0,\dots,0,a_{k+1},\dots,a_n))\cdots)\Big) \ , \tag{2.117}$$

so that by the same argument

$$T_* \frac{\partial}{\partial r_i} = Z_i \quad , \quad i \in \underline{k} \ . \tag{2.118}$$

Hence T^{-1} is the required coordinate transformation; the new coordinate map is given by $T^{-1}\circ\psi$. □

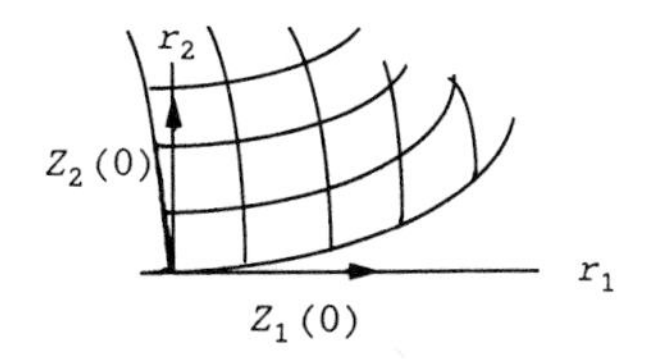

Fig. 2.9. The new coordinates constructed in Theorem 2.36 for n = k = 2.

2.2.2 Distributions, Frobenius' Theorem

Definition 2.37 *A distribution D on a manifold M is a map which assigns to each $p \in M$ a linear subspace $D(p)$ of the tangent space T_pM. D will be called a smooth distribution if around any point these subspaces are spanned by a set of smooth vectorfields, i.e. for each $p \in M$ there exists a neighborhood U of p and a set of smooth vectorfields X_i, $i \in I$, with I some (possibly infinite) index set, such that*

$$D(q) = \text{span}\{X_i(q);\ i \in I\}, \qquad q \in U\ . \tag{2.119}$$

In the sequel distribution will *always* mean smooth distribution. A vectorfield X is said to *belong* to (or is in) the distribution D (denoted $X \in D$) if $X(p) \in D(p)$ for any $p \in M$. The dimension of a distribution D at $p \in M$ is the dimension of the subspace $D(p)$. A distribution is called *constant dimensional* if the dimension of $D(p)$ does not depend on the point $p \in M$.

Lemma 2.38 *Let D be a constant dimensional distribution of dimension k. Then around any $p \in M$ there exist k independent vectorfields $X_1,\dots,X_k$ such that*

$$D(q) = \text{span}\{X_1(q),\dots,X_k(q)\}, \qquad q \text{ near } p. \tag{2.120}$$

Proof Since $\dim D(p) = k$ there exist k vectorfields from the index set I in (2.119), for simplicity denoted as $X_1,\dots,X_k$, such that

$$D(p) = \text{span}\{X_1(p),\dots,X_k(p)\} \tag{2.121}$$

Hence $X_1(p),\dots,X_k(p)$ are independent elements of T_pM. By continuity it follows that for q close to p the vectors $X_1(q),\dots,X_k(q)$ in T_qM are also independent, and hence, since $\dim D(q) = k$, span $D(q)$. □

The vectorfields $X_1,\dots,X_k$ above are called the local generators of D, since every vectorfield $X \in D$ can be written around p as $X(q) = \sum_{i=1}^{k} \alpha_i(q)X_i(q)$ for some smooth functions α_i, $i \in \underline{k}$.

Definition 2.39 *A distribution D is called involutive if* $[X,Y] \in D$ *whenever X and Y are vectorfields in D.*

Remark By Proposition 2.27(a) it follows that a distribution D given as in (2.119) is involutive if and only if $[X_i, X_j] \in D$ on U for $i,j \in I$. In particular if D is locally given as in (2.120) then we only have to check that $[X_i, X_j] \in D$ for $i,j \in k$, or said otherwise, $[X_i, X_j]$ has to be of the form $\sum_{\ell=1}^{k} c_{ij}^{\ell} X_{\ell}$ for some functions c_{ij}^{ℓ}.

Definition 2.40 *A submanifold P of M is an integral manifold of a distribution D on M if*

$$T_qP = D(q) \quad , \quad \text{for every } q \in P. \tag{2.122}$$

(Recall that since $P \subset M$ we have $T_qP \subset T_qM$, for all $q \in P$.)

We have

Proposition 2.41 *Let D be a distribution on M such that through each point of M there passes an integral manifold of D. Then D is involutive.*

Proof Let $X,Y \in D$ and $p \in D$. Let P be an integral manifold of D through p. Then for every $p \in P$

$$X(p) \in D(p) \in T_pP, \; Y(p) \in D(p) \in T_pP \; .$$

Since P is a submanifold around any $p \in P$ there is a coordinate chart $(U, x_1, \ldots, x_n)$ for M such that

$$U \cap P = \{q \in U \mid x_i(q) = x_i(p), \; i = k+1, \ldots, n\}. \tag{2.123}$$

Writing out X and Y in the basis $\frac{\partial}{\partial x_1}, \ldots, \frac{\partial}{\partial x_n}$ it follows that the last $n-k$ components of X and Y in $p \in P$ are zero. Hence by (2.94) the last $n-k$ components of $[X,Y]$ are also zero, and so $[X,Y](p) \in T_pP = D(p)$ for any $p \in P$. □

We say that a distribution D on M is *integrable* if through any point of M there passes an integral manifold of D. In Proposition 2.41 we saw that involutivity of D is a necessary condition for integrability; the next theorem shows that for constant dimensional distributions it is also sufficient.

Theorem 2.42 (Frobenius' Theorem) *Let D be an involutive distribution of constant dimension k on M. Then for any $p \in M$ there is a coordinate chart $(U,\varphi) = (U,x_1,\ldots,x_n)$ with*

$$\begin{aligned} &\varphi(p) = 0 \\ &\varphi(U) = (-\epsilon,\epsilon) \times \ldots \times (-\epsilon,\epsilon) \quad , \quad \epsilon > 0 , \end{aligned} \tag{2.124}$$

such that for each $a_{k+1},\ldots,a_n$, smaller in absolute value than ϵ, the submanifold

$$\{q \in U \mid x_{k+1}(q) = a_{k+1},\ldots,x_n(q) = a_n\} \tag{2.125}$$

is an integral manifold of D. Moreover every integral manifold is of this form.

Actually we usually need the following equivalent form of Theorem 2.42.

Corollary 2.43 (Frobenius) *Let D be an involutive constant dimensional distribution on M. Then around any $p \in M$ there exists a coordinate chart $(U,x_1,\ldots,x_n)$ such that*

$$D(q) = \text{span}\ \{\frac{\partial}{\partial x_1}\Big|_q,\ldots,\frac{\partial}{\partial x_k}\Big|_q\}, \qquad q \in U . \tag{2.126}$$

D as in (2.126) is called a flat distribution, and usually we will simply write (2.126) as $D = \text{span}\ \{\frac{\partial}{\partial x_1},\ldots,\frac{\partial}{\partial x_k}\}$.

Proof (of Theorem 2.42 and Corollary 2.43) Take a coordinate chart (U',φ') about p, with $\varphi'(p) = 0$. Mapping everything onto $\varphi'(U)$ using φ' we may as well assume that we are in $\mathbb{R}^n$ with $p = 0$. Moreover we can assume that $D(0) \subset T_0\mathbb{R}^n$ is spanned by

$$\frac{\partial}{\partial r_1}\Big|_0,\ldots,\frac{\partial}{\partial r_k}\Big|_0 \tag{2.127}$$

Let $\pi\colon \mathbb{R}^n \to \mathbb{R}^k$ be the projection onto the first k factors. Then $\pi_{*0}\colon T_0\mathbb{R}^n \to T_0\mathbb{R}^k$ is an isomorphism when restricted to $D(0) \subset T_0\mathbb{R}^n$. By continuity $\pi_{*q}\colon T_q\mathbb{R}^n \to T_{\pi(q)}\mathbb{R}^k$ is an isomorphism when restricted to $D(q)$ for q close to 0. So for q near 0 we can choose unique vectors

$$X_1(q),\ldots,X_k(q) \in D(q) \ ,$$

such that

$$\pi_{*q} X_i(q) = \frac{\partial}{\partial r_i}\Big|_{\pi(q)} \quad , \quad i \in \underline{k} \; . \tag{2.128}$$

It follows that the vectorfields X_i (defined on a neighborhood of $0 \in \mathbb{R}^n$) and $\frac{\partial}{\partial r_i}$ (on $\mathbb{R}^k$) are π-related. By Proposition 2.30 we get

$$\pi_{*q}[X_i, X_j]_q = \left[\frac{\partial}{\partial r_i}, \frac{\partial}{\partial r_j}\right]_{\pi(q)} = 0 \; . \tag{2.129}$$

By involutivity of D we have $[X_i, X_j]_q \in D(q)$, and since π_{*q} is one-one when restricted to $D(q)$ we therefore have $[X_i, X_j] = 0$, $i, j \in \underline{k}$. Hence by Theorem 2.36 we can choose a local coordinate chart $(U, x_1, \ldots, x_n)$, with $U \subset U'$, such that

$$X_i = \frac{\partial}{\partial x_i} \quad , \quad i \in \underline{k}, \text{ on } U \; . \tag{2.130}$$

Hence Corollary 2.43 is proved. Integral manifolds of D in these coordinates clearly are given as in (2.121). □

The totality of submanifolds (2.125) parametrized by a_i, $|a_i| < \epsilon$, $i = k+1, \ldots, n$, is called a *foliation* of the open submanifold $U \subset M$, and each submanifold (2.125) is called a *leaf* of this foliation. Hence Frobenius' theorem says that an involutive constant dimensional distribution on M locally generates a foliation of M, whose leaves are integral manifolds of the distribution.

Remark It is also possible to prove that M *as a whole* is foliated by integral manifolds of D, if we generalize the definition of integral manifold by allowing for *immersed* submanifolds, see the text above Fig. 2.4.

Example 2.44 Consider on $M = \{(x_1, x_2, x_3)^T \in \mathbb{R}^3 \mid x_i > 0, \; i = 1,2,3\}$ the distribution $D(x) = \text{span}\{X_1(x), X_2(x)\}$, where

$$X_1(x) = x_1 \frac{\partial}{\partial x_1} + x_2 \frac{\partial}{\partial x_2} + x_3 \frac{\partial}{\partial x_3} \; , \; X_2(x) = \frac{\partial}{\partial x_3} \; . \tag{2.131}$$

(These are the input vectorfields in the model of a mixed-culture bioreactor as treated in Example 1.4.) Since $[X_1, X_2] = X_2$ it follows that D is involutive, as well as (note the definition of M) constant dimensional. In order to apply Theorem 2.42 we consider the set of partial differential equations

$$X_1(\varphi)(x) = x_1 \frac{\partial\varphi}{\partial x_1}(x) + x_2 \frac{\partial\varphi}{\partial x_2}(x) + x_3 \frac{\partial\varphi}{\partial x_3}(x) = 0 ,$$
$$X_2(\varphi)(x) = \frac{\partial\varphi}{\partial x_3}(x) = 0 , \tag{2.132}$$

in $\varphi(x_1,x_2,x_3)$. A possible solution is

$$\varphi(x_1,x_2,x_3) = \ell n \frac{x_1}{x_2}. \tag{2.133}$$

Denote $z_1 := \varphi(x)$, $z_2 := x_2, z_3 := x_3$, then it is checked that z_1,z_2,z_3 are a set of new coordinates for M, in which the integral manifolds of D are of the form $\{(z_1,z_2,z_3)^T \mid z_1 = \text{constant}\}$, while $D = \text{span}\,\{\frac{\partial}{\partial z_2}, \frac{\partial}{\partial z_3}\}$. Note that the choice of coordinates is by no means unique. In particular arctan (x_2/x_1) also solves (2.132), and thus we can take $\tilde{z}_1 := \arctan(x_2/x_1)$. □

The classical version of the Frobenius' Theorem is at first sight quite different from Theorem 2.42.

Theorem 2.45 (Classical Frobenius' Theorem) *Consider the set of partial differential equations*

$$\frac{\partial k}{\partial r}(r,t) = b\big(r,k(r,t)\big) \tag{2.134}$$

with $r \in \mathbb{R}^m$, $t \in \mathbb{R}^n$ and b: $\mathbb{R}^m \times \mathbb{R}^n \to M(n,m)$ (n×m-matrices) in the unknown $k : \mathbb{R}^m \times \mathbb{R}^n \to \mathbb{R}^n$. Then locally there exists a solution k if and only if the matrix component functions $b_{i\alpha}(r,s)$, $s \in \mathbb{R}^n$, $i \in \underline{n}$, $\alpha \in \underline{m}$, satisfy

$$\frac{\partial b_{i\beta}}{\partial r_\gamma} - \frac{\partial b_{i\gamma}}{\partial r_\beta} + \sum_{j=1}^{n} \left(\frac{\partial b_{i\beta}}{\partial s_j} b_{j\gamma} - \frac{\partial b_{i\gamma}}{\partial s_j} b_{j\beta} \right) = 0, \ i \in \underline{n},\ \gamma,\beta \in \underline{m} . \tag{2.135}$$

Furthermore we can ensure that the solution $k(r,t)$ satisfies

$$\text{rank}\,\frac{\partial k}{\partial t}(r,t) = n. \tag{2.136}$$

The connection between Theorem 2.45 and Theorem 2.42 is as follows. Define the vectorfields

$$Z_\alpha = \frac{\partial}{\partial r_\alpha} + \sum_{j=1}^{n} b_{j\alpha} \frac{\partial}{\partial s_j}, \qquad \alpha \in \underline{m} , \tag{2.137}$$

and the constant dimensional distribution D spanned by $Z_1,\ldots,Z_m$. It is easily checked (see Exercise 2.12) that D is involutive if and only if (2.135) is satisfied; in fact (2.135) implies that $[Z_i,Z_j] = 0$, $i,j \in \underline{m}$. Hence by Theorem 3.36 we can find local coordinates for $\mathbb{R}^m \times \mathbb{R}^n$ in which D

is a flat distribution. By the special form of the vectorfields Z_α in (2.137) it follows that we may leave the coordinates $r_1,\dots,r_m$ unchanged, while coordinates $s_1,\dots,s_n$ can be transformed to new coordinates $t_1,\dots,t_n$ depending on r and s in such a way that in the coordinates (r,t) the distribution D is given as span $\{\frac{\partial}{\partial r_\alpha}, \alpha \in \underline{m}\}$. Denote $t = h(r,s)$, and define the inverse map $k(r,t)$ satisfying (2.136) and

$$h(r,k(r,t)) = t \ . \tag{2.138}$$

Differentation of (2.138) with respect to r_α, $\alpha \in \underline{m}$, yields

$$\frac{\partial h}{\partial r_\alpha}(r,s) + \sum_{j=1}^{n} \frac{\partial h}{\partial s_j}(r,s)\, \frac{\partial k_j}{\partial r_\alpha}(r,t) = 0, \tag{2.139}$$

where $s = k(r,t)$, and thus

$$0 = Z_\alpha(t) = Z_\alpha(h(r,s)) = \frac{\partial h}{\partial s_\alpha}(r,s) + \sum_{j=1}^{m} b_{j\alpha}(r,s)\, \frac{\partial k^{-1}}{\partial s_j}(r,s) =$$

$$= - \sum_{j=1}^{n} \frac{\partial h}{\partial s_j}(r,s)\, \frac{\partial k_j}{\partial r_\alpha}(r,t) + \sum_{j=1}^{n} b_{j\alpha}(r,s)\, \frac{\partial h}{\partial s_j}(r,s) \ . \tag{2.140}$$

By non-singularity of the $n\times n$-matrix $\frac{\partial h}{\partial s}(r,s)$ this immediately yields that $k(r,t)$ is a solution of (2.134).

Finally for later convenience, we define the *sum* and *intersection* of distributions. Let D_1 and D_2 be two smooth distributions on M. Then their *sum* $D_1 + D_2$ is defined as the smooth distribution given in any $q \in M$ as

$$(D_1+D_2)(q) = \text{span}\{X_1(q) + X_2(q) \mid X_1 \text{ smooth vectorfield in } D_2 \text{ and } X_2 \text{ smooth vectorfield in } D_2\}. \tag{2.141}$$

The *intersection* $D_1 \cap D_2$ is the smooth distribution given in $q \in M$ as

$$(D_1 \cap D_2)(q) = \text{span}\{X(q) \mid X \text{ smooth vectorfield contained in } D_1 \textit{ and } D_2\} \tag{2.142}$$

Note that for two involutive smooth distributions the intersection $D_1 \cap D_2$ is again involutive. It follows that for any smooth distribution D we can define the smallest *involutive* smooth distribution *containing* D (because if D_1 and D_2 are involutive smooth distributions containing D, then so is $D_1 \cap D_2$). This distribution is called the *involutive closure* of D, and is denoted by $\bar{D}$:

$$\bar{D} = \text{smallest involutive smooth distribution containing } D \ . \tag{2.143}$$

2.2.3 Cotangent Bundle, Differential One-Forms, Codistributions

Let M be a manifold, and let T_pM be its tangent space in a point p. Since T_pM is a linear space we can consider the *dual* space of T_pM, denoted T_p^*M, called *cotangent space* of M in p. (Recall that the dual V^* of a linear space V is the set of all linear functions on V.) Elements of T_p^*M are called cotangent vectors. Let $\{\frac{\partial}{\partial x_1}|_p,\ldots,\frac{\partial}{\partial x_n}|_p\}$ be a basis for T_pM corresponding to local coordinates $x_1,\ldots,x_n$ on M, then we denote the *dual* basis of T_p^*M by $dx_1|_p,\ldots,dx_n|_p$. By definition

$$dx_i|_p\left(\frac{\partial}{\partial x_j}|_p\right) = \delta_{ij} \quad , \quad i,j \in \underline{n} \ . \tag{2.144}$$

Any cotangent vector $\sigma_p \in T_p^*M$ can be written as $\sum_{i=1}^{n} \alpha_i dx_i|_p$ for some coefficients α_i, and is also denoted as a row-vector $(\alpha_1,\ldots,\alpha_n)$.

A function $f\colon M \to \mathbb{R}$ defines in every point p an element of T_p^*M, denoted as df_p or $df(p)$, by the formula

$$df(p)(X_p) = X_p(f), \qquad X_p \in T_pM \ . \tag{2.145}$$

We call $df(p)$ the *differential* of f at p. If we interpret $dx_i|_p$ in (2.144) as the differential of the coordinate function x_i then (2.144) and (2.145) are consistent. The differential $df(p)$ in the basis $dx_1|_p,\ldots,dx_n|_p$, is given as

$$df(p) = \frac{\partial f}{\partial x_1}(p)dx_1|_p + \cdots + \frac{\partial f}{\partial x_n}(p)dx_n|_p \ . \tag{2.146}$$

(In order to check (2.146) compute $df(p)\left(\frac{\partial}{\partial x_j}|_p\right)$, $j \in \underline{n}$.)

Let $z_1,\ldots,z_n$ be another set of local coordinates around p, with $z = S(x)$. Let $\sigma_p \in T_p^*M$ be represented as

$$\sigma_p = \sum_{i=1}^{n} \alpha_i dx_i|_p, \text{ and as } \sigma_p = \sum_{i=1}^{n} \beta_i dz_i|_p \ , \tag{2.147}$$

then the coefficients α_i and β_i are related by the formula

$$\alpha_i = \sum_j \frac{\partial S_j}{\partial x_i}\big(x(p)\big)\beta_j \ , \tag{2.148}$$

or, with $\alpha := (\alpha_1,\ldots,\alpha_n)$ and $\beta := (\beta_1,\ldots,\beta_n)$,

$$\alpha = \beta\,\frac{\partial S}{\partial x}\big(x(p)\big) \ . \tag{2.149}$$

(One says that cotangent vectors transform in a *covariant* fashion, contrary to tangent vectors which transform in a contravariant way.)

For $F\colon M_1 \to M_2$ we have defined the tangent map $F_{*p} : T_pM_1 \to T_{F(p)}M_2$ by

$$(F_{*p}X_p)(f) = X_p(f \circ F), \qquad f\colon M_2 \to \mathbb{R} . \tag{2.150}$$

The *adjoint map* of F_{*p} will be denoted by F_p^*. Thus dually to F_{*p} we have

$$F_p^*\colon T^*_{F(p)}M_2 \to T_p^*M_1 , \tag{2.151}$$

with the defining property ($\sigma_{F(p)} \in T^*_{F(p)}M_2$, $X_p \in T_pM_1$)

$$F_p^*\sigma_{F(p)}(X_p) = \sigma_{F(p)}(F_{*p}X_p) . \tag{2.152}$$

In local coordinates $x_1,\ldots,x_{n_1}$ for M_1, and $z_1,\ldots,z_{n_2}$ for M_2 the linear map F_p^* is given by the Jacobian matrix $\frac{\partial F}{\partial x}\big(x(p)\big)$ of the local representative of F, in the sense that if $\sigma_{F(p)}$ is expressed as a row-vector then $F_p^*\sigma_{F(p)}$ is given by the row-vector (compare with (2.149))

$$\sigma_{F(p)}\frac{\partial F}{\partial x}\big(x(p)\big) . \tag{2.153}$$

The *cotangent bundle* of a manifold M is defined as

$$T^*M = \bigcup_{p \in M} T_p^*M . \tag{2.154}$$

There is the natural projection $\pi\colon T^*M \to M$ taking a cotangent vector $\sigma_p \in T_p^*M \subset T^*M$ to $p \in M$. As in the case of a tangent bundle, the cotangent bundle can be given a manifold structure. Given local coordinates $x_1,\ldots,x_n$ on M we obtain natural local coordinates for T^*M by letting a cotangent vector $\sigma_p = \sum_{i=1}^{n}\alpha_i dx_i\big|_p$ correspond to the coordinate values

$$\big(x_1(p),\ldots,x_n(p),\alpha_1,\ldots,\alpha_n\big) \in \mathbb{R}^{2n}.$$

Now we define the dual object of a vectorfield.

Definition 2.46 *A smooth differential one-form σ, briefly smooth one-form, on a smooth manifold M is defined as a smooth map*

$$\sigma\colon M \to T^*M , \tag{2.155}$$

satisfying (with π the natural projection $T^*M \to M$)

$$\pi \circ \sigma = \textit{identity on } M . \tag{2.156}$$

Replacing "smooth" throughout by C^ω, resp. C^k, the one-form is called C^ω, resp. C^k. In the sequel differential one-form will always mean smooth differential one-form.

Hence a one-form σ is a map which assigns to each $p \in M$ a cotangent vector $\sigma(p) \in T_p^*M$. Let $(U, x_1, \ldots, x_n) = (U, \varphi)$ be a local coordinate chart for M about p, resulting in the basis $dx_1|_p, \ldots, dx_n|_p$ for T_p^*M, then we can write

$$\sigma(p) = \sum_{i=1}^{n} \sigma_i(p) dx_i|_p \tag{2.157}$$

for certain smooth functions $\sigma_i : U \to \mathbb{R}$, $i \in \underline{n}$. Letting $\hat{\sigma}_i := \sigma_i \circ \varphi^{-1}: \varphi(U) \to \mathbb{R}$ be the local representatives of σ_i, $i \in \underline{n}$, and omitting the carets, we write σ in local coordinates as the row-vector (compare with (2.71))

$$\sigma(x_1, \ldots, x_n) = \big(\sigma_1(x_1, \ldots, x_n), \ldots, \sigma_n(x_1, \ldots, x_n)\big) , \tag{2.158}$$

or, abusing notation by writing dx_i for dr_i, $i \in \underline{n}$ (the natural basis for $T^*_{\varphi(p)}\mathbb{R}^n$) as

$$\sigma = \sum_{i=1}^{n} \sigma_i(x) dx_i \quad . \tag{2.159}$$

Since one-forms are the dual objects of vectorfields, they act in a natural way upon vectorfields (with σ a one-form and X a vectorfield)

$$\sigma(X)(p) = \sigma(p)\big(X(p)\big) \in \mathbb{R} \quad . \tag{2.160}$$

Hence $\sigma(X)$ is a smooth function on M. Any function f defines a one-form, denoted as df, by letting $df(p)$ be defined as in (2.145). Notice that we have the equality

$$df(X) = X(f) = L_X f \quad . \tag{2.161}$$

Not every one-form can be written as df for a certain function f. In fact it follows from (2.146) that

$$df = \frac{\partial f}{\partial x_1} dx_1 + \cdots + \frac{\partial f}{\partial x_n} dx_n \quad , \tag{2.162}$$

and hence, since $\dfrac{\partial^2 f}{\partial x_i \partial x_j} = \dfrac{\partial^2 f}{\partial x_i \partial x_j}$, a necessary condition for a one-form $\sigma = \sum_{i=1}^{n} \sigma_i(x) dx_i$ to be of the form df is that

$$\frac{\partial \sigma_i}{\partial x_j} = \frac{\partial \sigma_j}{\partial x_i} \qquad i,j \in \underline{n}. \tag{2.163}$$

Conversely one can prove that condition (2.163) is sufficient for the *local* existence of a function f such that $\sigma_i = \frac{\partial f}{\partial x_i}$. One-forms df are called *exact*, and one-forms satisfying (2.163) are called *closed*.

Finally let $F: M_1 \to M_2$ be a smooth map, then for σ_2 being a one-form on M_2, we define a one-form σ_1 on M_1, denoted as $\sigma_1 = F^*\sigma_2$, by letting

$$\sigma_1(p)(X_p) = \sigma_2(F(p))(F_{*p}X_p) \ . \tag{2.164}$$

Notice that $F^*\sigma_2$ is *always* a well-defined one-form on M_1. On the other hand recall that if X is a vectorfield on M_1 then F_*X_1 need not be a well-defined vectorfield on M_2. (Of course if F is a *diffeomorphism*, then F_*X is always well-defined.) It is easily checked that F^* maps exact one-forms on M_2 into exact one-forms on M_1, in fact for any $X_p \in T_pM$ we have $\big(F^*(df)\big)(p)(X_p) = df(p)(F_{*p}X_p) = X_p(f \circ F)$ and so

$$F^*(df) = d(f \circ F), \qquad f: M_2 \to \mathbb{R} \ . \tag{2.165}$$

Since closed one-forms are locally of the form df it follows that also exact one-forms on M_2 are mapped by F^* onto exact one-forms on M_1.

One may also define the notion of the *Lie derivative* of a one-form σ along a vectorfield X. In fact we define $L_X\sigma$ as the one-form

$$L_X\sigma = \lim_{h\to 0} \frac{1}{h}\big((X^h)^*\sigma - \sigma\big) \ . \tag{2.166}$$

If σ is given in local coordinates $x_1,\dots,x_n$ as the row vector $\big(\sigma_1(x),\dots,\sigma_n(x)\big)$, and X as the column vector $\big(X_1(x),\dots,X_n(x)\big)^T$, then it may be checked (see Exercise 2.13) that $L_X\sigma$ is given as the row-vector

$$L_X\sigma = (X_1,\dots,X_n)\begin{pmatrix} \frac{\partial \sigma_1}{\partial x_1} & \cdots & \frac{\partial \sigma_n}{\partial x_1} \\ \vdots & & \vdots \\ \frac{\partial \sigma_1}{\partial x_n} & \cdots & \frac{\partial \sigma_n}{\partial x_n} \end{pmatrix} + (\sigma_1,\dots,\sigma_n)\begin{pmatrix} \frac{\partial X_1}{\partial x_1} & \cdots & \frac{\partial X_1}{\partial x_n} \\ \vdots & & \vdots \\ \frac{\partial X_n}{\partial x_1} & \cdots & \frac{\partial X_n}{\partial x_n} \end{pmatrix}, \tag{2.167}$$

where everything is taken in $x = (x_1,\dots,x_n) \in \mathbb{R}^n$. If σ is exact, i.e. $\sigma = df$, then (2.166) reduces to

$$L_X df = \lim_{h\to 0} \frac{1}{h}\big[(X^h)^*df - df\big] = d\big(\lim_{h\to 0}\big[f \circ X^h - f\big]\big) = d(L_X f) \ . \tag{2.168}$$

We thus see that the Lie derivative of a one-form is the generalization of the Lie derivative of a function.

Finally we give the following interesting "product" formula (X,Y vectorfields, σ one-form), which can be verified using the local coordinate expressions (2.96) and (2.167)

$$L_X\big(\sigma(Y)\big) = (L_X\sigma)(Y) + \sigma(L_X Y) \ . \tag{2.169}$$

For $\sigma = df$ this reduces to $L_X\big(Y(f)\big) = Y\big(X(f)\big) + df(L_X Y)$, or

$$X\big(Y(f)\big) - Y\big(X(f)\big) = [X,Y](f) \ , \tag{2.170}$$

which is just the definition of the Lie bracket $[X,Y]$.

The dual object of a distribution is a codistribution. A *codistribution* P on a manifold M is defined as a map which assigns to any $p \in M$ a linear subspace $P(p)$ of the cotangent space T_p^*M. P is called a *smooth* codistribution if around any point p there exists a neighborhood U of p and a set of smooth one-forms σ_i, $i \in I$, with I some (possible infinite) index set, such that (compare (2.119))

$$P(q) = \operatorname{span}\{\sigma_i(q);\ i \in I\}, \qquad q \in U \ . \tag{2.171}$$

In the sequel codistribution will always mean smooth codistribution. A one-form σ is said to *belong to* the codistribution P ($\sigma \in P$) if $\sigma(p) \in P(p)$ for any $p \in M$. The dimension of P at $p \in M$ is the dimension of the subspace $P(p)$. A codistribution P is called *constant dimensional* if the dimension of $P(p)$ does not depend on $p \in M$. It immediately follows (compare Lemma 2.38) that if P is a codistribution of constant dimension ℓ, then around any p there exist ℓ independent one-forms $\sigma_1,\dots,\sigma_\ell$ (called local generators of P) such that

$$P(q) = \operatorname{span}\{\sigma_1(q),\dots,\sigma_\ell(q)\}, \qquad q \text{ near } p \ . \tag{2.172}$$

Finally for any codistribution P we define $\ker P$ as the smooth distribution

$$(\ker P)(q) = \operatorname{span}\{X(q) \,|\, X \text{ vectorfield such that } \sigma(X) = 0,\ \forall \sigma \in P\} \tag{2.173}$$

Conversely for any distribution D we define its smooth annihilator $\operatorname{ann} D$ as the smooth codistribution

$$(\operatorname{ann} D)(q) = \operatorname{span}\{\sigma(q) \,|\, \sigma \text{ one-form such that } \sigma(X) = 0,\ \forall X \in D\} \tag{2.174}$$

It follows that if D and P are constant dimensional then $\operatorname{ann} D$, resp. $\ker D$ are constant dimensional. By definition $D \subset \ker(\operatorname{ann} D)$ and $P \subset \operatorname{ann}(\ker P)$, but in general equality does not hold. However if D and P are *constant dimensional* then it follows from Lemma 2.38, (2.172), and a dimensionality argument that equality does hold, i.e.

$$D = \ker(\operatorname{ann} D), \text{ respectively } P = \operatorname{ann}(\ker P) \ . \tag{2.175}$$

For convenience we call a codistribution P *involutive* if $\ker P$ is an involutive distribution. If P is locally generated by exact one-forms, i.e.

$$P(q) = \operatorname{span}\{df_1(q),\ldots,df_\ell(q)\}, \qquad q \text{ near } p \ , \tag{2.176}$$

then $\ker P$ is always involutive. Indeed, let $X_1, X_2 \in \ker P$, then by the definition of the Lie bracket

$$\begin{aligned} &df_i(q)\big([X_1,X_2](q)\big) = \\ &\big([X_1,X_2](f_i)\big)(q) = \big(X_1(X_2(f_i))\big)(q) - \big(X_2(X_1(f_i))\big)(q) = 0 \ , \end{aligned} \tag{2.177}$$

since $X_2(f_i) = X_1(f_i) = 0$, $i \in \underline{\ell}$. Conversely, let P be a constant-dimensional codistribution such that $D = \ker P$ is involutive. Then by Frobenius' Theorem (Corollary 2.43) there exist local coordinates $x_1,\ldots,x_n$ such that $D(q) = \operatorname{span}\{\frac{\partial}{\partial x_1}\big|_q,\ldots,\frac{\partial}{\partial x_k}\big|_q\}$. Since in view of (2.175) $P = \operatorname{ann} D$, it immediately follows that $P(q) = \operatorname{span}\{dx_{k+1}(q),\ldots,dx_n(q)\}$, $q \in M$, or abbreviated

$$P = \operatorname{span}\{dx_{k+1},\ldots,dx_n\} \tag{2.178}$$

As in the case of distributions (cf.(2.41)) we let the *sum* of two smooth codistributions P_1 and P_2 be the smooth codistribution $P_1 + P_2$ defined in every $q \in M$ as

$$\begin{aligned} (P_1+P_2)(q) = \operatorname{span}\{\sigma_1(q) + \sigma_2(q) \mid &\sigma_1 \text{ smooth one-form in } P_1, \\ &\sigma_2 \text{ smooth one-form in } P_2\}. \end{aligned} \tag{2.179}$$

The *intersection* $P_1 \cap P_2$ is the smooth codistribution defined in any $q \in M$ as (compare (2.142))

$$(P_1 \cap P_2)(q) = \operatorname{span}\{\sigma(q) \mid \sigma \text{ smooth one-form contained in } P_1 \textit{ and } P_2\} \tag{2.180}$$

Finally let $F : M \to N$, and let P be a codistribution on N. Then we define the codistribution F^*P on M as

$$F^*P(q) = \operatorname{span}\{(F^*\sigma)(q) \mid \sigma \text{ one-form in } P\} \ , \ q \in M \ . \tag{2.181}$$

2.3 Summary of Section 2.2

1. In local coordinates $x = (x_1, \ldots, x_n)$ a vectorfield X is given as $\sum_{i=1}^{n} X_i(x)\frac{\partial}{\partial x_i}$ or as a vector $X(x) = \big(X_1(x_1,\ldots,x_n),\ldots,X_n(x_1,\ldots,x_n)\big)^T$, and corresponds to the set of differential equations

$$\dot{x}_i = X_i(x_1,\ldots,x_n)\ ,\ i \in \underline{n}\ ,$$

abbreviated as $\dot{x} = X(x)$.

2. The Lie derivative $L_X f = X(f)$ of a function f along the vectorfield X equals in local coordinates

$$L_X f(x) = \sum_{i=1}^{n} \frac{\partial f}{\partial x_i}(x)X_i(x) = \lim_{h\to 0} \frac{f\big(X^h(x)\big)-f(x)}{h}\ ,$$

with $X^t\colon M \to M$ the time-t integral (flow) of X.

3. Let $X(p) \neq 0$, then there exist local coordinates $x_1,\ldots,x_n$ around p such that $X = \frac{\partial}{\partial x_1}$, or in vector notation $X = (1\ 0\ \ldots\ 0)^T$.

4. Let $X = (X_1,\ldots,X_n)^T$ and $Y = (Y_1,\ldots,Y_n)^T$ be two vectorfields. In local coordinates x the Lie bracket $[X,Y] = L_X Y$ is the vectorfield given by the vector

$$[X,Y](x) = \frac{\partial Y}{\partial x}(x)X(x) - \frac{\partial X}{\partial x}(x)Y(x).$$

5. Vectorfields on M with the Lie bracket form the Lie algebra $V^\infty(M)$; that is: $[X,Y]$ is bilinear, anti-symmetric, and Jacobi's identity holds

$$\big[[X,Y],Z\big] + \big[[Y,Z],X\big] + \big[[Z,X],Y\big] = 0.$$

6. Suppose that $F_* X_i = Y_i$, $i = 1,2$, for vectorfields X_1,X_2 and Y_1,Y_2 on M respectively N, with $F\colon M \to N$. Then

$$F_*[X_1,X_2] = [Y_1,Y_2].$$

7. $[X,Y](p) = \lim_{h\to 0} \frac{1}{h}\big[(X_*^{-h}Y)(p) - Y(p)\big]$.

8. $[X,Y] = 0$ if and only if $X^t \circ Y^s = Y^s \circ X^t$ for all s,t.

9. Let $X_1,\dots,X_k$ be linearly independent vectorfields with $[X_i,X_j] = 0$, $i,j \in \underline{k}$, then there exist local coordinates $x_1,\dots,x_n$ such that $X_i = \frac{\partial}{\partial x_i}$, $i \in \underline{k}$.

10. A distribution D is given in any $q \in M$ as

$$D(q) = \operatorname{span}\{X_i(q) \mid X_i,\ i \in I\} ,$$

for some vectorfields X_i and index set I.
D is involutive if, whenever $X,Y \in D$, also $[X,Y] \in D$.

11. Let D be an involutive distribution of constant dimension k, then there exist local coordinates $x_1,\dots,x_n$ such that $D = \operatorname{span}\{\frac{\partial}{\partial x_1},\dots,\frac{\partial}{\partial x_k}\}$.

12. For any function f on M the exact one-form df is defined as $df(p)(X_p) = X_p(f)$, $X_p \in T_pM$.

13. In local coordinates a one-form σ is given $\sum_{i=1}^{n} \sigma_i(x)dx_i$ or as a row vector $\sigma(x) = \big(\sigma_1(x_1,\dots,x_n),\dots,\sigma_n(x_1,\dots,x_n)\big)$, and an exact one-form df is given as $\big(\frac{\partial f}{\partial x_1}(x),\dots,\frac{\partial f}{\partial x_n}(x)\big)$.

14. For any vectorfield X and one-form σ we have

$$\sigma(X)(p) = \sigma(p)\big(X(p)\big) \in \mathbb{R}$$

and in local coordinates $\sigma(X)(x) = \sum_{i=1}^{n} \sigma_i(x)X_i(x)$.

15. For any map $F\colon M \to N$ and any one-form σ on N we define $F^*\sigma$ by $(F^*\sigma)(p)(X_p) = \sigma\big(F(p)\big)(F_{*p}X_p)$.

16. $F^*(df) = d(f \circ F)$.

17. The Lie derivative of a one-form σ along a vectorfield X is defined as $L_X\sigma = \lim_{h\to 0} \frac{1}{h}\big[(X^h)^*\sigma - \sigma\big]$ and equals in local coordinates the row vector

$$\Big[\big(\frac{\partial \sigma^T}{\partial x}(x)\big)X(x)\Big]^T + \sigma(x)\frac{\partial X}{\partial x}(x) \quad .$$

18. $L_X df = dL_X f$.

19. For any one-form σ, and vectorfields X,Y we have

$$L_X(\sigma(Y)) = (L_X\sigma)(Y) + \sigma([X,Y]).$$

20. A codistribution P is given in any $q \in M$ as

$$P(q) = \text{span}\{\sigma_i(q) \mid \sigma_i,\ i \in I\}.$$

for some one-forms σ_i and index set I.

21. Let D a distribution, and P a codistribution. Then

$$(\text{ann } D)(q) = \text{span}\{\sigma(q) \mid \sigma \text{ one-form such that } \sigma(X) = 0 \text{ for all } X \in D\},$$
$$(\ker P)(q) = \text{span}\{X(q) \mid X \text{ vectorfield such that } \sigma(X) = 0 \text{ for all } \sigma \in P\}.$$

P is called involutive if $\ker P$ is involutive.

22. $D \subset \ker(\text{ann } D)$, $P \subset \text{ann}(\ker P)$, and equality holds if D and P have constant dimension. If P is involutive and constant dimensional then there exist local coordinates $x_1,\ldots,x_n$ such that $P = \text{span}\{dx_{k+1},\ldots,dx_n\}$.

Notes and References

The material treated in this chapter is quite standard, and is adequately covered in many textbooks, such as [Bo], [Sp], [Wa], (see also [AM]), and we have made extensively use of these sources.

For more details on immersed submanifolds we refer to [Bo], [Sp]. The definition of tangent space as given here (Definition 2.21) is the most common one; see however [AM] for an alternative definition (see also [BJ]). The proof of Frobenius' Theorem (Theorem 2.42) given here is taken from [So]. A more constructive proof can be found e.g. in [Bo], [Wa]. For a global version of Frobenius' theorem and global foliations we refer to [Bo]. An important extension of Frobenius' theorem is the Hermann-Nagano theorem for *analytic* distributions with no constant dimension, see [He], [Na]. This was further generalized in [Su]. For more details concerning properties of distributions and codistributions we refer to [Is].

We only treated here differential one-forms. For general differential forms, the d-operator and Lie derivatives of differential forms we refer to [AM], [Bo], [Sp], [Wa]. Furthermore, involutivity of codistributions can be defined independently of distributions, and Frobenius' theorem can be equivalently stated for involutive codistributions using differential forms, see for instance [B], [S].

[AM] R.A. Abraham, J.E. Marsden, **Foundations of Mechanics**, Benjamin/Cummings, Reading, 1978.

[Bo] W.A. Boothby, **An Introduction to Differentiable Manifolds and Riemannian Geometry**, Academic, New York, 1975.

[BJ] T. Bröcker, K. Jänich, **Einführung in die Differentialtopologie**, Springer, Berlin, 1973.

[He] R. Hermann, "The differential geometry of foliations", J. Math. and Mech. 11, pp. 302-306, 1962.

[Is] A. Isidori, **Nonlinear Control Systems: An Introduction**, Lect. Notes Contr. Inf. Sci. 72, Springer, Berlin, 1985.

[Na] T. Nagano, "Linear differential systems with singularities and applications to transitive Lie algebras", J. Math. Soc. Japan, 18, pp. 398-404, 1966.

[Sp] M. Spivak, **A comprehensive introduction to differential geometry**, Vol I, Publish or Perish, Boston, 1970.

[Su] H. Sussmann, "Orbits of families of vectorfields and integrability of distributions", Trans. Amer. Math. Soc., 180, pp. 171-188, 1973.

[Wa] F.W. Warner, **Foundations of differentiable manifolds and Lie groups**, Scott, Foresman, Glenview, 1970.

Exercises

2.1 Consider the topological space $\mathbb{R}$ with coordinate charts $(\mathbb{R},\varphi)$ and $(\mathbb{R},\psi)$ with $\varphi(x) = x$, $\psi(x) = x^3$. Show that these charts are *not* C^∞-compatible. On the other hand show that $\mathbb{R}$ with differentiable structure defined by the atlas $(\mathbb{R},\varphi)$, and $\mathbb{R}$ with differentiable structure defined by the atlas $(\mathbb{R},\psi)$ are *diffeomorphic*.

2.2 Show that $f : \mathbb{R} \to \mathbb{R}$ given by

$$f(x) = 0 \ , \ x \leq 0 \ ,$$
$$f(x) = \exp(-\tfrac{1}{x}) \ , \ x > 0 \ ,$$

is C^∞ but not analytic.

2.3 Prove that the rank of the map $f(A) = A^TA$ in Example 2.15 is indeed $\frac{1}{2}n(n+1)$ in points of $O(n)$.

2.4 With the aid of the implicit function theorem prove the following

(a) Let $A(x)$ be a $p\times m$-matrix, and $b(x)$ a p-vector, with x in some open set U of $\mathbb{R}^n$. Suppose that for some $x_0 \in U$

$$\operatorname{rank} A(x_0) = p.$$

Then there exists a neighborhood $V \subset U$ of x_0 and a smooth map $\alpha : V \to \mathbb{R}^m$ such that

$$A(x)\ \alpha(x) = b(x) \quad , \ x \in V \ .$$

(b) Let $A(x)$ be a $p\times m$-matrix, and $b(x)$ a p-vector, with x in some neighborhood U of a point $x_0 \in \mathbb{R}^n$. Suppose that

$$\operatorname{rank} A(x) = r ,$$

for every $x \in U$. Then there exists a neighborhood $V \subset U$ of x_0 and a smooth map $\alpha : V \to \mathbb{R}^m$ such that

$$A(x)\ \alpha(x) - b(x) = \begin{pmatrix} 0 \\ \varphi(x) \end{pmatrix}^{r} , \quad x \in V ,$$

for some $(p - r)$-vector $\varphi(x)$, depending smoothly on x.

(c) Let $A(x)$ be a $p \times m$-matrix, with x in some neighborhood U of a point $x_0 \in \mathbb{R}^n$. Suppose that rank $A(x) = r$ for every $x \in U$. Then there exists a neighborhood $V \subset U$ of x_0 and a smooth map $\beta : V \to G\ell(m)$ (with $G\ell(m)$ the invertible $m \times m$-matrices) such that

$$A(x)\ \beta(x) = \left[\begin{array}{c|c} I_r & 0 \\ \hline \psi(x) & 0 \end{array}\right] , \quad x \in V ,$$

with $\psi(x)$ a $(p-r)xr$ matrix, depending smoothly on x.

2.5 Prove Proposition 2.20 in the following way:

(a) Take arbitrary coordinate charts $(U, \tilde{x}_1, \ldots, \tilde{x}_{n_1})$ about p_1 and $(V, \tilde{z}_1, \ldots, \tilde{z}_{n_2})$ about $p_2 = F(p_1)$, and write F in these coordinates as

$$F(\tilde{x}) = (F_1(\tilde{x}), \ldots, F_{n_2}(\tilde{x}))^T$$

Use Proposition 2.17 in order to show that, possibly after a permutation of the coordinates $\tilde{z}_1, \ldots, \tilde{z}_{n_2}$, the functions $x_1 = F_1, \ldots, x_{n_1} = F_{n_1}$ serve as new local coordinates system for M about p_1. Prove that in the coordinates x and $\tilde{z}$ F takes the form

$$(a_1, \ldots, a_{n_1}) \mapsto (a_1, \ldots, a_{n_1}, \psi_{n_1+1}(a), \ldots, \psi_{n_2}(a))$$

(b) Define the following coordinate transformation on N

$$z_i = \tilde{z}_i \quad , \ i = 1, \ldots, n_1 ,$$

$$z_i = \tilde{z}_i - \psi_i(\tilde{z}_1, \ldots, \tilde{z}_{n_1}) \quad , \ i = n_1 + 1, \ldots, n_2 .$$

Show that in the coordinates (x,z) F takes the form (2.38).

2.6 Let X, Y be vectorfields on a manifold M. Define for any smooth function $f : M \to \mathbb{R}$ the object $[[X,Y]]_p(f)$, $p \in M$, as satisfying

$$[[X,Y]]_p(f) = X_p(Y(f)).$$

Show that in general $[[X,Y]]$ is *not* a vectorfield.

2.7 Consider the vectorfields (on $\mathbb{R}^2$)

$$X = x_1 \frac{\partial}{\partial x_1} + x_2 \frac{\partial}{\partial x_2} \quad , \quad y = x_2 \frac{\partial}{\partial x_1} - x_1 \frac{\partial}{\partial x_2} .$$

Show that $[X,Y] = 0$. Interpret this geometrically.

Define on $\mathbb{R}^2 \backslash \{(x_1, x_2) \mid x_1 \geq 0, x_2 = 0\}$ the coordinate transformation (see Example 2.2) $x_1 = r \cos \varphi$, $x_2 = r \sin \varphi$, and compute X and Y in the new coordinates (r, φ). Check the vanishing of the Lie bracket in the new coordinates.

2.8 Apply Theorem 2.36 to the vectorfields X, Y as given in Exercise 2.7.

2.9 Show that the linear space of $n\times n$-matrices with bracket operation

$$[A,B] = AB - BA \;, \qquad A,B \;\; n\times n\text{-matrices} \;,$$

is indeed a Lie algebra.

2.10 Define the $2n\times 2n$-matrix $J = \begin{pmatrix} 0_n & -I_n \\ I_n & 0_n \end{pmatrix}$.

(0_n is $n\times n$ zero matrix, I_n is $n\times n$ identity matrix). A $2n\times 2n$-matrix A is called *Hamiltonian* if $A^T J + JA = 0$. Show that the linear space of $2n\times 2n$ Hamiltonian matrices is a Lie subalgebra of the Lie algebra of all $2n\times 2n$ matrices.

2.11 Consider on $\mathbb{R}^3$ the vectorfields

$$X_1(x) = \frac{\partial}{\partial x_2} - x_1 \frac{\partial}{\partial x_3} \text{ and } X_2(x) = \frac{\partial}{\partial x_1} .$$

Let $x_0 = (0,0,0)^T$. Compute that $X_2^{-h} \circ X_1^{-h} \circ X_2^{h} \circ X_1^{h} \circ (x_0) = h^2 [X_1, X_2](x_0)$.

2.12 Show that the distribution D spanned by the vectorfields Z_α, $\alpha \in \underline{m}$, as defined in (2.137) is involutive if and only if (2.135) holds.

2.13 Verify the local coordinate expression for the Lie derivative of a one-form as given in (2.167). (Hint : First verify the expression for the vectorfields $\frac{\partial}{\partial x_i}$, $i \in \underline{n}$.)

2.14 Let $F : M \to N$ and let P be a constant dimensional involutive co distribution on N. Prove that $F^* P$ (see (2.181)) is an involutive codistribution on M. (Hint: First write (cf. (2.178)) $P = \{dz_{k+1}, \ldots, dz_m\}$ in suitable local coordinates for N. Then compute $F^* P$.)

2.15 Let $V = \mathbb{R}^r$ be a Lie algebra. For any basis $\{v_1, \ldots, v_r\}$ of V there exist constants c_{ij}^k, $i,j,k \in \underline{r}$, called the structure constants, such that

$$[v_i, v_j] = \sum_{k=1}^{r} c_{ij}^k v_k , \qquad i,j \in \underline{r}.$$

Show that

(i) $c_{ij}^k = -c_{ji}^k , \qquad i,j,k \in \underline{r}$

(ii) $\sum_{k=1}^{r} c_{ij}^k c_{k\ell}^m + c_{\ell i}^k c_{kj}^m + c_{j\ell}^k c_{ki}^m) = 0, \qquad i,j,m \in \underline{r},$

Conversely, any set of constants satisfying (i), (ii), defines a Lie bracket.

3
Controllability and Observability, Local Decompositions

In the first two sections of this chapter we will give some basic concepts and results in the study of controllability and observability for nonlinear systems. Roughly speaking we will restrict ourselves to what can be seen as the nonlinear generalizations of the Kalman rank conditions for controllability and observability of linear systems. The reason for this is that in the following chapters we will not need so much the notions of nonlinear controllability and observability *per se*, but only the "structural properties" as expressed by these nonlinear "controllability" and "observability" rank conditions that will be obtained. In the last section of this chapter we will show how the geometric interpretation of reachable and unobservable subspaces for linear systems as invariant subspaces enjoying some maximality or minimality properties can be generalized to the nonlinear case, using the notion of *invariant* distributions. In this way we make contact with the nonlinear generalization of linear geometric control theory as dealt with in later chapters, where this last notion plays a fundamental role.

3.1 Controllability

In this chapter we consider smooth affine nonlinear control systems

$$\dot{x} = f(x) + \sum_{j=1}^{m} g_j(x)u_j, \qquad u = (u_1, \ldots, u_m) \in U \subset \mathbb{R}^m, \tag{3.1}$$

where $x = (x_1, \ldots, x_n)$ are local coordinates for a smooth manifold M (the state space manifold), and $f, g_1, \ldots, g_m$ are smooth vectorfields on M (see Chapter 2). f is called the *drift* vectorfield, and g_j, $j \in \underline{m}$, the *input* vectorfields. Given any point $x_0 \in M$ one may wonder what one can say about the set of points which can be reached from x_0 in finite time by a suitable choice of the input functions $u_j(\cdot)$, $j \in \underline{m}$. This is the general *controllability* (or *reachability*) problem.

Throughout we will make the following assumptions concerning the input space U and the class of admissible controls $\mathcal{U}$ (see Chapter 1).

Assumption 3.1 (a) The input space U is such that the set of associated vectorfields of the system (3.1)

$$\mathcal{F} = \{f + \sum_{j=1}^{m} g_j u_j \mid (u_1, \ldots, u_m) \in U\} \tag{3.2}$$

contains the vectorfields $f, g_1, \ldots, g_m$.

(b) $\mathcal{U}$ consists of the piecewise constant functions which are piecewise continuous from the right.

Remark It follows from standard results on the continuity of solutions of differential equations that if we *approximate* a more general control function $\bar{u}(\cdot)$ by piecewise constant functions in some suitable sense, then the solutions of (3.1) for these piecewise constant functions will approximate the solution of (3.1) for $\bar{u}(\cdot)$ (see also the references at the end of this chapter). In this sense many properties of systems with quite general control functions can be established by considering only the piecewise constant case.

Recall (see Chapter 1) that the unique solution of (3.1) at time $t \geq 0$ for a particular input (control) function $u(\cdot)$ and initial condition $x(0) = x_0$ is denoted as $x(t, 0, x_0, u)$, or simply as $x(t)$. Then we state

Definition 3.2 *The nonlinear system (3.1) is called* controllable *if for any two points x_1, x_2 in M there exists a finite time T and an admissible control function u: $[0,T] \to U$ such that $x(T, 0, x_1, u) = x_2$.*

One is especially interested in controllability because for *linear systems*

$$\dot{x} = Ax + Bu, \qquad x \in \mathbb{R}^n, \ u \in \mathbb{R}^m. \tag{3.3}$$

one knows that controllability is equivalent with the (easily verifiable) algebraic condition

$$\operatorname{rank}\left(B \mid AB \mid A^2B \mid \cdots \mid A^{n-1}B\right) = n \tag{3.4}$$

(*Kalman rank condition* for controllability). Furthermore in linear systems theory controllability is crucial, not only because of the concept of controllability per se, but also, or maybe especially, because of its importance for e.g. stabilizability, (optimal) control design and realization theory.

The simplest approach to study controllability of the nonlinear system (3.1) is to consider its *linearization*. In fact we have

Proposition 3.3 *Consider the nonlinear system (3.1), and let $x_0 \in M$*

satisfy $f(x_0) = 0$. *Furthermore let* U *contain a neighborhood* V *of* $u = 0$. *Suppose that the linearization of (3.1) in* x_0 *and* $u = 0$

$$\dot{z} = \frac{\partial f}{\partial x}(x_0)z + \sum_{j=1}^{m} g_j(x_0)v_j, \qquad z \in \mathbb{R}^n, \ v \in \mathbb{R}^m, \tag{3.5}$$

is a controllable linear system. Then for every $T > 0$ *and* $\epsilon > 0$ *the set of points which can be reached from* x_0 *in time* T *using admissible control functions* $u(\cdot): [0,T] \to V$, *satisfying* $\|u(t)\| < \epsilon$, *contains a neighborhood of* x_0.

Proof Since the linearization (3.5) is controllable there exist input functions $v^1(\cdot),\dots,v^n(\cdot)$ defined on $[0,T]$ steering the origin $z = 0$ in time T to independent points $z^1,\dots,z^n \in \mathbb{R}^n$. Furthermore we can take these input functions to be piecewise constant, and so small that

$$u(t,\xi_1,\dots,\xi_n) := \xi_1 v^1(t) + \dots + \xi_n v^n(t) \tag{3.6}$$

satisfies $\|u(t,\xi)\| < \epsilon$ for $0 \le t \le T$ and $|\xi_i| < 1$, $i \in \underline{n}$. Denote by $x(t,\xi)$ the solution of (3.1) for the input (3.6) initiating at $x(0,\xi) = x_0$. (By taking ϵ small enough $x(t,\xi)$ will exist for all $0 \le t \le T$.) Consider now the map

$$\xi \mapsto x(T,\xi), \qquad \xi \text{ near } 0. \tag{3.7}$$

We shall show that the matrix

$$Z(t) = \frac{\partial x}{\partial \xi}(t,\xi)\big|_{\xi=0} \tag{3.8}$$

is non-singular at $t = T$, and then the conclusion follows from the Inverse Function Theorem applied to the map (3.7) (see Proposition 2.16). Since

$$\frac{\partial}{\partial t}x(t,\xi) = f\big(x(t,\xi)\big) + \sum_{j=1}^{m} g_j\big(x(t,\xi)\big)u_j(t,\xi), \tag{3.9}$$

we can differentiate (3.9) with respect to ξ in $\xi = 0$, so as to obtain

$$\dot{Z}(t) = AZ(t) + B\big(v^1(t) \vdots \cdots \vdots v^n(t)\big), \qquad Z(0) = 0, \tag{3.10}$$

where $A = \frac{\partial f}{\partial x}(x_0)$ and $B = \big(g_1(x_0) \vdots \cdots \vdots g_m(x_0)\big)$. By definition of v^i it follows that the columns of $Z(T)$ are independent. □

Remark 3.4 (see Exercise 3.2) Similar statements can be derived for the points which can be steered in small time to x_0 using small controls, noting that controllability of (3.5) is equivalent to controllability of the linearization of the *time-reversed* system $\dot{x} = -f(x) - \sum_{j=1}^{m} g_j(x)u_j$.

However the linearization approach is often not satisfactory. Already in Chapter 1 we have seen that by linearizing a nonlinear system we may loose much of the structure of the system, see e.g. Example 1.2. In particular a nonlinear system can be controllable while its linearization is *not*:

Example 3.5 Consider the simplified model of maneuvering a car, as dealt with in Example 2.35. If the speed of rolling and rotation can be directly controlled then we obtain the nonlinear system

$$\frac{d}{dt}\begin{pmatrix} x_1 \\ x_2 \\ x_3 \end{pmatrix} = \begin{pmatrix} \sin x_3 \\ \cos x_3 \\ 0 \end{pmatrix} u_1 + \begin{pmatrix} 0 \\ 0 \\ 1 \end{pmatrix} u_2 . \tag{3.11}$$

Clearly, its linearization in any point $x \in \mathbb{R}^3$ is uncontrollable. However, as we will see, this model of driving a car *is* controllable (in accordance with most people's experience of driving a car!). □

So let us consider again the nonlinear system (3.1). We ask ourselves in which directions we can steer from a given point x_0, i.e. which are the points which can be reached from x_0 in arbitrarily small time. First we consider the situation where the drift vectorfield f is absent, and for simplicity we take only two input vectorfields while M is equal to $\mathbb{R}^n$, i.e.

$$\dot{x} = u_1 g_1(x) + u_2 g_2(x) \qquad x \in \mathbb{R}^n, \ (u_1, u_2) \in \mathbb{R}^2. \tag{3.12}$$

Clearly from any point $x_0 \in \mathbb{R}^n$ we may steer directly in all directions contained in the subspace of the tangent space $T_{x_0}\mathbb{R}^n \cong \mathbb{R}^n$ given by

$$G(x_0) = \text{span}\{g_1(x_0), g_2(x_0)\}, \tag{3.13}$$

by using constant inputs. Can we steer, maybe in an indirect way, in other directions as well? The answer is yes and the key idea for achieving this is to *switch* between the vectorfields g_1 and g_2 by choosing appropriate piecewise constant input functions u_1 and u_2.

Recall from Chapter 2 (Lemma 2.34) that in case $[g_1, g_2] = 0$ then the integral flows of g_1, g_2 commute, i.e. $g_1^t \circ g_2^s = g_2^s \circ g_1^t$ for all t, s, and thus we cannot steer into a direction outside $G(x_0)$ by switching between g_1 and g_2. However, if $[g_1, g_2](x_0) \notin G(x_0)$ we *can* steer into a direction outside

$G(x_0)$ as follows from

Proposition 3.6 *Consider for the nonlinear system (3.12) the control function* $u(t) = (u_1(t), u_2(t))$ *given as*

$$u(t) = \begin{cases} (1,0) & t \in [0,h) \\ (0,1) & t \in [h,2h) \\ (-1,0) & t \in [2h,3h) \\ (0,-1) & t \in [3h,4h) \end{cases}, \qquad h > 0. \tag{3.14}$$

Then the solution $x(t) = x(t,0,x_0,u)$ *satisfies*

$$x(4h) = x_0 + h^2[g_1,g_2](x_0) + O(h^3). \tag{3.15}$$

Proof By Taylor expansion we have

$$x(h) = x(0) + h\dot{x}(0) + \frac{1}{2}h^2\ddot{x}(0) + \ldots =$$

$$x_0 + hg_1(x_0) + \frac{1}{2}h^2 \frac{\partial g_1}{\partial x}(x_0)g_1(x_0) + \ldots .$$

Similarly (only writing terms up to order h^2)

$$x(2h) = x(h) + hg_2\big(x(h)\big) + \frac{1}{2}h^2 \frac{\partial g_2}{\partial x}\big(x(h)\big)g_2\big(x(h)\big) + \ldots =$$

$$x_0 + h\big(g_1(x_0) + g_2(x_0)\big) +$$

$$h^2\Big(\frac{1}{2}\frac{\partial g_1}{\partial x}(x_0)g_1(x_0) + \frac{\partial g_2}{\partial x}(x_0)g_1(x_0) + \frac{1}{2}\frac{\partial g_2}{\partial x}(x_0)g_2(x_0)\Big) + \ldots$$

where we have used the fact that $g_2(x_0+hx_1) = g_2(x_0) + h\frac{\partial g_2}{\partial x}(x_0) + \ldots$. Next we compute (to keep notation down: all functions are evaluated at x_0, unless stated otherwise)

$$x(3h) = x(2h) - hg_1\big(x_0 + h(g_1+g_2)+..\big) + \frac{1}{2}h^2\frac{\partial g_1}{\partial x}(x_0+..)g_1(x_0+..) + \ldots$$

$$= x_0 + h(g_1+g_2) + h^2\Big(\frac{1}{2}\frac{\partial g_1}{\partial x}g_1 + \frac{\partial g_2}{\partial x}g_1 + \frac{1}{2}\frac{\partial g_2}{\partial x}g_2\Big) -$$

$$- h\Big(g_1 + h\frac{\partial g_1}{\partial x}(g_1+g_2)\Big) + \frac{1}{2}h^2\frac{\partial g_1}{\partial x}g_1 + \ldots$$

$$= x_0 + hg_2 + h^2\Big(\frac{\partial g_2}{\partial x}g_1 - \frac{\partial g_1}{\partial x}g_2 + \frac{1}{2}\frac{\partial g_2}{\partial x}g_2\Big) + \ldots .$$

Finally we obtain

$$x(4h) = x(3h) - hg_2(x_0+hg_2+..) + \tfrac{1}{2}h^2 \frac{\partial g_2}{\partial x}(x_0+..)g_2(x_0+..) + \ldots =$$

$$= x_0 + hg_2 + h^2\left(\frac{\partial g_2}{\partial x}g_1 - \frac{\partial g_1}{\partial x}g_2 + \frac{1}{2}\frac{\partial g_2}{\partial x}g_2\right) -$$

$$- h\left(g_2 + h\frac{\partial g_2}{\partial x}g_2\right) + \tfrac{1}{2}h^2\frac{\partial g_2}{\partial x}g_2 + \ldots$$

$$= x_0 + h^2\left(\frac{\partial g_2}{\partial x}g_1 - \frac{\partial g_1}{\partial x}g_2\right) + O(h^3). \qquad \square$$

Example 3.5 (continued) In Example 2.35 we have directly computed (cf. 2.111) that for g_1 and g_2 as given in (3.11)

$$x(4h) = x_0 + h^2(-\cos x_{30}, \sin x_{30}, 0)^T + O(h^3),$$

where $(-\cos x_{30}, \sin x_{30}, 0)^T = [g_1, g_2](x_0)$. $\square$

Formula (3.15) implies that, at least approximately, we can steer the system (3.12) from x_0 into the direction given by the vector $[g_1,g_2](x_0)$; in particular if $[g_1,g_2](x_0) \notin G(x_0)$ we can steer into a direction outside $G(x_0)$. However, this is not the end of the story. By choosing more and more elaborate switchings for the inputs it is also possible to move in directions given by the higher-order brackets of g_1 and g_2, i.e. terms such as $[g_2,[g_1,g_2]]$, $[[g_1,g_2],[g_2,[g_1,g_2]]$, etc. (In fact already in the expansion (3.15) these higher-order brackets are present in the remainder $O(h^3)$.) In case a drift term f is present in (3.1) we also have to consider the Lie brackets involving the vectorfield f. For instance the system

$$\dot{x} = f(x) + ug(x) \tag{3.16}$$

can be regarded as a special case of (3.12) with $u_1 = 1$. Although we cannot go back and forth along the vectorfield f, by making switchings only in u the evolution of x still can be steered in directions involving the brackets of f and g. (Although generally we can now only steer along the positive or negative directions of these brackets.)

Motivated by the foregoing discussion we give

Definition 3.7 *Consider the nonlinear system (3.1). The* accessibility algebra $\mathcal{C}$ *is the smallest subalgebra of $V^\infty(M)$ (the Lie algebra of vectorfields on M, cf. Chapter 2) that contains $f, g_1, \ldots, g_m$.*

Remark The smallest subalgebra of $V^\infty(M)$ containing a set of vectorfields is well-defined, since the intersection of two subalgebras is again a subalgebra.

The following characterization of $\mathcal{C}$ is sometimes useful.

Proposition 3.8 Every element of $\mathcal{C}$ is a linear combination of repeated Lie brackets of the form

$$[X_k, [X_{k-1}, [\cdots, [X_2, X_1]\cdots]]] \tag{3.17}$$

where X_i, $i \in \underline{k}$, is in the set $\{f, g_1, \ldots, g_m\}$ and $k = 0,1,2,\ldots$.

Proof Denote the linear subspace of $V^\infty(M)$ spanned by the expressions (3.17) by $\mathcal{L}$. By definition of $\mathcal{C}$ we have $\mathcal{L} \subset \mathcal{C}$. In order to show $\mathcal{L} = \mathcal{C}$ we only have to prove that $\mathcal{L}$ is a subalgebra. Let the length of an expression (3.17) be the number of Lie brackets in it, i.e. k. Consider two arbitrary expressions (3.17) of length j, resp. ℓ

$$\begin{aligned} Z &= [Z_j, [Z_{j-1}, [\cdots, [Z_2, Z_1]\cdots]]] \\ Y &= [Y_\ell, [Y_{\ell-1}, [\cdots, [Y_2, Y_1]\cdots]]] \end{aligned} \tag{3.18}$$

By induction we will prove that $[X,Y] \in \mathcal{L}$ for any j and ℓ. For assume that $[Z,Y] \in \mathcal{L}$ for all Y, ℓ arbitrary, and for all Z with $j \leq k$. (This is clearly true for $k = 1$.) Now take $j = k+1$ in (3.17). Then by the Jacobi-identity (Proposition 2.27 (c))

$$[Z,Y] = -[Z^1, [Z_j, Y]] + [Z_j, [Z^1, Y]], \tag{3.19}$$

with $Z^1 = [Z_{j-1}, [\cdots [Z_2, Z_1]\cdots]]$. Since the length Z^1 equals $j-1 = k$, it follows by the induction assumption that the first term on the right-hand side is in $\mathcal{L}$, and $[Z^1, Y] \in \mathcal{L}$, so that also the second term is in $\mathcal{L}$. □

Now let us define the *accessibility distribution* C as the distribution generated by the accessibility algebra $\mathcal{C}$:

$$C(x) = \text{span}\{X(x) \mid X \text{ vectorfield in } \mathcal{C}\}, \qquad x \in M. \tag{3.20}$$

Since $\mathcal{C}$ is a subalgebra, it immediately follows that C is *involutive*. Furthermore, let $R^V(x_0, T)$ be the reachable set from x_0 at time $T > 0$, following trajectories which remain for $t \leq T$ in the neighborhood V of x_0, i.e.

$$R^V(x_0,T) = \{x \in M \mid \textit{there exists an admissible input } u\colon [0,T] \to U \textit{ such that the evolution of (3.1) for } x(0) = x_0 \textit{ satisfies } x(t) \in V,\ 0 \le t \le T, \textit{ and } x(T) = x\}, \tag{3.21}$$

and denote

$$R_T^V(x_0) = \bigcup_{\tau \le T} R^V(x_0,\tau). \tag{3.22}$$

We have the following basic theorem

Theorem 3.9 *Consider the system (3.1). Assume that*

$$\dim C(x_0) = n. \tag{3.23}$$

Then for any neighborhood V of x_0 and $T > 0$ the set $R_T^V(x_0)$ contains a non-empty open set of M.

Proof By continuity there exists a neighborhood $W \subset V$ of x_0 such that $\dim C(x) = n$, for any $x \in W$. We construct a sequence of submanifolds N_j in W, $\dim N_j = j$, $j \in \underline{n}$, in the following way. Let $\mathcal{F}$ be the set of associated vectorfields of the system (3.1), cf. (3.2). For $j = 1$ choose $X_1 \in \mathcal{F}$ such that $X_1(x_0) \neq 0$. Then by the Flow-box Theorem (Theorem 2.26) for sufficiently small $\epsilon_1 > 0$

$$N_1 = \{X_1^{t_1}(x_0) \mid 0 < t_1 < \epsilon_1\} \tag{3.24}$$

is a submanifold of M of dimension 1, contained in W. Let us now construct N_j for $j > 1$ by induction. Assume that we have constructed a submanifold $N_{j-1} \subset W$ of dimension j-1 defined as

$$N_{j-1} = \{X_{j-1}^{t_{j-1}} \circ X_{j-2}^{t_{j-2}} \circ \cdots \circ X_1^{t_1}(x_0) \mid 0 \le \sigma_i < t_i < \epsilon_i,\ i \in \underline{j-1}\} \tag{3.25}$$

where X_i, $i \in \underline{j-1}$, are vectorfields in $\mathcal{F}$, and $\sum_{i=1}^{j-1} \sigma_i$ is arbitrarily small. If $j-1 < n$ then we can find $X_j \in \mathcal{F}$ and $q \in N_{j-1}$ such that

$$X_j(q) \notin T_q N_{j-1}. \tag{3.26}$$

For if this was not possible then $X(q) \in T_q N_{j-1}$ for any $X \in \mathcal{F}$ and $q \in N_{j-1}$. However, in view of Proposition 2.41, this would mean that this holds for any $X \in \mathcal{C}$, so that $\dim C(q) < n$ for every $q \in N_{j-1} \subset W$, which is in contradiction with the definition of W. It also follows that we may take q in (3.26) arbitrarily close to x_0. Therefore the map

$$(t_j,\ldots,t_1) \mapsto X_j^{t_j} \circ X_{j-1}^{t_{j-1}} \circ \cdots \circ X_1^{t_1}(x_0) \tag{3.27}$$

has rank equal to j on some set $0 \leq \sigma_i < t_i < \epsilon_i$, $i \in \underline{j}$. Hence by Proposition 2.20 the image of this map for ϵ_i, $i \in \underline{j}$, sufficiently small is a submanifold $N_j \subset W$ of dimension j. Finally we conclude that N_n is the desired open set contained in $R_T^V(x_0)$. □

Motivated by this we give

Definition 3.10 *The system (3.1) is* locally accessible *from* x_0 *if* $R_T^V(x_0)$ *contains a non-empty open set of* M *for all neighborhoods* V *of* x_0 *and all* $T > 0$. *If this holds for any* $x_0 \in M$ *then the system is called* locally accessible.

Corollary 3.11 *If* dim $C(x) = n$ *for all* $x \in M$ *then the system is locally accessible.*

We call (3.23) the *accessibility rank condition* at x_0. If (3.23) holds for any $x \in M$ then we say that the system satisfies the accessibility rank condition. (The relation with the controllability rank condition (3.4) for linear systems will be explained soon.)

What can we say if dim $C(x_0) < n$ for some x_0? In case the distribution C is constant dimensional about x_0 we have

Proposition 3.12 *Suppose that* C *has constant dimension* k *less than* n *about* x_0. *By Frobenius' Theorem (Theorem 2.42) we can find a neighborhood* W *of* x_0 *and local coordinates* $x_1,\ldots,x_n$ *such that the submanifold*

$$S_{x_0} = \{q \in W \mid x_i(q) = x_i(x_0),\ i = k+1,\ldots,n\} \tag{3.28}$$

is an integral manifold of C. *Then for any neighborhood* $V \subset W$ *of* x_0 *and for all* $T > 0$, $R_T^V(x_0)$ *is contained in* S_{x_0}. *Furthermore* $R_T^V(x_0)$ *contains a non-empty open set of the integral manifold* S_{x_0}. *Hence the system restricted to* S_{x_0} *is locally accessible.*

Proof Since $f(x) + \sum_{j=1}^{m} g_j(x)u_j \in C(x)$ for any $(u_1,\ldots,u_m) \in U$ and $x \in M$ the system (3.1) for $x(0) = x_0$ can be restricted (for sufficiently small time) to S_{x_0}, where dim S_{x_0} = dim $C(x_0)$. Now apply Theorem 3.9 to this restricted system. □

Corollary 3.13 *If the system (3.1) is locally accessible then* $\dim C(x) = n$ *for* x *in an open and dense subset of* M.

Proof First, for any x_0 such that $\dim C(x_0) = n$ there exists a neighborhood of x_0 such that on this neighborhood $\dim C(x) = n$. Hence the set of x for which $\dim C(x) = n$ is always open (but possibly empty). Now suppose there is an open set $\bar{V} \neq \emptyset$ of M where $\dim C(x) < n$ for all $x \in \bar{V}$. Then there is also an open set $V \neq \emptyset$ with $\dim C(x) = k < n$ for all $x \in V$. Now use Proposition 3.12 for the system restricted to V. Then it follows that the system is not locally accessible, which contradicts the assumption. Hence the set of x for which $\dim C(x) = n$ is dense. □

Usually the property of local accessibility is far from controllability, as shown by the following example.

Example 3.14 Consider the system on $\mathbb{R}^2$

$$\dot{x}_1 = x_2^2, \qquad \dot{x}_2 = u. \tag{3.29}$$

The accessibility algebra $\mathcal{C}$ is spanned by the vectorfields $f = x_2^2 \frac{\partial}{\partial x_1}$, $g = \frac{\partial}{\partial x_2}$, and their Lie brackets $[f,g] = -2x_2 \frac{\partial}{\partial x_1}$, $\big[[f,g],g\big] = 2 \frac{\partial}{\partial x_1}$. Clearly $\dim C(x) = 2$ everywhere, and so the system is locally accessible. However since $x_2^2 \geq 0$ the x_1-coordinate is always non-decreasing. Hence the reachable sets look like in Fig. 3.1, and the system is not controllable. □

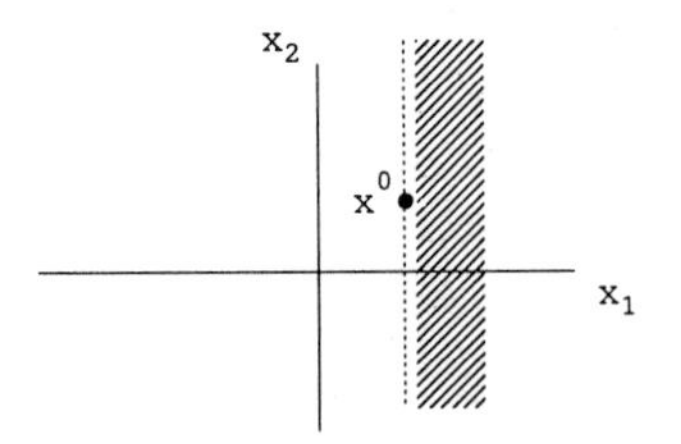

Fig. 3.1. Reachable set from x^0.

However in case the drift term f in (3.1) is absent the accessibility rank condition *does* imply controllability.

Proposition 3.15 *Suppose* $f = 0$ *in (3.1), and let* $\mathcal{F}$ *be symmetric, i.e. for any* $X \in \mathcal{F}$ *also* $-X \in \mathcal{F}$. *Then*

(a) If $\dim C(x_0) = n$ *then* $R_T^V(x_0)$ *contains a neighborhood of* x_0 *for all*

neighborhoods V of x_0 and $T > 0$.

(b) *If* $\dim C(x) = n$ *for all* $x \in M$ *and* M *is connected, then (3.1) is controllable.*

Proof (a) Go back to the proof of Theorem 3.9, and consider the map

$$(t_n, \ldots, t_1) \mapsto X_n^{t_n} \circ X_{n-1}^{t_{n-1}} \circ \cdots \circ X_1^{t_1}(x_0), \quad 0 \leq \sigma_i < t_i < \epsilon_i, \tag{3.30}$$

with $X_i \in \mathcal{F}$, of which the image is N_n. Now let $(s_1, \ldots, s_n)$ satisfy $\sigma_i < s_i < \epsilon_i$, $i \in \underline{n}$, and consider the map

$$(t_n, \ldots, t_1) \mapsto (-X_1)^{s_1} \circ (-X_2)^{s_2} \circ \cdots \circ (-X_n)^{s_n} \circ X_n^{t_n} \circ X_{n-1}^{t_{n-1}} \circ \cdots \circ X_1^{t_1}(x_0),$$

$$\sigma_i < t_i < \epsilon_i. \tag{3.31}$$

Since $(-X_i)^{s_i} = X_i^{-s_i}$ it follows that the image of this map is an open set of M *containing* x_0, and the result follows from symmetry of $\mathcal{F}$.

(b) (see Figure 3.2) Let $R(x_0) := \bigcup_{\tau > 0} R^M(x_0, \tau)$, i.e. the reachable set from x_0. By (a) $R(x_0)$ is open. Now suppose that $R(x_0)$ is strictly contained in M. Take a point z on the boundary of $R(x_0)$. By (a), $R(z)$ contains a neighborhood of z, and hence intersects non-trivially with $R(x_0)$. Hence z can not be a boundary point of $R(x_0)$, which is a contradiction. □

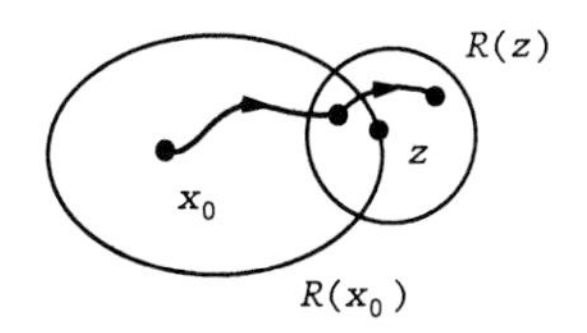

Fig. 3.2. Illustrating the proof of Proposition 3.15(b).

Remark 3.16 It can be easily seen that Proposition 3.15 also holds if $f(x) \in \operatorname{span}\{g_1(x), \ldots, g_m(x)\}$ for all $x \in M$.

Example 3.5 (continued) Since

$$[\sin x_3 \frac{\partial}{\partial x_1} + \cos x_3 \frac{\partial}{\partial x_2}, \frac{\partial}{\partial x_3}] = -\cos x_3 \frac{\partial}{\partial x_1} + \sin x_3 \frac{\partial}{\partial x_2}$$

(cf. (2.109)), we have $\dim C(x) = 3$ for every $x \in \mathbb{R}^3$. Hence by Proposition 3.15 $R_T^V(x)$ contains a neighborhood of x for every x (and every neighborhood V of x and all T), and the system is controllable, as alluded to before. □

Now let us apply the theory developed above to a linear system (3.3), written as

$$\dot{x} = Ax + \sum_{i=1}^{m} b_i u_i , \qquad x \in \mathbb{R}^n, \tag{3.32}$$

where $b_1, \ldots, b_m$ are the columns of the matrix B. First let us compute the accessiblility algebra $\mathcal{C}$ in this case. Clearly the Lie brackets of the constant input vectorfields given by the vectors $b_1, \ldots, b_n$ are all zero:

$$[b_i, b_j] = 0, \qquad i,j \in \underline{m}. \tag{3.33}$$

The Lie bracket of the drift vectorfield Ax with an input vectorfield b_i yields the constant vectorfield

$$[Ax, b_i] = -Ab_i . \tag{3.34}$$

The Lie brackets of Ab_i with Ab_j or b_j are zero, while

$$[Ax, -Ab_i] = A^2 b_i . \tag{3.35}$$

Continuing in this way we conclude that $\mathcal{C}$ is spanned by all constant vectorfields b_i, Ab_i, A^2b_i, ..., $i \in \underline{m}$, together with the linear drift vectorfield Ax. Therefore by Cayley-Hamilton

$$\mathcal{C} = \operatorname{span}\{Ax, b_i, Ab_i, \ldots, A^{n-1}b_i,\ i \in \underline{m}\} \tag{3.36}$$

and

$$C(x) = \operatorname{Im}\left(B \,\vdots\, AB \,\vdots\, \cdots \,\vdots\, A^{n-1}B\right) + \operatorname{span}\{Ax\}. \tag{3.37}$$

We see that the accessibility rank condition (3.23) at $x_0 = 0$ coincides with the Kalman rank condition for controllability (3.4). Hence if we would not have known anything special about linear systems, then at least it follows from Theorem 3.9 that a linear system which satisfies the rank condition (3.4) is *locally accessible*. (Of course we know from linear systems theory that (3.4) is equivalent with *controllability*. This stronger equivalence apparently is due to the linear structure. Notice that Proposition 3.15 does not really apply to linear systems; in fact the extra directions in which we can steer outside Im B are precisely due to Lie brackets with the drift term Ax.)

Remark 3.17 Consider the nonlinear system (3.1) and its linearization (3.5) in x_0 with $f(x_0) = 0$. Denote $A = \frac{\partial f}{\partial x}(x_0)$ and $b_j = g_j(x_0)$. It is easily verified (see Exercise 3.4) that

$$(-1)^k A^k b_j = \overbrace{[f,[f,[\dots[f}^{k\text{-times } f},g_j]\dots]]](x_0) \tag{3.38}$$

It thus follows from Proposition 3.3 that if the subspaces of $T_{x_0}M$ spanned by all repeated Lie brackets of the form given in the right-hand side of (3.38) for $j \in \underline{m}$ and $k = 0,1,\dots,$ has dimension n, then $R_T^V(x_0)$ contains a neighborhood of x_0 for all $T > 0$. Notice that the brackets in the right-hand side of (3.38) belong to a very special subclass of all brackets appearing in $\mathcal{C}$, cf. (3.17). This has motivated the search for stronger rank conditions than the one given in (3.23) guaranteeing stronger types of controllability than local accessibility; we refer to the references cited at the end of this chapter.

Notice that the term span$\{Ax\}$ in (3.35) is not present in the controllability rank condition (3.4) for linear systems. Furthermore for a linear system we know that not only the sets $R_T^V(x_0)$ contain a non-empty set but even the sets $R^V(x_0,T)$ for $T > 0$, i.e. the points that we reach exactly in time T with trajectories contained in V. This motivates the following definitions.

Definition 3.18 *Consider a nonlinear system (3.1). The system is said to be* locally strongly accessible *from* x_0 *if for any neighborhood* V *of* x_0 *the set* $R^V(x_0,T)$ *contains a non-empty open set for any* $T > 0$ *sufficiently small.*

Definition 3.19 *Let* $\mathcal{C}$ *be the accessibility algebra of (3.1). Define* $\mathcal{C}_0$ *as the smallest subalgebra which contains* $g_1,\dots,g_m$ *and satisfies* $[f,X] \in \mathcal{C}_0$ *for all* $X \in \mathcal{C}_0$. *Define the corresponding involutive distribution*

$$C_0(x) = \text{span}\{X(x) \mid X \textit{ vectorfield in } \mathcal{C}_0\}.$$

$\mathcal{C}_0$ *and* C_0 *are called the strong accessibility algebra, respectively strong accessibility distribution.*

Remark It can be immediately checked that for a linear system (3.32)

$$\begin{aligned} \mathcal{C}_0 &= \text{span}\{b_i, Ab_i, \dots, A^{n-1}b_i,\ i \in \underline{m}\} \\ C_0(x) &= \text{Im}\big(B \,\vdots\, AB \,\vdots\, \dots\dots \,\vdots\, A^{n-1}B\big) \end{aligned} \tag{3.39}$$

Analogously to Proposition 3.8 we give the following characterization of $\mathcal{C}_0$.

Proposition 3.20 *Every element of $\mathcal{C}_0$ is a linear combination of repeated Lie brackets of the form*

$$[X_k,[X_{k-1},[\cdots,[X_1,g_j]\cdots]]], \qquad j \in \underline{m},\ k = 0,1,\dots, \tag{3.40}$$

where X_i, $i \in \underline{k}$, is in the set $\{f,g_1,\dots,g_m\}$.

Proof See the proof of Proposition 3.8. □

We have the following extension of Theorem 3.9.

Theorem 3.21 *Consider the system (3.1). Suppose that*

$$\dim C_0(x_0) = n, \tag{3.41}$$

then the system is locally strongly accessible from x_0.

Proof The proof can be reduced to the proof of Theorem 3.9 by making use of the following trick. *Augment* the state space equations (3.1) by the equation $\dot{t} = 1$, t being the time variable, so that we have the augmented system

$$\tilde{\Sigma} \begin{cases} \dot{x} = f(x) + \sum_{j=1}^{m} g_j(x)u_j \\ \dot{t} = 1 \end{cases} \tag{3.42}$$

defined on $\tilde{M} = M \times \mathbb{R}$ with state $\tilde{x} = (x,t)$, drift vectorfield $\tilde{f}(x,t) = f(x)\frac{\partial}{\partial x} + \frac{\partial}{\partial t}$ and input vectorfields $\tilde{g}_j(x,t) = g_j(x)\frac{\partial}{\partial x}$. From the form of the vectorfields $\tilde{f}$ and $\tilde{g}_j$, $j \in \underline{m}$, it immediately follows that the control algebra $\tilde{C}$ of the augmented system satisfies for any t_0

$$\dim \tilde{C}(x_0,t_0) = \dim C_0(x_0) + 1. \tag{3.43}$$

By (3.41) and (3.43) we have $\dim \tilde{C}(x_0,0) = n+1$, and hence the augmented system is locally accessible from $(x_0,0)$. Hence for any $T > 0$ and V neighborhood of x_0 the reachable set $R_T^{\tilde{V}}\big((x_0,0)\big)$, with $\tilde{V} = V \times (-\epsilon,T+\epsilon)$, $\epsilon > 0$, contains a non-empty open set of $M \times \mathbb{R}$. Hence, there exists a non-empty open set $W \subset M$, and an interval (a,b), $0 < a < b \le T$, such that $W \times (a,b) \subset R_T^{\tilde{V}}\big((x_0,0)\big)$. Let $\tau \in (a,b)$. Since $W \times \{\tau\} \subset R_T^{\tilde{V}}\big((x_0,0)\big)$ we conclude that $W \subset R^{V}(x_0,\tau)$. Let $X \in \mathcal{F}$, then the mapping $x \mapsto X^{T-\tau}(x)$ maps W onto an open set $\tilde{W}$ which is contained in $R^{U}(x_0,T)$ for some neighborhood U of x_0. By choosing T small enough the intersection of $\tilde{W}$ with $R^{V}(x_0,T)$ will contain a non-empty open set of M. Hence the system is locally strongly accessible. □

We call (3.41) the *strong accessibility rank condition* at x_0. In case $\dim C_0(x_0) < n$, but C_0 has constant dimension around x_0, we have the following analogue of Proposition 3.12.

Proposition 3.22 *Suppose that C_0 has constant dimension $k < n$ around x_0. By Frobenius' Theorem there is a coordinate chart $(U,x_1,\dots,x_n)$ around x_0 such that the submanifolds $S = \{q \in U \mid x_{k+1}(q) = a_{k+1},\dots,x_n(q) = a_n\}$ for $|a_j| < \epsilon$, $j = k+1,\dots,n$, are integral manifolds of C_0 and the integral manifold S_{x_0} through x_0 is given by $a_j = 0$, $j = k+1,\dots,n$.*

There are now two possibilities:

(i) If $f(x_0) \in C_0(x_0)$, then $f(q) \in C_0(q)$ for all $q \in S_{x_0}$ and $R^{U}_{T}(x_0) \subset S_{x_0}$ for all $T > 0$. In this case the system restricted to S_{x_0} is locally strongly accessible.

(ii) If $f(x_0) \notin C_0(x_0)$, then by continuity $f(q) \notin C_0(q)$ for all $q \in \tilde{U}$, $\tilde{U} \subset U$ neighborhood of x_0, and $\dim C(q) = \dim C_0(q) + 1$ for all $q \in \tilde{U}$. In this case we can adapt the coordinates $x_{k+1},\dots,x_n$ on $\tilde{U}$ to coordinates $\tilde{x}_{k+1},\dots,\tilde{x}_n$ in such a way that as above

$$S_{x_0} = \{q \in U \mid \tilde{x}_{k+1}(q) = \tilde{x}_{k+2}(q) = \cdots = \tilde{x}_n(q) = 0\}$$

and if we let

$$S^{T}_{x_0} = \{q \in \tilde{U} \mid \tilde{x}_{k+1}(q) = T,\ \tilde{x}_{k+2}(q) = \cdots = \tilde{x}_n(q) = 0\} \tag{3.44}$$

then $R^{\tilde{U}}(x_0,T)$ is contained in $S^{T}_{x_0}$ for any $T > 0$ and moreover $R^{\tilde{U}}(x_0,T)$ contains a non-empty open set of $S^{T}_{x_0}$ for any $T > 0$ sufficiently small.

Proof From the definition of C_0 it immediately follows that for any vectorfield X contained in C_0 we have $[f,X] \in C_0$. Since in the above local coordinates $C_0 = \operatorname{span}\{\frac{\partial}{\partial x_1},\dots,\frac{\partial}{\partial x_k}\}$ the local coordinate expression for f takes the form (see also Proposition 3.42)

$$\begin{bmatrix} f_1(x_1,\dots,x_n) \\ \vdots \\ f_k(x_1,\dots,x_n) \\ f_{k+1}(x_{k+1},\dots,x_n) \\ \vdots \\ f_n\quad(x_{k+1},\dots,x_n) \end{bmatrix} \tag{3.45}$$

(i) If $f(x_0) \in C_0(x_0)$ then it immediately follows from (3.45) that $f(q) \in C_0(q)$ for all $q \in S_{x_0}$. Now apply Theorem 3.21 to the system restricted to S_{x_0}.
(ii) Since $C(q) = C_0(q) + \text{span}\{f(q)\}$ the equality $\dim C(q) = \dim C_0(q) + 1$ for all $q \in \tilde{U}$ immediately follows. By (3.45) we can define a vectorfield

$$\bar{f}(x_{k+1},\dots,x_n) = \begin{bmatrix} f_{k+1}(x_{k+1},\dots,x_n) \\ \vdots \\ f_n\quad(x_{k+1},\dots,x_n) \end{bmatrix} \tag{3.46}$$

living on an open part of R^{n-k} with coordinates $(x_{k+1},\dots,x_n)$. By assumption $\bar{f}(0) \neq 0$, and hence by Theorem 2.26 there exist coordinates $\tilde{x}_{k+1},\dots,\tilde{x}_n$ such that $\bar{f} = \dfrac{\partial}{\partial \tilde{x}_{k+1}}$. Then it follows that $R^{\tilde{U}}(x_0,T)$ is contained in $S^T_{x_0}$ for $T > 0$, and by the proof of Theorem 3.21 it follows that $R^{\tilde{U}}(x_0,T)$ contains a non-empty open set of $S^T_{x_0}$ for any $T > 0$ sufficiently small. □

Finally we give the following corollary; its proof parallels the proof of Corollary 3.13.

Corollary 3.23 *If the system (3.1) is locally strongly accessible, then* $\dim C_0(x) = n$ *for* x *in an open and dense subset of* M.

Example 3.24 Consider the equations of a spacecraft with gas jet actuators (Example 1.2). We only consider the equations describing the dynamics of the angular velocities $\omega_1,\omega_2,\omega_3$ (called Euler equations). Since the inertia matrix J is positive definite we can diagonalize it as $\text{diag}(a_1,a_2,a_3)$ to obtain the equations

$$\begin{aligned} a_1\dot{\omega}_1 &= \omega_2\omega_3(a_2-a_3) + \sum_{j=1}^{3} b_j^1 u_j \\ a_2\dot{\omega}_2 &= \omega_1\omega_3(a_3-a_1) + \sum_{j=1}^{3} b_j^2 u_j \\ a_3\dot{\omega}_3 &= \omega_2\omega_1(a_1-a_2) + \sum_{j=1}^{3} b_j^3 u_j \end{aligned} \tag{3.47}$$

where $b_i = (b_i^1, b_i^2, b_i^3)^T$, $i = 1,2,3$, are vectors in $\mathbb{R}^3$. We distinguish between three cases.

I. b_1, b_2, b_3 are independent. Clearly in this case the system is controllable.

II. $\dim \operatorname{span}\{b_1, b_2, b_3\} = 2$. Without loss of generality we may assume that $b_3 = 0$, so that in fact we only have two inputs u_1, u_2.

First consider the simple case $b_1 = (1\ 0\ 0)^T$, $b_2 = (0\ 1\ 0)^T$, so that the torques are around the first two principal axes. Rewrite the system as

$$\begin{aligned} \dot{\omega}_1 &= A_1 \omega_2 \omega_3 + \alpha_1 u_1 \\ \dot{\omega}_2 &= A_2 \omega_1 \omega_3 + \alpha_2 u_2 \\ \dot{\omega}_3 &= A_3 \omega_2 \omega_1 \end{aligned} \tag{3.48}$$

with $A_1 = (a_2 - a_3)a_1^{-1}$, $A_2 = (a_3 - a_1)a_2^{-1}$, $A_3 = (a_1 - a_2)a_3^{-1}$ and $\alpha_1 = a_1^{-1}$, $\alpha_2 = a_2^{-1}$. Denote $f(\omega) = (A_1\omega_2\omega_3, A_2\omega_1\omega_3, A_3\omega_2\omega_1)^T$, $g_1(\omega) = (\alpha_1\ 0\ 0)^T$, $g_2(\omega) = (0\ \alpha_2\ 0)^T$. Compute

$$\begin{aligned} [g_1, f](\omega) &= \begin{pmatrix} 0 & A_1\omega_3 & A_1\omega_2 \\ A_2\omega_3 & 0 & A_2\omega_1 \\ A_3\omega_2 & A_3\omega_1 & 0 \end{pmatrix} \begin{pmatrix} \alpha_1 \\ 0 \\ 0 \end{pmatrix} = \begin{pmatrix} 0 \\ \alpha_1 A_2 \omega_3 \\ \alpha_1 A_3 \omega_2 \end{pmatrix} \\ [g_2, f](\omega) &= \begin{pmatrix} 0 & A_1\omega_3 & A_1\omega_2 \\ A_2\omega_3 & 0 & A_2\omega_1 \\ A_3\omega_2 & A_3\omega_1 & 0 \end{pmatrix} \begin{pmatrix} 0 \\ \alpha_2 \\ 0 \end{pmatrix} = \begin{pmatrix} \alpha_2 A_1 \omega_3 \\ 0 \\ \alpha_2 A_3 \omega_1 \end{pmatrix} \end{aligned} \tag{3.49}$$

On the other hand

$$[g_2, [g_1, f]](\omega) = \begin{pmatrix} 0 & 0 & 0 \\ 0 & 0 & \alpha_1 A_2 \\ 0 & \alpha_1 A_3 & 0 \end{pmatrix} \begin{pmatrix} 0 \\ \alpha_2 \\ 0 \end{pmatrix} = \begin{pmatrix} 0 \\ 0 \\ \alpha_1 \alpha_2 A_3 \end{pmatrix} \tag{3.50}$$

which also equals $[g_1, [g_2, f]](\omega)$. Hence the vectors

$$\begin{pmatrix} \alpha_1 \\ 0 \\ 0 \end{pmatrix}, \begin{pmatrix} 0 \\ \alpha_2 \\ 0 \end{pmatrix}, \begin{pmatrix} 0 \\ 0 \\ \alpha_1 \alpha_2 A_3 \end{pmatrix} \tag{3.51}$$

are contained in $C_0(0)$, and thus if $A_3 \neq 0$, or equivalently $a_1 \neq a_2$, $\dim C_0(0) = 3$, and the system is locally strongly accessible from $\omega = 0$. Furthermore the condition $a_1 \neq a_2$ is also necessary for local strong accessibility, since if we would take $a_1 = a_2$ in (3.48) then we obtain

$$\dot{\omega}_1 = A_1\omega_2\omega_3 + \alpha_1 u_1$$
$$\dot{\omega}_2 = A_2\omega_1\omega_3 + \alpha_2 u_2 \qquad (3.52)$$
$$\dot{\omega}_3 = 0$$

which is clearly not accessible since ω_3 is constant. Therefore, *(3.48) is locally strongly accessible if and only if* $a_1 \neq a_2$.

For the general location of gas jet actuators the computations become more involved. Without proof we give the result

$$\begin{aligned}&(3.48) \text{ is locally strongly accessible} \Leftrightarrow \\ &\dim \operatorname{span}\{b_1, b_2, S(\omega)J^{-1}\omega;\ \omega \in \operatorname{span}\{b_1, b_2\}\} = 3\end{aligned} \qquad (3.53)$$

III. $\dim \operatorname{span}\{b_1, b_2, b_3\} = 1$. Without loss of generality we may assume that $b_2 = b_3 = 0$, so that in fact we have only one input u. For simplicity we only consider the case $a_1 = a_2$, so that the system becomes

$$\dot{\omega}_1 = A\omega_2\omega_3 + \alpha u$$
$$\dot{\omega}_2 = -A\omega_1\omega_3 + \beta u \qquad (3.54)$$
$$\dot{\omega}_3 = \gamma u$$

with $A = (a_1-a_3)a_1^{-1}$. Computing the algebra $\mathcal{C}_0$ for $f = A(\omega_2\omega_3\ -\omega_1\omega_3\ 0)^T$ and $g = (\alpha\ \beta\ \gamma)^T$ yields

$$[f,g] = -A\begin{pmatrix}\beta\omega_3 + \omega_2\gamma \\ -\alpha\omega_3 - \omega_1\gamma \\ 0\end{pmatrix},$$

$$\bar{g}_2 := [g,[f,g]] = -2A\gamma\begin{pmatrix}\beta \\ -\alpha \\ 0\end{pmatrix}, \qquad \bar{g}_3 := [\bar{g}_2,[f,g]] = -2A^2\gamma^2\begin{pmatrix}\alpha \\ \beta \\ 0\end{pmatrix}. \qquad (3.55)$$

Now g, $\bar{g}_2$, $\bar{g}_3$ span $\mathbb{R}^3$ for all $\omega \in \mathbb{R}^3$ if and only if

$$\det\begin{pmatrix}\alpha & A\gamma\beta & A^2\gamma^2\alpha \\ \beta & -A\gamma\alpha & A^2\gamma^2\beta \\ \gamma & 0 & 0\end{pmatrix} = A^3\gamma^4(\alpha^2+\beta^2) \neq 0. \qquad (3.56)$$

Therefore, if $\gamma \neq 0$ and $A \neq 0$, and not both α and β are zero, then the system is locally strongly accessible. These conditions are also necessary as can be checked as follows. If $A = 0$ then the system (3.54) is clearly not accessible. If $\gamma = 0$ then ω_3 is constant, and so the system is not accessible. Finally if $\alpha = \beta = 0$ then

$$\frac{d}{dt}\left(\frac{1}{2}\,\omega_1^2 + \frac{1}{2}\,\omega_2^2\right) = A(\omega_1\omega_2\omega_3 - \omega_1\omega_2\omega_3) = 0, \tag{3.57}$$

and so ω is constrained to lie in the surface $\frac{1}{2}\,\omega_1^2 + \frac{1}{2}\,\omega_2^2 = \text{constant}$. Hence the system is not accessible. □

Let us finally study controllability for a particular class of nonlinear systems, namely the *bilinear systems*

$$\dot{x} = Ax + \sum_{j=1}^{m} (B_j x)u_j, \qquad x \in \mathbb{R}^n, \tag{3.58}$$

where $A, B_1, \ldots, B_m$ are $n \times n$ matrices. First observe that the reachable set from the *origin* contains only the origin. What can we say about the reachable sets from other points? Let us first compute the accessibility algebra $\mathcal{C}$. The bracket of the drift vectorfield Ax and an input vectorfield $B_j x$ yields by the coordinate expression (2.94) of the Lie bracket

$$[Ax, B_j x] = B_j Ax - AB_j x = -[A, B_j]x, \tag{3.59}$$

where $[A, B_j] = AB_j - AB_j$ is now the *commutator* of the matrices A and B_j (cf. 2.99). Taking the Lie bracket of this linear vectorfield with say Ax yields

$$[Ax, -[A, B_j]x] = [A, [A, B_j]]x. \tag{3.60}$$

Continuing in this way we obtain

Proposition 3.25 *Consider the bilinear system (3.58). Let $\mathcal{M}$ be the smallest subalgebra in $G\ell(n)$ (the Lie algebra of $n \times n$ matrices with bracket $[A,B] = AB - BA$) containing the matrices $A, B_1, \ldots, B_m$. Then the accessibility algebra $\mathcal{C}$ is given as*

$$\mathcal{C} = \{\text{all linear vectorfields on } \mathbb{R}^n \text{ of the form } Mx, \text{ with } M \in \mathcal{M}\}. \tag{3.61}$$

Since $\mathcal{M}$ is contained in $G\ell(n) \cong \mathbb{R}^{n^2}$ it follows that $\mathcal{M}$ and hence $\mathcal{C}$ is a *finite-dimensional* Lie algebra. Furthermore, as in Proposition 3.8 it follows that every element of $\mathcal{M}$ can be written as a linear combination of elements of the form

$$[D_k, [D_{k-1}, [\cdots [D_2, D_1] \cdots]]], \tag{3.62}$$

with D_i, $i \in \underline{k}$, in the set $\{A, B_1, \ldots, B_m\}$.

The analysis of the subalgebra $\mathcal{C}_0$ is completely similar. In fact let $\mathcal{M}_0$ be the ideal in $\mathcal{M}$ generated by the matrices $B_1,\dots,B_m$, then vectorfields in $\mathcal{C}_0$ are of the form Mx with $M \in \mathcal{M}_0$.

Example 3.26 Consider again the spacecraft example (Example 1.2); in particular the equations describing the orientation of the rigid body

$$\dot{R}(t) = -R(t)S\big(\omega(t)\big) \tag{3.63}$$

where $R(t) = \big(r_1(t), r_2(t), r_3(t)\big) \in SO(3)$ and $r_i(t)$ describes the direction of the i-th axis of the spacecraft (with respect to an inertial frame). Let $N(t) := R^{-1}(t) = R^T(t)$, i.e. the columns of $N(t)$ describe the position of the axes of the inertial frame with respect to the moving frame given by the axes of the spacecraft. Since $R(t)N(t) = I$ we obtain

$$0 = \dot{R}(t)N(t) + R(t)\dot{N}(t) = -R(t)S\big(\omega(t)\big)N(t) + R(t)\dot{N}(t),$$

and hence

$$\dot{N}(t) = S\big(\omega(t)\big)N(t). \tag{3.64}$$

Now let us consider the time-evolution of a single column of $N(t)$, i.e.

$$\dot{x}(t) = \begin{pmatrix} 0 & \omega_3(t) & -\omega_2(t) \\ -\omega_3(t) & 0 & \omega_1(t) \\ \omega_2(t) & -\omega_1(t) & 0 \end{pmatrix} x(t), \qquad x \in \mathbb{R}^3, \tag{3.65}$$

and let us assume that we can control the angular velocities $\omega_1, \omega_2, \omega_3$ (w.r.t. the axes in the spacecraft) *directly*. Hence $u_i = \omega_i$, $i = 1,2,3$ are controls, and we obtain the bilinear system

$$\begin{aligned} \dot{x} &= \begin{pmatrix} 0 & 0 & 0 \\ 0 & 0 & 1 \\ 0 & -1 & 0 \end{pmatrix} x \cdot u_1 + \begin{pmatrix} 0 & 0 & -1 \\ 0 & 0 & 0 \\ 1 & 0 & 0 \end{pmatrix} x \cdot u_2 + \begin{pmatrix} 0 & 1 & 0 \\ -1 & 0 & 0 \\ 0 & 0 & 0 \end{pmatrix} x \cdot u_3 \\ &=: (B_1 x)u_1 + (B_2 x)u_2 + (B_3 x)u_3 . \end{aligned} \tag{3.66}$$

One computes

$$[B_1, B_2] = -B_3, \qquad [B_3, B_1] = -B_2, \qquad [B_2, B_3] = -B_1. \tag{3.67}$$

Hence $\mathcal{C}_0 = \mathcal{C} = \mathrm{span}\{B_1 x, B_2 x, B_3 x\}$ and

$$\mathcal{C}_0(x) = C(x) = \mathrm{span}\begin{pmatrix} 0 & -x_3 & x_2 \\ x_3 & 0 & -x_1 \\ -x_2 & x_1 & 0 \end{pmatrix} \tag{3.68}$$

so that $\dim C_0(0) = 0$, and $\dim C_0(x) = 2$ for all $x \neq 0$. By Proposition

3.12 it follows that the reachable sets from $x_0 \neq 0$ contain two-dimensional submanifolds of $\mathbb{R}^3$. In fact, it is easily seen that $R_T(x_0)$ is contained in the sphere in $\mathbb{R}^3$ with radius $r = \|x_0\|$, and by Proposition 3.15 is equal to this sphere. Of course, this expresses the fact that the columns of $N(t) \in SO(3)$ are vectors of unit length. If we have only two inputs, say $u_3 = 0$, then it follows from (3.67) that the controllability properties of (3.66) remain unchanged. □

3.2 Observability

Let us consider the same smooth affine control system (3.1) as before, but now together with an output map

$$\dot{x} = f(x) + \sum_{j=1}^{m} g_j(x)u_j, \qquad u = (u_1,\dots,u_m) \in U \subset \mathbb{R}^m,$$
$$y_i = h_i(x), \qquad i \in \underline{p}, \tag{3.69}$$

where $h = (h_1,\dots,h_p)^T\colon M \to Y = \mathbb{R}^p$ is the smooth output map of the system. The notion of *observability* we will deal with for these systems is defined as follows. Recall that $y(t,0,x_0,u) = h\big(x(t,0,x_0,u)\big)$ denotes the output of (3.69) for $u(\cdot)$ and initial state $x(0) = x_0$.

Definition 3.27 *Two states $x_1,x_2 \in M$ are said to be* indistinguishable *(denoted x_1Ix_2) for (3.69) if for every admissible input function u the output function $t \mapsto y(t,0,x_1,u)$, $t \geq 0$, of the system for initial state $x(0) = x_1$, and the output function $t \mapsto y(t,0,x_2,u)$, $t \geq 0$, of the system for initial state $x(0) = x_2$, are identical on their common domain of definition. The system is called* observable *if x_1Ix_2 implies $x_1 = x_2$.*

Notice that this definition of observability does not imply that *every* input function distinguishes points of M. However, if the output is the sum of a function of the initial state and a function of the input (as it is for linear systems) then it is easily seen that if some input distinguishes between two initial states then every input will do.

Since our aim is to replace the Kalman rank condition for observability of linear systems by a nonlinear observability rank condition (which inherently will be a local condition), we localize Definition 3.27 in the following way. Let $V \subset M$ be an open set containing x_1 as well as x_2. We say that x_1 and x_2 are V - indistinguishable, denoted as $x_1I^Vx_2$, if for every admissible constant control $u\colon [0,T] \to U$, $T > 0$ arbitrary, with the property that the solutions $x(t,0,x_1,u)$, and $x(t,0,x_2,u)$ both remain in V for $t \le T$, the output functions $y(t,0,x_1,u)$, respectively $y(t,0,x_2,u)$ are the same for $t \le T$ on their common domain of definition.

Definition 3.28 *The system (3.69) is called* locally observable *at* x_0 *if there exists a neighborhood* W *of* x_0 *such that for every neighborhood* $V \subset W$ *of* x_0 *the relation* $x_0 I^V x_1$ *implies that* $x_1 = x_0$. *If the system is locally observable at each* x_0 *then it is called* locally observable.

Roughly speaking a system is locally observable if every state x_0 can be distinguished from its neighbors by using system trajectories remaining close to x_0.

Recall that for studying local accessibility the *accessibility algebra* of the system was shown to be essential. Analogously, for local observability the *observation space* will prove to be instrumental.

Definition 3.29 *Consider the nonlinear system (3.69). The* observation space $\mathcal{O}$ *of (3.69) is the linear space (over* $\mathbb{R}$*) of functions on* M *containing* $h_1,\dots,h_p$, *and all repeated Lie derivatives*

$$L_{X_1}L_{X_2}\cdots L_{X_k}h_j, \qquad j \in \underline{p},\ k = 1,2,\dots \tag{3.70}$$

with X_i, $i \in \underline{k}$, *in the set* $\{f,g_1,\dots,g_m\}$.

The following propositions give equivalent characterizations of $\mathcal{O}$.

Proposition 3.30 $\mathcal{O}$ is also given as the linear space of functions on M containing $h_1,\dots,h_p$, and all repeated Lie derivatives

$$L_{Z_1}L_{Z_2}\cdots L_{Z_k}h_j, \qquad j \in \underline{p},\ k = 1,2,\dots \tag{3.71}$$

with Z_i, $i \in \underline{k}$, of the form

$$Z_i(x) = f(x) + \sum_{j=1}^{m} g_j(x)u_j^i, \tag{3.72}$$

for some point $u^i = (u_1^i,\dots,u_m^i) \in U$, i.e., $Z_i \in \mathcal{F}$.

Proof We use the facts that $L_{X_1+X_2}H = L_{X_1}H + L_{X_2}H$, and $L_X(H_1 + H_2) = L_XH_1 + L_XH_2$ for any vectorfields X,X_1,X_2 and functions H,H_1,H_2. Since Z_i is a linear combination of the vectorfields $f,g_1,\dots,g_m$, it immediately follows that expressions (3.71) are contained in $\mathcal{O}$. Conversely, all vectorfields $f, g_1, \ldots, g_m$ can be written as linear combinations of Z_i. In fact $f = Z_i$ for $u^i = 0$, and $g_j = \frac{1}{2}(f+g_j) - \frac{1}{2}(f-g_j)$. □

Remark Proposition 3.30 yields the following interpretation of the observation space $\mathcal{O}$: it contains the output functions and all derivatives of the output functions along the system trajectories. In particular, for an autonomous system (i.e. no inputs) $\mathcal{O}$ is constructed by taking $y_j = h_j(x)$ together with all repeated time derivatives $\dot{y}_j = L_f h_j(x)$, $\ddot{y}_j = L_f L_f h_j(x), \ldots$, $j \in \underline{p}$.

Proposition 3.31 *The definition of $\mathcal{O}$ is not changed if we allow the vectorfields X_i, $i \in \underline{k}$, in (3.70) to belong to the accessibility algebra $\mathcal{C}$.*

Proof Let X_1, X_2 be vectorfields. Then by definition of the Lie bracket

$$L_{[X_1,X_2]} h_i = L_{X_1}\big(L_{X_2}(h_i)\big) - L_{X_2}\big(L_{X_1}(h_i)\big). \tag{3.73}$$

Hence if X_1 and X_2 are in the set $\{f, g_1, \ldots, g_m\}$ then $L_{[X_1,X_2]} h_i$, $i \in \underline{p}$, belongs to $\mathcal{O}$, and similarly if h_i is replaced by any function (3.70). □

The observation space $\mathcal{O}$ defines the *observability codistribution*, denoted as $d\mathcal{O}$, by setting

$$d\mathcal{O}(q) = \operatorname{span}\{dH(q) \mid H \in \mathcal{O}\}, \qquad q \in M. \tag{3.74}$$

Since $d\mathcal{O}$ is generated by exact one-forms it follows that the codistribution $d\mathcal{O}$ is involutive (see (2.166)).

The main theorem concerning local observability reads as follows.

Theorem 3.32 *Consider the system (3.69) with* $\dim M = n$. *Assume that*

$$\dim d\mathcal{O}(x_0) = n, \tag{3.75}$$

then the system is locally observable at x_0.

Proof Since $\dim \mathcal{O}(x_0) = n$ there exist n functions $H_1, \ldots, H_n \in \mathcal{O}$ such that $dH_1(x_0), \ldots, dH_n(x_0)$ are linearly independent. Define the map $\Phi: M \to \mathbb{R}^n$ as

$$\Phi(x) = \big(H_1(x), \ldots, H_n(x)\big)^T. \tag{3.76}$$

It follows that the Jacobian matrix of Φ in x_0 is non-singular, and therefore by Proposition 2.16 there exists a neighborhood W of x_0 such that $\Phi: W \to \Phi(W)$ is a diffeomorphism. Now let $V \subset W$ be a neighborhood of x_0, and suppose that $x_0 I^V x_1$ for some $x_1 \in V$. Then for any $i \in \underline{p}$ and $k \geq 0$, and for small $t_1, \ldots, t_k$ we have

$$h_i\left(Z_k^{t_k} \circ Z_{k-1}^{t_{k-1}} \circ \cdots \circ Z_1^{t_1}(x_0)\right) = h_i\left(Z_k^{t_k} \circ Z_{k-1}^{t_{k-1}} \circ \cdots \circ Z_1^{t_1}(x_1)\right) \tag{3.77}$$

with Z_i, $i \in \underline{k}$, of the form (3.72). Differentiating of both sides with respect to $t_k, t_{k-1}, \ldots, t_1$ (in this order) at respectively $t_k = 0$, $t_{k-1} = 0, \ldots, t_1 = 0$ yields

$$L_{Z_1} L_{Z_2} \cdots L_{Z_k} h_i(x_0) = L_{Z_1} L_{Z_2} \cdots L_{Z_k} h_i(x_1) \tag{3.78}$$

for all Z_j, $j \in \underline{k}$, of the form (3.72). By Proposition 3.30 it follows that $H(x_0) = H(x_1)$ for all $H \in \mathcal{O}$. In particular $H_i(x_0) = H_i(x_1)$, $i \in \underline{n}$, and by injectivity of Φ on W this yields $x_0 = x_1$. □

We call (3.75) the *observability rank condition*. The system is said to satisfy the observability rank condition if (3.75) holds for any $x_0 \in M$.

Corollary 3.33 *Assume that (3.69) satisfies the observability rank condition, then it is locally observable.*

What can be said about the case $\dim d\mathcal{O}(x_0) < n$? In case the codistribution $d\mathcal{O}$ has constant dimension around x_0 we have (compare Proposition 3.12):

Proposition 3.34 *Suppose that $d\mathcal{O}$ has constant dimension $k < n$ around x_0. By Frobenius' Theorem (Theorem 2.42) we can find a coordinate chart $(U, x_1, \ldots, x_n)$ around x_0 such that the submanifold*

$$S_{x_0} = \{q \in U \mid x_i(q) = x_i(x_0),\ i = n-k+1, \ldots, n\} \tag{3.79}$$

is an integral manifold through x_0 of the involutive distribution $\ker d\mathcal{O}$. *There exists a neighborhood $W \subset U$ of x_0 such that for any neighborhood $V \subset W$ of x_0 we have*

$$\{x \mid x I^V x_0\} = S_{x_0} \cap V. \tag{3.80}$$

Proof As in the proof of Theorem 3.32 there exist k functions $H_1, \ldots, H_k \in \mathcal{O}$ such that $dH_1(x_0), \ldots, dH_k(x_0)$ are linearly independent. Therefore by Proposition 2.18 we can take $H_1, \ldots, H_k$ as partial coordinates on a neighborhood $W \subset U$ of x_0. By definition of S_{x_0} as an integral manifold of $\ker d\mathcal{O}$ we may as well assume that the coordinates $x_{n-k+1},\dots,x_n$ in (3.79) restricted to W are equal to $H_1,\dots,H_k$. Now let $V \subset W$ be a neighborhood of x_0. Suppose that $x_0 I^V x_1$ for some $x_1 \in V$. As in the proof of Theorem 3.32 it follows that

$$H_i(x_0) = H_i(x_1), \qquad i \in \underline{k}, \tag{3.81}$$

and so $x_1 \in S_{x_0} \cap V$. Therefore $\{x \mid x I^V x_0\} \subset S_{x_0} \cap V$.

For the converse inclusion, we note that by definition of $\mathcal{O}$ and $H_1,\dots,H_k$ we have for $q \in V$, $i \in \underline{p}$, $j \in \underline{m}$,

$$\begin{aligned} dL_f H_i(q) &\in \operatorname{span}\{dH_1(q),\dots,dH_k(q)\}, \\ dL_{g_j} H_i(q) &\in \operatorname{span}\{dH_1(q),\dots,dH_k(q)\}. \end{aligned} \tag{3.82}$$

Since $H_1,\dots,H_k$ are the coordinates $x_{n-k+1},\dots,x_n$, this yields

$$\begin{aligned} df_{n-k+i}(q) &\in \operatorname{span}\{dx_{n-k+1},\dots,dx_n\}(q), \\ d(g_j)_{n-k+i}(q) &\in \operatorname{span}\{dx_{n-k+1},\dots,dx_n\}(q), \end{aligned} \tag{3.83}$$

where the subscript n-k+i denotes the (n-k+i)-th component of the vectorfield. Hence f and g_j, $j \in \underline{m}$, are of the form

$$f(x) = \begin{pmatrix} f_1(x_1,\dots,x_n) \\ \vdots \\ f_{n-k}(x_1,\dots,x_n) \\ f_{n-k+1}(x_{n-k+1},\dots,x_n) \\ \vdots \\ f_n(x_{n-k+1},\dots,x_n) \end{pmatrix}, \quad g_j(x) = \begin{pmatrix} (g_j)_1(x_1,\dots,x_n) \\ \vdots \\ (g_j)_{n-k}(x_1,\dots,x_n) \\ (g_j)_{n-k+1}(x_{n-k+1},\dots,x_n) \\ \vdots \\ (g_j)_n(x_{n-k+1},\dots,x_n) \end{pmatrix} \tag{3.84}$$

Furthermore since $h_i \in \mathcal{O}$, it follows that h_i only depends on $x_{n-k+1},\dots,x_n$, $i \in \underline{p}$. Now let $\bar{x}_0 \in S_{x_0} \cap V$. It follows from the form of the vectorfields f and g_j, $j \in \underline{m}$, as displayed in (3.84) that for any $u \in \mathcal{U}$ the time t - integrals $x(t,0,x_0,u)$ and $x(t,0,\bar{x}_0,u)$, starting from x_0, respectively $\bar{x}_0$, will satisfy (for t small) $H_i\big(x(t,0,x_0,u)\big) = H_i\big(x(t,0,\bar{x}_0,u)\big)$, $i \in \underline{k}$, and thus $H\big(x(t,0,x_0,u)\big) = H\big(x(t,0,\bar{x}_0,u)\big)$ for all $H \in \mathcal{O}$. □

Corollary 3.35 *Assume that the system (3.69) is locally observable. Then* $\dim d\mathcal{O}(x) = n$ *for* x *in an open and dense subset of* M.

Proof (Compare Corollary 3.13). First, the set of points x for which $\dim d\mathcal{O}(x) = n$ is open. Assume that there exists a non-empty set V where $\dim d\mathcal{O}(x) < n$. By making V smaller we may assume that $\dim d\mathcal{O}(x) = k < n$

for $x \in V$. Now use Proposition 3.34 to conclude that on V the system is not locally observable. □

In general local observability does *not* imply observability, as is illustrated by

Example 3.36 Consider the nonlinear system

$$\begin{aligned} \dot{x} &= u, \qquad x \in \mathbb{R}, \\ y_1 &= \sin x, \qquad y_2 = \cos x. \end{aligned} \tag{3.85}$$

The observation space $\mathcal{O}$ consists of the two functions $\sin x$, $\cos x$. Clearly $\dim d\mathcal{O}(x) = \dim \operatorname{span}\{\cos x\, dx, \sin x\, dx\}$ is one for all $x \in \mathbb{R}$. Hence the system is locally observable. However the system is not observable since points x_0 and x_1 with $x_0 - x_1$ a multiple of 2π are not distinguishable. □

Now let us consider a linear system, written in accordance with (3.69) as

$$\begin{aligned} \dot{x} &= Ax + \sum_{j=1}^{m} b_j u_j, \qquad x \in \mathbb{R}^n, \\ y_i &= c_i x \qquad i \in \underline{p}, \end{aligned} \tag{3.86}$$

where c_i, $i \in \underline{p}$, are the rows of the observation matrix C. Let us compute the observation space $\mathcal{O}$. First

$$L_{b_j} c_i x = c_i b_j \in \mathbb{R}. \tag{3.87}$$

Hence the Lie derivative of a linear function along the constant vectorfield b_j yields a *constant* function, which will not contribute to the dimension of the observability co-distribution $d\mathcal{O}$. Furthermore

$$L_{Ax} c_i x = c_i Ax. \tag{3.88}$$

Again $L_{b_j}(c_i Ax)$ is constant, so we continue with

$$L_{Ax} L_{Ax} c_i x = L_{Ax}(c_i Ax) = c_i A^2 x. \tag{3.89}$$

In general we have

$$\underbrace{L_{Ax} L_{Ax} \cdots L_{Ax}}_{\leftarrow\ k\text{-times}\ \rightarrow} c_i x = c_i A^k x. \tag{3.90}$$

Hence by Cayley-Hamilton

$$\mathcal{O} = \operatorname{span}\{c_i x, c_i Ax, \ldots, c_i A^{n-1}x,\ i \in \underline{p}\} + \text{constant functions}. \tag{3.91}$$

It follows that the system satisfies the observability rank condition in an arbitrary point x_0 if and only if

$$rank \begin{bmatrix} C \\ CA \\ \vdots \\ CA^{n-1} \end{bmatrix} = n. \tag{3.92}$$

(Recall that $d(c_i x) = c_i$, $d(c_i Ax) = c_i A$, etc.). This is exactly the Kalman rank condition for observability. Of course we know from linear systems theory that (3.92) not only implies local observability but even (global) observability.

Corollary 3.35 suggests that, even if a nonlinear system is locally observable, the codistribution $d\mathcal{O}$ may have singular points q with $\dim d\mathcal{O}(q) < n$. Indeed this may easily happen, as shown by the following

Example 3.7 Consider the system

$$\dot{x} = 0, \qquad y = x^3, \qquad x \in \mathbb{R}, \tag{3.93}$$

which is clearly observable. However $\mathcal{O}$ consists of the single function x^3, and so $\dim d\mathcal{O}(0) = 0$. □

On the other hand the following proposition shows that in case the system is *locally accessible* and is *analytic* the codistribution $d\mathcal{O}$ always has constant dimension.

Proposition 3.38 *Let (3.69) be a locally accessible and analytic system. Furthermore, suppose that M is connected. Then the codistribution $d\mathcal{O}$ is constant-dimensional. In particular, (3.69) is locally observable if and only if it satisfies the observability rank condition.*

Proof As in the proof of Proposition 3.15 (b) it follows that for any x_0 and x_1 in M we can find $X_1, \ldots, X_n \in \mathcal{F}$ and $t_1, \ldots, t_n \in \mathbb{R}$ such that

$$x_1 = X_n^{t_n} \circ X_{n-1}^{t_{n-1}} \circ \cdots \circ X_1^{t_1}(x_0). \tag{3.94}$$

By (2.154), $(X^t)^*_{x_0}(dH) = d(H \circ X^t)(x_0)$. Furthermore

$$\frac{d^k}{dt^k}(H \circ X^t)(x)\Big|_{t=0} = L_X^k H(x) \tag{3.95}$$

(with $L_X^k H$ inductively defined by $L_X^0 H = H$, and $L_X^k H = L_X(L_X^{k-1} H)$, $k \geq 1$). By analyticity we therefore have the Taylor expansion

$$H \circ X^t(x) = \sum_{k=0}^{\infty} \frac{t^k}{k!} L_X^k H(x), \tag{3.96}$$

and so

$$(X^t)_{x_0}^*(dH) = \sum_{k=0}^{\infty} \frac{t^k}{k!}(dL_X^k H)(x_0). \tag{3.97}$$

It follows that (since $(X^t)_{x_0}^*: T_{x_1}^* M \to T_{x_0}^* M$)

$$(X^t)_{x_0}^* \, d\mathcal{O}(x_1) \subset d\mathcal{O}(x_0), \tag{3.98}$$

and so $\dim d\mathcal{O}(x_1) \leq \dim d\mathcal{O}(x_0)$. In the same manner we obtain $(X^{-t})_{x_1}^* d\mathcal{O}(x_0) \subset d\mathcal{O}(x_1)$, and so $\dim d\mathcal{O}(x_1) \geq \dim d\mathcal{O}(x_0)$, yielding the required equality. The last statement follows from Corollaries 3.33 and 3.35. □

Remark 3.39 In a similar way the following refinement of Corollary 3.13 can be proved: if the system (3.1) is analytic and lives on a connected state space M then local accessibility implies that $\dim C(x) = n$ for *all* $x \in M$.

Example 3.40 Consider the two-link rigid robot manipulator from Example 1.1. Assume for simplicity that $m_1 = m_2 = 1$, $\ell_1 = \ell_2 = 1$. Take as state the vector $x = (\theta_1, \theta_2, \dot{\theta}_1, \dot{\theta}_2) \in T(S^1 \times S^1)$. (Note that θ_1, θ_2 are angles, and so both are properly defined on the unit circle S^1.) Then the system is given as

$$\frac{d}{dt}\begin{bmatrix} \theta_1 \\ \theta_2 \\ \dot{\theta}_1 \\ \dot{\theta}_2 \end{bmatrix} = \underbrace{\begin{bmatrix} \dot{\theta}_1 \\ \dot{\theta}_2 \\ \\ -M^{-1}(\theta)C(\theta,\dot{\theta}) - M^{-1}(\theta)k(\theta) \end{bmatrix}}_{f} + \underbrace{\begin{bmatrix} 0 & 0 \\ 0 & 0 \\ \\ & M^{-1}(\theta) \end{bmatrix}}_{(g_1 \,\vdots\, g_2)} \begin{bmatrix} u_1 \\ u_2 \end{bmatrix}. \tag{3.99}$$

Suppose that the output is given as the angle of the first joint, i.e.

$$y = h(\theta_1, \theta_2, \dot{\theta}_1, \dot{\theta}_2) = \theta_1. \tag{3.100}$$

We will prove that the system is locally observable. For the computation of the observation space $\mathcal{O}$ we note that $L_f h = \dot\theta_1$ and $L_{f+gu}\dot\theta_1 = \ddot\theta_1$. Computing

$$M^{-1}(\theta) = \begin{pmatrix} 3+2\cos\theta_2 & 1+\cos\theta_2 \\ 1+\cos\theta_2 & 1 \end{pmatrix}^{-1} = \frac{1}{1+\sin^2\theta_2}\begin{pmatrix} 1 & -1-\cos\theta_2 \\ -1-\cos\theta_2 & 3+2\cos\theta_2 \end{pmatrix} \tag{3.101}$$

we obtain

$$\begin{aligned} L_f^2 h = {} & \frac{1}{1+\sin^2\theta_2}\left(\sin\theta_2\cdot\dot\theta_2(2\dot\theta_1+\dot\theta_2) + (1+\cos\theta_2)\sin\theta_2\cdot\dot\theta_1^2\right) \\ & + \frac{1}{1+\sin^2\theta_2}\left(-2g\sin\theta_1 - g\sin(\theta_1+\theta_2) + (1+\cos\theta_2)g\sin(\theta_1+\theta_2)\right). \end{aligned} \tag{3.102}$$

It follows that $L_f^2 h$ depends in a nontrivial way on the angle θ_2 in almost all points $(\theta_1,\theta_2,\dot\theta_1,\dot\theta_2)$. Since $L_f\theta_2 = \dot\theta_2$ it follows that in these points $L_f^3 h = L_f(L_f^2 h)$ will depend non-trivially on $\dot\theta_2$. Hence for almost all points we have proved that $\dim d\mathcal{O} = 4$. In order to prove that $\dim d\mathcal{O} = 4$ everywhere, and so the system is locally observable, we use Proposition 3.38. Clearly the system is analytic, and $M = T(S^1\times S^1)$ is connected. To show local accessibility we compute

$$[f,g_1](x) = -\begin{pmatrix} (M^{-1}(\theta))_1 \\ * \\ * \end{pmatrix}, \quad [f,g_2](x) = -\begin{pmatrix} (M^{-1}(\theta))_2 \\ * \\ * \end{pmatrix}, \tag{3.103}$$

with $(M^{-1}(\theta))_i$ the i-th column of $M^{-1}(\theta)$, $i = 1,2$, and where $*$ denote some unspecified functions of the state. Therefore the dimension of the span of g_1, g_2, $[f,g_1]$, $[f,g_2]$ equals

$$\dim\begin{pmatrix} \begin{matrix} 0 & 0 \\ 0 & 0 \end{matrix} & M^{-1}(\theta) \\ M^{-1}(\theta) & \begin{matrix} * & * \\ * & * \end{matrix} \end{pmatrix} \tag{3.104}$$

and clearly is 4 since $M^{-1}(\theta)$ has rank 2. Therefore by Corollary 3.11 the system is locally accessible. (In fact, the system (3.92) is controllable, since it is *feedback linearizable* to a controllable linear system, cf. Chapters 5,6.) □

3.3 Invariant Distributions; Local Decompositions

Let us first consider the linear system

$$\begin{aligned} \dot x &= Ax + Bu, & x \in \mathbb{R}^n,\ u \in \mathbb{R}^m, \\ y &= Cx, & y \in \mathbb{R}^p. \end{aligned} \tag{3.105}$$

A subspace $V \subset \mathbb{R}^n$ is called *A-invariant* if $AV \subset V$. If we choose a basis $e_1,\dots,e_n$ for $\mathbb{R}^n$ such that $V = \mathrm{span}\{e_1,\dots,e_k\}$ then in this basis A takes the form

$$A = \begin{bmatrix} A_{11} & A_{12} \\ 0 & A_{22} \end{bmatrix} \begin{matrix} \updownarrow k \\ \\ \end{matrix} \quad \text{(columns: } \overset{\leftrightarrow}{k}\text{)} \tag{3.106}$$

with A_{11} a $k{\times}k$-matrix. Let $(x_1,\dots,x_n)$ be the corresponding linear coordinate functions on $\mathbb{R}^n$. For convenience we write

$$x^1 = (x_1,\dots,x_k), \qquad x^2 = (x_{k+1},\dots,x_n). \tag{3.107}$$

Then the system dynamics (3.105) can be written as

$$\begin{aligned} \dot{x}^1 &= A_{11}x^1 + A_{12}x^2 + B_1u, \\ \dot{x}^2 &= \phantom{A_{11}x^1 +} A_{22}x^2 + B_2u. \end{aligned} \tag{3.108}$$

Now let us consider the *foliation* of $\mathbb{R}^n$ given by the affine subspaces

$$F_V = \{x + V \mid x \in \mathbb{R}^n\}. \tag{3.109}$$

or in the above coordinates, $F_V = \{x = (x^1,x^2) \mid x^2 = \text{constant}\}$. Consider two points q_1 and q_2 on a same leaf, i.e. $q_1 - q_2 \in V$, or equivalently $x^2(q_1) = x^2(q_2)$. Then it follows from (3.108) that for any $u \in U$ the solutions $x(t,0,q_1,u)$ and $x(t,0,q_2,u)$ at every time instant $t \geq 0$ will be again on a same leaf (depending on t).

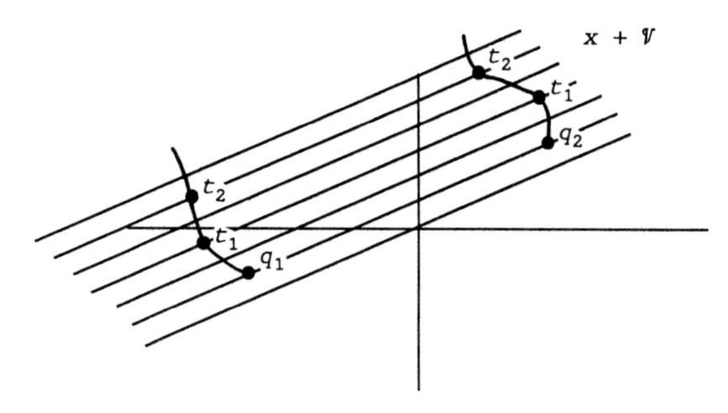

Fig. 3.3. Invariant foliation.

In fact the system dynamics (3.108) *project* to the dynamics on $\mathbb{R}^n/V$ given by $\dot{x}^2 = A_{22}x^2 + B_2u$. We say that the foliation F_V is *invariant* under the system dynamics $\dot{x} = Ax + Bu$, and we conclude that this is the case if and only if V is A-invariant.

Let us now generalize this idea to a nonlinear system (3.1). We only consider foliations F whose leaves are smooth submanifolds (or at least immersed submanifolds, see Chapter 2). Moreover we will assume that all leaves have the same dimension. Then by considering the tangent spaces of the leaves in any point $p \in M$ we obtain a constant dimensional distribution D

$$D(p) = \text{tangent space at } p \text{ of the leaf of } F \text{ through } p \tag{3.110}$$

which is involutive (see Proposition 2.41). Instead of working with the foliation F we will work with its associated constant dimensional involutive distribution. First, for a *general* distribution we define

Definition 3.41 *A (smooth) distribution D on M is invariant for the dynamics $\dot{x} = f(x)$, f being a vectorfield on M, if*

$$[f,X] \in D, \quad \text{for any vectorfield } X \in D, \tag{3.111}$$

or succinctly, $[f,D] \subset D$.

Analogously to the linear case we have the following geometric interpretation of invariance in case D is involutive and constant dimensional.

Proposition 3.42 *Let D be a constant dimensional involutive distribution on M, which is invariant for $\dot{x} = f(x)$. By Frobenius' Theorem we can take local coordinates $x = (x_1,\dots,x_n)$ on $V \subset M$ such that $D = \text{span}\{\frac{\partial}{\partial x_1},\dots,\frac{\partial}{\partial x_k}\}$. Denote $x^1 = (x_1,\dots,x_k)$, $x^2 = (x_{k+1},\dots,x_n)$, and correspondingly write $f = (f^1,f^2)$. Then in such local coordinates $\dot{x} = f(x)$ on V takes the form*

$$\begin{aligned} \dot{x}^1 &= f^1(x^1,x^2), \\ \dot{x}^2 &= f^2(x^2). \end{aligned} \tag{3.112}$$

Hence if $q_1, q_2 \in V$ are such that $x^2(q_1) = x^2(q_2)$ then $x^2(q_1(t)) = x^2(q_2(t))$ for all $t \geq 0$, with $q_i(t)$ denoting the solution within V of (3.112) for initial condition $q_i(0) = q_i$, $i = 1,2$.

Proof Since $D = \text{span}\{\frac{\partial}{\partial x_1},\dots,\frac{\partial}{\partial x_k}\}$, and

$$[f,\frac{\partial}{\partial x_i}] = -\sum_{j=1}^{n} \frac{\partial f_j}{\partial x_i} \frac{\partial}{\partial x_j} \tag{3.113}$$

with f_j denoting the j-th component of f, we obtain from (3.111)

$$\frac{\partial f_j}{\partial x_i}(x) = 0, \qquad i \in \underline{k},\ j = k+1,\dots,n, \tag{3.114}$$

which implies (3.112). The last statement immediately follows from the particular form of (3.112). □

Remark 3.43 In the linear case we associate with a subspace $\mathcal{V} \subset \mathbb{R}^n$ the distribution $D_{\mathcal{V}}$ on $\mathbb{R}^n$. In fact if $e_1,\dots,e_n$ is a basis for $\mathbb{R}^n$ such that $\mathcal{V} = \operatorname{span}\{e_1,\dots,e_k\}$, then in the corresponding linear coordinates $x_1,\dots,x_n$ for $\mathbb{R}^n$ the distribution $D_{\mathcal{V}}$ equals $\operatorname{span}\{\frac{\partial}{\partial x_1},\dots,\frac{\partial}{\partial x_k}\}$. It is immediately verified that $[Ax, D_{\mathcal{V}}] \subset D_{\mathcal{V}}$ if and only if $A\mathcal{V} \subset \mathcal{V}$.

Next we define invariance for nonlinear systems (3.1).

Definition 3.44 *A smooth distribution D on M is invariant for the nonlinear system (3.1) if*

$$[f, D] \subset D, \tag{3.115a}$$

$$[g_j, D] \subset D, \qquad j \in \underline{m}, \tag{3.115b}$$

(or, equivalently, if D is invariant for every vectorfield in $\mathcal{F}$; the set of associated vectorfields (cf. (3.2))).

As in Proposition 3.42 it follows that if D is constant dimensional and involutive, and if we choose coordinates $x = (x_1,\dots,x_k,x_{k+1},\dots,x_n) = (x^1,x^2)$, such that $D = \operatorname{span}\{\frac{\partial}{\partial x^1}\} := \operatorname{span}\{\frac{\partial}{\partial x_1},\dots,\frac{\partial}{\partial x_k}\}$, then invariance of D is equivalent with the following local representation of (3.1):

$$\begin{aligned} \dot{x}^1 &= f^1(x^1,x^2) + \sum_{j=1}^m g_j^1(x^1,x^2)u_j, \\ \dot{x}^2 &= f^2(x^2) + \sum_{j=1}^m g_j^2(x^2)u_j, \end{aligned} \tag{3.116}$$

and so the dynamics of x^2 are not influenced by x^1 (as in the linear case, cf. (3.108)).

Remark At first sight there is a difference with the linear case, since in that case, contrary to Definition 3.44, we did not impose any condition on the input matrix B. This is however explained by the fact that b_j, the columns of B, are in linear coordinates for $\mathbb{R}^n$ *constant* vectorfields, while also the distribution $D_{\mathcal{V}}$ associated with the subspace $\mathcal{V}$ is spanned by constant vectorfields. Since the Lie bracket of any two constant vectorfields is zero, it follows from Proposition 2.27 (a) that condition (3.115b) is automatically satisfied in the linear case.

We can also define invariance for the dual object of a distribution; a *codistribution*, cf. (2.171). We call a codistribution P invariant with respect to a vectorfield f if

$$L_f\sigma \in P, \qquad \text{for all one-forms } \sigma \in P. \tag{3.117}$$

Condition (3.117) will be abbreviated as $L_fP \subset P$.

Definition 3.45 A codistribution P on M is invariant for the system (3.1) if

$$L_fP \subset P, \tag{3.118a}$$

$$L_{g_j}P \subset P, \qquad j \in \underline{m}. \tag{3.118b}$$

The next proposition explains the relation with Definition 3.44.

Proposition 3.46

(a) Let the codistribution P be invariant for (3.1), then the distribution ker P *is invariant for (3.1).*

(b) Let the distribution D be invariant for (3.1), then the codistribution ann D *is invariant for (3.1).*

(c) Let D (P) be a constant dimensional (co-)distribution. Then D (P) is invariant for (3.1) if and only if ann D (ker P) *is invariant for (3.1).*

Proof The basic formula to be used is (2.169), with f and X vectorfields and σ a one-form:

$$L_f\big(\sigma(X)\big) = (L_f\sigma)(X) + \sigma([f,X]) \tag{3.119}$$

(a) Let $\sigma \in P$ and $X \in \ker P$. Then $\sigma(X) = 0$, and since $L_f\sigma \in P$ also $(L_f\sigma)(X) = 0$. Hence $\sigma([f,X]) = 0$ and thus $[f,X] \in \ker P$. The same holds for f replaced by g_j, $j \in \underline{m}$.

(b) Let $X \in D$ and $\sigma \in \text{ann } D$. Then $\sigma(X) = 0$, and since $[f,X] \in D$ also $\sigma([f,X]) = 0$. Hence $(L_f\sigma)(X) = 0$, and thus $L_f\sigma \in \text{ann } D$. The same is true for f replaced by g_j, $j \in \underline{m}$.

c. In case D has constant dimension, we have (cf. (2.175) $\ker(\text{ann } D) = D$. Now if D is invariant, then by (b) ann D is invariant. If ann D is invariant, then by (a) $D = \ker(\text{ann } D)$ is invariant. A similar reasoning holds for a constant dimensional codistribution P since (cf. (2.175)) $\text{ann}(\ker P) = P$. □

For a linear system (3.105) there is an appealing geometric characterization of $\mathcal{R} = \text{Im}\left(B \,\vdots\, AB \,\vdots\, \cdots \,\vdots\, A^{n-1}B\right)$, the reachable subspace, and of $\mathcal{N} = \bigcap_{i=0}^{n-1} \ker CA^{i-1}$, the unobservable subspace. Indeed, $\mathcal{R}$ is the *minimal* A-invariant subspace that contains Im B, and $\mathcal{N}$ is the *maximal* A-invariant subspace contained in ker C. For nonlinear systems we will now give a similar characterization of the strong accessibility distribution C_0 and the observability distribution ker $d\mathcal{O}$ (or the codistribution $d\mathcal{O}$).

Proposition 3.47 *Consider the nonlinear system (3.1) with distribution C_0 characterizing local strong accessibility, and the codistribution $d\mathcal{O}$ characterizing local observability.*
(a) C_0 is the smallest distribution that is invariant for (3.1) and contains the distribution $G(q) := \text{span}\{g_1(q), \ldots, g_m(q)\}$.
(b) $d\mathcal{O}$ is the smallest codistribution that is invariant for (3.1) and contains the codistribution $dh(q) := \text{span}\{dh_1(q), \ldots, dh_p(q)\}$.
(c) ker $d\mathcal{O}$ is the largest distribution that is invariant for (3.1) and is contained in the distribution ker *dh.*

Proof (a) By definition C_0 is invariant for (3.1) and contains G. Now let D be a smooth distribution, which is invariant for (3.1) and contains G. Then $g_j \in D$, $j \in \underline{m}$, and by invariance

$$[X_k, [X_{k-1}, [\cdots, [X_1, g_1]\cdots]]] \in D \tag{3.120}$$

for $X_i \in \{f, g_1, \ldots, g_m\}$, $i \in \underline{k}$. Hence (see Proposition 3.20) $X \in D$ for any $X \in \mathcal{C}_0$ and so $C_0(q) \subset D(q)$ for any $q \in M$. Therefore C_0 is the *smallest* distribution that is invariant and contains G.
(b) By definition $d\mathcal{O}$ is invariant (3.1) and contains dh. Let P be a codistribution that is invariant for (3.1) and contains dh. We prove that $d\mathcal{O}(q) \subset P(q)$ for any $q \in M$. Indeed, since $dh_j \in P$, $j \in \underline{p}$, it follows by invariance of P that

$$dL_{X_k} L_{X_{k-1}} \cdots L_{X_1} H_j \in P \tag{3.121}$$

for $X_i \in \{f, g_1, \ldots, g_m\}$, $i \in \underline{k}$ (since $L_{X_i} dH_j = dL_{X_i} H_j$, cf. (2.161)). Hence (see Definition 3.29) $dH(q) \in P(q)$ for any $H \in \mathcal{O}$.
(c) By Proposition 3.46 (a) the distribution ker $d\mathcal{O}$ is invariant for

(3.1). Furthermore $\ker d\mathcal{O}(q) \subset \ker dh(q)$ for any $q \in M$. Now let D be a distribution that is invariant for (3.1) and contained in $\ker dh$. Then

$$\text{ann } D \supset \text{ann}(\ker dh) \supset dh, \tag{3.122}$$

and by Proposition 3.46 (b) ann D is invariant for (3.1). Hence ann D is a codistribution that is invariant for (3.1) and contains dh. By part (b) this yields

$$\text{ann } D \supset d\mathcal{O}, \tag{3.123}$$

and therefore

$$D \subset \ker(\text{ann } D) \subset \ker d\mathcal{O}. \tag{3.124}$$

□

Remark 3.48 Part (a) also holds for the distribution C if we replace the distribution $G(q)$ by the distribution $\text{span}\{f(q), g_1(q), \ldots, g_m(q)\}$, $q \in M$.

In case C_0 and $d\mathcal{O}$ are *constant dimensional*, then by an application of Frobenius' Theorem we obtain the following local decompositions of a nonlinear system. These decompositions are quite analogous to the well-known decompositions of a linear system corresponding to its reachable subspace $\mathcal{R}$ and its unobservable subspace $\mathcal{N}$.

Theorem 3.49 *Consider the nonlinear system (3.1).*
(a) Let C_0 have constant dimension. By Frobenius' Theorem (Corollary 2.43) we can find local coordinates $x = (x^1, x^2) = (x_1, \ldots, x_k, x_{k+1}, \ldots, x_n)$ such that $C_0 = \text{span}\{\frac{\partial}{\partial x^1}\}$. In these coordinates the system takes the form

$$\dot{x}^1 = f^1(x^1, x^2) + \sum_{j=1}^{m} g_j^1(x^1, x^2)u_j, \tag{3.125a}$$

$$\dot{x}^2 = f^2(x^2). \tag{3.125b}$$

If we regard (3.125a) as a system with state x^1 parametrized by x^2 (everything locally), then for any value of x^2 the system (3.125a) satisfies the strong accessibility rank condition.
(b) Let $d\mathcal{O}$ have constant dimension. By Frobenius' Theorem we can find local coordinates $x = (x^1, x^2) = (x_1, \ldots, x_\ell, x_{\ell+1}, \ldots, x_n)$ such that $\ker d\mathcal{O} = \text{span}\{\frac{\partial}{\partial x^2}\}$. In these coordinates the system takes the form

$$\dot{x}^1 = f^1(x^1) + \sum_{j=1}^{m} g_j^1(x^1)u_j, \qquad y_i = h_i(x^1), \qquad i \in \underline{p}, \tag{3.126a}$$

$$\dot{x}^2 = f^2(x^1,x^2) + \sum_{j=1}^{m} g_j^2(x^1,x^2)u_j. \tag{3.126b}$$

If we regard (3.126a) as a system with state x^1 (locally defined), then (3.126a) satisfies the observability rank condition.

Proof (a) The form (3.125) follows from (3.116) and the fact that $g_j \in \text{span}\{\frac{\partial}{\partial x^1}\}$, $j \in \underline{m}$. From the characterization of $\mathcal{C}_0$ as in Proposition 3.20 it follows that in the above local coordinates all vectorfields in $\mathcal{C}_0$ have no components in $\frac{\partial}{\partial x^2}$ directions. Since $\dim C_0 = k = \dim x^1$ it follows that (3.125a) satisfies the strong accessibility rank condition for each x_2.
(b) The form (3.126) follows from (3.116) and the fact that $\ker dh_i \supset \text{span}\{\frac{\partial}{\partial x_2}\}$, $i \in \underline{p}$. It follows from (3.126) that every function in $\mathcal{O}$ does not depend on x^2. Since $\dim d\mathcal{O} = \ell = \dim x^1$ this implies that (3.126a) satisfies the observability rank condition. □

Remark The dynamics (3.125a) will be called the *strongly accessible dynamics* of the system. Similarly the dynamics (3.126b) will be called the *unobservable dynamics* of the system.

For a linear system (3.105) it is well-known that one can combine the above controllability and observability decomposition into a single decomposition (usually called the Kalman decomposition), where A, B, C take the form

$$A = \begin{pmatrix} A_{11} & 0 & A_{13} & 0 \\ A_{21} & A_{22} & A_{23} & A_{24} \\ 0 & 0 & A_{33} & 0 \\ 0 & 0 & A_{43} & A_{44} \end{pmatrix}, \quad B = \begin{pmatrix} B_1 \\ B_2 \\ 0 \\ 0 \end{pmatrix}, \quad C = (C_1 \;\; 0 \;\; C_3 \;\; 0), \tag{3.127}$$

in coordinates $x = (x^1,x^2,x^3,x^4)$ with $\mathcal{R} = \text{span}\{x^1,x^2\}$ and $\mathcal{N} = \text{span}\{x^2,x^4\}$. In order to generalize this to the nonlinear case, we need the following extension of Frobenius' Theorem (Theorem 2.42):

Proposition 3.50 *Let D_1 and D_2 be involutive constant dimensional distributions on a manifold M, such that also $D_1 + D_2$ is involutive and constant dimensional. Then about any $p \in M$ there exist local coordinates $x = (x^1,x^2,x^3,x^4)$ such that*

$$D_1 = \mathrm{span}\{\frac{\partial}{\partial x^1},\frac{\partial}{\partial x^2}\},$$
$$D_2 = \mathrm{span}\{\frac{\partial}{\partial x^2},\frac{\partial}{\partial x^3}\}. \tag{3.128}$$

Proof Recall the proof of Frobenius' Theorem. We may assume that we are in $\mathbb{R}^n$ with $p = 0$. Moreover we may assume that

$$D_1(0) = \mathrm{span}\{\frac{\partial}{\partial r^1}\Big|_0,\frac{\partial}{\partial r^2}\Big|_0\},$$
$$D_2(0) = \mathrm{span}\{\frac{\partial}{\partial r^2}\Big|_0,\frac{\partial}{\partial r^3}\Big|_0\}, \tag{3.129}$$

with $r = (r^1,r^2,r^3,r^4)$ natural coordinates for $\mathbb{R}^n$. (This can be always accomplished by a linear transformation of $\mathbb{R}^n$.) Let $\dim r^i = k_i$, $i = 1,2,3$. Let

$$\begin{aligned} &\pi_a : \mathbb{R}^n \to \mathbb{R}^{k_1+k_2} && \text{be the projection onto span}\{r^1,r^2\}, \\ &\pi_b : \mathbb{R}^n \to \mathbb{R}^{k_2+k_3} && \text{be the projection onto span}\{r^2,r^3\}, \\ &\pi_c : \mathbb{R}^n \to \mathbb{R}^{k_2} && \text{be the projection onto span}\{r^2\}. \end{aligned} \tag{3.130}$$

Since D_1,D_2 and $D_1 + D_2$ have constant dimension, also $D_1 \cap D_2$ has constant dimension. Then $\pi_{c*0} : T_0\mathbb{R}^n \to T_0\mathbb{R}^{k_2}$ is an isomorphism when restricted to $D_1(0) \cap D_2(0)$. By continuity π_{c*q} is one-one on $D_1(q) \cap D_2(q)$ for q close to 0. Hence near 0 we can find unique vectors $Z_1(q),\dots,Z_{k_2}(q) \in D_1(q) \cap D_2(q)$ such that

$$\pi_{c*q}Z_i(q) = \frac{\partial}{\partial z_i}\Big|_{\pi_c(q)}, \qquad \text{with } r^2 = (z_1,\dots,z_{k_2}). \tag{3.131}$$

Now consider π_a. By definition $\pi_{a*0} : T_0\mathbb{R}^n \to T_0\mathbb{R}^{k_1+k_2}$ is an isomorphism when restricted to $D_1(0)$. It follows that near 0 we can find unique vectors $X_1(q),\dots,X_{k_1+k_2}(q) \in D_1(q)$ such that

$$\pi_{a*q}X_i(q) = \frac{\partial}{\partial r_i}\Big|_{\pi_a(q)}, \qquad i = 1,\dots,k_1+k_2. \tag{3.132}$$

Furthermore since π_{a*q} coincides with π_{c*q} on $\mathrm{span}\{\frac{\partial}{\partial r^2}\Big|_q\}$ it follows that

$$X_{k_1+i}(q) = Z_i(q), \qquad i = 1,\dots,k_2. \tag{3.133}$$

In the same way by using π_b we obtain unique vectors

$$Y_{k_1+1}(q),\ldots,Y_{k_1+k_2}(q),X_{k_1+k_2+1}(q),\ldots,X_{k_1+k_2+k_3}(q),$$

such that

$$\begin{aligned}\pi_{a*q}Y_i(q) &= \frac{\partial}{\partial r_i}\Big|_{\pi_b(q)}, \qquad i = k_1+1,\ldots,k_1+k_2,\\ \pi_{b*q}X_i(q) &= \frac{\partial}{\partial r_i}\Big|_{\pi_b(q)}, \qquad i = k_1+k_2+1,\ldots,k_1+k_2+k_3.\end{aligned} \tag{3.134}$$

Since π_{b*q} coincides with π_{c*q} on $\mathrm{span}\{\frac{\partial}{\partial r^2}\big|_q\}$ it follows that

$$X_{k_1+i}(q) = Z_i(q) = Y_{k_1+i}(q), \qquad i = 1,\ldots,k_2. \tag{3.135}$$

In the same way as in the proof of Frobenius' Theorem it follows that

$$[X_i,X_j] = 0, \qquad i,j = 1,\ldots,k_1 + k_2 + k_3. \tag{3.136}$$

and hence by Theorem 2.36 we can find local coordinates

$$(x_1,\ldots,x_{k_1},x_{k_1+1},\ldots,x_{k_1+k_2},x_{k_1+k_2+1},\ldots,x_{k_1+k_2+k_3},x_{k_1+k_2+k_3+1},\ldots,x_n)$$
$$= (x^1,x^2,x^3,x^4)$$

such that (3.128) holds. □

Theorem 3.51 *Consider the nonlinear system (3.1). Assume that the distributions* C_0, $\ker d\mathcal{O}$ *and* $C_0 + \ker d\mathcal{O}$ *all have constant dimension. Then we can find local coordinates* $x = (x^1,x^2,x^3,x^4)$ *such that the system takes the form*

$$\dot{x}^1 = f^1(x^1,x^3) + \sum_{j=1}^m g_j^1(x^1,x^3)u_j \tag{3.137a}$$

$$\dot{x}^2 = f^2(x^1,x^2,x^3,x^4) + \sum_{j=1}^m g_j^2(x^1,x^2,x^3,x^4)u_j \tag{3.137b}$$

$$\dot{x}^3 = f^3(x^3) \tag{3.137c}$$

$$\dot{x}^4 = f^4(x^3,x^4) \tag{3.137d}$$

$$y_i = h_i(x^1,x^3), \qquad i \in \underline{p}, \tag{3.137e}$$

with $C_0 = \mathrm{span}\{\frac{\partial}{\partial x^1},\frac{\partial}{\partial x^2}\}$ *and* $\ker d\mathcal{O} = \mathrm{span}\{\frac{\partial}{\partial x^2},\frac{\partial}{\partial x^4}\}$.

Proof By Proposition 3.31 $\ker d\mathcal{O}$ is invariant for the vectorfields in $\mathcal{C}_0$.

Hence $\ker d\mathcal{O} + C_0$ is an involutive distribution. Now apply Proposition 3.50 and interchange x^3 with x^4. Then as in Theorem 3.49 the form (3.137) follows. □

Notice that the *input-output behavior* (cf. Chapter 4) of the system in local form (3.137) is completely described by the dynamics (3.137a,c), involving only the states x^1, x^3, together with the output equations (3.137e). Moreover, if the x^3-part of the initial state is an equilibrium for (3.137c) then x^3 remains constant, and thus also the subsystem (3.137c) does not contribute to the input-output behavior.

Notes and References

The idea of using Lie brackets in the study of accessibility or reachability can be traced back to [Ch]. It was further developed subsequently in the context of nonlinear control theory in e.g. [He], [Lo], [HH], [El], [SJ], [Su1], [Kr1], [HK] and [Su2], [Su3]. We have confined ourselves to a purely local treatment; for more global results we refer to especially [SJ], [Su1,2], [HK]. Our exposition is largely based on [SJ], [HK], and on the lecture notes [Cr1], [Is]. For more information concerning the remark after Assumption 3.1, we refer to [Su2]. The proof of Proposition 3.3 is taken from [LM]. More general expressions than the one given in Proposition 3.6 can be found in [He1]. The proof of Theorem 3.9 given here is due to [Kr1]. More information concerning the issues raised in Remark 3.17 can be found in e.g. [Her], [Su3], [Su4], [St]. Example 3.24 is taken from [Cr1,Cr3], and the final conditions for the two-input and one-input case can be found in [Ba]. Accessibility of bilinear systems was first studied in [Br].

The study of observability using the observation space as in Section 3.2 can be found in [HK], see also [So] for some modifications. In [Cr2] it is shown that an analytic system is observable if and only if the observation space distinguishes points in M. The relevance of the notion of invariant distributions in control theory, together with the resulting decomposition (3.116), was brought forward independently in [Hi], and [IKGM]. The relation with controllability and observability via the invariant distributions C_0 (or C) and $\ker d\mathcal{O}$ is due to [IKGM]. Previous work on local decompositions based on the accessibility *algebra* appeared in [Kr2], see for later work in this direction [Re]. The connection of controllability and observability with invariant distributions was particularly stressed and elaborated in [Is].

In this chapter we have confined ourselves to affine nonlinear systems. However, Theorem 3.9, Corollary 3.11, Proposition 3.12, Corollary 3.13 and Proposition 3.15 immediately carry over to general systems $\dot{x} = f(x,u)$ (see [HK]). For the extension of Theorem 3.21 to the general nonlinear case we refer to [SJ], [vdS1,2]. Also Theorem 3.32, Corollary 3.33, Proposition 3.34, Corollary 3.35 and Proposition 3.38 immediately carry over to systems $\dot{x} = f(x,u)$, $y = h(x)$. For systems $\dot{x} = f(x,u)$, $y = h(x,u)$ we refer to [vdS1,2], where also the relation with invariant distributions was treated. For more information concerning general nonlinear systems we refer to Chapter 13. Finally, in Chapter 12 some specializations of the theory of nonlinear controllability and observability to mechanical nonlinear control systems will be given.

[Ba] J. Baillieul, "Controllability and observability of polynomial dynamical systems", Nonlinear Anal., Theory, Meth. and Appl. 5, pp. 543-552, 1981.

[Br] R.W. Brockett, "System theory on group manifolds and coset spaces", SIAM J. Contr. 10, pp. 265-284, 1972.

[Ch] W.L. Chow, Uber Systemen von linearen partiellen Differentialgleichungen erster Ordnung, Math. Ann. 117, pp. 98-105, 1939.

[Cr1] P.E. Crouch, **Lecture Notes on Geometric Non Linear Systems Theory**, University of Warwick, Control Theory Centre, 1981.

[Cr2] P.E. Crouch, "Dynamical realizations of finite Volterra series", SIAM J. Contr. 19, pp 177-202, 1981.

[Cr3] P.E. Crouch, "Spacecraft attitude control and stabilization", IEEE Trans. Aut. Contr. AC-29, pp 321-331, 1984.

[El] D.L. Elliott, "A consequence of controllability", J. Diff. Eqns. 10, pp 364-370, 1970.

[GB] J.P. Gauthier, G. Bornard, "Observability for any $u(t)$ of a class of nonlinear systems", IEEE Trans. Autom. Contr. AC-26, pp. 922-926, 1981.

[He] R. Hermann, "On the accessibility problem in control theory", in **Int. Symp. on Nonlinear Differential Equations and Nonlinear Mechanics** (eds. J.P. La Salle, S. Lefschetz), pp. 325-332, Academic, New York, 1963.

[Hel] S. Helgason, **Differential geometry and symmetric spaces**, Academic, New York, 1962.

[Her] H. Hermes, "Control systems which generate decomposable Lie Algebras", J. Diff. Eqns. 44, pp. 166-187, 1982.

[HH] G.W. Haynes, H. Hermes, "Nonlinear controllability via Lie theory", SIAM J. Contr. 8, pp. 450-460, 1970.

[Hi] R.M. Hirschorn, "(A,B)-invariant distributions and disturbance decoupling", SIAM J. Contr. 19, pp. 1-19, 1981.

[HK] R. Hermann, A.J. Krener, "Nonlinear controllability and observability", IEEE Trans. Aut. Contr. AC-22, pp. 728-740, 1977.

[Is] A. Isidori, **Nonlinear Control Systems: An Introduction**, Lect. Notes Contr. Inf. Sci. 72, Springer, Berlin, 1985.

[IKGM] A. Isidori, A.J. Krener, C. Gori Giorgi, S. Monaco, "Nonlinear decoupling via feedback: a differential geometric approach", IEEE Trans. Aut. Contr. AC-26, pp. 331-345, 1981.

[Kr1] A.J. Krener, "A generalization of Chow's theorem and the Bang-bang theorem to nonlinear control systems", SIAM J. Contr. 12, pp. 43-52, 1974.

[Kr2] A.J. Krener, "A decomposition theory for differentiable systems", SIAM J. Contr. 15, pp. 289-297, 1977

[Kr3] A.J. Krener, "(Ad f,g), (ad f,g) and locally (ad f,g) invariant and controllability distributions", SIAM J. Contr. Optimiz. 23, pp. 523-549, 1985.

[LM] E.B. Lee, L. Markus, **Foundations of Optimal Control Theory**, Wiley, New York, 1967.

[Lo] C. Lobry, "Contrôlabilité des systèmes non linéaires", SIAM J. Contr. 8, pp. 573-605, 1979.

[Ne] E. Nelson, **Tensor analysis**, Princeton University Press, Princeton, 1967.

[Re] W. Respondek, "On decomposition of nonlinear control systems", Syst. & Contr. Lett. 1, pp. 301-308, 1982.

[So] E. Sonntag, "A concept of local observability", Systems Control Lett. 5, pp. 41-47, 1985.

[St] G. Stefani, "On the local controllability of a scalar input-control system", in **Theory and Applications of Nonlinear Control Systems** (eds. C.I. Byrnes, A. Lindquist), pp. 167-182, North-Holland, Amsterdam, 1986.

[Su1] H.J. Sussmann, "Orbits of families of vectorfields and integrability of distributions", Trans. American Math. Soc. 180, pp. 171-188, 1973.

[Su2] H.J. Sussmann, "Existence and uniqueness of minimal realizations of nonlinear systems", Math. Syst. Theory 10, pp. 263-284, 1977.

[Su3] H.J. Sussmann, "Lie brackets, real analyticity and geometric control", in **Differential Geometric Control Theory**, (eds. R.W. Brockett, R.S. Millman, H.J. Sussmann), pp. 1-116, Birkhäuser, Boston, 1983.

[Su4] H.J. Sussmann, "A general theorem on local controllability", SIAM J. Contr. Optimiz. 25, pp. 158-194, 1987.

[SJ] H.J. Sussmann, V. Jurdjevic, "Controllability of nonlinear systems", J. Diff. Eqns. 12, pp. 95-116, 1972.

[vdS1] A.J. van der Schaft, "Observability and controllability for smooth nonlinear systems", SIAM J. Contr. Optimiz. 20, pp. 338-354, 1982.

[vdS2] A.J. van der Schaft, **System theoretic descriptions of physical systems**, CWI Tract 3, CWI, Amsterdam, 1984.

Exercises

3.1 Consider the following more sophisticated model of a car (compare with Example 2.35).

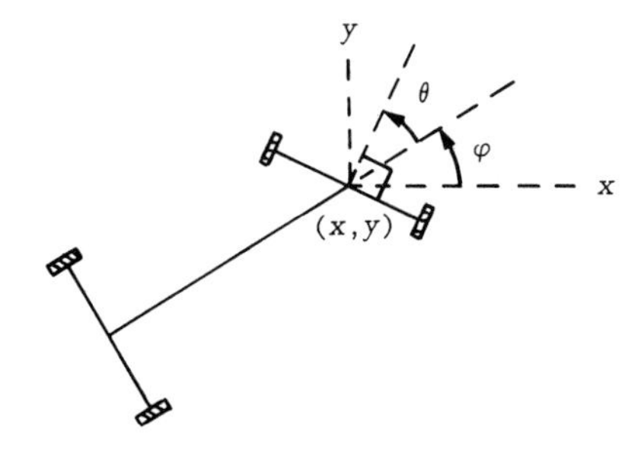

Fig. 3.4. Model of a car.

The configuration space is $M = \mathbb{R}^2 \times S^1 \times S^1$ parametrized by (x,y,φ,θ), where (x,y) are the Cartesian coordinates of the center of the front axis, the angle φ measures the direction in which the car is headed, and θ is the angle made by the front wheels with the car. (More realistically we take $-\theta_{max} < \theta < \theta_{max}$.)

There are two input vectorfields, called Steer and Drive. Clearly, Steer $= \frac{\partial}{\partial\theta}$, while after some analysis we see that, in the appropriate units, Drive $= \cos(\varphi+\theta)\frac{\partial}{\partial x} + \sin(\varphi+\theta)\frac{\partial}{\partial y} + \sin\theta\frac{\partial}{\partial\varphi}$.

(a) Prove that

$$[\text{Steer},\text{Drive}] = -\sin(\varphi+\theta)\frac{\partial}{\partial x} + \cos(\varphi+\theta)\frac{\partial}{\partial y} + \cos\theta\,\frac{\partial}{\partial\varphi} =: \text{Wriggle}.$$

(b) Define Slide $:= -\sin\varphi\,\frac{\partial}{\partial x} + \cos\varphi\,\frac{\partial}{\partial y}$. Prove that [Steer,Wriggle] = -Drive, and [Wriggle,Drive] = Slide. Compute furthermore the bracket of Slide with Steer, Drive and Wriggle.

(c) Compute the dimension of the distribution C spanned by $\mathcal{C}$, and show that the system is locally strongly accessible and controllable.

3.2 (see Remark 3.4). Show under the assumptions of Proposition 3.3 that the set of points which can be steered to x_0 in time $T > 0$ using admissible control satisfying $\|u(t)\| < \epsilon$, $\epsilon > 0$, contains a neighborhood of x_0.

3.3 ([Ba]) Prove condition (3.53).

3.4 (see also Remark 3.17) Consider a nonlinear system (3.69), with $f(x_0) = 0$. Let

$$(1) \quad \begin{aligned} \dot{z} &= Az + Bv, \qquad & z \in \mathbb{R}^n,\ v \in \mathbb{R}^m, \\ \bar{y} &= Cz, & \bar{y} \in \mathbb{R}^p, \end{aligned}$$

be its linearized system in x_0, $u = 0$, i.e., $A = \frac{\partial f}{\partial x}(x_0)$ and $B = (b_1,\ldots,b_m)$ with $b_j = g_j(x_0)$ and $C = \begin{pmatrix} c_1 \\ \vdots \\ c_p \end{pmatrix}$ with $c_i = dh_i(x_0)$.

(a) Prove equation (3.38).

(b) Show by using **(a)** that if (1) is controllable then $R^V(x_0,T)$ contains a non-empty open set for any neighborhood V of x_0, and any small $T > 0$. (This is a weaker version of Proposition 3.3.)

(c) Prove that

$$c_i A^k = dL_f^k h_i(x_0), \qquad k = 1,2,\ldots, \qquad i \in \underline{p},$$

and show that if (1) is observable then (3.69) is locally observable at x_0.

3.5 Consider a linear time-varying system

(1a) $\dot{x}(t) = A(t)x(t) + B(t)u(t), \quad x \in \mathbb{R}^n,\ u \in \mathbb{R}^m,$

(1b) $y(t) = C(t)x(t),\ y \in \mathbb{R}^p,$

where $A(t)$, $B(t)$, $C(t)$ are matrices of appropriate dimensions, whose elements are smooth functions of time t.

(a) Rewrite (1a) as a *nonlinear* time-invariant system

$$(2)\quad \begin{cases} \dot{x} = A(t)x + B(t)u \\ \dot{t} = 1 \end{cases}$$

with augmented state (x,t). Define for $k = 1,2,\ldots$ the matrices

$$(3)\quad D_k(t) = \left(B(t) \,\vdots\, (A(t) - \frac{d}{dt})B(t) \,\vdots\, \cdots \,\vdots\, (A(t) - \frac{d}{dt})^{k-1}B(t)\right).$$

Use the theory of local (strong) accessibility for nonlinear systems in order to prove that if for some k

$$(4)\quad \text{rank } D_k(t) = n, \qquad \text{for all } t > 0,$$

then $R^{\mathbb{R}^n}(x_0, t)$ contains a non-empty open set of $\mathbb{R}^n$ for all x_0 and all $T > 0$. (N.B. Using linear arguments one can even prove that (4) implies that $R^{\mathbb{R}^n}(x_0, t) = \mathbb{R}^n$, for all x_0, $T > 0$.)

(b) Can we always take k in (4) to be less or equal than n?

(c) Derive a condition similar to (4) for some sort of observability for the system (1), using the augmented system (2).

3.6 Consider a nonlinear system

$$(1)\quad \begin{aligned} \dot{x} &= f(x,u) \\ y &= h(x) \end{aligned}, \qquad x \in \mathbb{R}^n,\ u \in \mathbb{R}^m,\ y \in \mathbb{R}$$

We say that the system (1) is *uniformly* locally observable at x^0 ([GB]) if there exists a neighborhood W of x^0 such that for every neighborhood $V \subset W$ of x^0 the following holds. Let $x^1 \in V$ with $x^1 \neq x^0$. Then for every piecewise constant input function $u\colon [0,T] \to \mathbb{R}^m$, with $T > 0$ arbitrary, such that $x(t,x^0,u)$ and $x(t,x^1,u)$ remain in V for $t \in [0,T]$, we have that

$$y(t,x^0,u) \neq y(t,x^1,u), \qquad \text{for some } t \in [0,T].$$

Prove that (1) is uniformly locally observable at x^0 if in suitable local coordinates it is of the form

$$y = h(x_1), \qquad x = (x_1,\ldots,x_n),$$

$$(2)\quad \begin{aligned} \dot{x}_1 &= f_1(x_1,x_2,u) \\ \dot{x}_i &= f_i(x_1,x_2,\ldots,x_{i+1},u), \qquad i = 2,\ldots,n-1, \\ \dot{x}_n &= f_n(x_1,x_1,\ldots,x_n,u), \end{aligned}$$

with $\dfrac{\partial h}{\partial x_1}(x) \neq 0$, $\dfrac{\partial f_i}{\partial x_{i+1}}(x,u) \neq 0$ for every $u \in \mathbb{R}^m$ and every $x \in V$, $i = 1,2,\ldots,n-1$.

3.7 Consider, see Example 3.40, the two-link rigid robot manipulator (3.99) with single output (3.100). Linearize the system in $\theta_1 = \theta_2 = \dot{\theta}_1 = \dot{\theta}_2 = 0$ and $u_1 = u_2 = 0$ and show that the linearized system is observable for $g \neq 0$, while it is not observable for $g = 0$. On the other hand, show that dim $d\mathcal{O} = 4$ even in case $g = 0$.

3.8 ([So]) The system (3.69) is called *L-observable* at x_0 if for every neighborhood W of x_0 there exists a neighborhood $V \subset W$ of x_0 such that the relation $x_0 I^W x_1$ for $x_1 \in V$ implies that $x_1 = x_0$.

(a) Show that if (3.69) is locally observable at x_0 then it is also L-observable at x_0.

(b) (compare Corollary 3.35) Show that if the system is L-observable at every $x \in M$ then $\dim d\mathcal{O}(x) = n$ for every x in an open and dense subset of M.

(c) Assume that the system satisfies the accessibility rank condition and that M is connected. Show that if the system is L-observable at every x in an open and dense subset of M, then it is L-observable at every $x \in M$.

(d) (compare Proposition 3.38) Assume the system satisfies the accessibility rank condition and that M is connected. Prove: the system is L-observable at every $x \in M \Leftrightarrow$ the observability rank condition holds on an open and dense subset of M.

3.9 ([HK]) Prove Remark 3.39. (Hint: use Theorem 2.33.)

3.10 Prove Remark 3.48.

3.11 Let the accessibility distribution C for the nonlinear system (3.69) have constant dimension. Show, analogously to Theorem 3.49, that in local coordinates (x^1,x^2) such that $C = \text{span}\{\frac{\partial}{\partial x^1}\}$ the system takes the form (compare (3.126))

$$\dot{x}^1 = f^1(x^1,x^2) + \sum_{j=1}^{m} g_j^1(x^1,x^2)u_j, \qquad \dot{x}^2 = 0.$$

Assume that also the distributions $\ker d\mathcal{O}$ and $C + \ker d\mathcal{O}$ have constant dimension. Then show that in local coordinates (x^1,x^2,x^3,x^4) with $C = \text{span}\{\frac{\partial}{\partial x^1},\frac{\partial}{\partial x^2}\}$ and $\ker d\mathcal{O} = \text{span}\{\frac{\partial}{\partial x^2},\frac{\partial}{\partial x^4}\}$ the system takes the form (3.137) with (3.137c,d) replaced by $\dot{x}^3 = 0$ and $\dot{x}^4 = 0$.

3.12 **(a)** Show that if a distribution D is invariant under the nonlinear system (3.1), then $[Z,D] \subset D$ for every vectorfield $Z \in \mathcal{C}$ (the accessibility algebra).

(b) Let D be an involutive constant dimensional distribution on $\mathbb{R}^n$ with natural coordinates $(x_1,\ldots,x_n)$. Suppose that $[\frac{\partial}{\partial x_i},D] \subset D$, $i \in \underline{n}$. Prove that $D = D_{\mathcal{V}}$ for some subspace $\mathcal{V} \subset \mathbb{R}^n$ (cf. Remark 3.43).

(c) ([Kr3]) Let D be an involutive constant dimensional distribution for a controllable linear system on $\mathbb{R}^n$. Show that $D = D_{\mathcal{V}}$ for some subspace $\mathcal{V} \subset \mathbb{R}^n$.

4 Input-Output Representations

In this chapter we consider, as before, smooth nonlinear systems

$$\begin{aligned} \dot{x} &= f(x) + \sum_{j=1}^{m} g_j(x)u_j, \qquad u = (u_1,\ldots,u_m) \in u \subset \mathbb{R}^m, \\ y_i &= h_j(x), \qquad j \in \underline{p}, \qquad y = (y_1,\ldots,y_p) \in \mathbb{R}^p, \end{aligned} \tag{4.1}$$

where $x = (x_1,\ldots,x_n)$ are local coordinates for the state space manifold M and f, $g_1,\ldots,g_m$ are smooth vectorfields, while $h_1,\ldots,h_p$ are smooth functions. Our aim in this chapter is to derive explicit expressions relating the inputs u directly to the outputs y. Said otherwise, we want to eliminate in (4.1) the states x and obtain a direct connection between the inputs and outputs. Apart from the interest per se, this is potentially useful for various control purposes. Indeed, in linear systems theory it is well-known that some control or synthesis problems are easier formulated and/or solved for systems given in state space form, while other problems may be more naturally stated and/or solved for systems described in input-output form.

We recall that for linear systems there are several ways of representing a system

$$\begin{aligned} \dot{x} &= Ax + Bu, \qquad u \in \mathbb{R}^m, x \in \mathbb{R}^n, \\ y &= Cx, \qquad y \in \mathbb{R}^p, \end{aligned} \tag{4.2}$$

in input-output form. Of primary importance is the *impulse response matrix* representation

$$y(t) = \int_0^t G(t-s)u(s)ds + Ce^{At}x_0, \qquad t \geq 0, \tag{4.3}$$

with $G(\tau) = Ce^{A\tau}B$ the impulse response matrix, yielding the output $y(t)$, $t \geq 0$, as the sum of a convolution integral of the input $u(t)$, $t \geq 0$, and a term depending on the initial state $x(0) = x_0$. Laplace transformation of (4.3) for $x_0 = 0$ yields (with $\wedge$ denoting the Laplace transform).

$$\hat{y}(s) = \hat{G}(s)\hat{u}(s) \tag{4.4}$$

where the rational matrix $\hat{G}(s) = C(Is-A)^{-1}B$ is called the *transfer matrix*. It is well-known that $\hat{G}(s)$ can be alternatively written as a polynomial fraction

$$\hat{G}(s) = D^{-1}(s)N(s) \tag{4.5}$$

where $D(s)$ is a $p\times p$ polynomial matrix satisfying det $D(s) \neq 0$, and $N(s)$ is a $p\times m$ polynomial matrix. (The matrices $D(s)$ and $N(s)$ can be also directly obtained from the matrices A,B,C in (4.2).) Inverse Laplace transformation then yields the set of *higher-order differential equations* in the inputs and outputs

$$D\left(\frac{\mathrm{d}}{\mathrm{dt}}\right)y(t) = N\left(\frac{\mathrm{d}}{\mathrm{dt}}\right)u(t) \ , \tag{4.6}$$

which is called an *external differential* representation of the linear system (4.2).

In the first section of this chapter we study a nonlinear generalization of the impulse response matrix representation (4.3), known as the *Wiener-Volterra series* representation of (4.1). Briefly we also make contact with a different series expansion called the *Fliess functional expansion*. In the second section we deal with the nonlinear generalization of the *external differential* representation (4.6). We will also show how under constant rank assumptions the state x (in fact the *observable* part of the state) can be expressed in the inputs and outputs and their higher-order time-derivatives. Finally, in the third section we will study the particular problem when a given input component does not affect a certain output component. The conditions obtained will be instrumental to the solution of the disturbance decoupling and input-output decoupling problem as treated in Chapters 7, 8, 9 and 13.

In the present chapter we will *not* deal with the inverse problem of input-output representations; that is, how to obtain from an input-output representation (e.g. Wiener-Volterra or Fliess series representation, or external differential representation) a state space model (4.1). For this important *realization problem* we refer to the literature cited at the end of this chapter.

4.1 Wiener-Volterra and Fliess Series Expansion

For simplicity we first take the input and output in (4.1) to be scalar valued

$$\begin{aligned} \dot{x} &= f(x) + g(x)u, \quad x(0) = x_0 \ , & u \in U \subset \mathbb{R} \ , \\ y &= h(x), & y \in \mathbb{R} \ , \end{aligned} \tag{4.7}$$

where $x = (x_1,\ldots,x_n)$ are local coordinates for M, $f\colon \mathbb{R}^n \to \mathbb{R}^n$, $g\colon \mathbb{R}^n \to \mathbb{R}^n$ are the local coordinate expressions of two smooth vectorfields on M, and $h\colon \mathbb{R}^n \to \mathbb{R}$ is the local coordinate expression of a smooth function on M. We assume throughout that U is an open subset of $\mathbb{R}$ containing the origin.

First we fix the initial state x_0 contained in the above coordinate neighborhood. Let us conveniently denote by $\gamma_0(t,s,x) := x(t,s,x,u)$ the solution of (4.7) at time t for initial state $x(s) = x$ and input $u = 0$, i.e.

$$\begin{aligned} \frac{d}{dt}\gamma_0(t,s,x) &= f\big(\gamma_0(t,s,x)\big) \ , \\ \gamma_0(s,s,x) &= x \ . \end{aligned} \tag{4.8}$$

Since $\dot{x} = f(x)$ is time-invariant we clearly have

$$\gamma_0(t,s,x) = \gamma_0(t-s,0,x) \ . \tag{4.9}$$

For any given input function $u(t)$, say piecewise continuous, we denote by $\gamma_u(t,s,x) := x(t,s,x,u)$ the resulting solution of (4.7) for $x(s) = x$, i.e.

$$\begin{aligned} \frac{d}{dt}\gamma_u(t,s,x) &= f\big(\gamma_u(t,s,x)\big) + u(t)g\big(\gamma_u(t,s,x)\big) \ , \\ \gamma_u(s,s,x) &= x \ . \end{aligned} \tag{4.10}$$

In order to compare $h\big(\gamma_0(t,0,x_0)\big)$ and $h\big(\gamma_u(t,0,x_0)\big)$, i.e. the outputs for zero input and input function u, we introduce for fixed t the curve

$$\rho(s) = \gamma_0\big(t,s,\gamma_u(s,0,x_0)\big), \qquad 0 \le s \le t \ . \tag{4.11}$$

Note that

$$\rho(0) = \gamma_0(t,0,x_0) \quad , \quad \rho(t) = \gamma_u(t,0,x_0) \ , \tag{4.12}$$

implying the basic relation

$$h\big(\gamma_u(t,0,x_0)\big) - h\big(\gamma_0(t,0,x_0)\big) = h\big(\rho(t)\big) - h\big(\rho(0)\big) = \int_0^t \frac{d}{ds} h\big(\rho(s)\big)ds. \tag{4.13}$$

Proposition 4.1 *Let $\rho(s)$ be as defined in (4.11), then*

$$\frac{d}{ds} h\big(\rho(s)\big) = u(s) \left(\frac{\partial h\big(\gamma_0(t,s,x)\big)}{\partial x} g(x)\right)\Big|_{x = \gamma_u(s,0,x_0)} . \tag{4.14}$$

Proof By (4.9) and the chain rule

$$\frac{d}{ds} h\big(\rho(s)\big) = \frac{d}{ds} h\big(\gamma_0(t-s,0,\gamma_u(s,0,x_0))\big) = \tag{4.15}$$

$$\frac{\partial}{\partial s_1} h\big(\gamma_0(t-s_1,0,\gamma_u(s,0,x_0))\big)\Big|_{s_1 = s} + \frac{\partial}{\partial s_2} h\big(\gamma_0(t-s,0,\gamma_u(s_2,0,x_0))\big)\Big|_{s_2 = s}$$

The second term of the right-hand side of (4.15) equals, by (4.10),

$$\frac{\partial h(\gamma_0(t-s,0,x)}{\partial x}\big(f(x) + u(s)g(x)\big)\big|_{x = \gamma_u(s,0,x_0)} \tag{4.16}$$

The first term on the right-hand side of (4.15) equals, by (4.8),

$$-\frac{\partial h(x)}{\partial x} f(x)\big|_{x = \gamma_0(t-s,0,\gamma_u(s,0,x_0))} \tag{4.17}$$

Furthermore, for the first term in (4.16) we observe that by (4.8)

$$\begin{aligned}&\frac{\partial h(\gamma_0(t-s,0,z)}{\partial z} f(z) = \\ &\frac{\partial h(x)}{\partial x}\Big|_{x = \gamma_0(t-s,0,z)} \cdot \frac{\partial \gamma_0(t-s,0,z)}{\partial z} f(z) .\end{aligned} \tag{4.18}$$

We claim that since γ_0 is the flow of f

$$\frac{\partial \gamma_0(t-s,0,z)}{\partial z} f(z) = f\big(\gamma_0(t-s,0,z)\big) \tag{4.19}$$

Indeed $\gamma_0(\tau,0,\gamma_0(t-s,0,z)) = \gamma_0(t-s,0,\gamma_0(\tau,0,z))$ and so, by Corollary 2.32, we have $\gamma_0(t-s,0,z)_* f(z) = f\big(\gamma_0(t-s,0,z)\big)$ which is exactly (4.19). (Note that $\gamma_0(t-s,0,z) = f^{t-s}(z)$.) Therefore for $z = \gamma_u(s,0,x_0)$ (4.18) equals

$$\frac{\partial h(x)}{\partial x} f(x)\Big|_{x = \gamma_0(t-s,0,\gamma_u(s,0,x_0))} \tag{4.20}$$

and hence equals minus (4.17). Taking everything together we see that (4.15) equals the right-hand side of (4.14). □

Let us denote

$$\begin{aligned}w_0(t) &= h\big(\gamma_0(t,0,x_0)\big) ,\\ \bar{w}_1(t,s,x) &= \frac{\partial h\big(\gamma_0(t,s,x)\big)}{\partial x} g(x) .\end{aligned} \tag{4.21}$$

Then by (4.13) and (4.14) it follows that

$$h\big(\gamma_u(t,0,x_0)\big) = w_0(t) + \int_0^t u(s)\bar{w}_1\big(t,s,\gamma_u(s,0,x_0)\big)ds . \tag{4.22}$$

Repeating the same procedure as above for $h(\cdot)$ replaced by $\bar{w}_1(t,s,\cdot)$ we obtain

$$\bar{w}_1\big(t,s,\gamma_u(s,0,x_0)\big) = w_1(t,s) + \int_0^s u(r)\bar{w}_2(t,s,r,\gamma_u\big(r,0,x_0)\big)dr , \tag{4.23}$$

where we denote

$$w_1(t,s) = \bar{w}_1\big(t,s,\gamma_0(s,0,x_0)\big) ,$$
$$\bar{w}_2(t,s,r,x) = \frac{\partial\bar{w}_1\big(t,s,\gamma_0(s,r,x)\big)}{\partial x}\, g(x) . \tag{4.24}$$

Substition in (4.22) gives

$$h\big(\gamma_u(t,0,x_0)\big) = w_0(t) + \int_0^t w_1(t,s)u(s)ds + \int_0^t\int_0^s \bar{w}_2\big(t,s,r,\gamma_u(r,0,x_0)\big)u(s)u(r)drds \tag{4.25}$$

After r repetitions of this process we obtain the functional expansion of $y(t) = h\big(\gamma_u(t,0,x_0)\big)$ given by

$$\begin{aligned} y(t) &= w_0(t) + \int_0^t w_1(t,s)u(s)ds + \\ &\int_0^t\int_0^{s_1} w_2(t,s_1,s_2)u(s_1)u(s_2)ds_2ds_1 + \ldots\ldots + \\ &\int_0^t\int_0^{s_1}..\int_0^{s_{k-1}} w_k(t,s_1,..,s_k)u(s_1)u(s_2)..u(s_k)ds_k..ds_1 + \ldots. + \\ &\int_0^t\int_0^{s_1}..\int_0^{s_r} \bar{w}_{r+1}\big(t,s_1,..,s_{r+1},\gamma_u(s_{r+1},0,x_0)\big)\, u(s_1)..u(s_{r+1})ds_{r+1}..ds_1 \end{aligned} \tag{4.26}$$

where for $i = 2,3,\ldots,$ and $\bar{w}_1(t,s,x)$ as given in (4.21)

$$\bar{w}_k(t,s_k,\ldots,s_k,x) = \frac{\partial\bar{w}_{i-1}\big(t,s_1,\ldots,s_{k-1},\gamma_0(s_{k-1},s_k,x)\big)}{\partial x}\, g(x),$$
$$w_k(t,s_1,\ldots,s_k) = \bar{w}_i\big(t,s_1,\ldots,s_i,\gamma_0(s_i,0,x_0)\big), \tag{4.27}$$
$$t \geq s_1 \geq s_2 \geq \ldots \geq s_k \geq 0.$$

We call (4.26) the *r-th order Wiener-Volterra functional expansion* of $y(t)$, and $w_k(t,s_1,\ldots,s_k)$ as given in (4.27) the *k-th order Volterra kernel*. Notice that the kernels and therefore the expansion depend on the initial state x_0.

For *analytic* systems we can let r in (4.26) tend to infinity to obtain a convergent Wiener-Volterra series. (For the (simple) proof we refer to the literature cited at the end of the chapter.)

Theorem 4.2 *Let f,g, and h in (4.7) be analytic in a neighborhood of x_0. Then there exists $T > 0$ such that for each input function $u(t)$ on $[0,T]$ satisfying $|u(t)| < 1$ the Wiener-Volterra series*

$$y(t) = w_0(t) + \sum_{k=1}^{\infty} \int_0^t \int_0^{s_1} \dots \int_0^{s_{k-1}} w_i(t,s_1,\dots,s_k)u(s_1)\dots u(s_k)ds_k\dots ds_1 \quad (4.28)$$

with $w_k(t,s_1,\dots,s_k)$ as in (4.27), is uniformly absolutely convergent on $[0,T]$.

Remark For a linear single-input single-output system $\dot{x} = Ax + bu, y = cx$ we note that $\bar{w}_1(t,s,x)$ as in (4.21) is given by

$$\bar{w}_1(t,s,x) = \frac{\partial}{\partial x}\left(ce^{A(t-s)}x\right)b = ce^{A(t-s)}b\ , \quad (4.29)$$

and hence does not depend on x. Therefore by (4.24) and (4.27) we have $w_1(t,s) = ce^{A(t-s)}b$ and $w_i(t,s_1,\dots,s_i) = 0$ for $i > 1$, and the functional expansion reduces to

$$y(t) = ce^{At}x_0 + \int_0^t ce^{A(t-s)}bu(s)ds \quad (4.30)$$

which is just the impulse response representation (4.3).

We note that we can rephrase the definition of the *Volterra kernels* $w_k(t,s_i,\dots,s_k)$ as given in (4.27) in the following coordinate free way. Let $f^t\colon M \to M$ be the flow of the vectorfield f (i.e. $f^t(x) = \gamma_0(t,0,x)$). Then it can checked (see Exercise 4.1) that the kernels are also given as

$$\begin{aligned} &w_0(t) = h \circ f^t(x_0), \\ &w_k(t,s_1,\dots,s_k) = \\ &L_g(\dots(L_g(L_g(h\circ f^{(t-s_1)})\circ f^{(s_1-s_2)})\circ f^{(s_2-s_3)})\circ\dots\circ f^{(s_{k-1}-s_k)})\circ f^{(s_k)}(x_0), \\ &\qquad k = 1,2,\dots \end{aligned} \quad (4.31)$$

The Wiener-Volterra functional expansion for the *multi-input multi-output* case is completely similar. If (4.1) is an analytic system then there exists $T > 0$ such that if $|u_i(t)| < 1$, $t \in [0,T]$, $i \in \underline{p}$, then for any output component y_j, $j \in \underline{p}$, we have a uniformly absolutely convergent expansion

$$\begin{aligned} y_j(t) = w_0^j(t) &+ \sum_{k=1}^{\infty} \sum_{i_1\dots i_k=1}^{m} \int_0^t \int_0^{s_1} \dots \int_0^{s_k} \\ &w^j_{i_1\dots i_k}(t,s_1,\dots,s_k)u_{i_1}(s_1)\dots u_{i_k}(s_k)ds_k\dots ds_1, \end{aligned} \quad (4.32)$$

for certain Volterra kernels $w^j_{i_1\dots i_k}(t,s_1,\dots,s_k)$. In fact these Volterra kernels are given for $j \in \underline{p}$ as

$$w^j_0(t) = h_j \circ f^t(x_0),$$

$$w^j_{i_1\dots i_k}(t,s_1,\dots,s_k) = L_{g_{i_k}}(\dots L_{g_{i_2}}(L_{g_{i_1}}(h_j \circ f^{(t-s_1)}) \circ f^{(s_1-s_2)}) \circ f^{(s_2-s_3)}) \circ$$
$$\dots \circ f^{(s_{k-1}-s_k)}) \circ f^{(s_k)}(x_0), \quad i_1,\dots,i_k \in \underline{m}\ , \ k = 1,2.\dots \tag{4.33}$$

Example 4.3 Consider a *bilinear* system (cf. Chapter 3)

$$\begin{aligned} \dot{x} &= Ax + \sum_{j=1}^{m} (B_j x)u_j, \qquad x \in \mathbb{R}^n,\ u \in \mathbb{R}^m\ ,\quad x(0) = x_0\ , \\ y_j &= c_j x, \qquad j \in \underline{p}\ , \end{aligned} \tag{4.34}$$

with $A,B_1,\dots,B_m$ $n\times n$ matrices. In this case the flow f^t of the drift vectorfield is explicitly given as $f^t(x) = e^{At}x$. Therefore the Volterra kernels up to order two are given as

$$\begin{aligned} w^j_0(t) &= c_j e^{At}x_0, \\ w^j_i(t,s) &= c_j e^{A(t-s)}B_i e^{As}x_0, \\ w^j_{i_1 i_2}(t,s) &= c_j e^{A(t-s_1)}B_{i_1}e^{A(s_1-s_2)}B_{i_2}e^{As_2}x_0, \end{aligned} \tag{4.35}$$

and similar expressions hold for the higher-order kernels. □

Let us now briefly indicate how for *analytic* systems (4.1) we can deduce from the Wiener-Volterra functional expansion, as described above, another functional expansion, called the Fliess functional expansion, which reveals more clearly the underlying algebraic structure. For simplicity we first consider the single-input single-output case, and start from the Wiener-Volterra series (4.28) with its kernels given as in (4.31). Since f,g and h are assumed to be analytic all the kernels $w_j(t,s_1,\dots,s_j)$ are analytic functions of their arguments. The key idea is now to expand these kernels as Taylor series *not* in $t,s_1,\dots,s_j$, but in the variables $t-s_1,s_1-s_2,\dots,s_{j-1}-s_j,s_j$, i.e.

$$w_j(t,s_1,\dots,s_j) = \sum_{k_0,k_1,\dots,k_j=0}^{\infty} c_{j,k_0k_1\dots k_j}\cdot \frac{(t-s_1)^{k_0}.(s_1-s_2)^{k_1}\dots.(s_{j-1}-s_j)^{k_{j-1}}.(s_j)^{k_j}}{k_0!k_1!\dots k_j!} \tag{4.36}$$

for certain coefficients $c_{j,k_0k_1\ldots k_j}$ depending on x_0. Therefore by (4.28) we now have to consider the iterated integrals

$$\int_0^t \int_0^{s_1} \cdots \int_0^{s_{j-1}} \frac{(t-s_1)^{k_0}}{k_0!} u(s_1) \cdots \frac{(s_{j-1}-s_i)^{k_{j-1}}}{k_{i-1}!} u(s_j) \frac{s_j^{k_j}}{k_j!} ds_j \ldots ds_1 \tag{4.37}$$

which, however, have the following appealing structure. Define

$$\xi_0(t) = t \ , \ \xi_1(t) = \int_0^t u(s)ds \ , \tag{4.38}$$

and inductively let

$$\begin{aligned} &\int_0^t d\xi_i = \xi_i(t) \quad , \quad i = 0,1 \ , \\ &\int_0^t d\xi_{i_k} \ldots d\xi_{i_0} = \int_0^t d\xi_{i_k}(s) \int_0^s d\xi_{i_{k-1}} \ldots d\xi_{i_0} \end{aligned} \tag{4.39}$$

where $i_0,\ldots,i_k$ are 0 or 1. An easy computation shows that with these definitions (4.37) equals

$$\int_0^t (d\xi_0)^{k_0} \ d\xi_1 \ldots (d\xi_0)^{k_{j-1}} \ d\xi_1 (d\xi_0)^{k_j} \tag{4.40}$$

where $(d\xi_0)^k$ stands for k times $d\xi_0$. Furthermore the coefficients $c_{j,k_0k_1\ldots k_j}$ in (4.36) can be directly identified with the following expressions

$$c_{j,k_0k_1\ldots k_j} = L_f^{k_j} \ L_g \ L_f^{k_{j-1}} \ldots L_g \ L_f^{k_0} \ h(x_0) \ . \tag{4.41}$$

(One way to obtain this identity is to use Taylor expansions of (4.31) in the variables $t-s_1, s_1-s_2, \ldots, s_{k-1}-s_k, s_k$; for details we refer to the literature cited at the end of this chapter.) Combining (4.28), (4.36), (4.40) and (4.41) we arrive at the *Fliess functional expansion*

$$y(t) = h(x_0) + \sum_{k=0}^{\infty} \sum_{i_0,\ldots,i_k=0}^{m} L_{g_{i_0}} \ldots L_{g_{i_k}} \ h(x_0) \int_0^t d\xi_{i_k} \ldots d\xi_{i_0} \tag{4.42}$$

with $g_0 := f$. Similarly for the multi-input multi-output case we have for $j \in \underline{p}$ (compare (4.32))

$$y_j(t) = h_j(x_0) + \sum_{k=0}^{\infty} \sum_{i_0,\ldots,i_k=0}^{m} L_{g_{i_0}} \ldots L_{g_{i_k}} \ h_j(x_0) \int_0^t d\xi_{i_k} \ldots d\xi_{i_0} \tag{4.43}$$

with $g_0 := f$, and $\int_0^t d\xi_i = \xi_i(t) = \int_0^t u_i(s)ds, \qquad i \in \underline{m}$.

4.2. External Differential Representations

In this section we shall give an algorithm which, under constant rank assumptions, converts a system in state space form (4.1) into a set of higher-order differential equations in the inputs and outputs:

$$R_i(u,\dot{u},\ldots,u^{(k)},y,\dot{y},\ldots,y^{(k)}) = 0, \qquad i \in \underline{p}, \tag{4.44}$$

where $u^{(j)}$ and $y^{(j)}$ denote the j-th time derivative of the input function u, respectively output function y.

Let us first introduce the notion of a *prolongation* of a higher-order differential equation. Consider a higher-order differential equation

$$P(w,\dot{w},\ldots,w^{(k)}) = 0 \tag{4.45}$$

in the variables $w \in \mathbb{R}^q$. We will interpret (4.45) also as an *algebraic* equation in the *indeterminates* $w,\dot{w},\ldots,w^{(k)}$. The *prolonged equation* or *prolongation* of (4.45) is defined as

$$\dot{P}(w,\dot{w},\ldots,w^{(k)},w^{(k+1)}) := \frac{\partial P}{\partial w}\dot{w} + \frac{\partial P}{\partial \dot{w}}\ddot{w} + \ldots + \frac{\partial P}{\partial w^{(k)}} w^{(k+1)}, \tag{4.46}$$

where, for notational simplicity, $\dfrac{\partial P}{\partial w^{(j)}} w^{(j+1)} := \displaystyle\sum_{s=1}^{q} \frac{\partial P}{\partial w_s^{(j)}}(w,\ldots,w^{(k)}) w_s^{(j+1)}$.

The relation of (4.46) with (4.45) is as follows. Let $w(t)$, $t \in (a,b)$, be a smooth solution curve of (4.45), i.e.

$$P(w(t),\dot{w}(t),\ldots,w^{(k)}(t)) = 0, \qquad t \in (a,b), \tag{4.47}$$

then clearly for $t \in (a,b)$

$$\begin{aligned} 0 &= \frac{d}{dt} P(w(t),\dot{w}(t),\ldots,w^{(k)}(t)) = \\ &= \frac{\partial P}{\partial w}\dot{w}(t) + \frac{\partial P}{\partial \dot{w}}\ddot{w}(t) + \ldots + \frac{\partial P}{\partial w^{(k)}} w^{(k+1)}(t), \end{aligned} \tag{4.48}$$

and so $w(t)$, $t \in (a,b)$, is also a solution curve of the prolonged equation (4.46). Furthermore we note that

$$\frac{\partial \dot{P}}{\partial w^{(k+1)}} = \frac{\partial P}{\partial w^{(k)}}. \tag{4.49}$$

Now consider the nonlinear system (4.1), rewritten in implicit form as

$$P_i(x,\dot{x},u) = \dot{x}_i - f_i(x,u) = 0, \quad i = 1,\ldots,n, \tag{4.50a}$$

$$P_i(x,y) = y_{i-n} - h_{i-n}(x) = 0 \ , \ i = n+1,\ldots,n+p \ , \tag{4.50b}$$

with $f(x,u) := f(x) + \sum_{j=1}^{m} g_j(x)u_j$. Remember that our aim is to eliminate x and its derivatives in the equations (4.50). Roughly speaking, this will be achieved by successively differentiating the output equations along the system, and to solve from this set of equations for the state variables x. Mathematically this will be formalized by first prolonging the output equations (4.50b), to substitute these for some of the n equations $\dot{x}_i - f_i(x,u) = 0$ in (4.50a), and to replace $\dot{x}$ in these prolonged equations by $f(x,u)$. After doing this we obtain a system of equations of the same form as (4.50) where now, however, the number of equations involving $\dot{x}$ has been reduced. Then the same procedure is repeated.

Formally we have the following algorithm. Let $(\tilde{x}(t),\tilde{u}(t),\tilde{y}(t))$, $t \in (-\epsilon,\epsilon)$, be a smooth solution curve of (4.50). This yields a solution point

$$(\bar{x},\bar{u},\bar{y}) = (\tilde{x}(0),\dot{\tilde{x}}(0),..,\tilde{x}^{(n)}(0),\tilde{u}(0),\dot{\tilde{u}}(0),..,\tilde{u}^{(n)}(0),\tilde{y}(0),\dot{\tilde{y}}(0),..,\tilde{y}^{(n)}(0)) \tag{4.51}$$

of (4.50), regarded as a set of algebraic equations in the indeterminates $x,\dot{x},\ldots,x^{(n)},u,\dot{u},\ldots,u^{(n)},y,\dot{y},\ldots,y^{(n)}$.

Algorithm 4.4 (External representation algorithm)

Step 1 Assume that

$$\operatorname{rank} \left(\frac{\partial P_i}{\partial \dot{x}_j}(x,y) \right)_{\substack{i=n+1,\ldots,n+p \\ j=1,\ldots,n}} = s_1 \, , \ \text{around } (\bar{x},\bar{u},\bar{y}) \ . \tag{4.52}$$

Denote $p_1 = s_1 - s_0$, with $s_0 = 0$. If $p_1 = 0$ the algorithm terminates. If $p_1 > 0$ we proceed as follows. By (4.52) it follows that we can reorder the equations $P_1,\ldots,P_n$ in (4.50), and separately the equations $P_{n+1},\ldots,P_{n+p}$, in such a way that

$$\operatorname{rank} \begin{pmatrix} \left(\dfrac{\partial P_i}{\partial \dot{x}_j} \right)_{\substack{i=1,\ldots,n-p_1 \\ j=1,\ldots,n}} \\ \left(\dfrac{\partial P_i}{\partial x_j} \right)_{\substack{i=n+1,\ldots,n+p_1 \\ j=1,\ldots,n}} \end{pmatrix} = n, \ \text{around } (\bar{x},\bar{u},\bar{y}) \ . \tag{4.53}$$

Furthermore we re-order the variables $x_1,\ldots,x_n$ in the same way as we did for $P_1,\ldots,P_n$, so that still $P_i(x,\dot{x},u) = \dot{x}_i - f_i(x,u)$, $i \in \underline{n}$. Now consider the prolonged equations $\dot{P}_{n+1},\ldots,\dot{P}_{n+p_1}$ and replace (4.50) by the following set of equations

$$P_i(x,\dot{x},u) = \dot{x}_i - f_i(x,u) = 0 \;,\; i = 1,\ldots,n-p_1 \;, \tag{4.54a}$$

$$\dot{P}_i(x,f(x,u),u,y,\dot{y}) = 0 \;,\; i = n+1,\ldots,n+p_1 \;, \tag{4.54b}$$

$$P_i(x,y) = 0 \;,\; i = n+1,\ldots,n+p \;. \tag{4.54c}$$

Lemma 4.5 *Around $(\bar{x},\bar{u},\bar{y})$ the set of smooth solution curves of (4.54) equals that of (4.50).*

Proof Clearly if $(x(t),u(t),u(t))$ is a solution curve of (4.50), then it is also a solution curve of the prolonged equations, and since $\dot{x}_i - f_i(x,u) = 0$, $i \in \underline{n}$, we may substitute in these prolonged equations $f_i(x,u)$ for $\dot{x}_i$, so as to obtain (4.54b). For the converse we observe that by (4.59)

$$\operatorname{rank}\left(\frac{\partial P_i}{\partial x_k}\right)_{\substack{i=n+1,\ldots,n+p_1\\ k=n-p_1+1,\ldots,n}} = p_1\,,\ \text{around } (\bar{x},\bar{u},\bar{y}) \;. \tag{4.55}$$

Now let $(x(t),u(t),y(t))$ be a solution curve of (4.54) around $(\bar{x},\bar{u},\bar{y})$. Then it is a solution curve of (4.54c), and therefore of the prolonged equations $\dot{P}_{n+1} = \ldots = \dot{P}_{n+p_1} = 0$. Hence $(x(t),u(t),y(t))$ satisfies

$$\sum_{j=1}^{n-p_1} \frac{\partial P_i}{\partial x_j}\dot{x}_j + \sum_{k=n-p_1+1}^{n} \frac{\partial P_i}{\partial x_k}\dot{x}_k + \sum_{s=1}^{p} \frac{\partial P_i}{\partial y_s}\dot{y}_s = 0,\ i = n+1,..,n+p_1 \,. \tag{4.56}$$

Since $(x(t),u(t),y(t))$ satisfies (4.54a) it follows that

$$\sum_{j=1}^{n-p_1} \frac{\partial P_i}{\partial x_j} f_j(x,u) + \sum_{k=n-p_1+1}^{n} \frac{\partial P_i}{\partial x_k}\dot{x}_k + \sum_{s=1}^{p} \frac{\partial P_i}{\partial y_s}\dot{y}_s = 0 \;,\; i = n+1,\ldots,n+p_1 \,. \tag{4.57}$$

On the other hand $(x(t),u(t),y(t))$ satisfies (4.54b):

$$\sum_{j=1}^{n-p_1} \frac{\partial P_i}{\partial x_j} f_j(x,u) + \sum_{k=n-p_1+1}^{n} \frac{\partial P_i}{\partial x_k} f_k(x,u) + \sum_{s=1}^{p} \frac{\partial P_i}{\partial y_s}\dot{y}_s = 0 \;,\quad i = n+1,..,n+p_1 \;. \tag{4.58}$$

Comparing with (4.57) we see that

$$\sum_{k=n-p_1+1}^{n} \frac{\partial P_i}{\partial x_k} \dot{x}_k = \sum_{k=n-p_1+1}^{n} \frac{\partial P_i}{\partial x_k} f_k(x,u) \;, \qquad i = n+1,..,n+p_1 \;. \tag{4.59}$$

By (4.55) it now follows that

$$\dot{x}_k = f_k(x,u), \qquad k = n-p_1+1,...,n \;, \tag{4.60}$$

and hence $(x(t),u(t),y(t))$ is around $(\bar{x},\bar{u},\bar{y})$ a solution curve of (4.50). □

We rename equations (4.54b) by setting

$$P_i(x,u,y,\dot{y}) := \dot{P}_{i+p_1}(x,f(x,u),u,y,\dot{y}) \;, \; i = n-p_1+1,..,n \;. \tag{4.61}$$

Denote $n_1 := n$, and $n_2 := n_1 - p_1$ then (4.54) is rewritten as

$$P_i(x,\dot{x},u) = \dot{x}_i - f_i(x,u) = 0 \;, \; i = 1,...,n_2 \;, \tag{4.62a}$$

$$P_i(x,u,y,\dot{y}) = 0 \;, \; i = n_2+1,..,n+p \;. \tag{4.62b}$$

As a result of the first step of the algorithm we have transformed (4.50) into (4.62). Notice that (4.62) is of the same form as (4.50), but the number of equations involving $\dot{x}$ has decreased by p_1. Notice also that (4.62) satisfies

$$\operatorname{rank} \begin{bmatrix} \left(\frac{\partial P_i}{\partial \dot{x}_j}\right)_{\substack{i=1,...,n_2 \\ j=1,...,n}} \\ \left(\frac{\partial P_i}{\partial x_j}\right)_{\substack{i=n_2+1,...,n+p \\ j=1,...,n}} \end{bmatrix} = n, \text{ around } (\bar{x},\bar{u},\bar{y}) \;. \tag{4.63}$$

Step k of the algorithm Consider a system of equations

$$P_i(x,\dot{x},u) = \dot{x}_i - f_i(x,u) = 0 \;, \; i = 1,...,n_k \;, \tag{4.64a}$$

$$P_i(x,u,\dot{u},..,u^{(k-2)},y,\dot{y},..,y^{(k-1)}) = 0 \;, \; i = n_k+1,..,n+p \;, \tag{4.64b}$$

for which

$$\operatorname{rank} \begin{bmatrix} \left(\frac{\partial P_i}{\partial \dot{x}_j}\right)_{\substack{i=1,...,n_k \\ j=1,...,n}} \\ \left(\frac{\partial P_i}{\partial x_j}\right)_{\substack{i=n_k+1,...,n+p \\ j=1,...,n}} \end{bmatrix} = n, \text{ around } (\bar{x},\bar{u},\bar{y}) \;. \tag{4.65}$$

Now assume that

$$\operatorname{rank}\left[\frac{\partial P_i}{\partial x_j}\right]_{\substack{i=n_k+1,\dots,n+p\\ j=1,\dots,n}} = s_k\,, \text{ around } (\bar{x},\bar{u},\bar{y})\ . \tag{4.66}$$

Denote $p_k := s_k - s_{k-1}$. If $p_k = 0$ the algorithm terminates. If $p_k > 0$ we proceed as follows. By (4.49)

$$\left[\frac{\partial P_i}{\partial x_j}\right]_{\substack{i=n_k+1,\dots,n+p\\ j=1,\dots,n}} = \left[\frac{\partial \dot{P}_i}{\partial \dot{x}_j}\right]_{\substack{i=n_k+1,\dots,n+p\\ j=1,\dots,n}} \tag{4.67}$$

Furthermore in the $(k-1)$-th step we have assumed that

$$\operatorname{rank}\left[\frac{\partial P_i}{\partial x_j}\right]_{\substack{i=n_{k-1}+1,\dots,n+p\\ j=1,\dots,n}} = \operatorname{rank}\left[\frac{\partial \dot{P}_i}{\partial \dot{x}_j}\right]_{\substack{i=n_{k-1}+1,\dots,n+p\\ j=1,\dots,n}}$$

$$= s_{k-1}\,, \text{ around } (\bar{x},\bar{u},\bar{y})\ . \tag{4.68}$$

Now consider the prolongations of the equations obtained in the $(k-1)$-th step

$$\dot{P}_i(x,\dot{x},u,\dots,u^{(k-1)},y,\dots,y^{(k)}) = 0\ ,\ i = n_k+1,\dots,n_{k-1}\ . \tag{4.69}$$

By (4.68) there exist functions $\alpha_{i\ell}(x,u,\dots,u^{(k-2)},y,\dots,y^{(k-1)})$, $i = n_k+1,\dots,n_{k-1}$, $\ell = n_{k-1}+1,\dots,n+p$, such that if we define the following modifications of the equations (4.69)

$$S_i(x,\dot{x},u,\dots,u^{(k-1)},y,\dots,y^{(k)}) := \dot{P}_i(x,\dot{x},u,\dots,u^{(k-1)},y,\dots,y^{(k)}) + \tag{4.70}$$

$$\sum_{\ell=n_{k-1}+1}^{n+p} \alpha_{i\ell}(x,u,\dots,u^{(k-2)},y,\dots,y^{(k-1)})\ \dot{P}_i(x,\dot{x},u,\dots,u^{(k-1)},y,\dots,y^{(k)}),$$

$$i = n_k+1,\dots,n_{k-1}\,,$$

then

$$\frac{\partial S_i}{\partial \dot{x}_j} = 0\ ,\ i = n_k+1,\dots,n_{k-1}\,,\quad j = n_k+1,\dots,n\ . \tag{4.71}$$

By (4.66) we can now reorder $S_{n_k+1},\dots,S_{n_{k-1}}$ in such a way that

$$\operatorname{rank}\left[\frac{\partial S_i}{\partial \dot{x}_j}\right]_{\substack{i=n_k+1,\dots,n_k+p_k\\ j=1,\dots,n_k}} = p_k\,, \text{ around } (\bar{x},\bar{u},\bar{y})\ . \tag{4.72}$$

Then by (4.65) we can permute the equations $P_1,\dots,P_{n_k}$ in such a way that

$$\text{rank}\begin{pmatrix}\left(\dfrac{\partial P_i}{\partial \dot{x}_j}\right)_{\substack{i=1,\ldots,n_k-p_k\\ j=1,\ldots,n_k}}\\ \left(\dfrac{\partial S_i}{\partial \dot{x}_j}\right)_{\substack{i=n_k+1,\ldots,n_k+p_k\\ j=1,\ldots,n_k}}\end{pmatrix} = n_k\,,\ \text{around } (\bar{x},\bar{u},\bar{y}). \tag{4.73}$$

Furthermore we permute the variables $x_1,\ldots,x_{n_k}$ in the same way. Now consider instead of (4.64) the set of equations

$$P_i(x,\dot{x},u) = \dot{x}_i - f_i(x,u) = 0, \qquad i = 1,\ldots,n_k-p_k\ , \tag{4.74a}$$

$$S_i(x,f(x,u),u,\ldots,u^{(k-1)},y,\ldots,y^{(k)}) = 0, \qquad i = n_k+1,\ldots,n_k+p_k, \tag{4.74b}$$

$$P_i(x,u,\ldots,u^{(k-2)},y,\ldots,y^{(k-1)}) = 0, \qquad i = n_k+1,\ldots,n+p\ . \tag{4.74c}$$

Lemma 4.6 *Around $(\bar{x},\bar{u},\bar{y})$ the set of smooth solution curves of (4.74) equals that of (4.64).*

Proof Clearly any solution curve of (4.64) is a solution curve of (4.74) (compare Lemma 4.5). Let $(x(t),u(t),y(t))$ be a smooth solution curve of (4.74). We only have to prove that $(x(t),u(t),y(t))$ is a solution curve of

$$P_i(x,\dot{x},u) = \dot{x}_i - f_i(x,u) = 0, \qquad i = n_k-p_k+1,\ldots,n_k\ . \tag{4.75}$$

Clearly $(x(t),u(t),y(t))$ is also a solution curve of

$$S_i(x,\dot{x},u,\ldots,u^{(k-1)},y,\ldots,y^{(k)}) = 0, \qquad i = n_k+1,\ldots,n_k+p_k\ . \tag{4.76}$$

Using the fact that S_i is linear in $\dot{x}$, and $(x(t),u(t),y(t))$ is also a solution curve of (4.74b), it follows that

$$\sum_{j=1}^{n}\frac{\partial S_i}{\partial \dot{x}_j}\dot{x}_j = \sum_{j=1}^{n}\frac{\partial S_i}{\partial \dot{x}_j}f_j(x,u)\ , \qquad i = n_k+1,\ldots,n_k+p_k\ . \tag{4.77}$$

Furthermore since $(x(t),u(t),y(t))$ satisfies (4.74a) and (4.71) holds, (4.77) reduces to

$$\sum_{j=n_k-p_k+1}^{n_k}\frac{\partial S_i}{\partial \dot{x}_j}\dot{x}_j = \sum_{j=n_k-p_k+1}^{n_k}\frac{\partial S_i}{\partial \dot{x}_j}f_j(x,u), \qquad i = n_k+1,\ldots,n_k+p_k \tag{4.78}$$

and (4.75) follows by (4.72) and (4.73). □

We rename equations (4.74b) by setting

$$P_i(x,u,\ldots,u^{(k-1)},y,\ldots,y^{(k)}) := S_{i+p_k}(x,f(x,u),u,\ldots,u^{(k-1)},y,\ldots,y^{(k)}),$$

$$i = n_k - p_k + 1,\ldots,n_k. \qquad (4.79)$$

Denote $n_{k+1} := n_k - p_k$, then (4.74) is rewritten as

$$P_i(x,\dot{x},u) = \dot{x}_i - f_i(x,u) = 0, \qquad i = 1,\ldots,n_{k+1}, \qquad (4.80a)$$

$$P_i(x,u,\ldots,u^{(k-1)},y,\ldots,y^{(k)}) = 0, \qquad i = n_{k+1}+1,\ldots,n+p. \qquad (4.80b)$$

Clearly (4.80) satisfies (4.65) with n_k replaced by n_{k+1}, and is a system having p_k more equations not involving $\dot{x}$ than (4.64).

If in the above algorithm the constant rank assumption (4.66) is satisfied around $(\bar{x},\bar{u},\bar{y})$ for any $k = 1,2,\ldots,$ then we call $(\bar{x},\bar{u},\bar{y})$ a *regular point* for the algorithm.

It is clear that for a regular point the algorithm terminates after a finite number of steps, denoted by k^*, in the sense that $p_{k^*+1} = 0$. Moreover, necessarily $k^* \le n$. Denote $\bar{n} = n_{k^*+1}$, then after performing the algorithm we end up with a system

$$\dot{x}_i - f_i(x,u) = 0, \qquad i = 1,\ldots,\bar{n}, \qquad (4.81a)$$

$$P_i(x,u,\ldots,u^{(k^*-1)},y,\ldots,y^{(k^*)}) = 0, \qquad i = \bar{n}+1,\ldots,n+p, \qquad (4.81b)$$

Let us first consider the case $\bar{n} = 0$. Recall that at the k^*-th step we assumed that

$$\operatorname{rank}\left(\frac{\partial P_i}{\partial x_j}\right)_{\substack{i=n_{k^*}+1,\ldots,n+p\\ j=1,\ldots,n}} = s_{k^*}, \text{ around } (\bar{x},\bar{u},\bar{y}). \qquad (4.82)$$

Comparing the recursive definitions of n_k and p_k

$$n_{k+1} = n_k - p_k, \qquad n_1 = n, \qquad k = 1,2,\ldots \qquad (4.83)$$

$$p_{k+1} = s_{k+1} - s_k, \qquad s_0 = 0, \qquad k = 0,1,\ldots$$

we immediately obtain $n_k = n - s_{k-1}$, $k \ge 1$. Hence $\bar{n} = n_{k^*+1} = n - s_{k^*}$, and, because we assume $\bar{n} = 0$, $n = s_{k^*}$. Therefore by (4.82) we can reorder the equations $P_1,\ldots,P_{n+p}$ of (4.81) in such a way that

$$\operatorname{rank}\left(\frac{\partial P_i}{\partial x_j}\right)_{\substack{i=1,\ldots,n\\ j=1,\ldots,n}} = n \ (= s_{k^*}), \text{ around } (\bar{x},\bar{u},\bar{y}), \qquad (4.84)$$

where the equations $P_1,..,P_n$ are taken from the set $P_{n_k^*+1},..,P_{n+p}$. Then by the Implicit Function Theorem, we can solve, locally around $(\bar{x},\bar{u},\bar{y})$, from

$$P_i(x,u,\dot{u},..,u^{(k^*-2)},y,\dot{y},..,y^{(k^*-1)}) = 0, \qquad i \in \underline{n}, \tag{4.85}$$

for the variables $x_1,..,x_n$, i.e.

$$x_i = \psi_i(u,\dot{u},..,u^{(k^*-2)},y,\dot{y},..,y^{(k^*-1)}), \qquad i \in \underline{n}. \tag{4.86}$$

Substitution of (4.86) in the remaining equations $P_{n+1},..,P_{n+p}$ yields equations of the form

$$R_i(u,\dot{u},..,u^{(k^*-1)},y,\dot{y},..,y^{(k^*)}) = 0, \qquad i \in \underline{p}, \tag{4.87}$$

which finally constitute the *external differential representation* of the nonlinear system (4.1). Summarizing

Theorem 4.7 *Consider the smooth nonlinear system (4.1), and let $(\bar{x},\bar{u},\bar{y})$ be a regular point for the Algorithm 4.4. Suppose that $\bar{n} = 0$. Then (4.1) is transformed around $(\bar{x},\bar{u},\bar{y})$ into the equations (4.86) together with (4.87). The first equations (4.86) express the state x as function of u and y and their derivatives, up to order $k^*-1 \leq n-1$, while the last equations (4.87) form the external differential representation of the system.*

Example 4.8 Consider the bilinear system

$$0 = \dot{x}_1 - ux_2 \qquad =: P_1(x,\dot{x},u) \tag{4.88a}$$

$$0 = \dot{x}_2 - ux_1 \qquad =: P_2(x,\dot{x},u) \tag{4.88b}$$

$$0 = \dot{x}_3 - x_2 - ux_1 \qquad =: P_3(x,\dot{x},u) \tag{4.88c}$$

$$0 = y - x_3 \qquad =: P_4(x,y) \tag{4.88d}$$

Clearly rank $\frac{\partial P_4}{\partial x} = 1$ everywhere, and $\dot{P}_4 = \dot{y} - \dot{x}_3$. Hence at the first step of the algorithm we transform (4.88) into (4.88a,b,d) together with

$$0 = \dot{y} - x_2 - ux_1 =: P_3(x,u,y,\dot{y}). \tag{4.89}$$

Then

$$\text{rank}\begin{pmatrix} \frac{\partial P_3}{\partial x} \\ \frac{\partial P_4}{\partial x} \end{pmatrix} = \text{rank}\begin{pmatrix} -u & -1 & 0 \\ 0 & 0 & 1 \end{pmatrix} = 2, \tag{4.90}$$

and $\dot{P}_3 = \ddot{y} - \dot{x}_2 - u\dot{x}_1 - \dot{u}x_1$. Hence at the second step we transform the system into (4.88a,d), (4.89), together with

$$0 = \ddot{y} - ux_1 - u^2x_2 - \dot{u}x_1 =: P_2(x,u,\dot{u},y,y,\dot{y},\ddot{y}) . \tag{4.91}$$

For the third step we notice that

$$\begin{bmatrix} \dfrac{\partial P_2}{\partial x} \\ \dfrac{\partial P_3}{\partial x} \\ \dfrac{\partial P_4}{\partial x} \end{bmatrix} = \begin{bmatrix} -u-\dot{u} & -u^2 & 0 \\ -u & -1 & 0 \\ 0 & 0 & -1 \end{bmatrix} , \tag{4.92}$$

which has rank 3 for $u + \dot{u} - u^3 \neq 0$. Excluding this point we define

$$S_2 = \dot{P}_2 - u^2\dot{P}_3 = y^{(3)} - u^2\ddot{y} - 2u\dot{u}\ddot{y} - (u+\dot{u}-u^3)\dot{x}_1 - (\dot{u}+\ddot{u}-3u^2\dot{u})x_1 , \tag{4.93}$$

so that $\dfrac{\partial S_2}{\partial \dot{x}_1} = u + \dot{u} - u^3$. Outside $u + \dot{u} - u^3 = 0$ we transform the system finally into

$$\begin{aligned} 0 &= y^{(3)} - u^2\ddot{y} - 2u\dot{u}\ddot{y} - (u + \dot{u} - u^3)ux_2 - (\dot{u} + \ddot{u} - 3u^2\dot{u})x_1 \\ 0 &= \ddot{y} - ux_1 - u^2x_2 - \dot{u}\, x_1 \\ 0 &= \dot{y} - x_2 - u\, x_1 \\ 0 &= y - x_3 \end{aligned} \tag{4.94}$$

From the last three equations we can solve for x_1, x_2 and x_3, provided $u + \dot{u} - u^3 \neq 0$. Indeed

$$\begin{aligned} x_1 &= \frac{1}{u + \dot{u} - u^3} (\ddot{y} - u^2\dot{y}) , \\ x_2 &= \frac{1}{u + \dot{u} - u^3} (u\dot{y} + \dot{u}\dot{y} - u\ddot{y}) , \quad x_3 = y . \end{aligned} \tag{4.95}$$

Substitution of (4.95) in the first equation of (4.94) gives the external differential representation

$$\begin{aligned} (u + \dot{u} - u^3)y^{(3)} + (u^2\dot{y} - \ddot{y})\ddot{u} + (3u^2-1)\dot{u}\ddot{y} - \\ - 3u^2\dot{y}\dot{u} - 3u\dot{y}\dot{u}^2 + u^5\dot{y} - u^3\dot{y} = 0 . \end{aligned} \tag{4.96}$$

□

We shall now show how the property of $\bar{n}$ being equal to 0 immediately relates to the *local observability* of (4.1). In fact we will show that $\bar{n}$

is the dimension of the unobservable dynamics of the system (see Remark after Theorem 3.49).

Lemma 4.9 *The subspace of* $\mathbb{R}^n$ *given by*

$$\ker \left[\frac{\partial P_i}{\partial x_j}(x,u,..,u^{(k^*-1)}, y,..,y^{(k^*)}) \right]_{\substack{i = \bar{n}+1,..,n+p \\ j = 1,...,n}} \tag{4.97}$$

does not depend on $u,..,u^{(k^*-1)},y,..,y^{(k^*)}$.

Proof By definition of k^*, $s_{k^*+1} - s_{k^*} = p_{k^*+1} = 0$. Hence the subspace (4.97) is also given as

$$\ker \left[\frac{\partial P_i}{\partial x_j}(x,u,..,u^{(k^*-2)}, y,..,y^{(k^*-1)}) \right]_{\substack{i = n_{k^*}+1,..,n+p \\ j = 1,..,n}} \tag{4.98}$$

It follows that the subspace (4.98) will not change if we add to $P_i, i = n_{k^*}+1,..,n+p$, any function obtained by prolongation of some P_i, $i = n_{k^*}+1,..,n+p$, and substitution of $\dot{x} = f(x,u)$ herein. Now suppose that (4.97) and thereby (4.98) *does* depend on some variable $u,..,u^{(k^*-2)}$, $y,..,y^{(k^*-1)}$. Let $r \leq k^*-1$ be the highest derivative of u or y such that (4.98) depends on $u^{(r)}$ or $y^{(r)}$. Then prolong the equations $P_{n_{k^*}+1},..,P_{n+p}$ and substitute $\dot{x} = f(x,u)$, and join these equations to $P_{n_{k^*}+1},..,P_{n+p}$ in (4.98). It follows that (4.98) depends non-trivially on $u^{(r+1)}$ or $y^{(r+1)}$, which is a contradiction with the definition of r. □

It follows from Lemma 4.9 that (4.97) defines a *distribution* D on the coordinate chart of M, the state space manifold, that we are working in. Furthermore, since D is given as the intersection of the kernels of exact one-forms, the distribution *D is involutive*. Moreover, since we assume that $(\bar{x},\bar{u},\bar{y})$ is a regular point for Algorithm 4.4, D has constant dimension around $(\bar{x},\bar{u},\bar{y})$. Hence by Frobenius' Theorem (Theorem 2.42) we can choose local coordinates (x^1,x^2), with dim $x^1 = \bar{n}$, such that D = span $\{\frac{\partial}{\partial x^1}\}$. It follows that in such coordinates (4.81) takes the form

$$\dot{x}^1 - f^1(x^1,x^2) = 0 , \qquad x^1 = (x_1,..,x_{\bar{n}}) , \tag{4.99a}$$

$$P_i(x^2,u,..,u^{(k^*-1)}, y,..,y^{(k^*)}) = 0 , \qquad i = \bar{n}+1,..,n+p . \tag{4.99b}$$

We immediately see that D is *invariant* (Definition 3.44) for (4.1). Since the equations $y_i - h_i(x) = 0$, $i \in \underline{p}$, are contained in (4.99b) it follows that $h_i(x)$, $i \in \underline{p}$, only depends on x^2, and so D is contained in ker dh. As in the case $\bar{n} = 0$ we can solve from $s_{k^*}(= n-\bar{n})$ of the equations $P_i = 0$, $i = n_{k^*}+1,\ldots,n+p$, for the variables x^2 as functions of $u,\ldots,u^{(k^*-2)}$, $y,\ldots,y^{(k^*-1)}$. Substitution of x^2 in the remaining p equations of the set $P_{\bar{n}+1},\ldots,P_{n+p}$ yields the external differential representation of the system. Summarizing, we have obtained the following generalization of Theorem 4.7.

Theorem 4.10 *Consider the system (4.1), and let* $(\bar{x},\bar{u},\bar{y})$ *be a regular point for Algorithm 4.4. Then (4.1) is transformed around* $(\bar{x},\bar{u},\bar{y})$ *into a threefold set of equations*

$$\dot{x}^1 = f^1(x^1,x^2,u) \;, \qquad \dim x^1 = \bar{n} \;, \tag{4.100a}$$

$$x^2 = \psi(u,\ldots,u^{(k^*-2)}, y,\ldots,y^{(k^*-1)}) \;, \qquad \dim x^2 = n-\bar{n} \;, \tag{4.100b}$$

$$R_i(u,\dot{u},\ldots,u^{(k^*-1)}, y,\dot{y},\ldots,y^{(k^*)}) = 0 \;, \qquad i \in \underline{p} \;. \tag{4.100c}$$

Here (4.100a) are the dynamics of the unobservable part of the system, (4.100b) expresses the observable part of the state as function of u *and* y *and derivatives, and (4.100c) is the external differential representation of the system.*

Proof The only thing left to be shown is that $D = \operatorname{span}\{\frac{\partial}{\partial x^1}\}$ is the *largest* distribution contained in ker dh that is invariant for (4.1). However since x^2 is expressed in (4.100b) as function of u and y and their derivatives it follows that x^2 locally can be determined on the basis of knowledge of (any) input function and resulting output function. By the definition of local observability (Definition 3.28) this implies that, indeed, $D = \ker d\mathcal{O}$, with the $\mathcal{O}$ the observation space (Definition 3.29). □

4.3 Output Invariance

In this section we study the problem when in a nonlinear system (4.1) a certain input component u_j does not influence a particular output component y_i. For *linear* systems we know that this amounts to the transfer function (or impulse response function) from u_j to y_i being zero, and we look for a nonlinear generalization of this condition. In the linear case there also exists an equivalent *geometric* condition involving the existence of an invariant subspace containing the input vector b_j corresponding to u_j, and contained in the kernel of the output function $y_i = c_i x$. The nonlinear generalization of this last condition given here will be crucial for the developments in Chapters 7 and 9.

For simplicity of notation we consider instead of (4.1) the smooth system

$$\begin{aligned} \dot{x} &= f(x) + \sum_{j=1}^{m} g_j(x)u_j + e(x)v, \qquad (u_1,\ldots,u_m) \in U = \mathbb{R}^m, \ v \in \mathbb{R} \\ y &= h(x) \ , \end{aligned} \tag{4.101}$$

with $u_1,\ldots,u_m$ and v the scalar inputs, and y a scalar output, and we want to deduce conditions which ensure that the input v does not influence the output y for any initial state x_0 and any input functions $u_1,\ldots,u_m$. As in Chapter 3 we restrict ourselves throughout to *piecewise constant* (and piecewise continuous from the right) input functions $u_1,\ldots,u_m,v$.

Definition 4.11 *Consider the system (4.101) with the input functions $u_1,\ldots,u_m,v$ piecewise constant. We say that the output y is not affected by (or invariant under) the input v if for every initial state x_0, for every set of input functions $u_1,\ldots,u_m$ and for all $t \geq 0$*

$$y(t,x_0,u_1,\ldots,u_m,v^1) = y(t,x_0,u_1,\ldots,u_m,v^2) \ , \tag{4.102}$$

for every pair of functions v^1,v^2.

Remark 4.12 Suppose that (4.102) holds. If we now approximate more general input functions $u_1(t),\ldots,u_m(t),v(t)$ by piecewise constant functions it readily follows from standard results on differential equations that (4.102) also holds for these more general input functions (see also the corresponding remark after Definition 3.1).

Remark 4.13 We may, equivalently, replace (4.102) by the requirement that for all $x_0,u_1,\ldots,u_m$ and all $t \geq 0$

$$y(t,x_0,u_1,\ldots,u_m,v) = y(t,x_0,u_1,\ldots,u_m,0) \ , \tag{4.103}$$

for every function v.

We deduce the following necessary conditions for output invariance.

Proposition 4.14 *Consider the system (4.101) and suppose that the output* *y is invariant under v. Then for all $r \geq 0$ and any choice of vectorfields $X_1,\dots,X_r$ in the set $\{f,g_1,\dots,g_m\}$ we have*

$$L_e L_{X_1} L_{X_2} \dots L_{X_r} h(x) = 0, \qquad \text{for all } x . \tag{4.104}$$

Remark 4.15 Notice that condition (4.104) is not changed if we let $X_1,\dots,X_r$ belong to the extended set $\{f,g_1,\dots,g_m,e\}$.

Proof By Proposition 3.22, (4.104) is equivalent with the requirement

$$L_e L_{Z_1} \dots L_{Z_r} h(x) = 0 , \tag{4.105}$$

for all $r \geq 0$ and any choice of vectorfields $Z_1,\dots,Z_r$ of the form (see (3.65))

$$Z_i = f + \sum_{j=1}^{m} g_j(x)u_j^i , \quad i \in \underline{r} , \tag{4.106}$$

for some point $u^i = (u_1^i,\dots,u_m^i)^T \in U$. Now let y be invariant under v. Then by (4.103) for small $t_1,\dots,t_k$ and all x

$$h\left(Z_k^{t_k} \circ Z_{k-1}^{t_{k-1}} \circ \dots \circ Z_1^{t_1}(x)\right) = h\left(\tilde{Z}_k^{t_k} \circ \tilde{Z}_{k-1}^{t_{k-1}} \circ \dots \circ \tilde{Z}_1^{t_1}(x)\right) \tag{4.107}$$

where $Z_1,\dots,Z_k$ are of the form (4.106), and $\tilde{Z}_1,\dots,\tilde{Z}_k$ are given as

$$\tilde{Z}_i(x) = Z_i(x) + e(x), \qquad i \in \underline{k} , \tag{4.108}$$

(i.e. the left-hand side of (4.107) is the output for $v = 0$, while the right-hand side is the output for $v = 1$). Differentiating both sides of (4.107) with respect to $t_k, t_{k-1},\dots,t_1$ at respectively $t_k = 0$, $t_{k-1} = 0,\dots,t_1 = 0$, yields (compare (3.71))

$$L_{Z_1} L_{Z_2} \dots L_{Z_k} h(x) = L_{\tilde{Z}_1} L_{\tilde{Z}_2} \dots L_{\tilde{Z}_k} h(x) \tag{4.109}$$

for all $k \geq 0$. Now take $k = 1$, then

$$L_{Z_1} h(x) = L_{\tilde{Z}_1} h(x) . \tag{4.110}$$

Since $L_{\tilde{Z}_1} h(x) = L_{Z_1} h(x) + L_e h(x)$, this implies

$$L_e h(x) = 0 , \tag{4.111}$$

i.e. (4.105) for $r = 0$. In general we obtain

$$\begin{aligned} L_{Z_1} L_{Z_2} \dots L_{Z_k} h(x) &= L_{\tilde{Z}_1} L_{\tilde{Z}_2} \dots L_{\tilde{Z}_k} h(x) = \\ &= L_{\tilde{Z}_1} L_{Z_2} \dots L_{Z_k} h(x) , \end{aligned} \tag{4.112}$$

and using (4.108) this yields

$$L_e L_{Z_2} \cdots L_{Z_k} h(x) = 0 , \tag{4.113}$$

for all $Z_2, \ldots, Z_k$ of the form (4.106). □

If condition (4.104) is *not* satisfied, i.e. if there exists some choice of vectorfields $X_1, \ldots, X_r$ in the set $\{f, g_1, \ldots, g_m\}$ such that for some x

$$L_e L_{X_1} L_{X_2} \cdots L_{X_r} h(x) \neq 0 , \tag{4.114}$$

then we say that the input v *instantaneously affects* the output y. Indeed the proof of Proposition 4.14 shows that if we differentiate in this case $y(t)$ $(r+1)$-times with respect to time, along a suitably chosen trajectory of the system, then this $(r+1)$-th order derivative will depend non-trivially on v.

One may expect that condition (4.104) is also *sufficient* for output invariance, so that if v does not *instantaneously* affect y then v does not affect y at all. However for C^∞-systems this is, unfortunately, generally not true, as shown by the following example.

Example 4.16 Consider the following single-input single-output system on $\mathbb{R}$

$$\dot{x} = f(x) + e(x)v , \qquad x \in \mathbb{R} ,$$
$$y = h(x) , \tag{4.115}$$

where the smooth vectorfield $e(x)$ and the smooth function $h(x)$ are of the form

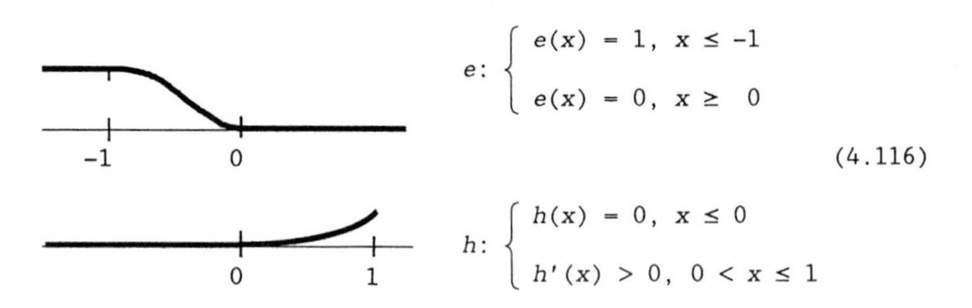

Fig. 4.1. Condition (4.104) is not sufficient for output invariance.

Furthermore we let $f(x) \equiv 1$. Condition (4.104) now amounts to

$$L_e L_f^r h(x) = 0 \tag{4.117}$$

for all $r \geq 0$, which is obviously satisfied by the definition of e and h.

Hence v does *not* instantaneously affect the output y. However v *does* affect the output y in the following, indirect, way. By (4.13) and (4.14) we have for any input function v

$$h(t,x_0,v) - h(t,x_0,0) = \int_0^t v(s)\left[\frac{\partial h(f^{t-s}(x))}{\partial x} e(x) \right]_{x = \gamma_v(s,0,x_0)} ds \tag{4.118}$$

where $\gamma_v(s,0,x_0)$ is the solution at time s of the differential equation $\dot{x} = f(x) + e(x)v$. By definition of f the term between brackets on the right-hand side of (4.118) equals

$$\frac{\partial h(x+t-s)}{\partial x} e(x) \ . \tag{4.119}$$

By the definition of h and e it follows that for $x < 0$ and $t - s$ satisfying $1 > x + t - s > 0$ the expression (4.119) is *not* equal to zero. Therefore if we take $v(s)$ in (4.118) equal to 1 then for some x_0 and t we will have $h(t,x_0,1) \neq h(t,x_0,0)$, and so v *does* affect the output y. □

Notice that in the above example the vectorfield $e(x)$ and the function $h(x)$ are smooth (C^∞) but *not* analytic. In fact for an analytic system condition (4.104) *does* imply output invariance:

Proposition 4.17 *Suppose that the system (4.101) is analytic, then y is invariant under v if and only if (4.104) is satisfied.*

Proof The "only if" direction has been proved in Proposition 4.14. The "if" direction is proved as follows. Let x_0 be the initial state. An arbitrary piecewise constant input sequence $u(t) = (u_1(t),\ldots,u_m(t))^T$ can be written as

$$\begin{aligned} u(t) &= (u_1^1,\ldots,u_m^1)^T \ , \quad t \in [t_0,t_1) \ , \ t_0 = 0 \ , \\ &= (u_1^2,\ldots,u_m^2)^T \ , \quad t \in [t_1,t_1+t_2) \ , \\ &\ \ \vdots \qquad\qquad\qquad\qquad \vdots \\ &= (u_1^r,\ldots,u_m^r)^T \ , \quad t \in [t_1 + \ldots + t_{r-1}, t_1 + \ldots + t_r). \end{aligned} \tag{4.120}$$

Consider two time instants s,t satisfying $0 \leq s \leq t \leq t_1 + \ldots + t_r$. Then we can write

$$\begin{aligned} s &= t_1 + \ldots + t_{p-1} + (t_p - \tau_p) \ , \\ t &= t_1 + \ldots + t_{\ell-1} + \tau_\ell \ , \end{aligned} \tag{4.121}$$

for some integers p,ℓ satisfying $0 \leq p \leq \ell \leq r$, and some τ_p and τ_ℓ

satisfying $0 \le \tau_p < t_p$, $0 \le \tau_\ell < \tau_\ell$. Denoting $Z_i = f + \sum_{j=1}^{m} g_j(x)u_j^i$, with $u^i = (u_1^i, \ldots, u_m^i)$, $i \in \underline{r}$, as in (4.120), then the solution $\gamma_0(t,s,x)$ of (4.101) for $v = 0$ and $x(s) = x$, and $u(t)$ defined by (4.120), is given as

$$\gamma_0(t,s,x) = Z_\ell^{\tau_\ell} \circ Z_{\ell-1}^{t_{\ell-1}} \circ \ldots \circ Z_{p+1}^{t_{p+1}}(x). \tag{4.122}$$

Let us denote the solution of (4.101) for the above input sequence $u(t) = (u_1(t), \ldots, u_m(t))^T$ and for arbitrary input $v(t)$ by $\gamma_v(t,s,x)$. Completely similar to Proposition 4.1 we have the following relation between the output $y(t,x_0,u_1,\ldots,u_m,0) = h(\gamma_0(t,0,x_0))$ of (4.101) for $v = 0$ and the output $y(t,x_0,u_1,\ldots,u_m,v) = h(\gamma_v(t,0,x_0))$ of (4.101) for arbitrary v, namely

$$y(t,x_0,u_1,\ldots,u_m,v) - y(t,x_0,u_1,\ldots,u_m,0) = \tag{4.123}$$

$$\int_0^t v(s) \left[\frac{\partial h(\gamma_0(t,s,x))}{\partial x} e(x) \right] \Big|_{x = \gamma_u(s,0,x_0)} ds$$

Now we will prove that the expression between brackets

$$\frac{\partial h(\gamma_0(t,s,x))}{\partial x} e(x) \tag{4.124}$$

is zero for all $t \ge s \ge 0$ and all x. By (4.121) we have that (4.124) equals

$$L_e h(Z_\ell^{\tau_\ell} \circ Z_{\ell-1}^{t_{\ell-1}} \circ \ldots \circ Z_{p+1}^{t_{p+1}} \circ Z_p^{\tau_p}(x)) . \tag{4.125}$$

Since (4.101) is assumed to be analytic we can write

$$h(Z_\ell^{\tau_\ell} \circ Z_{\ell-1}^{t_{\ell-1}} \circ \ldots \circ Z_p^{\tau_p}(x)) = \tag{4.126}$$

$$\sum_{i=0}^{\infty} \frac{\tau_\ell^i}{i!} L_{Z_\ell}^i h(Z_{\ell-1}^{t_{\ell-1}} \circ \ldots \circ Z_p^{\tau_p}(x))$$

(we take $t_1 + \ldots + t_r$ small enough so that (4.126) and all following Taylor expansions converge). In the same way we can expand for any i

$$L_{Z_\ell}^i h(Z_{\ell-1}^{t_{\ell-1}} \circ \ldots \circ Z_p^{\tau_p}(x)) =$$

$$\sum_{j=0}^{\infty} \frac{t_{\ell-1}^j}{j!} L_{Z_{\ell-1}}^j L_{Z_\ell}^i h(Z_{\ell-2}^{t_{\ell-2}} \circ \ldots \circ Z_p^{\tau_p}(x)). \tag{4.127}$$

Continuing in this way, and recalling that (4.104) is equivalent with (4.105), we immediately see that (4.125) equals zero, and hence the right-hand side of (4.123) is zero. □

Remark 4.18 Notice that actually we do *not* need analyticity of the vectorfield e. This observation can be useful e.g. for the disturbance decoupling problem (where e will denote the disturbance vectorfield, cf. Chapter 7).

Remark 4.19 In fact, Proposition 4.17 also follows from the Fliess functional expansion. If we let $g_0 = f$ and $g_{m+1} = e$ then by (4.43) the Fliess functional expansion of $y(t)$ equals

$$y(t) = h(x_0) + \sum_{k=0}^{\infty} \sum_{i_0,\ldots,i_k=0}^{m+1} L_{g_{i_0}} \ldots L_{g_{i_k}} h(x_0) \int_0^t d\xi_{i_k} \ldots d\xi_{i_0} \tag{4.128}$$

The iterated integral $\int_0^t d\xi_{i_k} \ldots d\xi_{i_0}$ depends on v $(= u_{m+1})$ only if one of the indices $i_0,\ldots,i_k$ is equal to $m+1$. However by (4.104) the expressions $L_{g_{i_0}} \ldots L_{g_{i_k}} h(x_0)$ are *zero* whenever one of the indices $i_0,\ldots,i_k$ equals $m+1$. Hence any iterated integral in (4.128) involving v is premultiplied by zero, and therefore $y(t)$ does not depend on v.

Conditions (4.104), which are necessary, and in the analytic case also sufficient, conditions for output invariance can be given the following geometric interpretation.

Proposition 4.20 *Consider the nonlinear system (4.101). Conditions (4.104) are satisfied if and only if there exists a distribution D with the following properties*

(i) *D is invariant for $\dot{x} = f(x) + \sum_{j=1}^{m} g_j(x)u_j$ (see Definition 3.44),*

(ii) *$e \in D$,*

(iii) *$D \subset \ker dh$.*

Proof (only if) Consider the unobservability distribution $\ker d\mathcal{O}$ for $\dot{x} = f(x) + \sum_{j=1}^{m} g_j(x)u_j$, $y = h(x)$. By Proposition 3.47(c) $\ker d\mathcal{O}$ is invariant and is contained in $\ker dh$. By (4.104) $L_e H(x) = 0$ for all $H \in \mathcal{O}$, and hence $e \in \ker d\mathcal{O}$.

(if) Let D satisfy (i), (ii), (iii). Let X be a vectorfield contained in D. By (i) and (iii) $[f,X] \in D \subset \ker dh$. Hence

$$L_f L_X h - L_X L_f h = 0 . \tag{4.129}$$

Since $X \in D \subset \ker dh$ this yields $L_X L_f h = 0$, and so

$$X \in \ker dL_f h. \tag{4.130}$$

In the same way $X \in \ker dL_{g_j} h$, $j \in \underline{m}$. Continuing in this way we have for any $X \in D$ that $X \in \ker d\mathcal{O}$. In particular we may take X equal to $e \in D$. Then $e \in \ker d\mathcal{O}$, or equivalently $L_e H(x) = 0$ for all $H \in \mathcal{O}$. Hence (4.104) is satisfied. □

Remark 4.21 We may assume without loss of generality that the distribution D in Proposition 4.20 *is involutive*. Indeed if D satisfies (i), (ii), (iii) then $\bar{D}$, the involutive closure of D (see (2.143)) also satisfies (i), (ii) and (iii). The fact that $\bar{D}$ satisfies (ii) is trivial, while $\bar{D}$ satisfies (iii) because $\ker dh$ is involutive. To prove that $\bar{D}$ satisfies (i) we use the Jacobi-identity (Proposition 2.27(c)): let $X_1, X_2 \in D$, then $[f,[X_1,X_2]] = [[f,X_1],X_2] + [X_1,[f,X_2]]$. Now $[[f,X_1],X_2] \in [X_2,D] \in \bar{D}$, and similarly $[X_1,[f,X_2]] \in \bar{D}$. The same holds for f replaced by g_j, $j \in \underline{m}$.

Example 4.22 Consider the linear system

$$\begin{aligned} \dot{x} &= Ax + Bu + ev, \qquad && x \in \mathbb{R}^n,\ u \in \mathbb{R}^m,\ v \in \mathbb{R}, \\ y &= cx, && y \in \mathbb{R}. \end{aligned} \tag{4.131}$$

Then $\ker d\mathcal{O} = \bigcap_{i=0}^{n-1} \ker cA^i =: \mathcal{N}$, and condition (4.104) amounts to

$$cA^i e = 0, \qquad i = 0,1,\ldots , \tag{4.132}$$

and is satisfied if and only if $e \in \mathcal{N}$. Since a linear system is trivially analytic this is equivalent with output invariance. □

Up to now we have not yet obtained a constructive necessary and sufficient condition for output invariance of C^∞ systems. However by adding a *constant* rank assumption the above geometric interpretation together with Frobenius' Theorem yields (compare with Proposition 4.20 and Remark 4.21):

Proposition 4.23 *Consider the smooth nonlinear system (4.101). The output y is invariant under v if there exists an involutive distribution D of constant dimension with the properties*

(i) *D is invariant for* $\dot{x} = f(x) + \sum_{j=1}^{m} g_j(x)u_j$,

(ii) $e \in D$,

(iii) $D \subset \ker dh$.

In particular if ker $d\mathcal{O}$ *has constant dimension then* y *is invariant under* v *if and only if (4.104) holds.*

Proof Let D be a distribution as above. By Frobenius' Theorem (Corollary 2.43) we can find local coordinates $(x^1,x^2) = (x_1,\dots,x_k,\ x_{k+1},\dots,x_n)$ such that $D = \text{span}\ \{\frac{\partial}{\partial x^1}\}$. By (i) and (ii) the system (4.101) in these coordinates takes the form (see (3.116))

$$\dot{x}^1 = f^1(x^1,x^2) + \sum_{j=1}^{m} g_j^1(x^1,x^2)u_j + e^1(x_1,x_2)v\ ,$$
$$\dot{x}^2 = f^2(x^2) + \sum_{j=1}^{m} g_j^2(x^2)u_j\ . \tag{4.133a}$$

Furthermore by (iii) the output equation equals

$$y = h(x^2)\ . \tag{4.133b}$$

Now it is clear from (4.133) that for solutions remaining in this coordinate chart the input v does *not* affect the output y for any choice of the input functions $u_1,\dots,u_m$. Since we can take such a coordinate chart around any point x this implies that y is invariant under v. In Proposition 4.20 we have shown that ker $d\mathcal{O}$ satisfies (i), (ii), (iii) if (4.104) is satisfied. Since ker $d\mathcal{O}$ is assumed to have constant dimension (and is by definition involutive) this shows that (4.104) implies that y is invariant under v. The converse has already been proved in Proposition 4.14. □

Notes and References

Since the fifties, Wiener-Volterra series have been used quite extensively in the study of nonlinear systems; we refer to the books [Ru] and [Sche] for a modern account. The Wiener-Volterra functional expansion for a nonlinear system in state space form as given in Section 4.1 is largely taken from [LK]. This paper on its turn generalises the results in [Br] and [Gi], and the Wiener-Volterra functional expansions for bilinear systems as given by in [BDK]. For realization theory of Wiener-Volterra series we refer to [Br] and [Cr].

The Fliess functional expansion has been developed in a series of papers, e.g. [Fl1,Fl3,Fl4] and the survey [FLL], and uses, among other things, the theory of formal power series in automata theory [Schü], and the theory of iterated integrals, cf. [Ch]. For realization theory of Fliess functional expansions we refer to [Fl3] and the detailed exposition in [Is]. For the proof of Theorem 4.2 we refer e.g. to [Is],[Fl4], and for the relation with the Volterra-Wiener expansion (4.36),(4.41) we refer to [Fl2],[Is]. Functional expansions for discrete-time nonlinear systems have been studied by [So] and [NC].

An alternative approach to realization theory starting with general input-output maps has been developed by [Ja],[Su].

The material of Section 4.2 has been taken almost verbatim from [vdSl]; for related treatments we refer to [CMP],[Gl] and [Re]. The importance of higher-order differential representations for nonlinear systems has been pointed out e.g. in [Wi],[Fl5] and [vdS2]. For realization theory of external differential representations we refer to [vdS2],[Gl]. Section 4.3 is largely based on [Is], which extends the results of [IKGM] and [H]. The fact that conditions (4.104) are *not* sufficient for output invariance in the smooth case (see Example 4.16) seems not to have been pointed out explicitly before.

[Br] R.W. Brockett, "Volterra series and geometric control theory", Automatica 12, pp. 167-176, 1976.

[BDK] C. Bruni, G. di Pillo, G. Koch, "On the mathematical models of bilinear systems", Ricerche di Automatica 2, pp. 11-26, 1971.

[Cr] P.E. Crouch, "Dynamical realizations of finite Volterra series", SIAM J. Contr. Optimiz. 19, pp. 177-202, 1981.

[Ch] K.T. Chen, "Iterated path integrals, Bull. Amer. Math. Soc. 83, pp. 831-879, 1977.

[CMP] G. Conte, C.M. Moog, A. Perdon, "Un théorème sur la répresentation entrée-sortie d'un système nonlinéaire", C.R. Acad. Sci. Paris, Sér. I, 307, pp. 363-366, 1988.

[Fl1] M. Fliess, "Sur la réalisation des systèmes dynamique bilinéaires", C.R. Acad. Sci. Paris, A-277, pp. 923-926, 1973.

[Fl2] M. Fliess, "A note on Volterra series expansions for nonlinear differential systems", IEEE Trans. Aut. Contr. 25, pp. 116-117, 1980.

[Fl3] M. Fliess, "Fonctionelles causales non linéaires et indéterminées non commutatives", Bull. Soc. Math. France 109, pp.3-40, 1981.

[Fl4] M. Fliess, "Réalisation des systémes non linéaires, algébres de Lie filtrées transitives et series génératrices non commutatives", Invent. Math. 71, pp. 521-537, 1983.

[Fl5] M. Fliess, "Automatique et corps différentiels", Forum Math. 1, pp. 227-238, 1989.

[FLL] M. Fliess, M. Lamnabhi, F. Lamnabhi-Lagarrigue, "An algebraic approach to nonlinear functional expansions", IEEE Trans. Circ. Syst. CAS- 30, pp. 554-570, 1983.

[Gl] S.T. Glad, "Nonlinear state space and input-output descriptions using differential polynomials" in **New Trends in Nonlinear Control Theory** (eds. J. Descusse, M. Fliess, A. Isidori, D. Leborgne), Lect. Notes Contr. Inf. Sci. 122, pp. 182-190, 1989.

[Hi] R.M. Hirschorn, "(A,B)-invariant distributions and disturbance decoupling of nonlinear systems", SIAM J. Contr. Optimiz. 19, pp. 1-19, 1981.

[Is] A. Isidori, **Nonlinear Control Systems: An Introduction**, Lect. Notes Contr. Inf. Sci. 72, Springer, Berlin, 1985.

[IKGM] A. Isidori, A.J. Krener, C. Gori Giorgi, S. Monaco, "Nonlinear decoupling via feedback: a differential geometric approach", IEEE Trans. Auto. Contr. AC-26, pp 331-345, 1981.

[Ja] B. Jakubczyk, "Existence and uniqueness of realizations of nonlinear systems", SIAM J. Contr. Optimiz. 18, pp. 455-471, 1980.

[LK] C.Lesiak, A.J. Krener, "The existence and uniqueness of Volterra series for nonlinear systems", IEEE Trans. Aut. Contr. AC-23, pp. 1090-1095, 1978.

[NC] D. Normand-Cyrot, **Théorie et pratique des systèmes nonlinéaires en temps discret**, Thèse d' Etat, Université de Paris Sud, 1983.

[Ru] W.J. Rugh, **Nonlinear System Theory: the Volterra-Wiener approach**, Johns Hopkins Press, Baltimore, 1981.

[Re] W. Respondek, "From state-space representation to differential equations in inputs and outputs", paper presented at the IFAC Symposium on Nonlinear Control Systems Design, Capri, Italy, 1989.

[Sche] M. Schetzen, **The Volterra and Wiener theories of nonlinear systems**, Wiley, New York, 1980.

[Schü] M.P. Schützenberger, "On the definition of a family of automata", Inform. Control, 4, pp. 245-270, 1961.

[So] E. Sontag, **Polynomial response maps**, Lect. Notes Contr. Inf. Sci. 13, Springer, Berlin, 1979.

[Su] H.J. Sussmann, "Existence and uniqueness of minimal realizations of nonlinear systems", Math. Syst. Theory 10, pp. 263-284, 1977.

[vdS1] A.J. van der Schaft, "Representing a nonlinear state space system as a set of higher-order differential equations in the inputs and outputs", Systems Control Lett. 812, pp. 151-160, 1989.

[vdS2] A.J. van der Schaft, "On realization of nonlinear systems described by higher-order differential equations", Math. Syst. Theory 19, pp. 239-275, 1987; Correction 20, pp. 305-306, 1987.

[Wi] J.C. Willems, "System theoretic models for the analysis of physical systems", Ricerche di Automatica 10, pp. 71-106, 1979.

Exercises

4.1 Verify that the k-th order Volterra kernel for a single-input single-output system is given as in (4.31).

4.2 Consider the nonlinear system (4.7). Define the functions $O_t = h \circ f^t$, and the vectorfields $P_t = (f^{-t})_* g$. Show that the k-th order Volterra kernel is alternatively given as ([Is])

$$w_k(t,s_1,\dots,s_k) = (L_{P_{s_k}} L_{P_{s_{k-1}}} \dots L_{P_{s_1}} O_t)(x_0) \ .$$

4.3 Consider the nonlinear system (4.7). Denote for clarity the k-th order Volterra kernel corresponding to initial state x_0 as $w_k(t,s_1,\dots,s_k,x_0)$. Show that for any $\tau \geq 0$

$$w_k(t,s_1,\dots,s_k,x_0) = w_k(t-\tau,s_1-\tau,\dots,s_k-\tau,f^{\tau}(x_0)) \ .$$

In particular, if $f(x_0) = 0$ we obtain *stationary* kernels:

$$w_k(t,s_1,\dots,s_k,x_0) = w_k(t-\tau,s_1-\tau,\dots,s_k-\tau,x_0), \ \forall \tau \geq 0.$$

Verify this for the Volterra kernels of a bilinear system with $x_0 = 0$ (Example 4.3).

4.4 Denote as in (4.3) the k-th order Volterra kernel corresponding to initial state x_0 as $w_k(t,s_1,\dots,s_k,x_0)$. The Wiener-Volterra expansion at x_0 is said to have length p if $w_{p+j}(t,s_1,\dots,s_{p+j},x_0) = 0$ for every $j \geq 1$.

(a) Prove that the Wiener-Volterra expansion has length p if $w_p(t,s_1,\dots,s_p,x_0)$ does not depend on x_0.

(b) Suppose the system satisfies the strong accessibility rank condition. Prove that if the Wiener-Volterra expansion at some x_0 has length p, then $w_p(t,s_1,\dots,s_p,x_0)$ does not depend on x_0 ([Cr]).

4.5 Consider the nonlinear system (4.7).

(a) Show that the first-order Volterra kernel can be written as

$$w_1(t,s,x_0) = \sum_{k_0,k_1=0}^{\infty} L_f^{k_1} L_g L_f^{k_0} h(x_0) \frac{(t-s)^{k_0}}{k_0!} \cdot \frac{s^{k_1}}{k_1!}$$

(b) Prove that $w_1(t,s,x_0)$ is independent of x_0 and only dependent on $t-s$ if and only if

(1) $L_g L_f^k h(x)$ is independent of x for all $k \geq 0$.

(c) Show that the output of (4.7) for $x(0) = x_0$ can be written as

$$y(t) = Q(t,x_0) + \int_0^t g(t-s)u(s)ds$$

for some functions Q and $g(\tau)$ (i.e., the input-output response of the system is *linear*) if and only if condition (1) is satisfied.

4.6 Consider the bilinear system

$$\begin{aligned} \dot{x}_1 &= -x_2 - x_3 u, \qquad y = x_1, \\ \dot{x}_2 &= x_3 - x_2 u, \\ \dot{x}_3 &= x_1 . \end{aligned}$$

(a) Use Algorithm 4.4 to determine an external differential representation.

(b) Show that the mapping $(u(0),\dot{u}(0),\ddot{u}(0),y(0),\dot{y}(0),\ddot{y}(0)) \longmapsto x_0$ is not injective for those input functions that satisfy $\dot{u} = -1 + u^2$.

4.7 Show that the integers $p_1,\dots,p_{k^*}$ defined in Algorithm 4.4 satisfy

$$p_1 \geq p_2 \geq \dots \geq p_{k^*} > 0 \ , \ p_1 + p_2 + \dots + p_{k^*} = n - \bar{n}.$$

Define k_i as the cardinality of the set $\{p_j \mid p_j \geq i\}$, $i \in \underline{p}$. Show that

$$k_1 \geq k_2 \geq \dots \geq k_p \geq 0 \ , \ k_1 + k_2 + \dots + k_p = n - \bar{n}$$

(The integers k_i are called observability indices, cf. [vdS1]).

4.8 Consider the Hamiltonian system (see also Chapter 12)

$$\dot{q}_i = \frac{\partial H}{\partial p_i} \ , \ \dot{p}_i = -\frac{\partial H}{\partial q_i} + u_i \ , \ y_i = q_i \ , \ i \in \underline{n} \ ,$$

where $H(q,p) = \frac{1}{2}p^T G(q)p + V(q)$, with $G(q)$ a positive definite $n \times n$ matrix for every $q \in \mathbb{R}^n$.

(a) Use Algorithm 4.4 to obtain an external differential representation.

(b) Show that the external differential system given by the Euler-Lagrange equations

$$\frac{d}{dt}\left(\frac{\partial L}{\partial \dot{y}_i}\right) - \frac{\partial L}{\partial y_i} = u_i \quad , \; i \in \underline{n},$$

where $L(y,\dot{y}) = \frac{1}{2}\,\dot{y}^T G^{-1}(y)\dot{y} - V(y)$, is equivalent to the external differential system obtained in **(a)** (i.e. the set of input-output trajectories is the same).

4.9 Prove Remark 4.13.

4.10 (see Remark 4.21) Consider a distribution D, locally given as (cf. 2.119) $D(q) = \text{span}\{X_i(q)\ ;\ i \in I\}$, $q \in V$, V open set in M. Show that $\overline{D}$, the involutive closure of D (cf. 2.143), on V is given as the distribution spanned by all vectorfields of the form

$$[X_k, [X_{k-1}, [\ldots, [X_2, X_1]\ldots]]],$$

where X_j , $j \in \underline{k}$, is in the set $\{X_i,\ i \in I\}$ and $k \in \mathbb{N}$ (compare with Proposition 3.5). Fill in the details of Remark 4.21.

5
State Space Transformation and Feedback

This chapter deals with some preliminairies which are basic to controller and observer design for nonlinear systems. In particular we discuss the possibility of linearizing a system by state space transformations and we introduce various types of nonlinear feedback.

5.1 State Space Transformations and Equivalence to Linear Systems

Consider a continuous time smooth affine nonlinear control system

$$\dot{x} = f(x) + \sum_{i=1}^{m} g_i(x)u_i, \tag{5.1}$$

where $x = (x_1, \ldots, x_n)$ are local coordinates for a smooth manifold M and $f, g_1, \ldots, g_m$ are smooth vectorfields on M. Together with the dynamics (5.1) we consider an output equation

$$y = h(x), \tag{5.2}$$

where $h\colon M \to \mathbb{R}^p$ is a smooth mapping from the state space M into the p-dimensional output space and $y = (y_1, \ldots, y_p)$ is the output of the system. (Without any problem the (local) results that will be obtained can be extended to more general output functions $h\colon M \to N$, by interpreting (5.2) as a local description of the output map.) There is a particular class of systems having a structure as in (5.1) and (5.2) for which various design problems are relatively easy to handle, namely the class of linear systems, i.e. systems having linear dynamics:

$$\dot{\tilde{x}} = A\tilde{x} + B\tilde{u}, \tag{5.3}$$

with A an $n \times n$ matrix and B an $n \times m$ matrix and with a linear output map

$$\tilde{y} = C\tilde{x}, \tag{5.4}$$

for a $p \times n$ matrix C. We may ask ourselves the question when the nonlinear system (5.1,2) is in some way *equivalent* to a linear system (5.3,4). To make this explicit we have to specify what is meant by the equivalence of two systems. We will do first this for the notion of equivalence under *state space transformations* (see Chapter 6 for the extension to feedback transformations). Since (5.1) (and 5.2)) describe a nonlinear system in local coordinates x on $\mathbb{R}^n$ we will consider *local* state space transformations. Recall, see Chapter 2, that a coordinate transformation $z = S(x)$ from x-coordinates to z-coordinates around a point $x_0 \in \mathbb{R}^n$ is a diffeomorphism $S: V \to S(V) \subset \mathbb{R}^n$ for some neighborhood $V \subset \mathbb{R}^n$ of x_0. Consider now the dynamics (5.1) with some initial state

$$x(0) = x_0, \tag{5.5}$$

and let $S: V \to S(V)$ be a diffeomorphism on a neighborhood V of x_0. In the new coordinates

$$z = S(x), \tag{5.6}$$

we have that

$$\dot{z} = \frac{\partial S}{\partial x}(x)\dot{x}, \tag{5.7}$$

and therefore

$$\dot{z} = \frac{\partial S}{\partial x}(x)\Big(f(x) + \sum_{i=1}^{m} g_i(x)u_i\Big). \tag{5.8}$$

Because $S: V \to S(V)$ is a diffeomorphism, the mapping S has a smooth inverse $S^{-1}: S(V) \to V$ with (see (5.6))

$$x = S^{-1}(z). \tag{5.9}$$

Substituting (5.9) into (5.8) yields

$$\dot{z} = \frac{\partial S}{\partial x}\big(S^{-1}(z)\big)f\big(S^{-1}(z)\big) + \sum_{i=1}^{m} \frac{\partial S}{\partial x}\big(S^{-1}(z)\big)g_i\big(S^{-1}(z)\big)u_i, \tag{5.10}$$

or briefly, see Chapter 2 (equation (2.85))

$$\dot{z} = \big(S_*f\big)(z) + \sum_{i=1}^{m} \big(S_*g_i\big)(z)u_i, \tag{5.11a}$$

which is again a system of the form (5.1) but now described in the new coordinates z and with initial state

$$z(0) = S(x_0). \tag{5.11b}$$

Obviously S maps trajectories $x(t,0,x_0,u)$ contained in V into trajectories $z(t,0,S(x_0),u)$ in $S(V)$ of (5.10). Let us see the effect of such a local state space transformation in a simple example.

Example 5.1 Consider on $\mathbb{R}^+ \times \mathbb{R}^+$ the system

$$\begin{aligned} \dot{x}_1 &= x_1 \ln x_2 \\ \dot{x}_2 &= - x_2 \ln x_1 + x_2 u \end{aligned} \tag{5.12}$$

and let $x_0 = (1,1)^T$. Introduce the coordinate transformation $S\colon \mathbb{R}^+ \times \mathbb{R}^+ \to \mathbb{R} \times \mathbb{R}$ via

$$\begin{aligned} z_1 &= S_1(x_1,x_2) = \ln x_1, \\ z_2 &= S_2(x_1,x_2) = \ln x_2. \end{aligned} \tag{5.13}$$

Then in the new coordinates (z_1,z_2) the system (5.12) takes the linear form

$$\begin{aligned} \dot{z}_1 &= z_2, \\ \dot{z}_2 &= - z_1 + u, \end{aligned} \tag{5.14}$$

while $\big(z_1(0),z_2(0)\big) = \big(S_1(1,1),S_2(1,1)\big) = (0,0)$. Observe that in this case the transformation (5.13) is even globally defined on $\mathbb{R}^+ \times \mathbb{R}^+$. □

The above example shows the great effect a coordinate change can have on a nonlinear control system. Apparently the system (5.12) is intrinsically a linear system.

In what follows we will assume that the point x_0 is such that $f(x_0) = 0$ and $S(x_0) = 0$. Without these assumptions all results that are given still hold under the modification that in the linear dynamics and output equation constant vectors ν and η are added, i.e. $\dot{z} = Az + Bu + \nu$, $y = Cz + \eta$. We now formulate our first equivalence problem.

Problem 5.2 (Coordinate transformation into a linear system) *Consider the nonlinear system (5.1) around a point x_0 with $f(x_0) = 0$. When does there exist a coordinate transformation $z = S(x)$ with $S(x_0) = 0$, which transforms the nonlinear dynamics (5.1) into linear dynamics, i.e. (5.11a) is linear?*

The following theorem (partially) answers this question by exploiting differential geometric tools as Lie-brackets of vectorfields, involutive distributions and so on. The answer is *partial* in that it only decides when the nonlinear dynamics is equivalent to a *controllable* linear system.

For any two vectorfields f and g we will define the repeated Lie bracket $ad_f^k g$, $k = 0,1,2,\dots$, inductively as $ad_f^k g = [f, ad_f^{k-1} g]$, $k \geq 1$, with $ad_f^0 g = g$.

Theorem 5.3 *Consider the nonlinear system (5.1) around an equilibrium point x_0, i.e. $f(x_0) = 0$. There exists a coordinate transformation of (5.1) into a controllable linear system if and only if the following two conditions hold in a neighborhood $\tilde{V}$ of x_0.*

$$\text{(i)} \quad \dim(\text{span}\{ad_f^j g_i(x),\ i \in \underline{m},\ j = 0,..,n-1\}) = n, \ \forall x \in \tilde{V}, \tag{5.15}$$

$$\text{(ii)} \quad [ad_f^k g_i, ad_f^\ell g_j](x) = 0, \text{ for all } i,j \in \underline{m},\ k,\ell \geq 0,\ \forall x \in \tilde{V}. \tag{5.16}$$

Proof First suppose that (5.1) with $f(x_0) = 0$ is transformed via $z = S(x)$, $S(x_0) = 0$ into a controllable linear system

$$\dot{z} = Az + Bu = Az + \sum_{i=1}^{m} b_i u_i, \tag{5.17}$$

where $b_1, \ldots, b_m$ are the columns of the matrix B. We have to show that (5.15) and (5.16) are satisfied. Observe that for a diffeomorphism $S: V \to S(V)$ and arbitrary smooth vectorfields f_1 and f_2 on V we have the identity (see Proposition 2.30)

$$S_*[f_1, f_2](z) = [S_* f_1, S_* f_2](z). \tag{5.18}$$

In the present situation we have (see (5.11) and (5.17))

$$S_* f(z) = Az, \tag{5.19a}$$

$$S_* g_i(z) = b_i \quad , \ i \in \underline{m}. \tag{5.19b}$$

So for $i \in \underline{m}$, we have

$$S_*\big(ad_f g_i\big)(z) = S_*[f, g_i](z) = [S_* f, S_* g_i](z) = [Az, b_i] = -Ab_i, \tag{5.20}$$

and similarly for $i \in \underline{m}$ and $\ell = 0,1,\ldots$

$$\begin{aligned} S_*\big(ad_f^{\ell+1} g_i\big)(z) &= S_*[f, ad_f^\ell g_i](z) = [S_* f, S_* ad_f^\ell g_i](z) \\ &= [Az, (-1)^\ell A^\ell b_i] = (-1)^{\ell+1} A^{\ell+1} b_i. \end{aligned} \tag{5.21}$$

Now, from the fact that (5.17) is controllable we know that

$$\dim(\text{span}\{B, AB, \ldots, A^{n-1}B\}) = n, \tag{5.22}$$

and so using (5.19-21) we may conclude that

$$\dim(\text{span}\{ad_f^j g_i(x),\ i \in \underline{m},\ j = 0,\ldots,n-1\}) = n, \ \forall x \in V. \tag{5.23}$$

Note that S_* maps the distribution $\text{span}\{ad_f^j g_i(x),\ i \in \underline{m},\ j = 0,\ldots,n-1\}$

into the flat distribution span$\{B, AB, \ldots A^{n-1}B\}$. To see that (5.16) hold, observe that

$$S_*[ad_f^k g_i, ad_f^\ell g_j](z) = [S_*(ad_f^k g_i), S_*(ad_f^\ell g_j)](z)$$

$$= (-1)^{k+\ell}[A^k b_i, A^\ell b_j] = 0. \tag{5.24}$$

Because S_* is a linear isomorphism for all points $x \in V$ we see that (5.16) holds. To prove the converse we assume that (5.15) an (5.16) hold on a neighborhood $\tilde{V}$ of x_0 and we have to show the existence of a diffeomorphism $S: V \to S(V)$ on some neighborhood V of x_0 which transforms (5.1) into the controllable system (5.17). Consider the set of vectorfields

$$\{ad_f^k g_i(x),\ i \in \underline{m},\ k = 0, \ldots, n-1\}. \tag{5.25}$$

From (5.15) an (5.16) we conclude that in a neighborhood $\bar{V} \subset \tilde{V}$ of x_0 we can select n vectorfields from (5.25), say $X_1, \ldots, X_n$, which satisfy

$$\dim(\text{span}\{X_1(x), \ldots, X_n(x)\}) = n,\ \forall x \in \bar{V}, \tag{5.26a}$$

$$[X_i, X_j](x) = 0,\ i, j \in \underline{n},\ \forall x \in \bar{V}. \tag{5.26b}$$

Thus we may apply Theorem 2.36 and find a coordinate transformation S defined on a neighborhood $V \subset \bar{V} \subset \tilde{V}$ which is such that $S(x_0) = 0$ and

$$S_* X_i = \frac{\partial}{\partial z_i}, \quad i \in \underline{n}. \tag{5.27}$$

Next we compute $S_* f$ and $S_* g_i$, $i \in \underline{m}$. First observe that, using (5.16),

$$[S_* g_i, \frac{\partial}{\partial z_j}](z) = [S_* g_i, S_* X_j](z) = S_*[g_i, X_j](z) = 0,\ i \in \underline{m},\ j \in \underline{n}. \tag{5.28}$$

Therefore $S_* g_i(z) = b_i$, for some constant vector b_i, $i \in \underline{m}$. Furthermore we have

$$[[\frac{\partial}{\partial z_i}, S_* f], \frac{\partial}{\partial z_j}](z) = S_*[[X_i, f], X_j](z) = 0,\ \text{for all } i, j \in \underline{n}, \tag{5.29}$$

(use (5.16) again), and so the vectorfield $S_* f$ is linear in the z coordinates. Because $f(x_0) = 0$ we conclude that $S_* f(z) = Az$ for some $n \times n$ matrix A. It is straightforward to see that the so constructed linear system (5.17) is indeed controllable. □

Example 5.4 Consider the nonlinear system

$$\frac{d}{dt}\begin{pmatrix} x_1 \\ x_2 \\ x_3 \end{pmatrix} = \begin{pmatrix} x_2 - 2x_2 x_3 + x_3^2 \\ x_3 \\ 0 \end{pmatrix} + \begin{pmatrix} 4x_2 x_3 \\ -2x_3 \\ 1 \end{pmatrix} u. \tag{5.30}$$

A straightforward computation shows that

$$\text{span}\{g(x), ad_f g(x), ad_f^2 g(x)\} = \text{span}\left\{ \begin{pmatrix} 4x_2x_3 \\ -2x_3 \\ 1 \end{pmatrix}, \begin{pmatrix} 2x_2 \\ -1 \\ 0 \end{pmatrix}, \begin{pmatrix} 1 \\ 0 \\ 0 \end{pmatrix} \right\},$$

which is 3-dimensional for all $x \in \mathbb{R}^3$. Therefore the condition (5.15) is satisfied. It is readily verified that also (5.16) is fullfilled and so Theorem 5.3 applies to the system (5.30). In order to find the linearizing coordinate transformation we need to find a mapping $S\colon V \subset \mathbb{R}^3 \to S(V) \subset \mathbb{R}^3$, satisfying $S(0) = 0$ and the condition (5.27). An easy inspection shows that

$$S_*(x) = \begin{pmatrix} 1 & 2x_2 & 0 \\ 0 & 1 & 2x_3 \\ 0 & 0 & 1 \end{pmatrix}, \tag{5.31}$$

does the job, and thus we may take as the (globally defined) coordinate change

$$z = S(x) = \begin{pmatrix} x_1 + x_2^2 \\ x_2 + x_3^2 \\ x_3 \end{pmatrix} \tag{5.32}$$

yielding the linear system in z-coordinates

$$\begin{cases} \dot{z}_1 = z_2 \\ \dot{z}_2 = z_3 \\ \dot{z}_3 = u \end{cases} \tag{5.33}$$

□

The importance of Theorem 5.3 lies in the fact that for a nonlinear system satisfying (5.15) and (5.16) we may use — after transforming the system into a linear one — linear control design.

Remark 5.5 Notice that the condition (5.15) alone implies that the nonlinear system (5.1) is strongly accessible in a neighborhood of x_0. Of course it is not equivalent to strong accessibility because no Lie-brackets of the form $[ad_f^k g_i, ad_f^\ell g_j](x)$ are involved in condition (5.15). Furthermore, in order to check (5.15), it suffices to compute (5.15) at x_0. When the condition (5.15) is satisfied at x_0, then it is by continuity also satisfied in a neighborhood of x_0.

In order to verify (5.16) in principle one needs to compute an infinite number of Lie brackets. That this is not the case, provided (5.15) holds, can be seen after a bit more analysis of both (5.15) and (5.16). Namely, if (5.15) and (5.16) are satisfied and if $\dim(\mathrm{span}\{g_1,\ldots,g_m\}) = m$, then, after a reordering of the vectorfields $g_1,\ldots,g_m$, there exist integers $\kappa_1 \geq \kappa_2 \geq \ldots \geq \kappa_m \geq 1$ with $\sum_{i=1}^{m} \kappa_i = n$ such that the set of vectorfields

$$\{ad_f^j g_i,\ j = 0,\ldots,\kappa_i - 1,\ i \in \underline{m}\} \tag{5.34}$$

satisfies (5.15). (In fact, the vectorfields given in (5.34) form a basis for the Lie algebra generated by the vectorfields $\{ad_f^k g_i,\ i \in \underline{m},\ k \geq 0\}$, that is, each vectorfield $ad_f^k g_i$ is a linear combination of the vectorfields of (5.30).) If $(\kappa_1,\ldots,\kappa_m)$ is the smallest m-tuple (with respect to the lexicographic ordering) with the property that the set of vectorfields (5.30) is n-dimensional, then these integers are called the *controllability indices* of the corresponding linear system (5.17). Now it is straightforward to verify that (5.15) and (5.16) can be restated as follows.

Corollary 5.6 *Consider the nonlinear system (5.1) with* $f(x_0) = 0$ *and assume* $\dim(\mathrm{span}\{g_1,\ldots,g_m\}) = m$. *There exists a coordinate transformation of (5.1) into a controllable linear system if and only if there exist controllability indices* $\kappa_1 \geq \kappa_2 \geq \ldots\ldots \geq \kappa_m \geq 1$, $\sum_{i=1}^{m} \kappa_i = n$ *such that the following two conditions are satisfied*

$$\text{(i)}\quad \dim(\mathrm{span}\{ad_f^j g_i(x_0),\ j = 0,\ldots,\kappa_i - 1,\ i \in \underline{m}\}) = n, \tag{5.35}$$

$$\text{(ii)}\quad [ad_f^k g_i, ad_f^\ell g_j](x) = 0,\ k + \ell = 0,\ldots,\kappa_i + \kappa_j - 1,\ i,j \in \underline{m},\ \forall x \in \tilde{V}. \tag{5.36}$$

As an illustration we show what conditions are needed for a single-input nonlinear system, i.e. $m = 1$. In that case (5.17) is a single-input linear system which has one controllability index, namely $\kappa = n$, and so (5.35) yields

$$\dim(\mathrm{span}\{g(x_0), ad_f g(x_0),\ldots, ad_f^{n-1} g(x_0)\}) = n, \tag{5.37}$$

and (5.36) reduces to

$$[ad_f^k g, ad_f^\ell g](x) = 0 \quad , \quad k+\ell = 0,\ldots,2n-1, \quad \forall x \in \tilde{V}, \tag{5.38}$$

which by using the Jacobi-identity (see Proposition 2.27(c)) can be simplified to

$$[g, ad_f^k g](x) = 0 \quad , \quad k = 1,3,5,\ldots,2n-1, \quad \forall x \in \tilde{V}. \tag{5.39}$$

Remark 5.7 By itself the foregoing results are of a local nature. The global equivalence problem can be addressed by requiring that the state space transformation S is defined on the whole state space M. An obvious requirement for the solvability of the global problem is that M is diffeomorphic to $\mathbb{R}^n$, and that conditions (5.15) and (5.16) hold on the whole state space. However one has to impose further conditions on the vectorfields $f, g_1, \ldots, g_m$ for the solvability of the global equivalence problem (see the references at the end of this chapter).

Problem 5.3 only addresses the question whether the nonlinear dynamics (5.1) are equivalent via a state space transformation to the controllable linear dynamics (5.17), and so the output equation (5.2) is not taken into account. Of course, when we want to control the system on the basis of the outputs, we have to calculate the effect of a coordinate change on the output equation as well. Let $z = S(x)$ be some coordinate transformation around x_0, then besides the new dynamics

$$\dot{z} = \big(S_* f\big)(z) + \sum_{i=1}^{m} \big(S_* g_i\big)(z) u_i \,, \tag{5.11}$$

we obtain the new output equation

$$y = h(S^{-1}(z)). \tag{5.40}$$

The equations (5.11) and (5.40) are again a system of the form (5.1,2), now described in the new coordinates z. The obvious extension of Problem 5.3 can now be formulated.

Problem 5.8 (Coordinate transformation into a linear system with outputs) *Consider the nonlinear system (5.1,2) around a point x_0 with $f(x_0) = 0$ and $h(x_0) = 0$. When does there exist a coordinate transformation $z = S(x)$ with $S(x_0) = 0$, which transforms the nonlinear system (5.1) with outputs (5.2) into a linear system with outputs, i.e. (5.11) and (5.40) are both linear?*

Before we can solve the above problem we recall the following facts about Lie derivatives of functions. Given a smooth vectorfield X on $\mathbb{R}^n$ and a smooth function h on $\mathbb{R}^n$ the Lie derivative of h with respect to X is the function $L_X h(x) = X(h)(x) = \frac{\partial h}{\partial x}(x) \cdot X(x)$. Similarly the functions $L_X^k h$ are defined as follows. By convention we set $L_X^0 h(x) = h(x)$ and inductively for $k \geq 1$, $L_X^k h(x) = L_X(L_X^{k-1} h)(x)$. Analogously for a smooth mapping $h\colon \mathbb{R}^n \to \mathbb{R}^p$ we define $L_X^k h$ componentwise, i.e. $L_X^k h = (L_X^k h_1, \ldots, L_X^k h_p)^T$ where $h = (h_1, \ldots, h_p)^T$. Returning to the nonlinear system (5.1,2) we introduce (in the local coordinates x) the mapping

$$W^k(x) = \begin{pmatrix} h(x) \\ L_f h(x) \\ \vdots \\ L_f^{k-1} h(x) \end{pmatrix} \tag{5.41}$$

which maps $\mathbb{R}^n$ into $\mathbb{R}^{kp}$, for $k = 1,2,\ldots$. The following theorem answers Problem 5.8 in case that the obtained linear system in the z-coordinates is minimal (i.e. controllable and observable).

Theorem 5.9 *Consider the nonlinear system (5.1,2) around a point x_0 with $f(x_0) = 0$ and $h(x_0) = 0$. There exists a coordinate transformation of the system (5.1,2) into a minimal linear system if and only if the following three conditions hold on a neighborhood $\tilde{V}$ of x_0.*

(i) $\dim(\text{span}\{ad_f^j g_i(x), j = 0,\ldots,n-1,\ i \in \underline{m}\}) = n,\ \forall x \in \tilde{V}$, (5.15)

(ii) $\text{rank}\ W^{n-1}(x) = n,\quad \forall x \in \tilde{V}$, (5.42)

(iii) $L_{X_1} \ldots L_{X_k} h_j(x) = 0,\quad \forall x \in \tilde{V},\ j \in \underline{p},\ k \geq 2$ and $X_1,\ldots,X_k \in \{f,g,\ldots,g_m\}$, with at least two X_i's different from f. (5.43)

Proof For simplicity we first assume $p = 1$. First we show that the conditions of Theorem 5.3 are satisfied, so we have to prove (5.16). From (5.42) we see that the functions $h, L_f h,\ldots,L_f^{n-1}h$ form a set of local coordinates on a neighborhood of x_0. So we may define the coordinate transformation S on a neighborhood $V \subset \tilde{V}$ of x_0 as $S(x) = W^{n-1}(x)$ with $S(x_0) = W^{n-1}(x_0) = 0$ because $h(x_0) = 0$ and $f(x_0) = 0$. By using (5.43) now for fixed k,ℓ and $i,j \in \underline{m}$ we observe that

$$[ad_f^k g_i,\ ad_f^\ell g_j] L_f^s h(x) = 0,\ \forall x \in V,\ s = 0,\ldots,n-1. \tag{5.44}$$

Considering $h,\ldots,L_f^{n-1}h$ as the coordinate functions we see that (5.16) holds. Therefore we may apply Theorem 5.4 and conclude that the transformation $z = S(x) = W^{n-1}(x)$ makes the dynamics (5.1) linear. It remains to show that h is also linear with respect to the z-coordinates. As in the proof of Theorem 5.3 we may select n independent vectorfields $X_1,\ldots,X_n$ from the set $\{ad_f^j g_i,\ j = 0,\ldots,n-1,\ i \in \underline{m}\}$ which are such that $[X_i, X_j] = 0$ for $i,j \in \underline{n}$. Now h is linear in the above coordinates if

$$L_{X_i} L_{X_j} h = 0 \quad \text{for all } i,j \in \underline{n}\ . \tag{5.45}$$

But (5.45) follows immediately from (5.43) and the fact that $X_i = ad_f^k g_q$

and $X_j = ad_f^{\ell} g_r$ for certain q,r,k and ℓ. As far as necessity of the conditions (5.15), (5.42) an (5.43) concerns, we notice that it is straightforward to check that for a minimal linear system these conditions hold. Finally we note that for $p > 1$ a similar procedure for deriving (5.16) may be used, by selecting a set of local coordinates from the functions $h_1,\dots,h_p,L_f h_1,\dots,L_f h_p,\dots$. That we can do so follows of course from (5.42). □

Remark 5.10 As mentioned in Remark 5.5, the condition (5.15) implies strong accessibility of the system (5.1) in a neighborhood of x_0. Similarly (5.42) implies that the system (5.1,2) is locally observable in a neighborhood of x_0. In order that the conditions (5.15) an (5.42) hold it suffices to check that (5.15) and (5.42) are satisfied at the point x_0, compare with Remark 5.5.

At a first glance condition (5.43) involves the computation of an infinite number of functions. Provided that (5.42) holds, this is not the case. The argument to see this is completely analogous to the one given in Corollary 5.6. From (5.42) we see that after a possible reordering of the outputs $y_1,\dots,y_p$ there exists integers $\mu_1 \geq \mu_2 \geq \dots \geq \mu_p \geq 1$ with $\sum_{i=1}^{p} \mu_i = n$ such that

$$\dim(\mathrm{span}\{L_f^j dh_i(x),\ j = 0,\dots,\mu_i-1,\ i \in \underline{p}\}) = n,\ \forall x \in \tilde{V}\ , \tag{5.46}$$

or, equivalently, the n functions $\{L_f^j h_i\ ,\ j = 0,\dots,\mu_i-1,\ i \in \underline{p}\}$ form a set of local coordinates around x_0. If $(\mu_1,\dots,\mu_p)$ is the smallest p-tuple with respect to the lexicographic ordering with the property that (5.46) holds, these integers are called the *observability* indices of the corresponding linear system. One can directly verify that (5.15), (5.42) and (5.43) can be reformulated as: There exist controllability indices $\kappa_1 \geq \kappa_2 \geq \dots \geq \kappa_m \geq 1$, with $\sum_{i=1}^{m} \kappa_i = n$, and observability indices $\mu_1 \geq \mu_2 \geq \dots \geq \mu_p \geq 1$, with $\sum_{i=1}^{p} \mu_i = n$, such that

$$\dim(\mathrm{span}\{ad_f^j g_i(x_0),\ j = 0,\dots,\kappa_i-1,\ i \in \underline{m}\}) = n\ , \tag{5.35}$$

$$\dim(\mathrm{span}\{L_f^j dh_i(x_0),\ j = 0,\dots,\mu_i-1,\ i \in \underline{p}\}) = n\ , \tag{5.47}$$

$$L_{g_j} L_f^k dh_i(x) = 0,\ \text{for } k = 0,\dots,\kappa_j+\mu_j-1,\ i \in \underline{m},\ i \in \underline{p}. \tag{5.48}$$

So we have obtained the following result.

Corollary 5.11 *Consider the nonlinear system (5.1,2) around x_0 satisfying $f(x_0) = 0$ and $h(x_0) = 0$. There exists a coordinate transformation of the system (5.1,2) into a minimal linear system if and only if there exist controllability indices $\kappa_1 \geq \kappa_2 \geq \ldots \geq \kappa_m \geq 1$, $\sum_{i=1}^{m} \kappa_i = n$, and observability indices $\mu_1 \geq \mu_2 \geq \ldots \geq \mu_p \geq 1$, $\sum_{i=1}^{p} \mu_i = n$, such that (5.35), (5.47) and (5.48) are satisfied.*

It is straightforward to see what the necessary and sufficient conditions (5.35), (5.47) and (5.48) amount to in the single-input single-output case, i.e. $m = p = 1$. The resulting single-input single-output linear system has controllability index $\kappa = n$ and observability index $\mu = n$ and thus (5.35) yields

$$\dim(\text{span}\{g(x_0), ad_f g(x_0), \ldots, ad_f^{n-1} g(x_0)\}) = n \,, \tag{5.37}$$

whereas (5.47) and (5.48) may be reduced to (see also (5.38))

$$[g, ad_f^k g](x) = 0 \;,\; k = 1,3,5,\ldots,2n-1, \quad \forall x \in \tilde{V}, \tag{5.39}$$

and

$$L_g L_f^j dh(x) = 0 \;,\; j = 0,\ldots,n-1, \;\forall x \in \tilde{V}. \tag{5.49}$$

Problem 5.8 deals with the question when a nonlinear system is in essence a transformed minimal linear system, thereby allowing for linear controller *and* observer design. On the other hand it does not say anything about a nonlinear system without inputs. However coordinate transformations can also be useful in this case. Consider

$$\dot{x} = f(x), \tag{5.50a}$$

around a point x_0, together with an output equation

$$y = h(x). \tag{5.50b}$$

As before we see that a coordinate change $z = S(x)$ results in the equations

$$\dot{z} = (S_* f)(z), \tag{5.51}$$

$$y = h(S^{-1}(z)), \tag{5.40}$$

and obviously we are interested in the question when (5.51) and (5.40) are linear in z. We will call (5.50) an *autonomous system with outputs*.

Problem 5.12 (Coordinate transformation of an autonomous system with outputs into a linear system) *Consider the autonomous system with outputs (5.50) around a point x_0 with $f(x_0) = 0$ and $h(x_0) = 0$. When does there exist a coordinate transformation $z = S(x)$ with $S(x_0) = 0$ which transforms (5.50) into a linear system with outputs, i.e. (5.51) and (5.40) are linear?*

The following theorem answers Problem 5.12 in case the resulting linear system is observable.

Theorem 5.13 *Consider the autonomous system with outputs (5.50) with $f(x_0) = 0$ and $h(x_0) = 0$. There exists a coordinate transformation of (5.50) into an (autonomous) observable linear system if and only if there exist observability indices $\mu_1 \geq \mu_2 \geq \ldots \geq \mu_p \geq 1$ with $\sum_{i=1}^{p} \mu_i = n$ such that the following two conditions are satisfied on a neighborhood $\tilde{V}$ of x_0.*

$$\text{(i)}\quad \dim(\operatorname{span}\{L_f^j dh_i(x),\ j = 0,\ldots,\mu_i-1,\ i \in \underline{p}\}) = n,\ \forall x \in \tilde{V}, \tag{5.46}$$

$$\text{(ii)}\quad L_f^{\mu_k} dh_k(x) = \sum_{i=1}^{p} \sum_{j=0}^{\mu_i-1} c_{ij}^k\, L_f^j dh_i(x),$$

$$\forall x \in \tilde{V},\ k \in p, \text{ for some constants } c_{ij}^k \in \mathbb{R}. \tag{5.52}$$

Proof First suppose (5.46) and (5.52) hold. As in the proof of Theorem 5.9 we may introduce a coordinate transformation S around x_0 by setting

$$S(x) = \big(h_1(x), L_f h_1(x), \ldots, L_f^{\mu_1-1} h_1(x), \ldots,$$
$$h_p(x), L_f h_p(x), \ldots, L_f^{\mu_p-1} h_p(x)\big). \tag{5.53}$$

Clearly, because $f(x_0) = 0$ and $h(x_0) = 0$, we have $S(x_0) = 0$. Moreover it is immediate that with respect to the new coordinates $z = S(x)$ the output map (5.40) is linear, namely

$$y = \begin{pmatrix} z_1 \\ \vdots \\ z_{\mu_1+1} \\ \vdots \\ z_{\mu_1+\mu_2+1} \\ \vdots \\ \vdots \\ z_{\mu_1+\ldots+\mu_{p-1}+1} \end{pmatrix}. \tag{5.54}$$

It remains to show that the vectorfield f is linear with respect to the z-coordinates. Let us compute $S_* f$. From (5.53) we have

$$(z_1,\ldots,z_n) = \left(h_1(x),\ldots,L_f^{\mu_p-1}h_p(x)\right). \tag{5.55}$$

Therefore we have

$$\dot{z}_1 = \frac{d}{dt}\left(h_1(x)\right) = \frac{\partial h_1}{\partial x}(x)\dot{x} = \frac{\partial h_1}{\partial x}(x)f(x) = L_f h_1(x) = z_2, \tag{5.56}$$

and similarly

$$\dot{z}_i = z_{i+1}, \quad i = 1,2,\ldots,\mu_1-1. \tag{5.57}$$

Now, using (5.52) we obtain

$$\dot{z}_{\mu_1} = \frac{d}{dt}\left(L_f^{\mu_1-1}h_1(x)\right) = L_f^{\mu_1}h_1(x) = \sum_{i=1}^{p}\sum_{j=0}^{\mu_1-1} c_{ij}^1 L_f^j h_1(x) \tag{5.58}$$

and clearly the right-hand side of (5.58) is a linear combination of the coordinates $(z_1,\ldots,z_n)$. The equations (5.56-58) show the linearity of the first μ_1 components of the vector field S_*f. In a similar way as above one may proceed to show the linearity of all components of S_*f. In the new coordinates S_*f takes the form

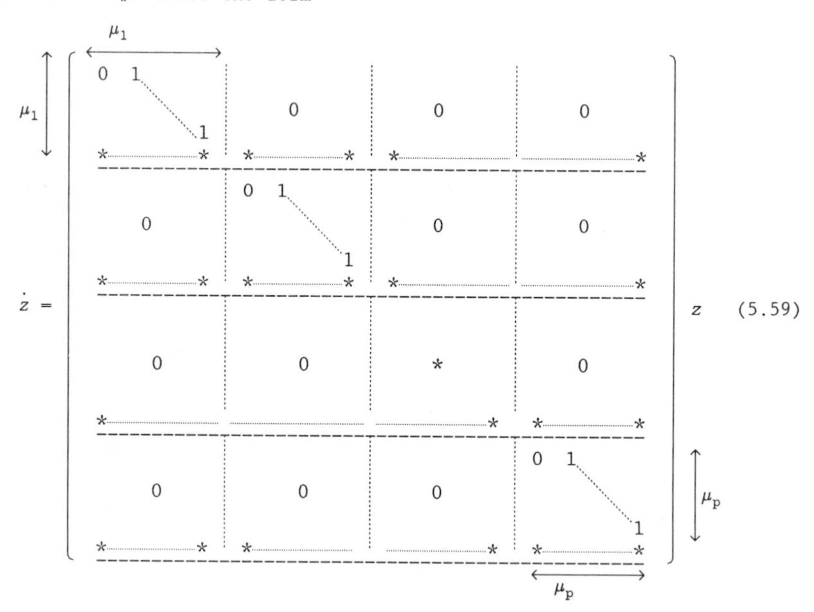

Using (5.54) and (5.59) it is immediate that the resulting linear system is observable. In fact (5.59), (5.54) form a linear system in the *observability canonical form* (without inputs). As far as the necessity of the conditions (5.46) and (5.52) concerns, we note that a linear system which is observable always can be put via a *linear* change of coordinates into the observability canonical form (5.59) and (5.54). The linear system (5.59) satisfies (5.46) and (5.52), which proves their necessity. □

It is emphasized that for the autonomous nonlinear systems (5.50) meeting the requirements of Theorem 5.13 we can use standard linear observer techniques. For instance, in order to construct a Luenberger observer, we first transform the system into the corresponding linear system (5.59,54), which for simplicity will be written as

$$\dot{z} = Az \ , \tag{5.60a}$$

$$y = Cz \ , \tag{5.60b}$$

and an observer for the state z of (5.60) is designed as the system

$$\dot{\hat{z}} = (A - KC)\hat{z} + Ky \ , \tag{5.61}$$

where K is chosen so that A-KC has all its eigenvalues in the open left half plane. In that case the error $e = z - \hat{z}$ satisfies

$$\dot{e} = (A - KC)e \ , \tag{5.62}$$

and thus $e(t)$ converges to zero when $t \to \infty$. This shows that $\hat{x} := S^{-1}(\hat{z})$ converges to the state x of the system (5.50). We notice that the above construction of an observer for the linear system (5.60) can be extended to a system which contains nonlinearities depending on the observations. Specifically consider the system

$$\dot{z} = Az + P(y) \ , \tag{5.63a}$$

$$y = Cz \ , \tag{5.63b}$$

where A and C are again as defined in (5.59) and (5.54) and $P(y) = (P_1(y),\dots,P_n(y))^T$ is some smooth vectorfield depending on y. In this case we replace the observer (5.61) by

$$\dot{\hat{z}} = (A - KC)\hat{z} + Ky + P(y) \ , \tag{5.64}$$

with again K chosen such that $A - KC$ has all its eigenvalues in the open left half plane. Then the *error* $e = z - \hat{z}$ also satisfies (5.62), which shows that $\hat{x} = S^{-1}(\hat{z})$, with $\hat{z}$ given by (5.64), yields an observer for the state x of the original nonlinear system. Motivated by this, we define

Problem 5.14 (Coordinate transformation of an autonomous system with outputs into linearizable error dynamics) *Consider the nonlinear system (5.50) around a point x_0 with $f(x_0) = 0$ and $h(x_0) = 0$. When does there*

exist a coordinate transformation $z = S(x)$ with $S(x_0) = 0$ which transforms (5.50) into the form (5.63)?

We will address here Problem 5.14 only for the single-output case, i.e. $p = 1$. The general case when $p > 1$ is in fact analogous, but needs much more analysis. Before we can state the solution of Problem 5.14 for $p = 1$ we need the following result, which reformulates the conditions of Theorem 5.13.

Proposition 5.15 *Consider the single output nonlinear system (5.50) with $f(x_0) = 0$ and $h(x_0) = 0$. There exists a coordinate transformation of (5.50) into an observable linear system if and only if the following two conditions hold on a neighborhood $\tilde{V}$ of x_0.*

(i) $$\dim(\text{span}\{dh(x), L_f dh(x), \ldots, L_f^{n-1} dh(x)\}) = n, \quad \forall x \in \tilde{V}, \tag{5.65}$$

(ii) the vectorfield g defined on $\tilde{V}$ via

$$L_g L_f^j h(x) = \begin{cases} 0, & j = 0,\ldots,n-2, \quad \forall x \in \tilde{V}, \\ 1, & j = n-1, \quad \forall x \in \tilde{V}, \end{cases} \tag{5.66}$$

satisfies

$$[g, ad_f^k g](x) = 0, \quad k = 1,3,5,\ldots,2n-1, \quad \forall x \in \tilde{V}. \tag{5.67}$$

Proof First suppose that (5.65) is satisfied and that the vectorfield g that is uniquely defined via (5.66) satisfies (5.67). Then using (5.65) we obtain that

$$\dim(\text{span}\{g(x), ad_f g(x), \ldots, ad_f^{n-1} g(x)\}) = n, \quad \forall x \in \tilde{V}, \tag{5.68}$$

and using the Jacobi-identity (see Proposition 2.27) equation (5.67) yields

$$[ad_f^k g, ad_f^\ell g](x) = 0, \quad k+\ell = 0,\ldots,2n, \quad \forall x \in \tilde{V}. \tag{5.69}$$

Applying Theorem 2.36 we can find a coordinate transformation $z = S(x)$ with $S(x_0) = 0$ such that

$$S_*\left((-1)^j ad_f^j g\right) = \frac{\partial}{\partial z_{n-j}}, \quad j = 0,\ldots,n-1. \tag{5.70}$$

It is immediate from (5.66) and (5.70) that

$$y = z_1. \tag{5.71}$$

Now we compute $S_* f$. For $j = 0,\ldots,n-2$ we have

$$[S_*f, S_*(-1)^j ad_f^j g](z) = S_*\big((-1)^j ad_f^{j+1} g\big)(z) = -\frac{\partial}{\partial z_{n-j-1}}, \tag{5.72}$$

which yields that

$$S_*f(z) = \begin{bmatrix} \alpha_1(z_1) \\ \vdots \\ \alpha_n(z_1) \end{bmatrix} + \begin{bmatrix} z_2 \\ \vdots \\ z_n \\ 0 \end{bmatrix}, \tag{5.73}$$

for smooth functions $\alpha_1, \ldots, \alpha_n$. From (5.69) it follows that

$$\frac{\partial^2 \alpha_i}{\partial z_1^2}(z_1) = 0, \quad i \in \underline{n}, \tag{5.74}$$

and so indeed S_*f is a linear vectorfield in the z coordinates. On the other hand suppose the system (5.50) is transformed via $z = S(x)$ into the linear system

$$\dot{z} = Az, \tag{5.75a}$$

$$y = Cz. \tag{5.75b}$$

Define the n-vector b by

$$CA^j b = \begin{cases} 0, & j = 0,1,\ldots,n-2, \\ 1, & j = n-1, \end{cases} \tag{5.76}$$

and let $g(x) = \big(S_*^{-1} b\big)(x)$, then it is straightforward to verify that this vectorfield satisfies (5.66) and (5.67). □

We are now able to solve Problem 5.14 when $p = 1$.

Theorem 5.16 *Consider the single output nonlinear system (5.50) with $f(x_0) = 0$ and $h(x_0) = 0$. There exists a coordinate transformation $z = S(x)$, with $S(x_0) = 0$ which transforms (5.50) into a system of the form*

$$\dot{z} = Az + P(y), \tag{5.63a}$$

$$y = Cz, \tag{5.63b}$$

with (C,A) observable if and only if the following two conditions hold on a neighborhood $\tilde{V}$ of x_0.

(i) $\dim(\mathrm{span}\{dh(x), L_f dh(x), \ldots, L_f^{n-1} dh(x)\}) = n, \quad \forall x \in \tilde{V},$ (5.65)

(ii) the vectorfield g defined as in (5.66) satisfies

$$[g, ad_f^k g](x) = 0, \quad k = 1,3,5,\ldots,2n-3, \quad \forall x \in \tilde{V}. \tag{5.77}$$

Proof Suppose that the conditions (i) and (ii) are satisfied. As in the proof of Proposition 5.15 we see that the vectorfield g defined in (5.66) helps us to define a coordinate transformation $z = S(x)$. Namely we have

$$\dim(\text{span}\{g(x), ad_f g(x), \ldots, ad_f^{n-1} g(x)\}) = n, \ \forall x \in \tilde{V}, \tag{5.68}$$

$$[ad_f^k g, ad_f^\ell g](x) = 0, \quad k + \ell = 0,1,\ldots,2n-2, \quad \forall x \in \tilde{V}, \tag{5.78}$$

and so, see Theorem 2.36, we can define the transformation $z = S(x)$, with $S(x_0) = 0$ and

$$S_*(-1)^j ad_f^j g = \frac{\partial}{\partial z_{n-j}}, \quad j = 0,\ldots,n-1. \tag{5.70}$$

It is obvious that in the new coordinates

$$y = z_1, \tag{5.71}$$

while

$$\dot{z} = \begin{bmatrix} 0 & 1 & \cdots & 0 \\ \vdots & \ddots & \ddots & \vdots \\ \vdots & & \ddots & 1 \\ 0 & \cdots & \cdots & 0 \end{bmatrix} z + \begin{bmatrix} \alpha_1(z_1) \\ \vdots \\ \alpha_n(z_1) \end{bmatrix}, \tag{5.79}$$

(compare Proposition 5.15), which is a system of the form (5.63). Conversely, when a state space transformation $z = S(x)$ exists which brings (5.50) into the form (5.63) with (C,A) observable, we have to establish (5.65) and (5.77). That (5.65) is satisfied follows from the fact that the pair (C,A) is observable and the fact that the system (5.63) is locally observable. Namely using the notation $\tilde{A}(z) = Az + P(y)$ we have

$$\text{span}\{dCz,\ dL_{\tilde{A}}Cz, \ldots, dL_{\tilde{A}}^{n-1}Cz\} = \text{span}\{C, CA, \ldots, CA^{n-1}\} = n. \tag{5.80}$$

Therefore (5.65) holds true. Define the n-vector b via

$$CA^j b = \begin{cases} 0, & j = 0,1,\ldots,n-2, \\ 1, & j = n-1, \end{cases} \tag{5.76}$$

and let g be the vectorfield defined by

$$g(x) = \left(S_*^{-1} b\right)(x). \tag{5.81}$$

Clearly using (5.76) we see that this vectorfield satisfies the requirements (5.66) and a direct computation shows that (5.77) is satisfied. □

Remark 5.17 For a nonlinear system satisfying the conditions of Theorem 5.16 we obtain a system described by the equations (5.71) and (5.79). Even in case that the functions $\alpha_1, \ldots, \alpha_n$ in (5.79) are linear in z_1 this description differs from the one given in Theorem 5.13, see equations

(5.54) and (5.59). This is the difference between the observability canonical form (5.59,54) and the observer canonical form (5.79,71), which in the linear case are isomorphic, but not necessarily in the nonlinear case.

Remark 5.18 As explained before, in the observer design for a system satisfying the conditions of Theorem 5.13 or Theorem 5.16 it is essential to introduce output injection of the form Ky or $Ky + P(y)$, see (5.61) respectively (5.64). For linear systems the concepts of state feedback and output injection are dual. Without formalizing here the nonlinear concept of state feedback and output injection, we remark that in general for nonlinear systems such a duality is not immediate.

5.2 Static and Dynamic Feedback

So far we have discussed various versions of the question when a nonlinear system is equivalent under a change of state space coordinates to an (almost) linear system. The state space transformation is only an intermediate step in the controller and observer design. As will be clear, most nonlinear systems are not equivalent via a state space transformation to a linear one and thus the forementioned techniques will not be of much help to us. In the next chapters we will discuss various other ways of changing nonlinear control systems. The cornerstone in this is the notion of *feedback*. We will discuss in this section some different types of feedback.

Definition 5.19 *A strict static state feedback for the nonlinear dynamics (5.1) is defined as a map*

$$u = \alpha(x), \tag{5.82}$$

where $u = (u_1, \ldots, u_m)^T$ *and* $\alpha: M \to \mathbb{R}^m$ *is a smooth function.*

Strict static state feedback, or for short, when no confusion arises, strict feedback, can be represented as follows:

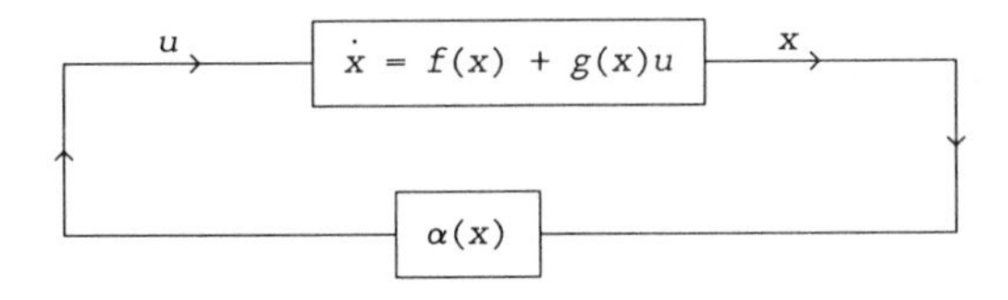

Fig. 5.1. Strict static state feedback.

So the actual state at time t, $x(t)$, is fed back via the function α, yielding the input at time t as $u(t) = \alpha(x(t))$. One of the applications of strict feedback is that of stabilizing the nonlinear system (5.1) around an equilibrium point x_0.

Example 5.20 (see Example 1.1) Consider again the dynamics of a rigid two-link robot manipulator with control torques u_1 and u_2 applied at the joints. Introducing the state space coordinates $\theta = (\theta_1, \theta_2)$, $\dot{\theta} = (\dot{\theta}_1, \dot{\theta}_2)$ we obtain the dynamics (see (1.8))

$$\frac{d}{dt}\begin{bmatrix}\theta \\ \dot{\theta}\end{bmatrix} = \begin{bmatrix}\dot{\theta} \\ -M(\theta)^{-1}C(\theta,\dot{\theta}) - M(\theta)^{-1}k(\theta) + M(\theta)^{-1}u\end{bmatrix}, \tag{5.83}$$

where $u = (u_1, u_2)$ and the matrices $M(\theta)$, $C(\theta,\dot{\theta})$ and $k(\theta)$ are as in Example 1.8. It is easily verified that each point $(\theta_0, 0) = (\theta_{10}, \theta_{20}, 0, 0)$ appears as an equilibrium point for (5.83) when setting u equal to

$$u_0 = C(\theta_0, 0) + k(\theta_0). \tag{5.84}$$

On the other hand let $\ell(\theta,\dot{\theta}) = (\ell_1(\theta,\dot{\theta}), \ell_2(\theta,\dot{\theta}))$ be an arbitrary vector linearly depending on $(\theta-\theta_0, \dot{\theta})$. Then the strict nonlinear state feedback

$$u = M(\theta)\ell(\theta,\dot{\theta}) + C(\theta,\dot{\theta}) + k(\theta), \tag{5.85}$$

yields the closed loop system

$$\frac{d}{dt}\begin{bmatrix}\theta \\ \dot{\theta}\end{bmatrix} = \begin{bmatrix}\dot{\theta} \\ \ell(\theta,\dot{\theta})\end{bmatrix}. \tag{5.86}$$

Note that (5.85) evaluated at $(\theta_0, 0)$ coincides with (5.84). Because the 2-vector $\ell(\theta,\dot{\theta})$ in (5.86) is arbitrarily linearly depending on θ an $\dot{\theta}$ we may choose $\ell(\theta,\dot{\theta})$ such that (5.86) becomes asymptotically stable. □

A second important type of state feedback is what will be called regular static state feedback.

Definition 5.21 *A regular static state feedback for the nonlinear dynamics (5.1) is defined as a relation*

$$u = \alpha(x) + \beta(x)v, \tag{5.87}$$

where $u = (u_1, \ldots, u_m)$, $\alpha: M \to \mathbb{R}^m$ *and* $\beta: M \to \mathbb{R}^{m\times m}$ *are smooth mappings with the property that the* $m\times m$ *matrix* $\beta(x)$ *is nonsingular for all* x *and* $v = (v_1, \ldots, v_m)$ *represents a new vector of control variables.*

Schematically regular static state feedback can be given as follows.

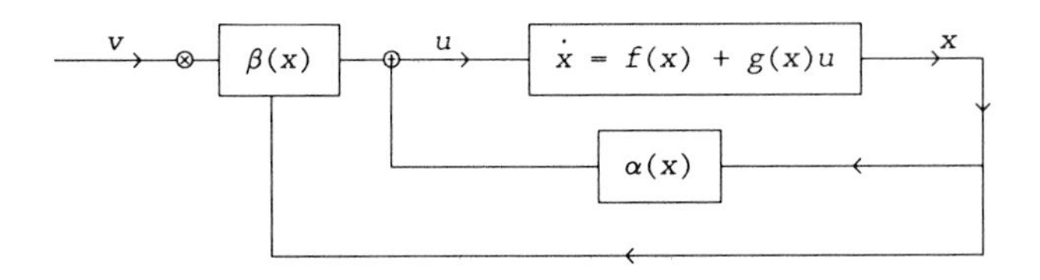

Fig. 5.2. Regular static state feedback.

Application of a feedback (5.87) to the system (5.1) yields the controlled dynamics

$$\dot{x} = f(x) + \sum_{i=1}^{m} g_i(x)\alpha_i(x) + \sum_{j=1}^{m}\left[\sum_{i=1}^{m} g_i(x)\beta_{ij}(x)\right]v_j, \tag{5.88}$$

which is again an affine control system with the newly defined inputs $(v_1,\ldots,v_m)$. We obtain the strict static state feedback of Definition 5.19 by setting $v = 0$. Note that for a linear system $\dot{x} = Ax + Bu$ strict static (linear) state feedback takes the form $u = Fx$, whereas linear regular state feedback is of the form $u = Fx + Gv$ with $|G| \neq 0$. Regular state feedback may enable us to meet certain design goals, where at the same time we keep as much control on the system as before applying the feedback. Because $\beta(x)$ is nonsingular for all x we have that

$$\text{span}\{g_1(x),\ldots,g_m(x)\} = \text{span}\{\sum_{i=1}^{m} g_i(x)\beta_{i1}(x),\ldots,\sum_{i=1}^{m} g_i(x)\beta_{im}(x)\}, \tag{5.89}$$

which shows that the input distributions of (5.1) and (5.88) are the same.

Remark 5.22 It is a straightforward exercise to show that regular static state feedback does not change the (strong) accessibility properties of a system.

Regular static state feedback, or for short, when no confusion arises, regular feedback, may be used in various design problems as will be shown in the next chapters. Here we will give a fairly simple illustration.

Example 5.23 Consider the two-dimensional single-input system

$$\begin{aligned} \dot{x}_1 &= x_2, \\ \dot{x}_2 &= f(x_1,x_2) + g(x_1,x_2)u, \end{aligned} \tag{5.90}$$

which models simple physical systems as for instance a pendulum or a cart rolling in one direction. Suppose $g(x_1,x_2) \neq 0$ for all (x_1,x_2). Then we may introduce the regular static state feedback

$$u = -\frac{f(x_1,x_2)}{g(x_1,x_2)} + \frac{1}{g(x_1,x_2)}v, \tag{5.91}$$

yielding the system

$$\begin{aligned} \dot{x}_1 &= x_2, \\ \dot{x}_2 &= v, \end{aligned} \tag{5.92}$$

which is simply a controllable linear system, that may be used for further control design. □

An important variation of strict static state feedback and regular static state feedback is when only use is made of the outputs (5.2) of the system.

Definition 5.24 *A strict static output feedback for the nonlinear system (5.1) with outputs (5.2) is defined as a relation*

$$u = \tilde{\alpha}(y), \tag{5.93}$$

where $u = (u_1,\ldots,u_m)$ *and* $y = (y_1,\ldots,y_p)$ *and* $\tilde{\alpha}\colon \mathbb{R}^p \to \mathbb{R}^m$ *is a smooth function.*

Definition 5.25 *A regular static output feedback for the system (5.1) with outputs (5.2) is defined as a relation*

$$u = \tilde{\alpha}(y) + \tilde{\beta}(y)v, \tag{5.94}$$

where $u = (u_1,\ldots,u_m)$, $y = (y_1,\ldots,y_p)$, $\tilde{\alpha}\colon \mathbb{R}^p \to \mathbb{R}^m$ *and* $\tilde{\beta}\colon \mathbb{R}^p \to \mathbb{R}^{m\times m}$ *are smooth mappings with the property that* $\tilde{\beta}(y)$ *is nonsingular for all* y *and* $v = (v_1,\ldots,v_m)$ *represents a vector of new control variables.*

Because $y = h(x)$ we observe that the output feedbacks (5.93) and (5.94) are special cases of the state feedbacks (5.82) and (5.87), and therefore in general less can be achieved.

Example 5.26 (see Example 5.23) Consider again the system

$$\begin{aligned} \dot{x}_1 &= x_2, \\ \dot{x}_2 &= f(x_1,x_2) + g(x_1,x_2)u, \end{aligned} \tag{5.90}$$

together with the output

$$y = x_1 . \tag{5.95}$$

Suppose that $g(x_1,x_2) \neq 0$ for all (x_1,x_2). In analogy with Example 5.23 we may try to apply a regular static output feedback to (5.90) which makes the overall system linear. Clearly this is possible only if the nonlinearities $f(x_1,x_2)$ and $g(x_1,x_2)$ depend on x_1 only. If so, we can apply the regular output feedback

$$u = -\frac{f(y)}{g(y)} + \frac{1}{g(y)} v, \tag{5.96}$$

again yielding the linear system (5.92). □

The last type of feedback we are going to discuss is in contrast with the previous ones, *dynamic*.

Definition 5.27 *A dynamic state feedback for the system (5.1) is defined as a relation*

$$\begin{cases} \dot{z} = \gamma(z,x) + \delta(z,x)v, \\ u = \alpha(z,x) + \beta(z,x)v, \end{cases} \tag{5.97}$$

where $z = (z_1,\ldots,z_q) \in \mathbb{R}^q$, and $\gamma\colon \mathbb{R}^q \times M \to \mathbb{R}^q$, $\delta\colon \mathbb{R}^q \times M \to \mathbb{R}^{q\times m}$, $\alpha\colon \mathbb{R}^q \times M \to \mathbb{R}^m$ *and* $\beta\colon \mathbb{R}^q \times M \to \mathbb{R}^{m\times m}$ *are smooth mappings, and* $v = (v_1,\ldots,v_m)$ *represents a new input vector.*

Dynamic state feedback can be viewed as the composition of the system (5.1) with the system

$$\begin{cases} \dot{z} = \gamma(z,\tilde{x}) + \delta(z,\tilde{x})v, \\ \tilde{u} = \alpha(z,\tilde{x}) + \beta(z,\tilde{x})v, \end{cases} \tag{5.98}$$

with the interconnections $\tilde{x} = x$ and $\tilde{u} = u$. Sometimes the system (5.98) itself is called a *compensator* and schematically we may represent dynamic state feedback as follows:

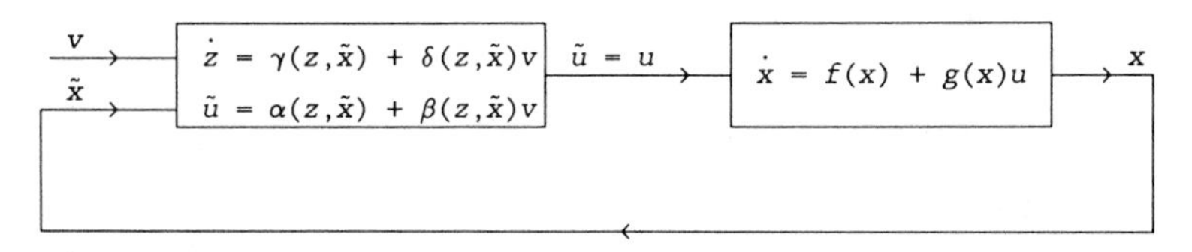

Fig. 5.3. Dynamic state feedback.

Obviously dynamic state feedback includes regular static state feedback (take $\dot{z} = 0$ in (5.98)). It also includes the idea of "adding integrators" to the system (5.1). For, consider (5.1) and suppose we set

$$\dot{u}_i = v_i \quad , \quad i \in \underline{m}. \tag{5.99}$$

with v_i, $i \in \underline{m}$, the new inputs. The same can be achieved by introducing the compensator

$$\begin{cases} \dot{z}_i = v_i, & i \in \underline{m}, \\ u_i = z_i, & i \in \underline{m}, \end{cases} \tag{5.100}$$

and adding this to the system (5.1), so as to obtain

$$\begin{cases} \dot{x}_i = f(x) + \sum_{i=1}^{m} g_i(x) z_i, \\ \dot{z}_i = v_i \qquad , \; i \in \underline{m}. \end{cases} \tag{5.101}$$

A mathematical example will illustrate the richness of dynamic state feedback.

Example 5.28 Consider on $\mathbb{R}^3$ the system

$$\begin{cases} \dot{x}_1 = u_1 \\ \dot{x}_2 = e^{x_1} u_1 + x_3 \\ \dot{x}_3 = u_2 \\ y_1 = x_1 \\ y_2 = x_2 \end{cases} \tag{5.102}$$

Because $\dot{y}_1 = u_1$ and $\dot{y}_2 = e^{x_1} u_1 + x_3$ we see that the control u_1 influences both outputs in (5.102). On the other hand, applying the dynamic state feedback

$$\begin{aligned} \dot{z} &= v_1 \; , \\ u_1 &= z \; , \\ u_2 &= - e^{x_1} z^2 - e^{x_1} v_1 + v_2 \; , \end{aligned} \tag{5.103}$$

to the system (5.102) yields:

$$\ddot{y}_1 = v_1,$$

$$\ddot{y}_2 = \frac{d}{dt}(e^{x_1} z + x_3) = e^{x_1} z^2 + e^{x_1} v_1 - e^{x_1} z^2 - e^{x_1} v_1 + v_2 = v_2,$$

from which we deduce that in the compensated system (5.102,103) the control v_1 only influences the output y_1 and the control v_2 only affects the other output y_2. □

In the following chapters it will be shown that dynamic state feedback may be used for various design purposes. In a similar way as has been done for static feedback we may introduce dynamic *output* feedback.

Definition 5.29 *A dynamic output feedback for the system (5.1,2) is defined as a relation*

$$\begin{cases} \dot{z} = \tilde{\gamma}(z,y) + \tilde{\delta}(z,y)v, \\ u = \tilde{\alpha}(z,y) + \tilde{\beta}(z,y)v, \end{cases} \tag{5.104}$$

where $z = (z_1,\ldots,z_q) \in \mathbb{R}^q$, and $\tilde{\gamma}: \mathbb{R}^q \times \mathbb{R}^p \to \mathbb{R}^q$, $\tilde{\delta}: \mathbb{R}^q \times \mathbb{R}^p \to \mathbb{R}^{q\times m}$, $\tilde{\alpha}: \mathbb{R}^q \times \mathbb{R}^p \to \mathbb{R}^m$ *and* $\tilde{\beta}: \mathbb{R}^q \times \mathbb{R}^p \to \mathbb{R}^{m\times m}$ *are smooth mappings, and* $v = (v_1,\ldots,v_m)$ *represents a new input vector.*

As in the static case, dynamic output feedbacks form a subclass of dynamic state feedbacks.

We conclude this chapter with some comments on nonlinear systems that are not necessarily affine.

Remark 5.30 Consider the nonlinear system on a manifold M of the form

$$\begin{aligned} \dot{x} &= f(x,u), \\ y &= h(x,u). \end{aligned} \tag{5.105}$$

Various questions and concepts introduced so far for affine control systems can be generalized to (5.105). Since many of the previous notions may be repeated (and are not changed) we will briefly concentrate on one of the differences. A further discussion is contained in Chapter 13. Regular static state feedback for the system (5.105) is defined as a relation of the form

$$u = \alpha(x,v) \ , \tag{5.106}$$

where $u = (u_1,\ldots,u_m)$, and $\alpha: M \times \mathbb{R}^m \to \mathbb{R}^m$ is a smooth mapping with the property that $\alpha(x,\cdot): \mathbb{R}^m \to \mathbb{R}^m$ is a diffeomorphism for all $x \in M$ and $v = (v_1,\ldots,v_m)$ denotes the new input. The difference with Definition 5.21 is that the relation (5.106) is not necessarily an x-dependent affine correspondence between u and v. The feedback (5.106) yields the feedback modified system

$$\dot{x} = f(x,\alpha(x,v)). \tag{5.107}$$

Often we will be concerned with a feedback (5.106) which is only locally defined in a neighborhood of a point (x_0, u_0). Regularity of the feedback $u = \alpha(x,v)$ is then *locally* guaranteed by the Implicit Function Theorem if the matrix $\frac{\partial\alpha}{\partial v}(x,v)$ is nonsingular. Henceforth we will refer to a mapping α, with $\frac{\partial\alpha}{\partial v}(x,v)$ nonsingular for all (x,v), as a *regular* static state feedback. In the foregoing we assumed the controls u to belong to $\mathbb{R}^m$, though one can also work with more general input manifolds U not necessarily being an Euclidean space. In that case a regular static state feedback α is defined as a mapping $\alpha : M \times U \to U$ with the requirement that $\frac{\partial\alpha}{\partial x}(x,v)$ is nonsingular for all (x,v).

Similarly one can introduce dynamic state feedback for the system (5.105) as a smooth relation of the not necessarily affine form

$$\begin{cases} \dot{z} = \gamma(z,x,v), \\ u = \alpha(z,x,v), \end{cases} \tag{5.108}$$

which differs from Definition 5.27 by the fact that the relation (5.108) is not necessarily an x-dependent affine correspondence between u and v.

Notes and References

In linear system theory the study of state space and feedback transformations has attracted a lot of attention. As a result, several relatively simple and useful canonical forms, like for instance the observability canonical form (5.59,54), have appeared, see e.g. [Ka] and references therein. In the area of nonlinear control, state space transformations were first studied by Krener [Kr1], see also [Re1],[Su] where some corrections on that paper are given. Theorem 4.4 can be found in [Kr1]. The extension including output functions (Theorem 4.13) comes from [Nij1] (see also [Nij2],[Re1],[Re2]). The study of linearizable error dynamics was initiated in [KI] and continued in [KR], see also [BZ]. The result in the single-input single-output case comes from [KI] and the multivariable case is treated in [KR], see also [XG]. The idea of transforming a nonlinear system via state feedback has been used for specific applications in [Br1],[Fr],[MC],[Po],[SR]. A general philosophy on state feedbacks may be found in [Br2],[Wi], see also [Ba].

[Ba] S.P. Banks, **Mathematical Theories of Nonlinear Systems**, Prentice Hall, Hertfordshire, 1988.

[Br1] R.W. Brockett, "Feedback invariants for nonlinear systems", Preprints 6th IFAC Congress, Helsinki, pp. 1115-1120, 1978.

[Br2] R.W. Brockett, "Global descriptions of nonlinear control problems; vector bundles and nonlinear control theory", Notes for a CBMS conference, manuscript, 1980.

[BZ] D. Bestle, M. Zeitz, "Canonical form observer design for nonlinear time-variable systems", Int. J. Control, 38, pp. 419-431, 1983.

[FK] M. Fliess, I. Kupka, "A finiteness criterion for nonlinear input-output differential systems", SIAM J. Contr. Optimiz. 21, pp. 721-728, 1983.

[Fr] E. Freund, "The structure of decoupled nonlinear systems", Int. J. Contr. 21, pp. 443-450, 1975.

[Ka] T. Kailath, **Linear Systems**, Prentice Hall, Englewood Cliffs, N.J., 1980.

[Kr] A.J. Krener, "On the equivalence of control systems and linearization of nonlinear systems", SIAM J. Contr. Optimiz. 11, pp. 670-676, 1973.

[KI] A.J. Krener, A. Isidori, "Linearization by output injection and nonlinear observers", Systems Control Lett. 3, pp. 47-52, 1983.

[KR] A.J. Krener, W. Respondek, "Nonlinear observers with linearizable error dynamics", SIAM J. Contr. Optimiz. 23, pp. 197-216, 1985.

[MC] G. Meyer, L. Cicolani, "A formal structure for advanced automatic flight-control systems", NASA Technical Note TND-7940, Ames Research Center, Moffett Field (Ca), 1975.

[Nij1] H. Nijmeijer, "State-space equivalence of an affine nonlinear system with outputs to a minimal linear system", Int. J. Contr. 39, pp. 919-922, 1984.

[Nij2] H. Nijmeijer, "Observability of a class of nonlinear systems: a geometric approach", Ricerche di Automatica 12, pp. 1-19, 1981.

[Po] W.A. Porter, "Decoupling and inverses for time-varying linear systems", IEEE Trans. Aut. Contr. 14, pp. 378-380, 1969.

[Re1] W. Respondek, "Geometric methods in linearization of control systems", in **Mathematical Control Theory** (eds. C. Olech, B. Jakubczyk, J. Zabczyk), Banach Center Publications, 14, pp. 453-467, 1985.

[Re2] W. Respondek, "Linearization, feedback and Lie brackets", in **Geometric theory of nonlinear control systems**, (eds. B. Jakubczyk, W. Respondek, K. Tchoń), Technical University of Wroclaw, Poland, pp. 131-166, 1985.

[SR] S.N. Singh, W.J. Rugh, "Decoupling in a class of nonlinear systems by state variable feedback", J. Dynamic Systems, Measurement and Control, pp. 323-329, 1972.

[Su] H.J. Sussmann, "Lie brackets, real analyticity and geometric control", in **Differential geometric control theory** (eds. R.W. Brockett, R.S. Millman, H.J. Sussmann), Birkhäuser, Boston, pp. 1-116, 1983.

[Wi] J.C. Willems, "System theoretic models for the analysis of physical systems", Ricerche di Automatica 10, pp. 71-106, 1979.

[XG] Xiao-Hua Xia, Wei-Bin Gao, "Nonlinear observer design by observer canonical forms", Int. J. Contr. 47, pp. 1081-1100, 1988.

Exercises

5.1 Prove Corollary 5.6.

5.2 Prove Remark 5.22.

5.3 Consider Example 1.1. Construct a regular static state feedback such that the closed loop system is linear.

5.4 Show that Problem 5.14 is solvable for the system

$$\dot{x}_1 = -x_1 - \frac{1}{4}x_2^2 - \frac{1}{4}x_1^2 + \frac{1}{2}x_1x_2,$$

$$\dot{x}_2 = x_1 + \frac{3}{4}x_2^2 + \frac{3}{4}x_1^2 - \frac{3}{2}x_1x_2,$$

$$y = \frac{1}{2}x_2 - \frac{1}{2}x_1.$$

5.5 Consider on $\mathbb{R}^n$ the smooth dynamics $\dot{x} = f(x)$ with $f(0) = 0$ and the smooth output function $y = h(x)$, where $h : \mathbb{R}^n \to \mathbb{R}$ and $h(0) = 0$. Suppose the system satisfies the observability rank condition about $x = 0$. Define a coordinate transformation $z = S(x)$ around $x = 0$ via $S(x) = (h(x), L_fh(x), \ldots, L_f^{n-1}h(x))$.

(a) Determine the system in the new coordinates $z = S(x)$, i.e. compute $\tilde{f}(z) = (S_*f)(z)$ and $\tilde{h}(z) = h(S^{-1}(z))$.

(b) Assume $n = 2$, and consider the system in the above z-coordinates. Determine under which conditions *on* $\tilde{f}$ *and* $\tilde{h}$ *only* the system can be transformed via a new transformation $\tilde{z} = \tilde{S}(z)$ into the form $\dot{\tilde{z}} = A\tilde{z} + P(y)$, $y = C\tilde{z}$.

(c) Express the conditions found in (b) in conditions on the original dynamics and output function.

(d) Repeat part (b) in case $n = 3$. Have you any idea what the conditions on $\tilde{f}$ and $\tilde{h}$ are for arbitrary n?

5.6 Consider on $\mathbb{R}^n$ a single input single output nonlinear system $\dot{x} = f(x) + g(x)u$, $y = h(x)$. Let $f(0) = 0$ and $h(0) = 0$ and assume that the map $S : \mathbb{R}^n \to \mathbb{R}^n$ defined as $S(x) = (h(x), L_fh(x), \ldots, L_f^{n-1}h(x))$ has rank n around $x = 0$. Suppose that $L_gL_f^kh(x) = 0$ for $k = 0,1,\ldots,n-2$.

(a) Determine the system with respect to the new local coordinates $z = S(x)$.

(b) Suppose additionally that $L_gL_f^{n-1}h(0) \neq 0$. Prove that there exists a regular static state feedback $u = \alpha(z) + \beta(z)v$ defined around $x = 0$ such that the closed loop system is linear. (See also Chapter 6 about feedback linearizability.)

5.7 Consider on $\mathbb{R}$ the system $\Sigma : \dot{x} = u$, $y = e^x + e^{2x}$.

(a) Show that the observation space (cf. Section 3.2) of the system Σ is given as $\{e^x + e^{2x}, e^x + 2e^{2x}\}$.

(b) Define $(z_1, z_2) = S(x) = (e^x + e^{2x}, e^x + 2e^{2x})$. Show that $\tilde{\Sigma} : \dot{z}_1 = z_2u$, $\dot{z}_2 = (3z_2 - 2z_1)u$, $y = z_1$, and conclude that the input-output trajectories of Σ are input-output trajectories of the bilinear system $\tilde{\Sigma}$.

5.8 ([FK]) Let $\Sigma : \dot{x} = f(x) + g(x)u$, $y = h(x)$ be a single input single-output nonlinear system on $\mathbb{R}^n$, and $\tilde{\Sigma} : \dot{\tilde{x}} = A\tilde{x} + (B\tilde{x})u + bu$, $\tilde{y} = C\tilde{x}$ a single-input single-output bilinear system on $\mathbb{R}^p$. The system Σ can be *immersed* into the bilinear system $\tilde{\Sigma}$ if there exists a mapping $S : \mathbb{R}^n \to \mathbb{R}^p$ such that each input-output trajectory $y(t,0,x_0,u)$ of Σ coincides with the input-output trajectory $\tilde{y}(t,0,S(x_0),u)$ of the bilinear system $\tilde{\Sigma}$. Prove that a necessary and sufficient condition that Σ can be immersed into a bilinear system $\tilde{\Sigma}$ is that the observation space C of Σ is finite dimensional (see also Exercise 5.7).

5.9 Consider the nonlinear system $\Sigma : \dot{x} = f(x) + \sum_{i=1}^{m} g_i(x)u_i$. Introduce the compensator $\dot{z}_i = v_i$, $u_i = z_i$, $i \in \underline{m}$, for the system Σ. Show that the system Σ satisfies the strong accessibility rank condition if and only if the precompensated system satisfies the strong accessibility rank condition.

5.10 Consider the nonlinear system with outputs $\Sigma : \dot{x} = f(x) + \sum_{i=1}^{m} g_i(x)u_i$, $y = h(x)$. Introduce the compensator $\dot{z}_i = v_i$, $u_i = z_i$, $i \in \underline{m}$. Show that Σ satisfies the observability rank condition if and only if the precompensated system satisfies the observability rank condition.

6
Feedback Linearization of Nonlinear Systems

In the previous chapter we have seen that by applying state space transformations $z = S(x)$ a nonlinear system given in local coordinates $x = (x_1, \ldots, x_n)$ as

$$\dot{x} = f(x) + \sum_{j=1}^{m} g_j(x) u_j, \qquad u = (u_1, \ldots, u_m) \in U \subset \mathbb{R}^m, \tag{6.1a}$$

$$y_i = h_i(x), \qquad i \in \underline{p}, \tag{6.1b}$$

transforms into a system which may look rather different from the system in the original coordinates. In fact, in Theorem 5.3 and Corollary 5.6 we have given necessary and sufficient conditions under which a locally strongly accessible nonlinear system (6.1a), which may be highly nonlinear in the original coordinates, can be transformed into a *linear* system

$$\dot{z} = Az + Bu \tag{6.2a}$$

where $(S_* f)(z) = Az$ and $(S_* g_j)(z) = b_j$, the j-th column of the matrix B. Furthermore in Theorem 5.9 conditions have been given which ensure that also the output equations are transformed into linear equations

$$y = Cz \tag{6.2b}$$

with $h_i\big(S^{-1}(z)\big) = c_i z$ and c_i the i-th row of the matrix C. Although the conditions as given in these theorems in principle only ensure the *local* existence of such a linearizing coordinate transformation $z = S(x)$ still these theorems are potentially very useful for control purposes since in this way the control of a seemingly highly nonlinear control system may be reduced to the control of a linear control system, for which we have many tools at our disposal.

Intuitively it is clear that the set of nonlinear systems (6.1) which can be transformed (locally) into a linear system forms a very thin subset of the set of all nonlinear systems. On the other hand in the preceding chapter we have also seen that application of a nonlinear feedback $u = \alpha(x) + \beta(x)v$ to (6.1a) results in the system (see(5.88))

$$\dot{x} = f(x) + \sum_{j=1}^{m} g_j(x)\alpha_j(x) + \sum_{j=1}^{m} \Big(\sum_{i=1}^{m} g_i(x)\beta_{ij}(x) \Big) v_j, \tag{6.3}$$

which may be quite different from the original system (6.1a). Therefore, a logical next step is to investigate when the dynamics (6.1a) can be transformed by a (local) state space transformation $z = S(x)$ *and* a regular static state feedback $u = \alpha(x) + \beta(x)v$ into a *linear* system (6.2a) where u is replaced by the new input vector v. This is called the *feedback linearization problem*. If the controlled dynamics (6.2a) *is* feedback linearizable in the above sense then in principle the control of the system can be split into two parts: a nonlinear feedback loop which renders the system (in suitable local coordinates) linear, and a super imposed linear control strategy for the obtained linear system. We have added the words "in principle", because the nonlinear feedback, required to make the system linear, may be very complicated, and very sensitive to parameter uncertainties. Furthermore, if the original controls $u_1,\ldots,u_m$ are bounded, say if $U = \{(u_1,\ldots,u_m) \mid |u_i| \le 1\}$, then the bounds for the new inputs $v_1,\ldots,v_m$ may become complicated functions of the state. Nevertheless, it is clear that feedback linearizability is a very enjoyable property for a nonlinear control system.

The feedback linearization problem is also important from a mathematical and system theoretic point of view, since it forms a first step in the *classification* of nonlinear systems (under state space and feedback transformations) by singling out the essentially linear systems. Finally we note that the problem of linearization of (6.1a) by state space transformations, or by state space transformations *and* feedback transformations, can be seen as the generalization of the problem of linearizing a *single* vectorfield

$$\dot{x} = f(x), \tag{6.4}$$

by means of coordinate transformations $z = S(x)$ in a neighborhood of an equilibrium point of f. This problem has a long history in mathematics, starting at least with the work of Poincaré, and is known to be hard in general. Surprisingly, we shall see that in a sense the addition of inputs makes the problem much easier, since by assuming local strong accessibility we add a lot of extra structure to the system.

This chapter is organized as follows. In the first section we shall give the geometric conditions for feedback linearizability of (6.1a). We will also show how these conditions nicely generalize to the solution of the problem of feedback linearizability of *general* controlled dynamics $\dot{x} = f(x,u)$, using feedback $u = \alpha(x,v)$. In the second section we will approach the problem from a more computational point of view. We will show that the geometric conditions as deduced in the first section contain a lot of redundancy, and moreover that often there is a simple recipe to construct the required state space and feedback transformation. This will become especially clear in the single-input case.

6.1 Geometric Conditions for Feedback Linearization

Consider a nonlinear affine system (without outputs)

$$\dot{x} = f(x) + \sum_{j=1}^{m} g_j(x)u_j \qquad (u_1,\ldots,u_m) \in U \subset \mathbb{R}^m, \tag{6.5}$$

where $x = (x_1,\ldots,x_n)$ are local coordinates for a smooth manifold M and $f, g_1,\ldots,g_m$ are smooth vector fields on M. For simplicity we first *throughout* take $U = \mathbb{R}^m$.

Definition 6.1 *Let* x_0 *be an equilibrium point for* f, *i.e.* $f(x_0) = 0$. *The system (6.5) is* feedback linearizable *around* x_0 *if there exists*

(i) *a coordinate transformation* $S\colon O \subset \mathbb{R}^n \to \mathbb{R}^n$ *defined on a neighborhood* O *of* x_0, *with* $S(x_0) = 0$,

(ii) *a regular static state feedback* $u = \alpha(x) + \beta(x)v$, *with* $\alpha(x)$ *an* $m\times 1$ *vector satisfying* $\alpha(x_0) = 0$ *and* $\beta(x)$ *an* $m\times m$ *invertible matrix, both defined on* O,

such that the feedback transformed system

$$\dot{x} = \tilde{f}(x) + \sum_{j=1}^{m} \tilde{g}_j(x)v_j, \tag{6.6}$$

with (see (5.88),(6.3))

$$\begin{aligned} \tilde{f}(x) &= f(x) + \sum_{i=1}^{m} g_i(x)\alpha_i(x), \\ \tilde{g}_j(x) &= \sum_{i=1}^{m} g_i(x)\beta_{ij}(x), \qquad j \in \underline{m}, \end{aligned} \tag{6.7}$$

transforms under $z = S(x)$ *into the linear system*

$$\dot{z} = Az + Bv \tag{6.8}$$

with $B = (b_1,\ldots,b_m)$ *and*

$$Az = (S_*\tilde{f})(z), \quad b_j = (S_*\tilde{g}_j)(z), \qquad j \in \underline{m}. \tag{6.9}$$

Remark 6.2 The above notion of linearizability should not be confused with the classical notion of linearizing the system (6.6) around the equilibrium point x_0 and $v = 0$. In this last case one writes (6.6) in a Taylor expansion

$$\dot{x} = \frac{\partial \tilde{f}}{\partial x}(x_0)(x-x_0) + \sum_{j=1}^{m} \tilde{g}_j(x_0)v_j + \text{higher-order terms in } x, \tag{6.10}$$

and the linearized system is defined by omitting all the higher order terms, yielding the linear system

$$\dot{\bar{x}} = \frac{\partial \tilde{f}}{\partial x}(x_0)\bar{x} + \sum_{j=1}^{m} \tilde{g}_j(x_0)v_j . \tag{6.11}$$

The relation of both notions of linearization is as follows. Assume (6.9) holds, then the matrices A and B in (6.9) are related to the matrices in (6.11) by

$$A = P\,\frac{\partial \tilde{f}}{\partial x}(x_0)\,P^{-1}, \qquad b_j = Pg_j(x_0), \qquad j \in \underline{m}, \tag{6.12}$$

where $P = \frac{\partial S}{\partial x}(x_0)$. Thus we conclude that linearization in the classical sense amounts to *neglecting* the higher-order terms in (6.10), while in feedback linearization the higher-order terms of (6.10) are *eliminated* by state space transformations (see also Exercise 6.1).

With every nonlinear system (6.5) we associate a nested sequence of distributions $D_1 \subset D_2 \subset D_3 \subset \dots$ as follows. Denote $ad_f^0 g_j = g_j$, and inductively $ad_f^k g_j = [f, ad_f^{k-1} g_j]$, $k = 1,2,..$, with $[f,g]$ the Lie bracket of vectorfields f and g. Then define

$$D_k(x) = \text{span}\{ad_f^r\, g_1(x),..,ad_f^r g_m(x) \mid r = 0,1,..,k-1\},\ k = 1,2,.. \tag{6.13}$$

The main theorem of this section reads as follows

Theorem 6.3 *Consider the nonlinear system (6.5) with $f(x_0) = 0$. Assume that the strong accessibility rank condition in x_0 (cf. (3.41)) is satisfied. Then the system is feedback linearizable around x_0 if and only if the distributions $D_1,..,D_n$ defined in (6.13) are all involutive and constant dimensional in a neighborhood of x_0. Moreover the resulting linear system (6.9) is controllable.*

First we prove the "only if" direction.

Proof of Theorem 6.3 ("only if") Suppose that the system is feedback linearizable around x_0, i.e. (6.9) holds for some regular feedback $u = \alpha(x) + \beta(x)v$ and local coordinate transformation $z = S(x)$ around x_0.

For the linear system (6.8) the distributions $D_1,..,D_n$ as in (6.13) are the following flat distributions (superscripts A,B indicate the dependence on the vectorfields Az and columns of B), defined for z around 0:

$$D_k^{A,B}(z) = \mathrm{Im}(B \vdots AB \vdots \cdots \vdots A^{k-1}B) \subset \mathbb{R}^n \cong T_z\mathbb{R}^n, \qquad k \in \underline{n}. \tag{6.14}$$

Clearly the distributions $D_1^{A,B},...,D_n^{A,B}$ are all involutive and of constant dimension. For the feedback transformed system (6.6) the corresponding distributions are given as (x in a neighborhood of x_0):

$$D_k^{\tilde{f},\tilde{g}}(x) = (S_{*x})^{-1} D_k^{A,B}(S(x)) \tag{6.15}$$

where $(S_{*x})^{-1} D_k^{A,B}(S(x)) = \{X(x) \mid S_{*x}X(x) \in D_k^{A,B}(S(x))\}$.

Also $D_1^{\tilde{f},\tilde{g}},..,D_n^{\tilde{f},\tilde{g}}$ are constant dimensional, and by Proposition 2.30 involutive because of the involutivity of $D_1^{A,B},..,D_n^{A,B}$. We will now show that the distributions $D_k = D_k^{f,g}$ defined in (6.13) for system (6.5) equal $D_k^{\tilde{f},\tilde{g}}$, $k \in \underline{n}$; in other words they are *feedback invariant*. First, for $k = 1$ we have

$$D_1^{\tilde{f},\tilde{g}}(x) = \mathrm{span}\,\{\tilde{g}_1(x),..,\tilde{g}_m(x)\} = \mathrm{span}\{g_1(x),..,g_m(x)\} = D_1(x), \tag{6.16}$$

since the matrix $\beta(x)$ is invertible. Now assume that for some k we have $D_k^{\tilde{f},\tilde{g}}(x) = D_k(x)$ for x near x_0, then we will prove that $D_{k+1}^{\tilde{f},\tilde{g}}(x) = D_{k+1}(x)$. Indeed take $ad_{\tilde{f}}^k \tilde{g}_j \in D_{k+1}^{\tilde{f},\tilde{g}}$, then

$$\begin{aligned} ad_{\tilde{f}}^k \tilde{g}_j &= [f + \sum_{i=1}^m g_i\alpha_i,\ ad_{\tilde{f}}^{k-1} \tilde{g}_j] = \\ &= [f,\ ad_{\tilde{f}}^{k-1} \tilde{g}_j] + \sum_{i=1}^m \alpha_i\ [g_i,\ ad_{\tilde{f}}^{k-1} \tilde{g}_j] - \sum_{i=1}^m ad_{\tilde{f}}^{k-1}\tilde{g}_j(\alpha_i)g_i. \end{aligned} \tag{6.17}$$

Now $ad_{\tilde{f}}^{k-1}\tilde{g}_j \in D_k^{\tilde{f},\tilde{g}} = D_k$, and thus $[f,\ ad_{\tilde{f}}^{k-1}\tilde{g}_j] \in D_{k+1}$ by definition of D_{k+1}. Furthermore since $D_k = D_k^{\tilde{f},\tilde{g}}$ is involutive and $g_i \in D_1 \subset D_k$ it follows that $[g_i, ad_{\tilde{f}}^{k-1}\tilde{g}_j] \in D_k$ and $ad_{\tilde{f}}^{k-1}\tilde{g}_j(\alpha_i)g_i \in D_k$. Therefore by (6.17) $ad_{\tilde{f}}^k \tilde{g}_j \in D_{k+1}$ and thus $D_{k+1}^{\tilde{f},\tilde{g}}(x) \subset D_{k+1}(x)$. For the converse inclusion we observe that

$$\begin{aligned} ad_f^k g_j &= [\tilde{f} - \sum_{i=1}^m g_i\alpha_i,\ ad_f^{k-1} g_j] = \\ &= [\tilde{f}, ad_f^{k-1}g_j] - \sum_{i=1}^m \alpha_i [g_i, ad_f^{k-1}g_j] + \sum_{i=1}^m ad_f^{k-1}g_j(\alpha_i)g_i \in D_{k+1}^{\tilde{f},\tilde{g}}. \end{aligned} \tag{6.18}$$

□

Remark The above proof shows that in general the distributions $D_1, D_2, \ldots$ are *not* feedback invariant, i.e. in general we do not have $D_k^{f,g} \neq D_k^{\tilde{f},\tilde{g}}$, $k = 1,2,\ldots$. In fact the feedback invariance of $D_1, D_2, \ldots$ in the above proof follows from the assumption of involutiveness. Therefore for systems which are *not* feedback linearizable these distributions are not the right tools for studying equivalence of the system under coordinate and feedback transformations. Instead we may consider the following distributions

$$
\begin{aligned}
\Delta_1(x) &= \text{span}\,\{g_1(x),\ldots,g_m(x)\} \quad (= D_1(x)) \\
\Delta_k(x) &= [f,\Delta_{k-1}](x) + \sum_{j=1}^{m} [g_j,\Delta_{k-1}](x), \qquad k = 2,3,\ldots
\end{aligned}
\tag{6.19}
$$

which *are* always feedback invariant (see Exercise 6.3). □

The key mathematical tool in the proof of the "if" part of Theorem 6.3 is the following generalization of Frobenius' Theorem (Theorem 2.42) (compare also Proposition 3.50):

Lemma 6.4 *Let $D_1 \subset D_2 \subset D_3 \subset \ldots \subset D_N$ be a sequence of involutive and constant dimensional distributions on M, with dimensions $m_1 \leq m_2 \leq \ldots m_N$. Then around any point $x_0 \in M$ there exist local coordinates*

$$
x = (x_1,\ldots,x_{m_1},x_{m_1+1},\ldots,x_{m_2},x_{m_2+1},\ldots,x_{m_N},x_{m_N+1},\ldots,x_n) =: (x^1,x^2,\ldots,x^N,\bar{x})
\tag{6.20}
$$

such that

$$
D_k = \text{span}\{\frac{\partial}{\partial x^1},\ldots,\frac{\partial}{\partial x^k}\}, \qquad k = 1,2,\ldots,N.
\tag{6.21}
$$

Proof First consider the distribution D_N. By Frobenius' Theorem (Theorem 2.42) there exists a partial set of coordinate functions $\bar{x} = (x_{m_N+1},\ldots,x_n)$ on a neighborhood U of x_0, such that the integral manifolds of D_N are given as

$$
\{q \in U \mid \bar{x}(q) = \bar{c}\},
\tag{6.22}
$$

for constant vectors $\bar{c} \in \mathbb{R}^{n-m_N}$. Next consider the distribution D_{N-1}. By Frobenius' Theorem there exist partial coordinate functions $y^{N-1} = (y_{m_{N-1}},\ldots,y_n)$ on a neighborhood of x_0 such that the integral manifolds of D_{N-1} are locally given as

$$
\{q \text{ near } x_0 \mid y^{N-1}(q) = d^{N-1}\},
\tag{6.23}
$$

for constant vectors $d^{N-1} \in \mathbb{R}^{n-m_{N-1}}$. Since $D_{N-1} \subset D_N$ it follows that every integral manifold (6.23) close to x_0 is contained in some integral manifold (6.22), i.e. the functions $\bar{x}$ are constant on the integral manifolds (6.23) of D_{N-1}. (Recall that $L_X \bar{x} = 0$ for any $X \in D_N$, and a fortiori for $X \in D_{N-1} \subset D_N$). Define $\rho_N := m_N - m_{N-1}$, then it follows that we can permute the coordinate functions y^{N-1} as

$$y^{N-1} = (x^N, \bar{y}), \tag{6.24}$$

with $\dim x^{N-1} = \rho_N$ and $\dim \bar{y} = n - m_N$ such that the functions x^N and $\bar{x}$ are independent. Then the integral manifolds (6.23) are also given as

$$\{q \text{ near } x_0 \,|\, x^N(q) = c^N,\ \bar{x}(q) = \bar{c}\}, \tag{6.25}$$

for constant vectors $c^N \in \mathbb{R}^N$, $\bar{c} \in \mathbb{R}^{n-m_N}$. Next consider D_{N-2}. By Frobenius' Theorem there again exist partial coordinate functions y^{N-2} such that the integral manifolds of D_{N-2} are given as the sets where all these functions are constant. However since $D_{N-2} \subset D_{N-1}$, all the functions $(x^N, \bar{x})$ are also constant on these integral manifolds. It follows that we can permute y^{N-2} as $y^{N-2} = (x^{N-1}, \bar{y}^{N-1})$ with $\dim(x^{N-1}) := \rho_{N-1} = m_{N-1} - m_{N-2}$ such that the integral manifolds of D_{N-2} are also given as

$$\{q \text{ near } x_0 \,|\, x^{N-1}(q) = c^{N-1},\ x^N(q) = c^N,\ \bar{x}(q) = \bar{c}\}. \tag{6.26}$$

Continuing in this way, we obtain local coordinates $x = (x^1, x^2, \ldots, x^N, \bar{x})$ such that for every $k = 1,..,N$ the integral manifolds of D_k are given as

$$\{q \text{ near } x_0 \,|\, x^{k+1}(q) = c^{k+1}, \ldots, x^N(q) = c^N, x(q) = \bar{c}\}, \tag{6.27}$$

for constant vectors $c^{k+1} \in \mathbb{R}^{\rho_{k+1}}, \ldots, c^N \in \mathbb{R}^{\rho_N}$, $\bar{c} \in \mathbb{R}^{n-m_N}$ where $\rho_k := m_k - m_{k-1}$, $k = N, N-1, \ldots$, with $m_0 = 0$. By Corollary 2.43 this is equivalent to (6.21). □

Remark 6.5 Notice that the description of the integral manifolds of D_k in (6.27), and therefore of the distributions D_k in (6.21), $k \in \underline{N}$, is invariant under any coordinate transformation $z = S(x)$, $z = (z^1, \ldots, z^N, \bar{z})$, of the "triangular" form

$$\begin{aligned}
\bar{z} &= \bar{S}(\bar{x}) \\
z^N &= S^N(x^N, \bar{x}) \\
z^{N-1} &= S^{N-1}(x^{N-1}, x^N, \bar{x}) \\
z^k &= S^{k-1}(x^{k-1}, \ldots, x^N, \bar{x}), \qquad k = N-2, \ldots, 2 \\
z^1 &= S^1(x^1, \ldots, x^N, \bar{x})
\end{aligned} \tag{6.28}$$

Proof of Theorem 6.3 "if" Consider the distributions $D_1 \subset D_2 \subset D_3 \ \ldots$ as defined in (6.13). Since they all have constant dimension and $\dim D_k \leq n$ for all k, it follows that there exists a least integer κ such that

$$D_\kappa = D_{\kappa+1}. \tag{6.29}$$

Clearly, $\kappa \leq n$. We claim that D_κ is *invariant* under the system dynamics (6.5), see Definition 3.44. Indeed D_κ is invariant under f by (6.29) and invariant under g_j, $j \in \underline{m}$, since $g_j \in D_1 \subset D_\kappa$, $j \in \underline{m}$, and D_κ is involutive. Since $g_j \in D_\kappa$, $j \in \underline{m}$, it follows by Proposition 3.47 that $D_\kappa = C_0$ and since (6.5) is strongly accessible it thus follows that $\dim D_\kappa = n = \dim M$. Thus by Lemma 6.4 there exist local coordinates $x = (x^1,..,x^\kappa)$ such that $x(x^0) = 0$, and

$$D_k = \operatorname{span}\{\frac{\partial}{\partial x^1},\ldots,\frac{\partial}{\partial x^k}\}, \qquad k = 1,\ldots,\kappa, \tag{6.30}$$

where

$$\dim(x^k) = \rho_k = m_k - m_{k-1}, \quad k = 1,\ldots,\kappa, \tag{6.31}$$

with $m_k = \dim D_k$, and $m_0 := 0$. By definition of D_k it follows that

$$[f, D_k] \subset D_{k+1}, \qquad k = 1,\ldots,\kappa-1. \tag{6.32}$$

Using (6.30) this yields for any $k = 1,\ldots,\kappa-1$

$$[f,\frac{\partial}{\partial x^j}](x) = -\frac{\partial f}{\partial x^j}(x) \in \operatorname{span}\{\frac{\partial}{\partial x^1},\ldots,\frac{\partial}{\partial x^{k+1}}\} \qquad j \in \underline{k}. \tag{6.33}$$

Writing f corresponding to $(x^1,..,x^\kappa)$ as $f = (f^1,..,f^\kappa)$ this yields

$$\frac{\partial f^i}{\partial x^j}(x) = 0, \qquad j = 1,2,\ldots,i-2, \qquad i = 3,\ldots,\kappa. \tag{6.34}$$

Thus the vectorfield f in such coordinates is of the form

$$f = \begin{pmatrix} f^1\ (x^1,\ldots,x^\kappa) \\ f^2\ (x^1,\ldots,x^\kappa) \\ f^3\ (x^2,\ldots,x^\kappa) \\ \vdots \\ f^i\,(x^{i-1},\ldots,x^\kappa) \\ \vdots \\ f^\kappa(\quad x^{\kappa-1},x^\kappa) \end{pmatrix} \tag{6.35}$$

Moreover, since $g_j \in D_1$, $j = 1,..,m$, the vectorfields g_j, $j \in \underline{m}$, are of the form

$$g_j = \begin{bmatrix} g_j^1(x^1,\dots,x^\kappa) \\ 0 \\ \vdots \\ 0 \end{bmatrix}. \tag{6.36}$$

Furthermore, by definition of D_k

$$D_{k+1}(x) = D_k(x) + \operatorname{span}\{[f,X](x) \mid X \in D_k\}. \tag{6.37}$$

Togehter with (6.34) this implies that

$$\operatorname{rank} \frac{\partial f^{k+1}}{\partial x^k}(x) = \rho_{k+1} = \dim(x^{k+1}) = \dim D_{k+1} - \dim D_k . \tag{6.38}$$

Hence for any $i = 2,\dots,\kappa$

$$\operatorname{rank} \frac{\partial f^i}{\partial x^{i-1}}(x^{i-1},x^i,\dots,x^\kappa) = \rho_i , \qquad x \text{ near } x_0 . \tag{6.39}$$

Since $\dim(x^{i-1}) = \rho_{i-1}$ it immediately follows that $\rho_1 \geq \rho_2 \geq \dots \geq \rho_\kappa > 0$.

Now we are going to define a coordinate transformation $z = S(x)$ with $S(0) = 0$ such that in the z-coordinates the system can be made linear by feedback. Furthermore this coordinate transformation will be of the form (6.28), i.e. (recall that $\dim D_\kappa = \dim M$)

$$\begin{aligned} z^\kappa &= S^\kappa(x^\kappa) \\ z^{\kappa-1} &= S^{\kappa-1}(x^{\kappa-1},x^\kappa) \\ &\ \vdots \\ z^i &= S^i(x^i,\dots,x^\kappa) \\ &\ \vdots \\ z^1 &= S^1(x^1,\dots,x^\kappa) \end{aligned} \tag{6.40}$$

and thus

$$D_k = \operatorname{span}\{\frac{\partial}{\partial z^1},\dots,\frac{\partial}{\partial z^k}\}, \qquad k \in \underline{\kappa}. \tag{6.41}$$

First we set $z^\kappa := x^\kappa$. For the definition of $z^{\kappa-1}$ we observe that by (6.39)

$$\operatorname{rank} \frac{\partial f^\kappa}{\partial x^{\kappa-1}}(x^{\kappa-1},z^\kappa) = \rho_\kappa = \dim(x^\kappa). \tag{6.42}$$

Now set

$$z^{\kappa-1} = \begin{bmatrix} f^\kappa(x^{\kappa-1},z^\kappa) \\ \tilde{x}^{\kappa-1} \end{bmatrix}, \tag{6.43}$$

where $\tilde{x}^{\kappa-1}$ are $\rho_{\kappa-1}-\rho_\kappa$ functions chosen from the set $x^{\kappa-1}$ in such a way that $\tilde{x}^{\kappa-1}$ and f^κ are independent functions of $x^{\kappa-1}$. Clearly this defines a coordinate transformation of the form (6.40), so that in the new coordinates $x^1,\dots,x^{\kappa-2},z^{\kappa-1},z^{\kappa}$ the distributions are still given as in (6.41). Therefore (6.34) and (6.39) still hold if we replace x^{κ} and $x^{\kappa-1}$ by z^{κ}, respectively $z^{\kappa-1}$. (Although the component functions $f^1,\dots,f^{\kappa}$ in the new coordinates are not the same as in the old coordinates!) Thus by (6.39) for $i = \kappa-1$ in these new coordinates we have

$$\operatorname{rank} \frac{\partial f^{\kappa-1}}{\partial x^{\kappa-2}}(x^{\kappa-2},z^{\kappa-1},z^{\kappa}) = \rho_{\kappa-1} = \dim(z^{\kappa-1}). \tag{6.43}$$

Then let

$$z^{\kappa-2} = \begin{bmatrix} f^{\kappa-1}(x^{\kappa-2},z^{\kappa-1},z^{\kappa}) \\ \tilde{x}^{\kappa-2} \end{bmatrix}, \tag{6.44}$$

where $\tilde{x}^{\kappa-2}$ are $\rho_{\kappa-2}-\rho_{\kappa}$ functions chosen from the set $x^{\kappa-2}$ in such a way that $\tilde{x}^{\kappa-2}$ and $f^{\kappa-1}$ are independent functions of $x^{\kappa-2}$. Clearly the coordinate transformation (6.44) is of the form (6.40), and thus (6.34) and (6.39) still hold in the new coordinates $x^1,\dots,x^{\kappa-3},z^{\kappa-2},z^{\kappa-1},z^{\kappa}$.

In this way we continue till we have introduced new coordinates $(z^1,\dots,z^{\kappa}) = S(x^1,\dots,x^{\kappa})$, defined inductively as

$$z^k = \begin{bmatrix} f^{k+1}(x^k,z^{k+1},\dots,z^{\kappa}) \\ \tilde{x}^k \end{bmatrix} =: \begin{bmatrix} \bar{z}^k \\ \tilde{x}^k \end{bmatrix}, \qquad k = \kappa-1,\dots,1. \tag{6.45}$$

Since $f(x_0) = 0$ it follows that $S(0) = 0$. In the new coordinates $z = S(x)$ the vectorfield f is given as

$$f(z) = \begin{bmatrix} f^1(z) \\ \bar{z}^1 \\ \vdots \\ \bar{z}^{\kappa-2} \\ \bar{z}^{\kappa-1} \end{bmatrix}, \qquad \dim(\bar{z}^k) = \rho_{k+1}, \qquad k = \kappa-1,\dots,1, \tag{6.46}$$

and thus is a linear vectorfield, except for the first part $f^1(z)$, where f^1 denotes some nonlinear function of z. However, since $\operatorname{span}\{g_1,\dots,g_m\} = D_1 = \operatorname{span}\{\frac{\partial}{\partial z^1}\}$ there exists a regular feedback $u = \alpha(z) + \beta(z)v = \alpha(S(x)) + \beta(S(x))v$ such that

$$f^1(z) + \sum_{i=1}^{m} g_i(z)\alpha_i(z) = 0, \tag{6.47a}$$

$$\bar{g}_j(z) = \sum_{i=1}^{m} g_i(x)\,\beta_{ij}(x) = \begin{cases} e_j, & j = 1,\dots,m_1, \\ 0, & j = m_1+1,\dots,m. \end{cases} \tag{6.47b}$$

Clearly the resulting closed loop system in the coordinates z is *linear*, and moreover it is controllable. □

Remark 6.6 It is easily seen that if $f(x_0) \neq 0$, then the same conditions as above guarantee that the nonlinear system (6.5) can be transformed into an "almost linear" system $\dot{z} = Az + Bv + f(x_0)$. Moreover if $f(x_0) \in \text{span}\{g_1(x_0),..,g_m(x_0)\}$ then by an additional feedback the constant drift term $f(x_0)$ can be also removed.

Finally we note that we can reformulate Theorem 6.3 in the following slightly different way.

Corollary 6.7 *The nonlinear system (6.5) with $f(x_0) = 0$ is feedback linearizable around x_0 to a controllable linear system if and only if the distributions $D_1,..,D_n$ defined in (6.13) are all involutive and constant dimensional in a neighborhood of x_0, and $D_n(x_0) = T_{x_0}M$.*

Proof The "if" part follows from the proof of the "if" part of Theorem 6.3 and the observation therein that $C_0 = D_n$. The "only if" part follows from the "only if" part of Theorem 6.3 plus the fact that for a controllable linear system $D_n = \mathbb{R}^n$. □

Let us study the resulting feedback linearized system defined by (6.46) and (6.47) in some more detail. Define for simplicity of notation $\sigma_i = \rho_{i-1}-\rho_i$, $i = 2,..,\kappa$, $\rho_0 = m$. Then the resulting linear system is given as $\dot{z} = Az + Bv$ where

$$A = \begin{pmatrix} & 0_{\rho_1 \times n} & & \\ \begin{matrix} I_{\rho_2} & 0_{\rho_2\times\sigma_2} \end{matrix} & & 0 & \\ & \begin{matrix} I_{\rho_3} & 0_{\rho_3\times\sigma_3} \end{matrix} & & \\ 0 & & \ddots & \\ & & & \begin{matrix} I_{\rho_\kappa} & 0_{\rho_\kappa\times\sigma_\kappa} \end{matrix} \end{pmatrix}, \quad B = \begin{pmatrix} I_{\rho_1} \quad 0_{\rho_1\times\sigma_1} \\ 0 \\ 0 \end{pmatrix} \tag{6.48}$$

By a permutation of the coordinates z the system can be put into the following form, known as the *Brunovsky normal form*. Indeed, define the controllability indices κ_i, $i \in \underline{m}$, as

$$\kappa_i := \text{number of integers in the set } \{\rho_1,..,\rho_\kappa\} \text{ which are} \geq i. \tag{6.49}$$

Thus $\kappa_1 + \kappa_2 + \ldots + \kappa_m = n$. Notice also that $\kappa_1 = \kappa$. Then by a simple permutation (6.48) takes the form (assume for simplicity that $m_1 = \rho_1 = m$)

$$\tilde{A} = \text{diag}\,(A_1,\ldots,A_m),\ \tilde{B} = \text{diag}\,(b_1,\ldots,b_m), \tag{6.50a}$$

with

$$A_i = \begin{bmatrix} 0 & 1 & & 0 \\ \vdots & & \ddots & 1 \\ 0 & \cdots & & 0 \end{bmatrix}_{\kappa_i \times \kappa_i}, \quad b_i = \begin{bmatrix} 0 \\ \vdots \\ 0 \\ 1 \end{bmatrix}_{\kappa_i \times 1} \tag{6.50b}$$

The proof of Theorem 6.3 ("if"-part) yields some useful information about the *structure* of any feedback linearizable system. Indeed it follows that a locally strongly accessible system (6.5) is feedback linearizable around x_0 if and only if there exists a coordinate system $x = (x^1,\ldots,x^\kappa)$ around x_0 such that span $\{g_1(x),\ldots,g_m(x)\}$ = span $\{\frac{\partial}{\partial x^1}\}$, and f satisfies (6.34) as well as (6.39). Therefore a locally strongly accessible system is feedback linearizable if and only if it has the following flow diagram structure, implied by equations (6.34) and the condition span $\{g_1(x),\ldots,g_m(x)\}$ = span $\{\frac{\partial}{\partial x^1}\}$,

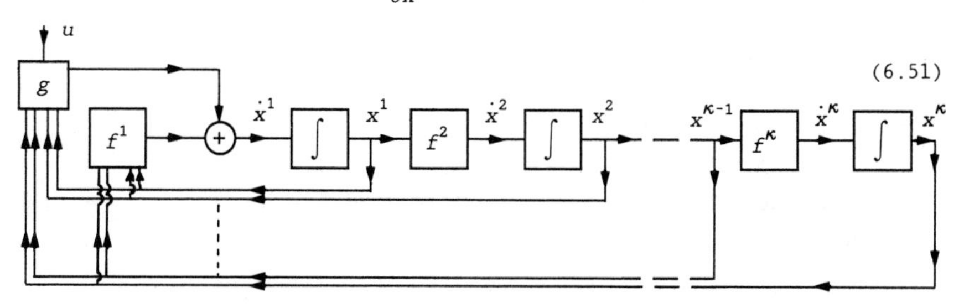

(6.51)

and moreover (6.39) is satisfied. Notice that in (6.51) u enters the central string of integrators, interlaced with the nonlinear mappings $f^2,\ldots,f^\kappa$, only at the beginning and that there are only "backward feedback loops". The system is brought into linear form by successively redefining $x^{\kappa-1},\ldots,x^1$, thereby successively eliminating the feedback loops in (6.51) from the right, and transforming $f^\kappa,\ldots,f^2$ into linear mappings. Finally the feedback loops influencing f^1 and g are removed by static state feedback.

Example 6.8 (see also Example 5.20) Consider the two-link rigid robot manipulator from Example 1.1 written in Euler-Lagrange form as

$$M(\theta)\ddot{\theta} + C(\theta,\dot{\theta}) + k(\theta) = u. \tag{6.52}$$

In this case a linearizing feedback is immediate, namely

$$u = C(\theta,\dot{\theta}) + k(\theta) + M(\theta)v, \tag{6.53}$$

with $v \in \mathbb{R}^2$ the new input. Substitution of (6.53) in (6.52) yields $M(\theta)\ddot{\theta} = M(\theta)v$, or equivalently, since $\det M(\theta) \neq 0$,

$$\ddot{\theta} = v, \tag{6.54}$$

which corresponds to the linear controllable dynamics

$$\frac{d}{dt}\begin{pmatrix} \theta \\ \dot{\theta} \end{pmatrix} = \begin{pmatrix} 0 & I_2 \\ 0 & 0 \end{pmatrix}\begin{pmatrix} \theta \\ \dot{\theta} \end{pmatrix} + \begin{pmatrix} 0 \\ I_2 \end{pmatrix} v. \tag{6.55}$$

Notice, however, that (6.55) is *not* a global linear system since $\theta \in S^1 \times S^1 \neq \mathbb{R}^2$. □

Example 6.9 Consider the controlled Euler equations from Example 1.2

$$J\dot{\omega} = S(\omega)J\omega + b_1u_1 + b_2u_2 + b_3u_3, \qquad \omega \in \mathbb{R}^3. \tag{6.56}$$

In case the vectors b_1, b_2, b_3 are independent the system is trivially feedback linearizable; simply set

$$S(\omega)J\omega + b_1u_1 + b_2u_2 + b_3u_3 = v, \tag{6.57}$$

with $v \in \mathbb{R}^3$ the new control vector, and solve for $u = (u_1, u_2, u_3)$.

If $\operatorname{rank}\,(b_1 b_2 b_3) = 2$ we effectively have two controls, and we may assume without loss of generality that $b_3 = 0$. In this case the distribution D_1 as defined in (6.13) is given by the flat distribution $\operatorname{span}\{b_1, b_2\}$ which is clearly involutive and of constant dimension. The computation of the distribution D_2 is more involved, and we restrict ourselves to the simplified situation (see Example 3.24, (3.48))

$$\begin{aligned} \dot{\omega}_1 &= A_1\omega_2\omega_3 + \alpha_1 u_1 \\ \dot{\omega}_2 &= A_2\omega_1\omega_3 + \alpha_2 u_2 \\ \dot{\omega}_3 &= A_3\omega_2\omega_1 \end{aligned} \tag{6.58}$$

with $J = \operatorname{diag}(a_1, a_2, a_3)$ and $A_1 = (a_2 - a_3)a_1^{-1}$, $A_2 = (a_3 - a_1)a_2^{-1}$, $A_3 = (a_1 - a_2)a_3^{-1}$, $\alpha_1 = a_1^{-1}$, $\alpha_2 = a_2^{-1}$. In Example 3.24 it has been computed that

$$[g_1, f](\omega) = \begin{pmatrix} 0 \\ \alpha_1 A_2 \omega_3 \\ \alpha_1 A_3 \omega_2 \end{pmatrix}, \quad [g_2, f](\omega) = \begin{pmatrix} \alpha_2 A_1 \omega_3 \\ 0 \\ \alpha_2 A_3 \omega_1 \end{pmatrix}, \tag{6.59}$$

and so the distribution D_2 equals

$$D_2(\omega) = \operatorname{span}\left\{ \begin{pmatrix} 1 \\ 0 \\ 0 \end{pmatrix}, \begin{pmatrix} 0 \\ 1 \\ 0 \end{pmatrix}, \begin{pmatrix} 0 \\ \alpha_1 A_2 \omega_3 \\ \alpha_1 A_3 \omega_2 \end{pmatrix}, \begin{pmatrix} \alpha_2 A_1 \omega_3 \\ 0 \\ \alpha_2 A_3 \omega_1 \end{pmatrix} \right\}. \tag{6.60}$$

It follows that $D_2(\omega) = T_\omega\mathbb{R}^3$ if and only if $A_3 \neq 0$ and $\omega_1 \neq 0$ or $\omega_2 \neq 0$. Hence if $a_1 \neq a_2$ then by Theorem 6.2 the system is feedback linearizable everywhere, except for the line $\omega_1^0 = \omega_2^0 = 0$. Outside this line the linearization is performed as follows. Set $z_3 := \omega_3$, and

$$z_2 := A_3\omega_2\omega_1 . \tag{6.61}$$

If $\omega_1^0 \neq 0$ then we set $z_1 = \omega_1$ and if $\omega_2^0 \neq 0$ then we set $z_1 = \omega_2$. In the first case we obtain in the new coordinates the equations

$$\begin{aligned}
\dot{z}_1 &= \frac{A_1}{A_3}\frac{z_2 z_3}{z_1} + \alpha_1 u_1 \\
\dot{z}_2 &= A_3 A_2 z_1^2 z_3 + \frac{A_1}{A_3}\frac{z_2^2 z_3}{z_1^2} + \alpha_2 A_3 z_1 u_2 + \frac{\alpha_1}{A_3}\frac{z_2}{z_1} u_1 \\
\dot{z}_3 &= z_2
\end{aligned} \tag{6.62}$$

since by (6.61) we have $\omega_2 = \dfrac{z_2}{A_3\omega_1} = \dfrac{z_2}{A_3 z_1}$. The first two equations of (6.62) can be linearized by setting the right-hand side of the first equation equal to v_1, and the right-hand side of the second equation equal to v_2, with $v = (v_1, v_2)$ the new input vector. Since $A_3 \neq 0$ this can be solved for u_1 and u_2 in all points for which $z_1 = \omega_1 \neq 0$.

For the one-input case we only consider the simplified situation treated in Example 3.24 (see equation (3.54))

$$\begin{aligned}
\dot{\omega}_1 &= A\omega_2\omega_3 + \alpha u \\
\dot{\omega}_2 &= -A\omega_1\omega_3 + \beta u \\
\dot{\omega}_3 &= \gamma u
\end{aligned} \tag{6.63}$$

with $A = (a_1 - a_3)a_1^{-1}$. Clearly the distribution D_1 equals $\operatorname{span}\{(\alpha,\beta,\gamma)^T\}$, and thus is trivially involutive and of constant dimension. In Example 3.24 it is computed that $[f,g] = -A(\beta\omega_3 + \omega_2\gamma, -\alpha\omega_3 - \omega_1\gamma, 0)^T$ so that

$$D_2(\omega) = \operatorname{span}\left\{ \begin{pmatrix} \alpha \\ \beta \\ \gamma \end{pmatrix}, A \begin{pmatrix} \beta\omega_3 + \omega_2\gamma \\ -\alpha\omega_3 - \omega_1\gamma \\ 0 \end{pmatrix} \right\} \tag{6.64}$$

Clearly D_2 does not have constant dimension (take $\omega = 0$!). Furthermore in general D_2 is not involutive, since

$$\left[\begin{pmatrix} \alpha \\ \beta \\ \gamma \end{pmatrix}, A \begin{pmatrix} \beta\omega_3 + \omega_2\gamma \\ -\alpha\omega_3 - \omega_1\gamma \\ 0 \end{pmatrix} \right] = 2A\gamma \begin{pmatrix} \beta \\ -\alpha \\ 0 \end{pmatrix} \notin D_2 \tag{6.65}$$

for general values of A, α, β, γ, and so the system (6.63) is not feedback linearizable. □

We shall now show how Theorem 6.3 immediately generalizes to *general* nonlinear systems $\dot{x} = f(x,u)$. For completeness we give (compare Definition 6.1).

Definition 6.10 *Consider a smooth nonlinear system*

$$\dot{x} = f(x,u) \qquad\qquad u = (u_1,\ldots,u_m) \in U \subset \mathbb{R}^m, \tag{6.66}$$

with equilibrium (x_0,u_0), *i.e.* $f(x_0,u_0) = 0$. *Throughout we assume that* U *is an open subset of* $\mathbb{R}^n$ *containing* u_0. *The system is feedback linearizable around* (x_0,u_0) *if there exists*

(i) *a coordinate transformation* $S\colon \mathbb{R}^n \to \mathbb{R}^n$ *defined on a neighborhood* O *of* x_0 *with* $S(x_0) = 0$,

(ii) *a regular static state feedback* $u = \alpha(x,v)$ *(see (5.106) satisfying* $\alpha(x_0,0) = u_0$ *and defined on a neighborhood* $O \times V$ *of* $(x_0,0)$, $V \subset U$, *with* $\frac{\partial\alpha}{\partial v}(x,v)$ *non-singular on* $O \times V$ *such that for suitable constant matrices* A *and* B

$$(S_*)_x\, f(x,\alpha(x,v)) = A\, S(x) + Bv, \qquad x \in O,\ v \in V \tag{6.67}$$

The conditions for feedback linearizability in this case rely on the notion of the *extended system* of a general nonlinear system (6.6) (compare with Definition 5.27 and (5.101)).

Definition 6.11 *The* extended system *of (6.66) is the* affine *nonlinear system*

$$\begin{aligned} \dot{x} &= f(x,u) \\ \dot{u} &= w \end{aligned} \tag{6.68}$$

with state $(x,u) \in M \times U$ *and input* $w \in \mathbb{R}^m$.

Theorem 6.12 *Consider the nonlinear system (6.66) with* $f(x_0,u_0) = 0$. *Suppose that the extended system (6.68) satisfies the strong accessibility rank condition in* (x_0,u_0). *Then the nonlinear system (6.66) is feedback linearizable around* (x_0,u_0) *if and only if the extended system (6.68) is feedback linearizable around* (x_0,u_0), *i.e. satisfies the conditions of Theorem 6.3.*

Proof (Only if) One can interpret the state space transformation $x = S^{-1}(z)$ together with the feedback $u = \alpha(x,v)$ for (6.66), as a state space transformation $(x,u) \to (z,v)$ for the extended system. In the new coordinates (z,v) the extended system has the form

$$\dot{z} = Az + Bv \tag{6.69a}$$

$$\dot{v} = G(z,v,w) \tag{6.69b}$$

for some function G. By the regularity of the feedback $u = \alpha(x,v)$ we can locally solve from the relation $u = \alpha(x,v)$ for v as function of x and u, i.e. $v = \beta(x,u)$ with $\alpha(x,\beta(x,u)) = u$. It follows that (6.69b) is given as

$$\dot{v} = \frac{\partial\beta}{\partial x}(S^{-1}(z),\alpha(S^{-1}(z),v))f(S^{-1}(z),\alpha(S^{-1}(z),v)) \tag{6.69b'}$$

$$+ \frac{\partial\beta}{\partial u}(S^{-1}(z),\alpha(S^{-1}(z),v))w$$

Setting the right-hand side of (6.69b') equal to a new control vector $\bar{w}$ defines the required linearizing feedback for the extended system.

(if) Let $D_1 \subset D_2 \subset \ldots$ be the sequence of distributions as defined in (6.13) for the extended system. By assumption they are all involutive and of constant dimension. Hence by Lemma 6.4 and local strong accessibility there exist local coordinates $q = (q^1,\ldots,q^\kappa)$ around $(x_0,u_0) \in M \times U$ such that $D_k = \text{span}\,\{\frac{\partial}{\partial q^1},\ldots,\frac{\partial}{\partial q^k}\}$, with $\dim D_k = m_k$ and $\dim q^k = m_k - m_{k-1}$, $k = 2,\ldots,N$. Since $D_1 = \text{span}\,\{\frac{\partial}{\partial u_1},\ldots,\frac{\partial}{\partial u_m}\}$ it follows we can take $q^1 = u := (u_1,\ldots,u_m)^T$ and $q^1 = m =: m_1$. Now consider the vectorfields $\frac{\partial}{\partial q_{m_1+1}},\ldots,\frac{\partial}{\partial q_{m_2}}$. In coordinates (x,u) for $M \times U$, where x are coordinates for M around x_0, and u coordinates for U around u_0, they are of the general form

$$A_j(x,u)\frac{\partial}{\partial x} + B_j(x,u)\frac{\partial}{\partial u}, \qquad j = m_1+1,\ldots,m_2. \tag{6.70}$$

By definition $[\frac{\partial}{\partial q_j},\frac{\partial}{\partial u_i}] = 0$, $j = m_1+1,\ldots,m_2$, $i \in \underline{m}$, and thus

$$0 = [A_j(x,u)\frac{\partial}{\partial x} + B_j(x,u)\frac{\partial}{\partial u}, \frac{\partial}{\partial u_i}] = -\frac{\partial A_j}{\partial u_i}\frac{\partial}{\partial x} + \frac{\partial B_j}{\partial u_i}\frac{\partial}{\partial u}. \tag{6.71}$$

Therefore $A_j(x,u)$ and $B_j(x,u)$ do not depend on u. Hence we may write in the (x,u)-coordinates

$$D_2 = D_1 + \text{span}\{A_j(x)\frac{\partial}{\partial x},\ j = m_1+1,\ldots,m_2\}. \tag{6.72}$$

In the same way D_k for any $k = 2,\ldots,N$ is of the form

$$D_k = D_1 + \text{span}\{A_j(x)\frac{\partial}{\partial x},\ j = m_1+1,\ldots,m_k\}. \tag{6.73}$$

Define the nested sequence of distributions $E_1,\ldots,E_{N-1}$ on M

$$E_{k-1} = \text{span}\{A_j(x)\frac{\partial}{\partial x},\ j = m_1 + 1,..,m_k\}, \qquad k = 2,..,N. \tag{6.74}$$

Since the distributions $D_1,..,D_N$ are all involutive and of constant dimension it immediately follows that the distributions $E_1,..,E_{N-1}$ on M also have these properties. (Note that E_{k-1} is the projection of D_k on M.) Hence by Lemma 6.4 there exist local coordinates $x = (x^1,..,x^{N-1})$ around x_0 for M such that

$$E_k = \text{span}\{\frac{\partial}{\partial x^1},..,\frac{\partial}{\partial x^k}\}, \qquad k \in \underline{N-1}, \tag{6.75}$$

implying that

$$D_k = \text{span}\ \{\frac{\partial}{\partial u}\} + \text{span}\{\frac{\partial}{\partial x^1},...,\frac{\partial}{\partial x^{k-1}}\}, \qquad k \in \underline{N}. \tag{6.76}$$

In the same way as in the proof of Theorem 6.3 (if-part) it now follows that in these coordinates $f(x,u)$ has the form

$$f(x,u) = \begin{pmatrix} f^1\ (u,x^1,..,x^{N-1}) \\ f^2\ (x^1,....,x^{N-1}) \\ \vdots \\ f^i\ (x^{i-1},..,x^{N-1}) \\ \vdots \\ f^{N-1}(\ x^{N-2},x^{N-1}) \end{pmatrix}, \tag{6.77}$$

and furthermore

$$\text{rank}\ \frac{\partial f^k}{\partial x^{k-1}} = \dim(x^k), \qquad k \in \underline{N-1}, \tag{6.78}$$

where we denote $x^0 := u$. We modify successively the coordinates $x^{N-1},..,x^1,u$ to coordinates $z^{N-1},..,z^1,v$ in the same way as in the proof of Theorem 6.3 (if-part). The thus defined transformation $(x^1,..,x^{N-1}) \mapsto (z^1,..,z^{N-1})$ is the required state space transformation $z = S(x)$. Furthermore the transformation from u to v is the required feedback $u = \alpha(x,v)$. Indeed, recall that the transformation from u to v is defined by setting (see (6.45))

$$v = \begin{pmatrix} f^1(u,x^1,..,x^{N-1}) \\ \tilde{u} \end{pmatrix} \tag{6.79}$$

where $\tilde{u}$ are $m - \dim(x^1)$ functions from the set $(u_1,..,u_m)$, taken in such a way that $\tilde{u}$ and $f^1(x,u)$ are independent functions of u. Hence by the Implicit Function Theorem we can solve u from (6.79) locally around (x_{01},u_0), yielding an explicit regular static state feedback

$$u = \alpha(x,v). \tag{6.80}$$

□

Remark Notice that the adaptation of the state space coordinates performed in every step is of the same nature as the final adaptation of the u-coordinates defining the required feedback transformation. Indeed consider (6.45), then by the Implicit Function Theorem we can solve for x^k as function of z^k and $z^{k+1},\ldots,z^k$, i.e.

$$x^k = \alpha^k(z^k,z^{k+1},\ldots,z^k). \tag{6.81}$$

This can be interpreted as "feedback" for a fictitious lower dimensional system with old input $u = x^k$, new input $v = z^k$ and state $(z^{k+1},\ldots,z^k)$.

Example 6.13 Consider the dynamics of a rocket outside the atmosphere. The forces which act on the rocket are the gravitational force and the force as delivered by the rocket motor.

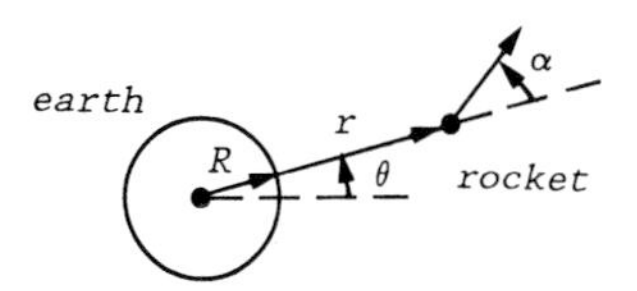

Fig. 6.1. Rocket outside the atmosphere.

The control variable is the angle α expressing the direction of the force as delivered by the rocket motor. Take state space variables $x_1 = r$, $x_2 = \theta$, $x_3 = \dot{r}$, $x_4 = \dot{\theta}$, thus $x \in T(\mathbb{R}^+ \times S^1)$. Then the dynamics are given as

$$\begin{aligned}
\dot{x}_1 &= x_3 \\
\dot{x}_2 &= x_4 \\
\dot{x}_3 &= -gR^2/x_1^2 + \frac{T}{m}\cos u + x_1x_4^2 \\
\dot{x}_4 &= -2x_3x_4/x_1 + \frac{T}{mx_1}\sin u
\end{aligned} \tag{6.82}$$

(with m the mass of the rocket; g the gravitational constant and R the radius of the earth). In order to check if the system is feedback linearizable (around an arbitrary equilibrium point) we consider the extended system, i.e. (6.82) together with the equation $\dot{u} = w$. The extended system is of the form $\dot{x}_e = f(x_e) + g(x_e)w$, where $x_e = (x,u)$ and

$$f(x,u) = \begin{bmatrix} x_3 \\ x_4 \\ -gR^2/x_1^2 + x_1x_4^2 + \frac{T}{m}\cos u \\ -2x_3x_4/x_1 + \frac{T}{mx_1}\sin u \\ 0 \end{bmatrix}, \quad g(x,u) = \begin{bmatrix} 0 \\ 0 \\ 0 \\ 0 \\ 1 \end{bmatrix}. \tag{6.83}$$

Hence

$$[f,g](x,u) = \begin{bmatrix} 0 \\ 0 \\ \frac{T}{m}\sin u \\ -\frac{T}{mx_1}\cos u \\ 0 \end{bmatrix}. \tag{6.84}$$

It follows that $D_2(x,u) = \text{span}\{g(x,u),[f,g](x,u)\}$ is *not* involutive, since

$$[g,[f,g]](x,u) = \begin{bmatrix} 0 \\ 0 \\ -\frac{T}{m}\cos u \\ -\frac{T}{mx_1}\sin u \\ 0 \end{bmatrix} \notin D_2(x,u). \tag{6.85}$$

Therefore by Theorem 6.12 the system is *not* feedback linearizable. □

6.2 Computational Aspects of Feedback Linearization

Consider a nonlinear system (6.5) with $f(x_0) = 0$ and satisfying the strong accessibility rank condition in x_0, for which the distributions $D_1,..,D_\kappa$ are all involutive and of constant dimension. Then Theorem 6.3 gives us the following recipe to construct a state space transformation $z = S(x)$ and a feedback $u = \alpha(x) + \beta(x)v$ which will transform the system into a linear one:

(a) Construct as in Lemma 6.4 coordinates $x = (x^1,\dots,x^\kappa)$ such that $D_k = \text{span}\{\frac{\partial}{\partial x^1},..,\frac{\partial}{\partial x^k}\}$, $k = 1,\dots,\kappa$.

(b) Adapt successively the coordinates $x^{\kappa-1},..,x^1$ to $z^{\kappa-1},..,z^1$ as in (6.45),using the vectorfield f.

(c) Apply the feedback (6.47).

In this section we shall show that in general this is not the most efficient way to perform feedback linearization. The crucial observation is that the sequence $D_1 \subset D_2 \subset .. \subset D_\kappa$ is not just an arbitrary nested sequence of distributions as in Lemma 6.4, but instead is constructed in a very special way. As a result, we do not have to construct *all* the coordinates $x^1,..,x^\kappa$ as in (a), but only a (possibly small) part of them, and still step (b) will provide us with a complete set of required coordinates $z^1,..,z^\kappa$. Also this will imply that there is a lot of redundancy in requiring that *all* distributions D_i are involutive and of constant dimension.

These considerations become most clear in the single-input case

$$\dot{x} = f(x) + g(x)u, \qquad f(x_0) = 0, \tag{6.85}$$

in which case the distributions D_i are simply given as

$$D_i(x) = \operatorname{span}\{g(x), ad_f g(x),..,ad_f^{i-1}g(x)\}, \tag{6.86}$$

and thus, since $\dim D_{i+1}(x) \leq \dim D_i(x)+1$, $D_n(x_0)$ is equal to $T_{x_0}M$ if and only if

$$\dim D_i(x_0) = i, \qquad i \in \underline{n}. \tag{6.87}$$

The main technical proposition reads as follows. For notational convenience we denote $L_X\varphi = d\varphi(X)$ as $\langle d\varphi, X\rangle$, for any vectorfield X and function φ.

Proposition 6.14 *Consider the single-input system (6.85). Suppose there exists a function φ such that*

$$\langle d\varphi, ad_f^k g\rangle(x) = 0, \qquad k = 0,1,..,n-2 \tag{6.88}$$

(or, equivalently, $\langle d\varphi, X\rangle = 0$ for $X \in D_i$, $i = 1,..,n-1$). Then the statement
(a) $\langle d\varphi, ad_f^{n-1}g\rangle(x_0) \neq 0$
implies the following two statements
(b) the functions $\varphi, L_f\varphi,..,L_f^{n-1}\varphi$ are independent around x_0.
(c) the vectorfields $g,[f,g],..,ad_f^{n-1}g$ are independent around x_0.
Conversely, if $g(x_0) \neq 0$ then (b) implies (a), and if $d\varphi(x_0) \neq 0$ then (c) implies (a).

Remark Note that conditions (6.88) and $\langle d\varphi, ad_f^{n-1}g\rangle(x_0) \neq 0$ are in a sense dual to the conditions for linearizable error dynamics, cf. Chapter 5; in this last case φ is known, while g has to be found.

In order to prove Proposition 6.14 we state the following auxiliary lemma.

Lemma 6.15 *Let f and g be vectorfields and φ a function on M. The following statements are equivalent*

(a) $\langle d\varphi, ad_f^k g\rangle = 0, \qquad k = 0,1,2,..,n-2,$

(b) $\langle dL_f^j\varphi, ad_f^i g\rangle = 0, \qquad i+j = 0,1,2,..,n-2,$

(c) $\langle dL_f^k\varphi, g\rangle = 0, \qquad k = 0,1,2,..,n-2.$

Furthermore if one of the above statements hold, then

(d) $\langle d\varphi, ad_f^{n-1} g\rangle = (-1)^k \langle dL_f^k\varphi, ad_f^{n-1-k} g\rangle, \qquad k = 0,1,...,n-1.$

Proof Clearly (b) implies (a) and (c). We prove that both (a) and (c) imply (b). Assume that (a) holds. From Chapter 2, cf. (2.170), we recall that $L_X\langle d\varphi, Y\rangle = \langle dL_X\varphi, Y\rangle + \langle d\varphi, ad_X Y\rangle$ for any vectorfields X,Y and function φ on M. Let k be such that $k+1 \leq n-2$. Then certainly $\langle d\varphi, ad_f^k g\rangle = 0$ and thus

$$0 = L_f\langle d\varphi, ad_f^k g\rangle = \langle dL_f g, ad_f^k g\rangle + \langle d\varphi, ad_f^{k+1} g\rangle. \tag{6.89}$$

Since $k+1 \leq n-2$ the second term on the right-hand side is zero, so that also $\langle dL_f g, ad_f^k g\rangle = 0$. Hence we have proved (b) for $j = 1$. Now suppose that (b) holds for a certain $j < n-2$. Then we prove that (b) also holds for $j+1$. Let k be such that $k+j+1 \leq n-2$. Then by the induction assumption

$$\langle dL_f^j\varphi,\ ad_f^k g\rangle = 0 = \langle dL_f^j\varphi,\ ad_f^{k+1} g\rangle. \tag{6.90}$$

Hence

$$\begin{aligned} 0 = L_f\langle dL_f^j\varphi,\ ad_f^k g\rangle &= \langle dL_f^{j+1}\varphi,\ ad_f^k g\rangle + \langle dL_f^j\varphi,\ ad_f^{k+1} g\rangle = \\ &= \langle dL_f^{j+1}\varphi,\ ad_f^k g\rangle. \end{aligned} \tag{6.91}$$

so that indeed (b) holds for j+1. Similarly (b) follows from (c) as follows. Suppose (b) holds for a certain i $< n-2$ (it does hold for $i = 0$), then we prove it also holds for $i+1$. Indeed let k be such that $k+i+1 \leq n-2$. Then by the induction assumption

$$\langle dL_f^k\varphi,\ ad_f^i g\rangle = 0 = \langle dL_f^{k+1}\varphi,\ ad_f^i g\rangle. \tag{6.92}$$

Hence

$$0 = L_f\langle dL_f^k\varphi,\ ad_f^i g\rangle = \langle dL_f^{k+1}\varphi, ad_f^i g\rangle + \langle dL_f^k\varphi,\ ad_f^{i+1} g\rangle = \langle dL_f^k\varphi,\ ad_f^{i+1} g\rangle, \tag{6.93}$$

so that indeed (b) holds for $i+1$. Finally (d) will be proved by induction to k. For k=0 the identity is trivial. Let (d) be satisfied for k-1. Then differentiate the function $\langle dL_f^{k-1}\varphi,\ ad_f^{n-1-k} g\rangle$ which is zero by (b), with respect to f:

$$0 = L_f\langle dL_f^{k-1}\varphi,\ ad_f^{n-1-k} g\rangle = \langle dL_f^k\varphi,\ ad_f^{n-1-k} g\rangle + \langle dL_f^{k-1}\varphi,\ ad_f^{n-k} g\rangle. \tag{6.94}$$

Hence $<dL_f^k\varphi, ad_f^{n-k}g> = -<dL_f^{k-1}\varphi, ad_f^{n-k}g> = -(-1)^{k-1}<d\varphi, ad_f^{n-1}g>$, so that (d) holds for k. □

Proof of Proposition 6.14 Consider the following product of two $n\times n$ matrices

$$\begin{bmatrix} d\varphi \\ dL_f\varphi \\ \vdots \\ dL_f^{n-1}\varphi \end{bmatrix} \left(g \,\vdots\, ad_f g \,\vdots \cdots \vdots\, ad_f^{n-1}g\right)(x_0) = \tag{6.95}$$

$$\begin{bmatrix} <d\varphi,g> & \cdots & <d\varphi,ad_f^{n-1}g> \\ \vdots & & \vdots \\ <dL_f^{n-1}\varphi,g> & \cdots & <dL_f^{n-1}\varphi,ad_f^{n-1}g> \end{bmatrix}(x_0)$$

which by (6.88) and Lemma 6.15(b) is of the form

$$\begin{bmatrix} 0 & \cdots & 0 & <d\varphi,ad_f^{n-1}g> \\ \vdots & ⋰ & ⋰ & \\ 0 & ⋰ & & \\ <dL_f^{n-1}\varphi,g> & & * & \end{bmatrix}(x_0) \tag{6.96}$$

Now suppose that (a) holds, then because of Lemma 6.15(d) the matrix (6.96) is nonsingular, and hence the two $n\times n$ matrices on the left-hand side of (6.95) are nonsingular, so that (b) and (c) follow. Conversely, suppose (c) holds and $d\varphi(x_0) \neq 0$ then by (6.88) necessarily (a) holds. Also, suppose (b) holds and $g(x_0) \neq 0$. By (6.88) and Lemma 6.15(b) it follows that $<dL_f^k\varphi,g> = 0$, $k = 0,1,2,\ldots,n-2$. Since $g(x_0) \neq 0$ necessarily $<dL_f^{n-1}\varphi,g>(x_0) \neq 0$ and thus by Lemma 6.15(d) (a) follows. □

We now obtain the following refined versions of Corollary 6.7 in the single-input case.

Proposition 6.16 *Consider the single-input system (6.85). Then the system is feedback linearizable around x_0 to a controllable linear system if and only if there exists a function φ such that*

$$<d\varphi,ad_f^k g>(x) = 0, \quad k = 0,1,\ldots,n-2, \ x \text{ in a neighborhood of } x_0, \tag{6.97a}$$

$$<d\varphi,ad_f^{n-1}g>(x_0) \neq 0. \tag{6.97b}$$

Furthermore for such a φ the distributions D_k are given as

$$D_i = \ker \operatorname{span}\{d\varphi, dL_f\varphi,\ldots,dL_f^{n-1-i}\varphi\}, \qquad i = 1,\ldots,n-1. \tag{6.98}$$

Corollary 6.17 *The single-input system (6.85) is feedback linearizable around x_0 to a controllable linear system if and only if*

$$\dim D_n(x_0) = n, \tag{6.99a}$$

$$D_{n-1} \text{ is involutive around } x_0. \tag{6.99b}$$

Proof of Proposition 6.16 and Corollary 6.17 If (6.97) holds then by Proposition 6.14 the functions $\varphi, L_f\varphi, .., L_f^{n-1}\varphi$ are independent around x_0. By Lemma 6.15(b) it follows from (6.97a) that $D_i \subset \ker \text{span} \{d\varphi, L_f\varphi, .., dL_f^{n-1-i}\varphi\}$. By Proposition 6.14(c) and (6.86) it follows that $\dim D_i = i$, and thus by the independency of $\varphi, .., L_f^{n-1}\varphi$, dim span $\{d\varphi, .., dL_f^{n-1-k}\varphi\} = n-k$, so that (6.98) follows. Hence the distributions $D_1, .., D_n$ are all involutive and constant dimensional and $D_n(x_0) = T_{x_0}M$. Then by Corollary 6.7 the system is feedback linearizable to a controllable linear system. Conversely if the system is feedback linearizable to a controllable linear system then by Corollary 6.7 (6.99) is satisfied. Finally if (6.99) is satisfied then by Frobenius' Theorem (Theorem 2.42) applied to the distribution D_{n-1} there exist local coordinates $(x_1, ..., x_n)$ such that the integral manifolds of D_{n-1} (which by (6.99a) and (6.87) have dimension $n-1$) are given as

$$\{q \text{ near } x_0 \mid x_n(q) = \text{constant}\}, \tag{6.100}$$

or equivalently, (6.97a) holds for $\varphi = x_n$. Moreover by (6.99a) necessarily (6.97b) holds for $\varphi = x_n$. □

We conclude that (6.99) implies the involutivity and constant dimensionality of *all* the distributions D_k, $k = 1, 2, .., n$. Moreover if we have found a function φ with $d\varphi(x_0) \neq 0$ and $\langle d\varphi, X\rangle(x) = 0$ for all $X \in D_{n-1}$ then by (6.98) the "rectifying" local coordinates as in Lemma 6.4 for the sequence of distributions $D_1 \subset D_2 \subset \ldots \subset D_{n-1}$ are directly given as $L_f^{n-1}\varphi, L_f^{n-2}\varphi, \ldots, L_f\varphi, \varphi$. Indeed if we define the coordinates $z = (z_1, .., z_n)$ by

$$z_i = L_f^{i-1}\varphi(x), \qquad i \in \underline{n}, \tag{6.101}$$

then it immediately follows from (6.98) that

$$D_i = \text{span} \{\frac{\partial}{\partial z_{n-i+1}}, \ldots, \frac{\partial}{\partial z_n}\}, \qquad i \in \underline{n}. \tag{6.102}$$

(Notice that for notational convenience we have reversed the ordering of $z_1, .., z_n$ in comparison with the ordering of the coordinates $x^1, .., x^N$ or

$z^1,\ldots,z^\kappa$ of Lemma 6.4, respectively Theorem 6.3.) Moreover the coordinates $z_1,\ldots,z_n$ are already the *linearizing* coordinates. If we assume without loss of generality, that $\varphi(x_0) = 0$, implying that $z_i(x_0) = L_f^{i-1}\varphi(x_0) = 0$, $i \in \underline{n}$. In fact, for any $i \in \underline{n-1}$

$$\dot{z}_i = L_{f+gu}(L_f^{i-1}\varphi)(x) = L_f^i\varphi(x) + uL_gL_f^{i-1}\varphi(x) = L_f^i\varphi(x) = z_{i+1}, \tag{6.103a}$$

since $L_gL_f^{i-1}\varphi(x) = \langle dL_f^{i-1}\varphi, g\rangle(x) = 0$ by Lemma 6.15(b), and

$$\dot{z}_n = L_f^n\varphi(x) + u\, L_gL_f^{n-1}\varphi(x). \tag{6.103b}$$

By (6.97b) and Lemma 6.15(d) $L_gL_f^{n-1}\varphi(x) \neq 0$ in a neighborhood of x_0 so that we can define the regular static state feedback

$$u = (L_gL_f^{n-1}\varphi(x))^{-1}\,(-\,L_f^n\varphi(x) + v), \tag{6.104}$$

resulting in the linear system

$$\dot{z} = \begin{bmatrix} 0 & 1 & & 0 \\ & & \ddots & \\ & & & 1 \\ 0 & \cdots & & 0 \end{bmatrix} z + \begin{bmatrix} 0 \\ \vdots \\ 0 \\ 1 \end{bmatrix} v. \tag{6.105}$$

Summarizing, we have obtained

Corollary 6.18 *Suppose the single-input system (6.85) satisfies (6.99). Then locally around* x_0 *there exists a function* φ *satifying (6.97) and* $\varphi(x_0) = 0$. *By defining the state space transformation* $z = S(x)$ *around* x_0 *as*

$$z_i = L_f^{i-1}\varphi(x), \qquad i \in \underline{n}, \tag{6.106a}$$

and the regular static state feedback $u = \alpha(x) + \beta(x)v$ *as*

$$\begin{aligned} \alpha(x) &= -(L_gL_f^{n-1}\varphi(x))^{-1}\, L_f^n\varphi(x) \\ \beta(x) &= (L_gL_f^{n-1}\varphi(x))^{-1} \end{aligned} \tag{6.106b}$$

the system around x_0 *is transformed into the linear system (6.105).*

Consequently the only possibly difficult step in performing feedback linearization lies in the computation of the function φ satisfying (6.97a) and $d\varphi(x_0) \neq 0$ (and therefore, assuming that $g(x_0)$, $ad_fg(x_0),\ldots$ $\ldots,ad_f^{n-1}g(x_0)$ are independent, also satisfying (6.97b) by Proposition 6.14), i.e. in finding a non-trivial solution of the n-dimensional set of partial differential equations

$$L_{X_i}\varphi(x) = 0, \qquad X_i := \mathrm{ad}_f^{i-1} g, \qquad i \in \underline{n}. \tag{6.107}$$

The treatment of the multi-input case follows the same lines as in the single-input case, and will be only sketched here, leaving the details to the reader. Suppose the distributions $D_1,..,D_{\kappa-1}$ (with κ the least integer such that $\dim D_\kappa = \dim M$) are all involutive and of constant dimension, and let $\rho_k = \dim D_k - \dim D_{k-1}$, $k \in \underline{\kappa}$, with $D_0 = 0$. Since $D_{\kappa-1}$ is involutive and constant dimensional there exist by Frobenius' Theorem ρ_κ independent functions φ_i such that $<d\varphi_i, D_{\kappa-1}> = 0$, $i \in \underline{\rho_\kappa}$. As in the proof of Lemma 6.15 it then follows $<dL_f\varphi_i, D_{\kappa-2}> = 0$, $i \in \underline{\rho_\kappa}$. Furthermore as in the proof of Proposition 6.14 it follows that the functions φ_i, $L_f\varphi_i$, $i \in \underline{\rho_\kappa}$, are independent. However, and at this point there is a major difference with the single-input case, in general the codimension of $D_{\kappa-2}$, which is $\rho_\kappa + \rho_{\kappa-1}$, is *larger* than the number of functions φ_i, $L_f\varphi_i$, $i \in \underline{\rho_\kappa}$, which is $2\rho_\kappa$ (since $\rho_0 \geq \rho_1 \geq \ldots \geq \rho_\kappa$, see the proof of Theorem 6.3). Therefore we have to use the assumption that $D_{\kappa-2}$ is involutive and constant dimensional to construct by an application of Frobenius' Theorem $\rho_{\kappa-1}-\rho_\kappa$ functions φ_i, $i = \rho_\kappa+1,\ldots,\rho_{\kappa-1}$, such that the functions φ_i, $i \in \underline{\rho_{\kappa-1}}$, and $L_f\varphi_i$, $i \in \underline{\rho_\kappa}$, are independent and annihilate $D_{\kappa-2}$. As in Lemma 6.15 and Proposition 6.14 it then follows that $<dL_f\varphi_i, D_{\kappa-3}> = 0$, $i \in \underline{\rho_{\kappa-1}}$, and $<dL_f^2\varphi_i, D_{\kappa-3}> = 0$, $i \in \underline{\rho_\kappa}$, and moreover the functions φ_i, $L_f\varphi_i$, $i \in \underline{\rho_{\kappa-1}}$, $L_f^2\varphi_i$, $i \in \underline{\rho_\kappa}$, are independent. Continuing in this way, we end up with n independent functions

$$\begin{array}{ll} \varphi_i, & i \in \underline{\rho_1}, \\ L_f\varphi_i, & i \in \underline{\rho_2}, \\ \vdots & \\ L_f^{\kappa-1}\varphi_i, & i \in \underline{\rho_\kappa}, \end{array} \tag{6.108}$$

which form the required coordinate system in which the system can be made linear by feedback. This feedback is given as the solution $u = \alpha(x) + \beta(x)v$ of (recall the definition of the controllability indices κ_i in (6.49))

$$v_j = L_f(L_f^{\kappa_j-1}\varphi_j)(x) + \sum_{\ell=1}^{m} L_{g_\ell}(L_f^{\kappa_j-1}\varphi_j)(x)u_\ell, \; j \in \underline{m}, \tag{6.109}$$

where we have supposed for simplicity that $\rho_1 = m$ (i.e. the vectorfields $g_1,..,g_m$ are independent around x_0). As in the proof of Lemma 6.15(d) the $m \times m$ matrix

$$A(x) = \left[\left(L_{g_i} L_f^{\kappa_j-1}\varphi_j\right)(x)\right]_{\substack{i=1,..,m \\ j=1,..,m}} \tag{6.110}$$

is nonsingular around x_0, and so (6.109) can be solved for u yielding the feedback (compare with (6.104))

$$u = -A^{-1}(x)\, b(x) + A^{-1}(x)v, \tag{6.111}$$

where $b(x) = (L_f^{\kappa_1}\varphi_1, \ldots, L_f^{\kappa_m}\varphi_m)^T(x)$, and v is the new input.

Remark 6.19 It is clear that like in the single-input case the assumption of involutivity and constant dimensionality of *all* the distributions D_i in Theorem 6.3 can be somewhat relaxed. In particular if for a certain $j < \kappa$, $\rho_j = \rho_{j+1}$ then the involutivity and constant dimensionality of D_j implies the same properties for the distribution D_{j-1}.

Example 6.20 Consider the model of a mixed-culture bioreactor as treated in Example 1.4

$$\frac{d}{dt}\begin{bmatrix} x_1 \\ x_2 \\ I \end{bmatrix} = \begin{bmatrix} \mu_1(S)x_1 \\ \mu_2(S,I)x_2 \\ -px_1 I \end{bmatrix} + \begin{bmatrix} -x_1 \\ -x_2 \\ -I \end{bmatrix} u_1 + \begin{bmatrix} 0 \\ 0 \\ 1 \end{bmatrix} u_2, \tag{6.112}$$

where $\mu_1(S)$ and $\mu_2(S,I)$ depend on x_1 and x_2, resp. x_1, x_2, I, through

$$\mu_1(S) = \frac{\overset{1}{\mu} S}{K+S}, \quad \mu_2(S,I) = \frac{\overset{2}{\mu} S}{K+S}\cdot\frac{K_I}{K+I}, \tag{6.113a}$$

$$S = S_f - x_1/Y_1 - x_2/Y_2 . \tag{6.113b}$$

As discussed in Example 1.4 we assume that there exists a point (x_1^0, x_2^0, I^0) in the positive orthant for which $\mu_1(S) = \mu_2(S,I) = \mu$. Then (x_1^0, x_2^0, I^0) is an equilibrium point of (6.112) for the constant inputs

$$u_1^0 = \mu, \tag{6.114a}$$

$$u_2^0 = (px_1^0 + \mu)I^0. \tag{6.114b}$$

We will now show that (6.112) can be feedback linearized around any point (x_1, x_2, I), $x_1 > 0$, $x_2 > 0$, $I > 0$, in the sense of Remark 6.6, i.e. the feedback linearized system will contain an extra constant drift term. Linearizing in the point (x_1^0, x_2^0, I^0) this extra drift term can be removed by first subtracting from the controls u_1, u_2 the constant terms u_1^0, resp. u_2^0. First we observe that the distribution $D_1 = \operatorname{span}\left\{\begin{bmatrix} -x_1 \\ -x_2 \\ -I \end{bmatrix}, \begin{bmatrix} 0 \\ 0 \\ 1 \end{bmatrix}\right\}$ is constant dimensional and involutive since

$$\left[\begin{pmatrix} -x_1 \\ -x_2 \\ -I \end{pmatrix}, \begin{pmatrix} 0 \\ 0 \\ 1 \end{pmatrix} \right] = \begin{pmatrix} 0 \\ 0 \\ 1 \end{pmatrix} \in D_1(x,I). \tag{6.115}$$

Furthermore

$$\left[\begin{pmatrix} \mu_1(S)x_1 \\ \mu_2(S,I)x_2 \\ -px_1 I \end{pmatrix}, \begin{pmatrix} 0 \\ 0 \\ 1 \end{pmatrix} \right] = \begin{pmatrix} 0 \\ \dfrac{\mu_2(S,I)x_2}{K_I+I} \\ -px_1 \end{pmatrix}, \tag{6.116}$$

and we conclude that $\dim D_2(x,I) = 3$ since

$$\det \begin{pmatrix} -x_1 & 0 & 0 \\ -x_2 & \dfrac{\mu_2 x_2}{K_I+I} & 0 \\ -I & px_1 & 1 \end{pmatrix} = -\mu_2 x_1 x_2/(K_I+I) \neq 0. \tag{6.117}$$

Therefore, by Theorem 6.3, the system is feedback linearizable in any point in the sense of Remark 6.6. The feedback linearization is performed as follows. First we have to find a function φ, with $d\varphi \neq 0$, such that $\langle d\varphi, g_1 \rangle = \langle d\varphi, g_2 \rangle = 0$, or equivalently

$$-x_1 \frac{\partial \varphi}{\partial x_1} - x_2 \frac{\partial \varphi}{\partial x_2} - I \frac{\partial \varphi}{\partial I} = 0, \tag{6.118a}$$

$$\frac{\partial \varphi}{\partial I} = 0. \tag{6.118b}$$

From (6.118b) it follows that φ only depends on the cell-densities x_1 and x_2. A possible solution to (6.118a) is then

$$\varphi(x_1,x_2) = \ln \frac{x_1}{x_2} =: z_1. \tag{6.119}$$

By a simple calculation

$$L_f\varphi(x_1,x_2,I) = \mu_1(S) - \mu_2(S,I) =: z_2. \tag{6.120}$$

The remaining new coordinate z_3 only has to satisfy the requirement that z_1,z_2,z_3 are independent; we simply take $z_3 := x_1$. In these new coordinates the system is described as

$$\begin{aligned} \dot z_1 &= z_2, \\ \dot z_2 &= L_f z_2 + u_1 L_{g_1} z_2 + u_2 L_{g_2} z_2, \\ \dot z_3 &= L_f z_3 + u_1 L_{g_1} z_3 + u_2 L_{g_2} z_3. \end{aligned} \tag{6.121}$$

It can be immediately checked that the matrix

$$A = \begin{pmatrix} L_{g_1}z_2 & L_{g_2}z_2 \\ L_{g_1}z_3 & L_{g_2}z_3 \end{pmatrix}, \tag{6.122}$$

is non-singular, so that the linearizing feedback is given as

$$\begin{pmatrix} u_1 \\ u_2 \end{pmatrix} = -A^{-1} \begin{pmatrix} L_f z_2 \\ L_f z_3 \end{pmatrix} + A^{-1} \begin{pmatrix} v_1 \\ v_2 \end{pmatrix}, \tag{6.123}$$

(everything depending on x_1, x_2, I). We note that the choice of coordinates is by no means unique. Certainly z_3 is not unique, and instead of (6.119) we could also take for example $\tilde{\varphi} = \arctan (x_2/x_1) =: \tilde{z}_1$. □

For completeness we will finally give the extension of Corollaries 6.17 and 6.18 to the case of a general single-input system

$$\dot{x} = f(x,u), \qquad f(x_0, u_0) = 0, \qquad u \in \mathbb{R}. \tag{6.124}$$

Corollary 6.21 *Consider the distributions $D_1, D_2, \ldots, D_{n+1}$ for the extended system (cf. Definition 6.11) of (6.126):*

$$\begin{aligned} \dot{x} &= f(x,u), \\ \dot{u} &= w, \end{aligned} \tag{6.125}$$

with state space coordinates $(x,u) \in \mathbb{R}^{n+1}$ and control w. Then (6.124) is feedback linearizable around (x_0, u_0) to a controllable linear system if and only if

$$\dim D_{n+1}(x_0, u_0) = n+1, \tag{6.126a}$$

$$D_n \text{ is involutive around } (x_0, u_0). \tag{6.126b}$$

The feedback linearization is performed by constructing a function $\varphi(x)$ with $\varphi(x_0) = 0$, such that $\langle d\varphi, X\rangle = 0$ for all vectorfields $x \in D_n$, and defining the state space transformation $z = S(x)$ as

$$z_i = L_f^{i-1} \varphi(x,u), \qquad i \in \underline{n}, \tag{6.127}$$

and solving locally u as a function $u = \alpha(x,v)$ from

$$v = L_f^n \varphi(x,u). \tag{6.128}$$

The resulting linear system is again given by (6.105).

Remark 6.22 Using Lemma 6.15 it is clear that the functions $L_f^{i-1}\varphi$, $i \in \underline{n}$, do not depend on u, while $\frac{\partial}{\partial u} L_f^n \varphi(x_0, u_0) \neq 0$.

As is well-known, the controllability of a linear system $\dot{x} = Ax + Bu$ guarantees the existence of a linear feedback $u = Kx$ such that the characteristic polynomial of the closed loop matrix $A + BK$ is equal to any desired monic polynomial of the same degree. It is therefore clear that if a nonlinear system is feedback linearizable to a controllable linear system $\dot{z} = Az + Bv$ then by an extra linear feedback $u = Kz$ the poles of the feedback linearized system can be arbitrarily assigned. Hence the nonlinear system can not only be transformed to a linear system in Brunovsky normal form (cf. (6.50)), but also to a linear system with an arbitrary characteristic polynomial (of degree n). Let us study this in more detail for a single-input affine system (6.85), which we assume to be feedback linearizable to (6.105). As explained in Corollary 6.18 the transformation to (6.105) is performed by taking coordinates $(z_1,\dots,z_n)$ as in (6.106a), thus defining a linear transformation $z = S(x)$, and by the feedback $u = \alpha(x) + \beta(x)v$ as given in (6.106b). It is clear that the *modified* feedback

$$u = -(L_g L_f^{n-1}\varphi(x))^{-1}(L_f^n\varphi + r_{n-1} L_f^{n-1}\varphi + \dots + r_0\varphi)(x) + (L_g L_f^{n-1}\varphi(x))^{-1}v \tag{6.129}$$

results in the linear system

$$\dot{z} = \begin{bmatrix} 0 & 1 & & \\ & & \ddots & \\ & & & 1 \\ -r_0 & \cdots & & -r_{n-1} \end{bmatrix} z + \begin{bmatrix} 0 \\ \vdots \\ 0 \\ 1 \end{bmatrix} v \tag{6.130}$$

with characteristic polynomial $r(\lambda)=\lambda^n+r_{n-1}\lambda^{n-1}+r_{n-2}\lambda^{n-2}+\dots+r_0$. Furthermore it is clear that we can always choose φ in such a way that (cf. Lemma 6.15)

$$L_g L_f^{n-1}\varphi = (-1)^{n-1} \langle d\varphi, ad_f^{n-1}g\rangle = 1. \tag{6.131}$$

Then (6.129) for $v = 0$ reduces to the static state feedback

$$u = -(L_f^n\varphi + r_{n-1} L_f^{n-1}\varphi + \dots + r_0\varphi)(x). \tag{6.132}$$

Remark 6.23 Assume that the nonlinear system (6.85) is already in linear form $\dot{x} = Ax + bu$, and let us see what the feedback (6.132) amounts to in this case. First, by (6.97) and (6.131) the function φ will be a linear function $\varphi(x) = kx$, with the $(1\times n)$-vector k satisfying

$$k\left(b \,\vdots\, Ab \,\vdots\, \cdots \,\vdots\, A^{n-1}b\right) = \left(0 \cdots 0 \;\, 1\right). \tag{6.133}$$

Secondly, the new system of coordinates $z = (z_1,\dots,z_n)$ as in (6.106a) is given as $z_i = L_{Ax}^{i-1}kx = kA^{i-1}x$, $i \in \underline{n}$. Said otherwise, if we apply to $\dot{x} = Ax + bu$ the linear basis transformation $z = Sx$ given as

$$S = \begin{bmatrix} k \\ kA \\ \vdots \\ kA^{n-1} \end{bmatrix}, \tag{6.134}$$

together with the feedback (cf.(6.20))

$$\begin{aligned} u &= -(L^n_{Ax} kx + r_{n-1} L^{n-1}_{Ax} kx + \ldots + r_0 kx) = \\ &= -k\,(A^n + r_{n-1} A^{n-1} + \ldots + r_0 I)x = -kr(A)x, \end{aligned} \tag{6.135}$$

where $r(\lambda) = \lambda^n + r_{n-1}\lambda^{n-1} + \ldots + r_0$, then the system $\dot{x} = Ax + bu$ is transformed into (6.130). Actually, the feedback expression $u = -kr(A)x$, assigning the characteristic polynomial $r(\lambda)$ to the closed loop matrix, is well-known in linear systems theory (Ackermann's formula). Furthermore, it can be seen that the i-th column of the inverse matrix S^{-1} is given as

$$A^{n-i}b + p_{n-1} A^{n-i-1}b + \ldots + p_i b, \tag{6.136}$$

with $p(\lambda)$ the characteristic polynomial of A, and it is well-known that these columns form a basis in which the system is in *controller canonical form*

$$\dot{z} = \begin{bmatrix} 0 & 1 & & \\ & & \ddots & \\ & & & 1 \\ -p_0 & \cdots & & -p_{n-1} \end{bmatrix} z + \begin{bmatrix} 0 \\ \vdots \\ 0 \\ 1 \end{bmatrix} u. \tag{6.137}$$

Notes and References

The problem of feedback linearization was first posed and treated in [Br] for the restricted class of feedback transformations $u = \alpha(x) + v$. Previous work in this area can be found e.g. in [Ko]. The necessary and sufficient conditions for feedback linearization were obtained in [JR], and in a slightly more elaborate way in [HS], [HSM1], see also [Su]. The extension to the general nonlinear case (Theorem 6.12) can be found in [vdS], see also [Su]. For additional results we refer to [Re1], [MBE], [MSH] and the survey [Cl]. The problem of partially linearizing the system was addressed in [IK], [KIR], [MBE], while the largest feedback subsystem was identified in [Ma2], see also [Re1], [Re3]. Linearization by dynamic state feedback has been dealt with in [CLM1,2], [IML]. The problem of global feedback linearization was studied in e.g. [HSM2], [Bo], [Re2], [CTI], [DWE]. Feedback linearization of the input-output map of the system was treated e.g. in [IR], [Is] (see also [Ru], [CIRT]), while feedback linearization of systems with outputs was studied in [CIRT]. A different approach to linearization by feedback was taken in e.g. [RC], [CMR], [WR]. Finally, the problem of approximate feedback linearization has been addressed in [Kr]. The non-genericity of feedback linearizable nonlinear systems, for n (= dim M) - m (= dim U) not too small, has been shown in [Tc].

[Bo] W.M. Boothby, "Some comments on global linearization of nonlinear systems", Syst. Control Lett., 4, pp. 143-147, 1984.

[Br] R.W. Brockett, "Feedback invariants for nonlinear systems", Proc. VIIth IFAC World Congress, Helsinki, pp. 1115-1120, 1978.

[CIRT] D. Cheng, A. Isidori, W. Respondek, T.J. Tarn, "Exact linearization of nonlinear systems with outputs", Math. Systems Theory, 21, pp. 63-83, 1988.

[Cl] D. Claude, "Everything you always wanted to know about linearization", in **Algebraic and Geometric Methods in Nonlinear Control Theory** (eds. M. Fliess, M. Hazewinkel), Reidel, Dordrecht, pp. 181-226, 1986.

[CLM1] B. Charlet, J. Levine, R. Marino, "Two sufficient conditions for dynamic feedback linearization of nonlinear systems", in **Analysis and Optimization of Systems** (eds. A. Bensoussan, J.L. Lions), Lect. Notes Contr. Inf. Sci., 111, Springer, Berlin, pp. 181-192, 1988.

[CLM2] B. Charlet, J. Levine, R. Marino, "On dynamic feedback linearization", Systems Control Lett., 13, pp. 143-151, 1989.

[CMR] C. Champetier, P. Mouyon, C. Reboulet, "Pseudo-linearization of multi-input nonlinear systems", Proc. 23rd IEEE Conf. on Decision and Control, Las Vegas, pp. 96-97, 1984.

[CTI] D. Cheng, T.J. Tarn, A. Isidori, "Global feedback linearization of nonlinear systems", Proc. 23rd IEEE Conf. on Decision and Control, Las Vegas, pp. 74-83, 1984.

[DWE] W. Dayawansa, W.M. Boothby, D.L. Elliott, "Global state and feedback equivalence of nonlinear systems", Systems Control Lett., 6, pp. 229-234, 1985.

[HS] L.R. Hunt, R. Su, "Linear equivalents of nonlinear time-varying systems", Proc. Int. Symposium on Math. Theory of Networks and Systems, Santa Monica, pp. 119-123, 1981.

[HSM1] L.R. Hunt, R. Su, G. Meyer, "Design for multi-input nonlinear systems", in **Differential Geometric Control Theory** (eds. R.W. Brockett, R.S. Millman, H.J. Sussmann), Birkhäuser, Boston, pp. 268-298, 1983.

[HSM2] L.R. Hunt, R. Su, G. Meyer, "Global transformations of nonlinear systems", IEEE Trans. Automat. Contr., AC-28, pp. 24-31, 1983.

[IK] A. Isidori, A.J. Krener, "On feedback equivalence of nonlinear systems", Systems Control Lett., 2, pp. 118-121, 1982.

[IML] A. Isidori, C. Moog, A. de Luca, "A sufficient condition for full linearizability via dynamic state-feedback", 25th IEEE Conf. Decision and Control, Athens, pp. 203-208, 1986.

[IR] A. Isidori, A. Ruberti, "On the synthesis of linear input-output responses for nonlinear systems", Systems Control Lett., 4, pp. 17-22, 1984.

[Is1] A. Isidori, "The matching of a prescribed linear input-output behavior in a nonlinear system", IEEE Trans. Automat. Contr., AC-30, pp. 258-265, 1985.

[Is2] A. Isidori, **Nonlinear Control Systems: An Introduction**, Lect. Notes Contr. Inf. Sci., 72, Springer, Berlin, 1985.

[KIR] A.J. Krener, A. Isidori, W. Respondek, "Partial and robust linearization by feedback", Proc. 22nd IEEE Conf. Decision and Control, San Antonio, pp. 126-130, 1983.

[Ko] W. Korobov, "Controllability, stability of some nonlinear systems", Differencialnyje Uravnienje, 9, pp. 466-469, 1973.

[Kr] A.J. Krener, "Approximate linearization by state feedback and coordinate change", Systems Control Lett., 5, pp. 181-185, 1984.

[Ma1] R. Marino, "Stabilization and feedback equivalence to linear coupled oscillators", Int. J. Control, 39, pp. 487-496, 1984.

[Ma2] R. Marino, "On the largest feedback linearizable subsystem", Systems Control Lett., 6, pp. 345-351, 1986.

[MBE] R. Marino, W.M. Boothby, D.L. Elliott, "Geometric properties of linearizable control systems", Math. Systems Theory, 18, pp. 97-123, 1985.

[RC] C. Reboulet, C. Champetier, "A new method for linearization nonlinear systems: the pseudo-linearization", Int. J. Control, 40, pp. 631-638, 1984.

[Re1] W. Respondek, "Geometric methods in linearization of control systems", in **Mathematical Control Theory** (eds. Cz. Olech, B. Jakubczyk, J. Zabczyk), Banach Center Publications, Polish Scientific Publishers, Warsaw, pp. 453-467, 1985.

[Re2] W. Respondek, "Global aspects of linearization, equivalence to polynomial forms and decomposition of nonlinear systems", in **Algebraic and Geometric Methods in Nonlinear Control Theory** (eds. M. Fliess, M. Hazewinkel), Reidel, Dordrecht, pp. 257-284, 1986.

[Re3] W. Respondek, "Partial linearizations, decompositions and fibre linear systems", in **Theory and Applications of Nonlinear Control Systems** (eds. C.I. Byrnes, A. Lindquist), North-Holland, Amsterdam, pp. 137-154, 1986.

[Ru] W.J. Rugh, "An input-output characterization for linearization by feedback", Systems Control Lett., 4, pp. 227-229, 1984.

[Su] R. Su, "On the linear equivalents of nonlinear systems", Systems Control Lett., 2, pp. 48-52, 1982.

[vdS] A.J. van der Schaft, "Linearization and input-output decoupling for general nonlinear systems", Systems Control Lett., 5, pp. 27-33, 1984.

[Tc] K. Tchon, "On some applications of transversality to system theory", Systems Control Lett., 4, pp. 149-156, 1984.

[WR] J. Wang, W.J. Rugh, "Feedback linearization families for nonlinear systems, IEEE Trans. Automat. Contr., AC-32, pp. 935-940, 1987.

[ZC] M. Zribi, J. Chiasson, "Exact linearization control of a PM stepper motor", Proc. American Control Conference, 1989, Pittsburgh, 1989.

Exercises

6.1 (see also Remark 6.2) **(a)** Consider the nonlinear system (6.5) with $f(x_0) = 0$. Denote its linearization around x_0 and $u = 0$ by $\dot{\bar{x}} = A\bar{x} + Bu$, with $A = \frac{\partial f}{\partial x}(x_0)$ and $B = \left(g_1(x_0) \vdots \cdots \vdots g_m(x_0)\right)$. Suppose the system is feedback linearizable around x_0. Show that around x_0 the system can be also transformed (using state space and feedback transformations) to the linear system $\dot{z} = Az + Bv$, with A and B as above. (In applications this may be a more sensible thing to do than to transform the system into Brunovsky normal form (6.50).)

(b) Consider a system $\dot{x} = f(x,u)$, $f(x_0,u_0) = 0$, which is feedback linearizable around (x_0,u_0). Show that the system can be also transformed into the linear system $\dot{z} = Az + Bv$, with $A = \frac{\partial f}{\partial x}(x_0,u_0)$, $B = \frac{\partial f}{\partial u}(x_0,u_0)$.

6.2 Consider the single-input nonlinear system $\dot{x} = f(x) + g(x)u$ on a $2n$-dimensional manifold M, with $f(x_0) = 0$ and satisfying the strong accessibility rank condition in x_0. Show that the system around x_0 can be transformed using state space transformations and feedback transformations into a system of n coupled linear oscillators with unit masses

$$\dot{z} = \begin{bmatrix} 0_{n\times n} & I_{n\times n} \\ -K & 0_{n\times n} \end{bmatrix} z + \begin{bmatrix} 0 \\ \vdots \\ 0 \\ 1 \end{bmatrix} v,$$

where

$$K = \begin{bmatrix} k_1 & k_{12} & 0 & \cdots & 0 \\ k_{12} & k_2 & k_{23} & & \vdots \\ 0 & k_{23} & k_3 & & 0 \\ \vdots & & & & k_{n-1,n} \\ 0 & \cdots & & k_{n-1,n} & k_n \end{bmatrix},$$

if and only if the system is feedback linearizable around x_0 ([Mal]).

6.3 Prove that the distributions Δ_k defined in (6.19) are feedback invariant.

6.4 ([ZC]) Consider the following nonlinear system (a model of a permanent magnet stepper motor)

$$\begin{aligned}
\dot{x}_1 &= -K_1 x_1 + K_2 x_3 \sin(K_5 x_4) + u_1 \\
\dot{x}_2 &= -K_1 x_2 + K_2 x_3 \cos(K_5 x_4) + u_2 \\
\dot{x}_3 &= -K_3 x_1 \sin(K_5 x_4) + K_3 x_2 \cos(K_5 x_4) - K_4 x_3 + K_6 \sin(4K_5 x_4) - \tau_L/J \\
\dot{x}_4 &= x_3
\end{aligned}$$

(Here x_1, x_2 denote currents, x_3 denotes the rotor speed, x_4 is the motor position, J is the rotor inertia, and τ_L is the load torque, which is assumed to be measurable.)

(a) Verify the conditions for feedback linearizability of the system (in the sense of Remark 6.6) in the point $x_1 = x_2 = x_3 = x_4 = 0$, and compute the controllability indices.

(b) Show that the coordinate transformation involved in the linearizing transformation is given as

$$\begin{aligned}
z_1 &= x_4/K_3 \\
z_2 &= x_3/K_3
\end{aligned}$$

$$z_3 = -x_1 \sin(K_4 x_4) + x_2 \cos(K_5 x_4) - K_4 x_3/K_3 + (K_6/K_3)\sin(4K_5 x_4) - \tau_L/(JK_3)$$

$$z_4 = x_1 \cos(K_5 x_4) + x_2 \sin(K_5 x_4)$$

and compute the corresponding linearizing feedback $u = \alpha(x) + \beta(x)v$.

6.5 Consider the following feedback linearizable system (motivated by the system (6.112) considered in Example 6.20)

$$\frac{d}{dt}\begin{pmatrix} x_1 \\ x_2 \\ x_3 \end{pmatrix} = \begin{pmatrix} x_1 \\ (1-\ln x_3)x_2 \\ -px_1x_3 \end{pmatrix} + \begin{pmatrix} -x_1 \\ -x_2 \\ -x_3 \end{pmatrix} u_1 + \begin{pmatrix} 0 \\ 0 \\ 1 \end{pmatrix} u_2$$

with $x_1 > 0$, $x_2 > 0$, $x_3 > 0$.

(a) Show that $z_1 = \ln(x_1/x_2)$, $z_2 = \ln x_3$, $z_3 = \ln x_1$ is the coordinate transformation involved in a linearizing transformation for this system. Show that the resulting closed-loop system is a *global* linear system on $\mathbb{R}^3$.

(b) Show that $z_1 = \arctan(x_2/x_1)$, $z_1 = -(1+x_1^2)^{-1}\ln x_3$, $z_3 = \ln x_1$ is also part of a linearizing transformation. Show, however, that the resulting closed-loop system is not a global linear system.

6.6 (see Remark 6.23) Show, using the Cayley-Hamilton theorem, that the i-th column of the inverse of the matrix S defined in (6.134) is given as in (6.136). Furthermore, show that the columns form a basis in which the system is in the form (6.137).

6.7 [Br] Consider the single-input system (6.85). Show that the system is feedback linearizable around x_0 to a controllable linear system using the restricted class of feedbacks $u = \alpha(x) + v$ (i.e. $\beta(x) = 1$), if and only if

(i) $\dim D_n(x_0) = n$

(ii) $[ad_f^i g, ad_f^j g](x) \in D_k(x)$, for every $0 \le i \le j \le k$ and $k \in \underline{n-1}$, x near x_0.

6.8 Consider the nonlinear system

$$\dot{x}_1 = \sin x_3 \qquad \dot{x}_4 = x_4 + x_5 u_2 - 1$$
$$\dot{x}_2 = x_1 x_2 \qquad \dot{x}_5 = -x_4 u_2$$
$$\dot{x}_3 = e^{x_1} u_1 + x_3$$

about the equilibrium $x_1 = 0$, $x_2 = 1$, $x_3 = 0$, $x_4 = 1$, $x_5 = 1$.

(a) Verify the conditions for feedback linearizability.

(b) Compute the linearization using Corollary 6.18, and the fact that the system is in "decoupled form".

6.9 Consider the Hamiltonian control system

$$\dot{q}_i = \frac{\partial H}{\partial p_i}(q,p) \qquad i \in \underline{n}$$
$$\dot{p}_i = -\frac{\partial H}{\partial q_i}(q,p) + u_i$$

where $H(q,p) = \frac{1}{2}p^T G(q)p + V(q)$ for some positive definite matrix $G(q)$, and $dV(q_0) = 0$. Check feedback linearizability about the point $(q_0, 0)$.

6.10 Consider the nonlinear system (6.5) with $f(x_0) = 0$, and satisfying the strong accessibility rank condition, together with its extended system

$$\dot{x} = f(x) + \sum_{j=1}^{m} g_j(x)u_j, \qquad u = (u_1, \ldots, u_m) \in \mathbb{R}^m,$$
$$\dot{u} = w.$$

(a) Prove, as a direct consequence of Theorem 6.12, that (6.5) is feedback linearizable around x_0 if and only if the extended system is feedback linearizable around $(x_0, 0)$.

(b) [CLM1] Consider the system

$$\dot{x}_1 = x_2, \qquad \dot{x}_3 = u_2,$$
$$\dot{x}_2 = u_1, \qquad \dot{x}_4 = x_3 - x_3 u_1.$$

Show that this system is not feedback linearizable around 0. Consider the *partial* extended system

$$\dot{x}_1 = x_2, \qquad \dot{x}_3 = u_2, \qquad \dot{u}_1 = w_1,$$
$$\dot{x}_2 = u_1, \qquad \dot{x}_4 = x_3 - x_3 u_1,$$

with state $(x_1, x_2, x_3, x_4, u_1)$ and inputs (w_1, u_2). Show that this system is feedback linearizable, and compute the linearizing transformation.

7
Controlled Invariant Distribution and the Disturbance Decoupling Problem

In this chapter, Section 7.1, we will introduce and discuss the concept of controlled invariance for nonlinear systems. Controlled invariant distributions play a crucial role in various synthesis problems like for instance the disturbance decoupling problem and the input-output decoupling problem. A detailed account of the disturbance decoupling problem together with some worked examples will be given in Section 7.2. Later, in Chapter 9, we will exploit controlled invariant distributions in the input-output decoupling problem.

7.1 Controlled Invariant Distributions

Consider the smooth nonlinear control system

$$\dot{x} = f(x) + \sum_{i=1}^{m} g_i(x)u_i \tag{7.1}$$

where $x = (x_1, \ldots, x_n)$ are local coordinates for a smooth manifold M and $f, g_1, \ldots, g_m$ are smooth vectorfields. Recall, see Definition 3.31, that a smooth distribution D is called *invariant* for the system (7.1) if

$$[f, D] \subset D, \tag{7.2a}$$

$$[g_i, D] \subset D, \quad i \in \underline{m}. \tag{7.2b}$$

Such invariant distributions play a central role in the output invariance of a nonlinear system, cf. Section 4.3. We generalize the notion of an invariant distribution for the system (7.1) by allowing for a regular static state feedback, i.e.

$$u = \alpha(x) + \beta(x)v \tag{7.3}$$

where $\alpha\colon M \to \mathbb{R}^m$ and $\beta\colon M \to \mathbb{R}^{m\times m}$ are smooth mappings with $\beta(x)$ nonsingular for all x in M, and where $v = (v_1, \ldots, v_m)$ denotes the new inputs. Applying (7.3) to (7.1) yields the feedback modified system

$$\dot{x} = \tilde{f}(x) + \sum_{i=1}^{m} \tilde{g}_i(x)v_i \tag{7.4}$$

where

$$\tilde{f}(x) = f(x) + \sum_{i=1}^{m} g_i(x)\alpha_i(x), \tag{7.5a}$$

$$\tilde{g}_i(x) = \sum_{j=1}^{m} g_j(x)\beta_{ji}(x), \quad i \in \underline{m}. \tag{7.5b}$$

We now define

Definition 7.1 *A smooth distribution D on M is called* controlled invariant *for the dynamics (7.1) if there exists a regular static state feedback (7.3) such that D is invariant for the feedback modified system (7.4), i.e.*

$$[\tilde{f},D] \subset D, \tag{7.6a}$$

$$[\tilde{g}_i,D] \subset D, \quad i \in \underline{m}. \tag{7.6b}$$

As we will see in the next section, this generalization of an invariant distribution will be instrumental in the solution of various synthesis problems. At this point we observe that it may be difficult to check if a given distribution D is controlled invariant, because this requires to test if there does exist some feedback (7.3) which makes D invariant. Before establishing convenient criteria on the distribution D and the original dynamics (7.1) which guarantee that D is controlled invariant, we will briefly discuss controlled invariance for a linear system.

Example 7.2 Consider the linear system

$$\dot{x} = Ax + Bu, \tag{7.7}$$

with $x \in \mathbb{R}^n$, $u \in \mathbb{R}^m$ and A and B matrices of appropriate size. In analogy with Section 3.3, a subspace $\mathcal{V} \subset \mathbb{R}^n$ is called controlled invariant (or sometimes called (A,B)-invariant) if there exists a linear feedback $u = Fx + Gv$, $|G| \neq 0$, which makes $\mathcal{V}$ invariant, thus

$$(A+BF)\mathcal{V} \subset \mathcal{V}. \tag{7.8}$$

A standard simple result from geometric linear system theory states that such a feedback matrix F exists if and only if

$$A\mathcal{V} \subset \mathcal{V} + \operatorname{Im} B. \tag{7.9}$$

As in Section 3.3 we can put this in a more differential geometric setting by identifying the subspace $\mathcal{V}$ with its corresponding flat distribution $D_{\mathcal{V}}$. Let $\{v_1,\ldots,v_r\}$ form a basis for $\mathcal{V}$, then $D_{\mathcal{V}}$ is the distribution generated by the constant vectorfields $v_1,\ldots,v_r$. The condition (7.8) then translates into

$$[(A+BF)x,v_\ell] \in D_{\mathcal{V}}(x), \quad \ell \in \underline{r},\ x \in \mathbb{R}^n, \tag{7.10}$$

which is the "linear" counterpart of (7.6a). Denoting $(BG)_i$ as the constant vectorfield formed by the i-th column of the matrix BG, $i \in \underline{m}$, we also obtain that

$$[(BG)_i,v_\ell] = 0 \in D_{\mathcal{V}}(x), \quad \ell \in \underline{r},\ i \in \underline{m},\ x \in \mathbb{R}^n, \tag{7.11}$$

which yields the counterpart of (7.6b). Since the condition (7.11) is automatically satisfied for a linear system, we obtain as a necessary and sufficient condition for the controlled invariance of $D_{\mathcal{V}}$ that, see (7.9),

$$[Ax,v_\ell] \in D_{\mathcal{V}}(x) + D_{\mathrm{Im}\,B}(x), \quad \ell \in \underline{r},\ x \in \mathbb{R}^n, \tag{7.12}$$

where $D_{\mathrm{Im}\,B}$ is the flat distribution corresponding to Im B. □

We next turn our attention to the question under which conditions a smooth distribution D is controlled invariant for the nonlinear system (7.1). First we identify a set of necessary conditions on D and the vectorfields $f,g_1,\ldots,g_m$ that should hold when D is controlled invariant under the feedback (7.3). Because $g_i(x) = \sum_{j=1}^m \tilde{g}_j(x)\beta_{ji}^{-1}(x)$, $i \in \underline{m}$, (see 7.5b), we obtain from (7.6b) that for any vectorfield $X \in D$

$$[g_i(x),X(x)] = [\sum_{j=1}^m \tilde{g}_j(x)\beta_{ji}^{-1}(x),X(x)] = \sum_{j=1}^m [\tilde{g}_j(x),X(x)]\beta_{ji}^{-1}(x) -$$

$$- \sum_{j=1}^m \tilde{g}_j(x)L_X\beta_{ji}^{-1}(x) \in D(x) + G(x), \quad i \in \underline{m}, \tag{7.13a}$$

where $G(x)$ is the distribution generated by the input vectorfields:

$$G(x) = \mathrm{span}\{g_1(x),\ldots,g_m(x)\} = \mathrm{span}\{\tilde{g}_1(x),\ldots,\tilde{g}_m(x)\}. \tag{7.14}$$

Similarly (7.6a) yields, using (7.13a)

$$[f(x),X(x)] = [f(x) + \sum_{i=1}^m g_i(x)\alpha_i(x),X(x)] - [\sum_{i=1}^m g_i(x)\alpha_i(x),X(x)] =$$

$$[\tilde{f}(x),X(x)] - \sum_{i=1}^m [g_i(x),X(x)]\alpha_i(x) + \sum_{i=1}^m g_i(x)L_X\alpha_i(x) \in D(x) + G(x). \tag{7.13b}$$

Summarizing, we have obtained the following necessary conditions for the controlled invariance of the distribution D

$$[f,D] \subset D + G, \tag{7.15a}$$

$$[g_i, D] \subset D + G, \qquad i \in \underline{m}. \tag{7.15b}$$

Assuming some regularity conditions we will see that the conditions (7.15a,b) are also sufficient for *local* controlled invariance.

Definition 7.3 *A smooth distribution D on M is called* locally controlled invariant *for the dynamics (7.1) if for each point $x_0 \in M$ there exists a neighborhood V of x_0 and a regular static state feedback (7.3) defined on V such that the feedback modified dynamics (7.4) defined on V satisfy (7.6a,b) on V.*

Remark 7.4 Notice that a locally controlled invariant distribution is in general not controlled invariant. The point is that the locally defined feedbacks of Definition 7.3 need not patch together into a globally defined smooth feedback which makes the distribution invariant.

As announced, the following theorem shows the sufficiency of (7.15a,b) for local controlled invariance, provided some constant dimension conditions are met.

Theorem 7.5 *Consider the smooth nonlinear system (7.1) and assume that the distribution G has constant dimension. Let D be an involutive distribution of constant dimension and assume $D \cap G$ has constant dimension. Then the distribution D is locally controlled invariant if and only if*

$$[f,D] \subset D + G, \tag{7.15a}$$

$$[g_i, D] \subset D + G, \qquad i \in \underline{m}. \tag{7.15b}$$

Proof As we already have shown the necessity of (7.15a,b), we only have to prove the sufficiency part of the theorem. Let $x_0 \in M$. We have to show the existence of a feedback (7.3) in a neighborhood V of x_0 such that (7.6a,b) hold true on V for the modified dynamics (7.4). Consider the constant dimensional distributions D, G and $D + G$ and assume first that $D \cap G = 0$. Let $\dim D = k$ and $\dim G = m$. In a neighborhood V_1 of x_0 we may choose $r = n-m-k$ vectorfields $X_1, \ldots, X_r$ such that

$$\dim(D+G+\text{span}\{X_1,\ldots,X_r\}) = \dim D + \dim G + \dim(\text{span}\{X_1,\ldots,X_r\}) = n \tag{7.16}$$

Also, by Corollary 2.43 (Frobenius), we can find local coordinates on a neighborhood V_2 of x_0, again denoted as $(x_1,\ldots,x_n)$ such that

$$D = \text{span}\{\frac{\partial}{\partial x_1},\ldots,\frac{\partial}{\partial x_k}\} \tag{7.17}$$

and thus

$$D + G + \text{span}\{X_1,\ldots,X_r\} = \text{span}\{\frac{\partial}{\partial x_1},\ldots,\frac{\partial}{\partial x_n}\}. \tag{7.18}$$

So we find that the distribution $G + \text{span}\{X_1,\ldots,X_r\}$ is spanned by $n-k$ vectorfields

$$\begin{cases} Z_1(x) = \dfrac{\partial}{\partial x_{k+1}} + \displaystyle\sum_{i=1}^{k} \eta_i^1(x)\frac{\partial}{\partial x_i} \\ \quad\vdots \\ Z_{n-k}(x) = \dfrac{\partial}{\partial x_n} + \displaystyle\sum_{i=1}^{k} \eta_i^{n-k}(x)\frac{\partial}{\partial x_i} \end{cases} \tag{7.19}$$

In the sequel we will use $G(x)$ to denote the distribution of input vectorfields (7.14) as well as the $n\times m$-matrix formed by the input vectorfields: $G(x) = \big(g_1(x),\ldots,g_m(x)\big)$. Define the $n\times(n-k)$-matrices $B(x)$ and $Z(x)$ by

$$\begin{cases} B(x)=\big(g_1(x),\ldots,g_m(x),X_1(x),\ldots,X_r(x)\big) \\ Z(x)=\big(Z_1(x),\ldots\ldots\ldots\ldots,Z_{n-k}(x)\big) \end{cases} \tag{7.20}$$

Letting $\bar{G}(x)$ and $\bar{Z}(x)$ be the $(n-k)\times m$-matrix, respectively $(n-k)\times(n-k)$-matrix which are obtained from the matrix $G(x)$, respectively $Z(x)$, by skipping the first k rows, we obtain from (7.15b) on the neighborhood $V = V_1 \cap V_2$ of x_0 that

$$\frac{\partial \bar{G}}{\partial x_i}(x) = \bar{G}(x)K_i(x), \qquad i \in \underline{k}, \tag{7.21}$$

for some $m\times m$-matrices $K_1(x),\ldots,K_k(x)$. Because $\text{Im}\, B(x) + D(x) = T_xM$ we have that

$$\frac{\partial \bar{B}}{\partial x_i}(x) = \bar{B}(x)\tilde{K}_i(x), \qquad i \in \underline{k}, \tag{7.22}$$

where $\bar{B}(x)$ is the $(n-k)\times(n-k)$-matrix obtained from $B(x)$ by deleting the first k rows and $\tilde{K}_1(x),\ldots,\tilde{K}_k(x)$ are some suitably chosen $(n-k)\times(n-k)$-matrices. From the special form of the matrix $B(x)$ we conclude from (7.21) and (7.22) that

$$\tilde{K}_i(x) = \begin{bmatrix} K_i(x) & * \\ 0 & * \end{bmatrix}, \qquad i \in \underline{k}. \tag{7.23}$$

As $\{g_1(x),\ldots,g_m(x),X_1(x),\ldots,X_r(x)\}$ and $\{Z_1(x),\ldots,Z_{n-k}(x)\}$ both span the same distribution $G + \text{span}\{X_1,\ldots,X_r\}$, there exists a nonsingular $(n-k)\times(n-k)$-matrix $M(x)$ such that

$$B(x) = Z(x)M(x). \tag{7.24}$$

Partitioning the matrix M as

$$M(x) = \left[\begin{array}{c|c} M^1(x) & M^2(x) \\ \hline M^3(x) & M^4(x) \end{array}\right] \begin{matrix} \updownarrow m \\ \\ \end{matrix}, \tag{7.25}$$

$$\underset{m}{\longleftrightarrow}$$

then we may assume, without loss of generality, that the $m\times m$-matrix $M^1(x)$ is nonsingular. For if this is not the case, then a permutation of columns of the matrix $Z(x)$ will produce a nonsingular $m\times m$-matrix in the upper left corner of $M(x)$. From (7.24) we obtain, using (7.20) and (7.19), that

$$\frac{\partial \bar{B}}{\partial x_i}(x) = \bar{Z}(x)\frac{\partial M}{\partial x_i}(x), \qquad i \in \underline{k}, \tag{7.26}$$

and so, from (7.22) and again (7.24), we derive

$$\bar{B}(x)\tilde{K}_i(x) = \bar{B}(x)\big(M(x)\big)^{-1}\frac{\partial M}{\partial x_i}(x), \qquad i \in \underline{k}, \tag{7.27}$$

which implies

$$\frac{\partial M}{\partial x_i}(x) = M(x)\tilde{K}_i(x), \qquad i \in \underline{k}, \tag{7.28}$$

since the matrix $\bar{B}(x)$ has full column rank. Using (7.25) and (7.23) we conclude that the nonsingular $m\times m$-matrix $M^1(x)$ satisfies

$$\frac{\partial M^1}{\partial x_i}(x) = M^1(x)K_i(x), \qquad i \in \underline{k}. \tag{7.29}$$

Define now the nonsingular $m\times m$-matrix $\beta(x)$ as

$$\beta(x) = \big(M^1(x)\big)^{-1}, \tag{7.30}$$

then this matrix $\beta(x)$ yields the desired change of input vectorfields. Namely using the identity

$$\frac{\partial M^1}{\partial x_i}(x)\big(M^1(x)\big)^{-1} + M^1(x)\frac{\partial}{\partial x_i}\big(M^1(x)\big)^{-1} = 0, \qquad i \in \underline{n}, \tag{7.31}$$

we obtain

$$\frac{\partial}{\partial x_i}\big(\bar{G}(x)\beta(x)\big) = \frac{\partial}{\partial x_i}\big(\bar{G}(x)\big(M^1(x)\big)^{-1}\big) = \frac{\partial \bar{G}}{\partial x_i}(x)\big(M^1(x)\big)^{-1} + \bar{G}(x)\frac{\partial}{\partial x_i}\big(M^1(x)\big)^{-1} =$$
$$= \bar{G}(x)K_i(x)\big(M^1(x)\big)^{-1} - \bar{G}(x)\big(M^1(x)\big)^{-1}M^1(x)K_i(x)\big(M^1(x)\big)^{-1} = 0, \qquad i \in \underline{k},$$

and thus

$$[G\beta, D] \subset D. \tag{7.32}$$

So far we have assumed that the distribution $D \cap G$ equals the zero distribution. In case the constant dimensional distribution $D \cap G$ has positive dimension, say q, we first construct an $m\times m$ transformation-matrix $\tilde{\beta}(x)$ such that for the transformed vectorfields $\tilde{g}_1,\dots,\tilde{g}_m$ defined via (7.5b) we have $D \cap G = \mathrm{span}\{\tilde{g}_1,\dots,\tilde{g}_q\}$. Obviously the vectorfields $\tilde{g}_1,\dots,\tilde{g}_q$ satisfy (7.6b), and for the other vectorfields $\tilde{g}_{q+1},\dots,\tilde{g}_m$ we may use a similar procedure as given in case $D \cap G = 0$.

Finally we have to show the existence of an m-vector $\alpha(x)$ such that the vectorfield $\tilde{f}(x) = f(x) + \sum_{i=1}^{m} g_i(x)\alpha_i(x)$ satisfies (7.6a). As we already have obtained the nonsingular $m\times m$-matrix $\beta(x)$ such that (7.6b) is fulfilled, we may as well construct an m-vector $\tilde{\alpha}(x)$ such that $\tilde{f}(x) = f(x) + \sum_{i=1}^{m} \tilde{g}_i(x)\tilde{\alpha}_i(x)$ satisfies (7.6a). (The vectors $\alpha(x)$ and $\tilde{\alpha}(x)$ are related via $\beta(x)\tilde{\alpha}(x) = \alpha(x)$.) In the above coordinates where D is given via (7.17), equation (7.15) yields the existence of m-vectors $\bar{\alpha}_1(x),\dots,\bar{\alpha}_k(x)$ and vectorfields $D_1(x),\dots,D_k(x)$ in D such that

$$\frac{\partial f}{\partial x_i}(x) = G^1(x)\bar{\alpha}_i(x) + D_i(x), \qquad i \in \underline{k}, \tag{7.33}$$

where $G^1(x) = [\tilde{g}_1(x),\dots,\tilde{g}_m(x)]$. Skipping again the first k rows in (7.33) yields an equation of the form

$$\frac{\partial \bar{f}}{\partial x_i}(x) = \bar{G}^1(x)\bar{\alpha}_i(x), \qquad i \in \underline{k}. \tag{7.34}$$

As the $(n-k)\times m$-matrix $\bar{G}^1(x)$ satisfies (see (7.32))

$$\frac{\partial \bar{G}^1}{\partial x_i}(x) = 0, \qquad i \in \underline{k}, \tag{7.35}$$

and

$$\frac{\partial^2 \bar{f}}{\partial x_j \partial x_i}(x) = \frac{\partial^2 \bar{f}}{\partial x_i \partial x_j}(x), \qquad \text{for all } i,j \in \underline{n}, \tag{7.36}$$

we obtain from (7.34) that

$$\frac{\partial \bar{\alpha}_i}{\partial x_j}(x) = \frac{\partial \bar{\alpha}_j}{\partial x_i}(x), \qquad i,j \in \underline{k}. \tag{7.37}$$

However, this is a well-known set of integrability conditions. In fact, define the m-vector $\tilde{\alpha}(x)$ by

$$\begin{aligned} \tilde{\alpha}(x) = & \int_0^{x_1} \bar{\alpha}_1(\tilde{x}_1, 0, \ldots, 0, x_{k+1}, \ldots, x_n)\, d\tilde{x}_1 + \\ & + \int_0^{x_2} \bar{\alpha}_2(x_1, \tilde{x}_2, 0, \ldots, 0, x_{k+1}, \ldots, x_n)\, d\tilde{x}_2 + \\ & + \ldots + \int_0^{x_k} \bar{\alpha}_k(x_1, \ldots, x_{k-1}, \tilde{x}_k, x_{k+1}, \ldots, x_n)\, d\tilde{x}_k, \end{aligned} \tag{7.38}$$

then it follows from (7.38) that this vector satisfies

$$\frac{\partial \tilde{\alpha}}{\partial x_i}(x) = \bar{\alpha}_i(x), \qquad i \in \underline{k}. \tag{7.39}$$

$\tilde{\alpha}(x)$ is the required feedback since it can be checked that the vectorfield $\tilde{f}(x) = f(x) + \sum_{i=1}^{m} \tilde{g}_i(x)\tilde{\alpha}_i(x)$ satisfies indeed (7.6b). This completes the proof. □

Remark 7.6 The underlying result of the if part of the proof is that the set of partial differential equations (7.29) has a locally defined solution $M^1(x)$. The necessary and sufficient condition for the existence of such a solution is that the matrices $K_i(x)$, $i \in \underline{k}$, satisfy

$$\frac{\partial K_i}{\partial x_j}(x) - \frac{\partial K_j}{\partial x_i}(x) + K_i(x)K_j(x) - K_j(x)K_i(x) = 0, \qquad i,j \in \underline{k}. \tag{7.40}$$

These equations are called the *integrability conditions* for (7.29). (Compare this with the classical version of the Frobenius' Theorem of Chapter 2, i.e. Corollary 2.45.) The necessity of (7.40) follows by assuming that a solution $M^1(x)$ of (7.29) exists. Then (7.40) follows by the fact that

$$\frac{\partial^2 M^1}{\partial x_i \partial x_j}(x) = \frac{\partial^2 M^1}{\partial x_j \partial x_i}(x), \qquad i,j \in \underline{k}. \tag{7.41}$$

On the other hand one obtains (7.40) by using (7.22) and

$$\frac{\partial^2 \bar{B}}{\partial x_j \partial x_i}(x) = \frac{\partial^2 \bar{B}}{\partial x_i \partial x_j}(x), \qquad i,j \in \underline{k}. \tag{7.42}$$

So the remaining thing to be shown is that the integrability conditions (7.40) are indeed sufficient for the existence of a solution of (7.29).

Theorem 7.5 gives a "geometric" proof of this.

The proof of Theorem 7.5 reveals that the conditions (7.15a,b) will only guarantee the local existence of a feedback which makes the distribution invariant. One needs further assumptions on the manifold M and the distribution D in order that a regular feedback on M exists, which renders D invariant. We shall not pursue the mathematical problems of local versus global controlled invariance here, but confine us to the local solutions as obtained in Theorem 7.5.

7.2 The Disturbance Decoupling Problem

In this section we study in detail the Disturbance Decoupling Problem for nonlinear control systems. Instrumental in the (local) solution of this problem will be the notion of controlled invariance as introduced in the previous section. As announced in Chapter 4 an essential role in the solution is played by the concept of output invariance, cf. Section 4.3. Consider the nonlinear dynamics

$$\dot{x} = f(x) + \sum_{i=1}^{m} g_i(x)u_i + \sum_{i=1}^{\ell} e_i(x)d_i \tag{7.43}$$

where $f, g_1, \ldots, g_m$ and $(u_1, \ldots, u_m)$ are as in Section 7.1, while $e_1, \ldots, e_\ell$ are smooth vectorfields on M and $d = (d_1, \ldots, d_\ell)$ is an arbitrary unknown time-function. The elements of the vector d can be interpreted as disturbances or unknown inputs acting on the system. Together with the dynamics (7.43), we consider the outputs

$$y = h(x) \tag{7.44}$$

where $h: M \to \mathbb{R}^p$ is a smooth map. From Proposition 4.16 we know that the disturbances d do not affect the outputs y if there exists a constant dimensional involutive distribution D on M with the following three properties

(i) $\quad [f, D] \subset D, \qquad$ (7.2a)

$\quad [g_i, D] \subset D, \quad i \in \underline{m}, \qquad$ (7.2b)

(ii) $\quad e_j \in D, \quad j \in \underline{\ell}, \qquad$ (7.45)

(iii) $\quad D \subset \ker dh. \qquad$ (7.46)

Obviously, these conditions for output invariance are usually not met and thus the disturbances d do influence the output. This leads to the Disturbance Decoupling Problem.

Problem 7.7 Disturbance Decoupling Problem (DDP) *Consider the nonlinear system (7.43,44). Under which conditions can we find a regular static state feedback (7.3) such that in the feedback modified dynamics*

$$\dot{x} = \tilde{f}(x) + \sum_{i=1}^{m} \tilde{g}_i(x)v_i + \sum_{i=1}^{\ell} e_i(x)d_i \tag{7.47}$$

the disturbances d do not influence the outputs (7.44)?

Completely analogous to section 4.3 we obtain the following result.

Proposition 7.8 *The Disturbance Decoupling Problem is solvable for the smooth system (7.43,44) if there exists a constant dimensional involutive distribution D which is controlled invariant and which satisfies the condition*

$$\text{span}\{e_1,\ldots,e_\ell\} \subset D \subset \bigcap_{j=1}^{p} \ker dh_j = \ker dh. \tag{7.48}$$

In case the system (7.43,44) is analytic *a necessary and sufficient condition for the solvability of the Disturbance Decoupling Problem is that there exists an analytic involutive controlled invariant distribution D satisfying (7.48).*

Proof The first part of the statement follows immediately from Proposition 4.16, whereas the second result is a consequence of Proposition 4.14. □

Proposition 7.8 completely solves the DDP for analytic systems and provides a sufficient condition for its solvability in case the system is smooth. However, in both cases this result is by itself not very useful as it may be difficult to check if there exists a controlled invariant distribution satisfying (7.48). To circumvent this difficulty, we approach the problem in a slightly different manner. We first search for the maximal controlled invariant distribution D^* in ker dh - provided such an object does exist - and then we check whether D^* contains the disturbance vectorfields $e_1,\ldots,e_\ell$. The following example shows that this approach indeed works for the linear DDP.

Example 7.9 Consider the linear system

$$\begin{cases} \dot{x} = Ax + Bu + Ed \\ y = Cx \end{cases} \tag{7.49}$$

with $x \in \mathbb{R}^n$, $u \in \mathbb{R}^m$, $y \in \mathbb{R}^p$, $d \in \mathbb{R}^\ell$, A, B, C and E matrices of appropriate size. In the *linear* Disturbance Decoupling Problem one searches for a linear state feedback $u = Fx + Gv$, $|G| \neq 0$, such that in the modified dynamics the disturbances d do not affect the output. The solvability of the linear DDP is known to be equivalent to (see the references cited at the end of this chapter) the existence of a controlled invariant subspace $\mathcal{V}$ which satisfies

$$\text{Im } E \subset \mathcal{V} \subset \ker C \tag{7.50}$$

which is the linear counterpart of the condition (7.48) stated in Proposition 7.8. On the other hand, given the subspace ker C, there exists a unique maximal controlled invariant subspace $\mathcal{V}^*$ for the dynamics of (7.49) contained in ker C, i.e. $\mathcal{V}^*$ is controlled invariant and contains any other controlled invariant subspace contained in ker C. Therefore, it immediately follows that the linear DDP is solvable for (7.49) if and only if

$$\text{Im } E \subset \mathcal{V}^*, \tag{7.51}$$

with $\mathcal{V}^*$ the maximal controlled invariant subspace in ker C. Provided (7.51) holds, a feedback $u = Fx + Gv$ which solves the linear DDP is given by an $m\times n$-matrix F such that $(A+BF)\mathcal{V}^* \subset \mathcal{V}^*$ and an arbitrary nonsingular $m\times m$-matrix G. □

In the sequel we will closely mimic the solution of the linear DDP as sketched in Example 7.9. The next observations show that similarly to the linear case, there exists a largest involutive distribution D^* contained in ker dh which satisfies (7.15a,b).

Proposition 7.10 *Let D be a distribution contained in* ker dh *satisfying (7.15a,b). Then also $\bar{D}$, the involutive closure of D, see (2.132), is contained in* ker dh *and satisfies (7.15a,b).*

Proof As ker dh is an involutive distribution, we immediately have that $D \subset \bar{D} \subset \ker dh$. Now let X_1 and X_2 be smooth vectorfields in D. Then by the Jacobi-identity

$$[f,[X_1,X_2]] = -[X_1,[X_2,f]] - [X_2,[f,X_1]] \in \bar{D} + G,$$

and also

$$[g_i,[X_1,X_2]] = -[X_1,[X_2,g_i]] - [X_2,[g_i,X_1]] \in \bar{D} + G, \; i \in \underline{m}.$$

Repeating this argument for iterated Lie brackets of vectorfields in D yields the desired conclusion. □

Proposition 7.11 *Let D_1 and D_2 be distributions in* ker dh *satisfying (7.15a,b). Then the distribution $D_1 + D_2$ is contained in* ker dh *and also satisfies (7.15a,b).*

Proof This follows immediately by observing that a smooth vectorfield X in $D_1 + D_2$ may be decomposed (locally) as the sum $X = X_1 + X_2$ with $X_1 \in D_1$ and $X_2 \in D_2$ and then writing out $[f,X]$ and $[g_i,X]$, $i \in \underline{m}$. □

Because the zero-distribution trivially is contained in ker dh and satisfies (7.15a,b) we have as a result:

Corollary 7.12 *There exists a unique involutive distribution in* ker dh *that satisfies (7.15a,b) and which contains all distributions in* ker dh *satisfying (7.15a,b). This distribution will be denoted as $D^*\left(f,g;\bigcap_{j=1}^{p}\ker dh_j\right)$ or, when no confusion arises, as D^*.*

Using the foregoing analysis we can effectively solve the DDP in a *local* way. That is, we will solve

Problem 7.13 (Local Disturbance Decoupling Problem) *Consider the nonlinear system (7.43,44). Under which conditions can we find for each point $x_0 \in M$ a regular static state feedback (7.3) defined on a neighborhood V of x_0 such that in the modified dynamics (7.47) defined on V the disturbances d do not influence the outputs?*

Using Corollary 7.12 we obtain a solution of Problem 7.13 in case that the distributions D^*, $D^* \cap G$ and G are constant dimensional.

Theorem 7.14 *Consider the nonlinear system (7.43,44). Suppose that the distributions D^*, $D^* \cap G$ and G are constant dimensional. Then the Local Disturbance Decoupling Problem is solvable if and only if*

$$\operatorname{span}\{e_1,\dots,e_\ell\} \subset D^*. \tag{7.52}$$

The effectiveness of Theorem 7.14 lies in the fact that there exists an algorithm which computes D^* in regular cases. Consider the algorithm (the

D^*-algorithm):

$$\begin{cases} D^0 = TM \\ D^{\mu+1} = \ker dh \cap \{X \in V(M) \mid [f,X] \in D^\mu + G,\ [g_i,X] \in D^\mu + G,\ i \in \underline{m}\} \end{cases} \tag{7.53}$$

where $V(M) = V^\infty(M)$ denotes the set of smooth vectorfields on M. Suppose the following holds.

Assumption 7.15 *For all $\mu \geq 0$ the distributions D^μ and $D^\mu \cap G$ as well as the distribution G have constant dimension on M (or equivalently $D^\mu + G$, $\mu \geq 0$ has constant dimension).*

Proposition 7.16 *Consider the algorithm (7.53) under Assumption 7.15. Then*

(i) $D^0 \supset D^1 \supset \cdots \supset D^\mu \supset D^{\mu+1} \supset \cdots$, (7.54)

(ii) *D^μ is involutive for $\mu \geq 0$*, (7.55)

(iii) $D^* = D^n$, (7.56)

(iv) *If $D \subset \ker dh$ is a distribution meeting the requirements of Theorem 7.5 then $D \subset D^*$.*

Proof (i) Clearly $D^0 \supset D^1$. Now suppose $D^\mu \supset D^{\mu+1}$, then

$$D^{\mu+2} = \ker dh \cap \{X \in V(M) \mid [f,X] \in D^{\mu+1} + G,\ [g_i,X] \in D^{\mu+1} + G,\ i \in \underline{m}\}$$

$$\subset \ker dh \cap \{X \in V(M) \mid [f,X] \in D^\mu + G,\ [g_i,X] \in D^\mu + G,\ i \in \underline{m}\} = D^{\mu+1}$$

which proves (i).

(ii) Clearly D^0 is involutive. Next suppose D^μ is involutive and let $X_1, X_2 \in D^{\mu+1}$. This implies $[f,X_k] \in D^\mu + G$ and $[g_i,X_k] \in D^\mu + G$, $k = 1,2$, $i \in \underline{m}$, as well $X_1, X_2 \in \ker dh$. Then $[X_1,X_2] \in \ker dh$ and moreover using the Jacobi identity one finds that $\big[f,[X_1,X_2]\big] \in [D^{\mu+1},D^\mu+G]$ and $\big[g_i,[X_1,X_2]\big] \in [D^{\mu+1},D^\mu+G]$, $i \in \underline{m}$. As D^μ is involutive we have $[D^{\mu+1},D^\mu+G] \subset D^\mu + G$, which proves the assertion.

(iii) From (i) and (ii) we conclude that the distributions $\{D^\mu\}$ form a decreasing sequence of involutive distributions which by Assumption 7.15 are of constant dimension. The only thing we need to prove is that the sequence stabilizes, i.e. if for some μ, $D^{\mu+1} = D^\mu$, then $D^{\mu+k} = D^\mu$ for all $k = 2,3,\ldots$. But this follows directly from the algorithm (7.53) as $D^{\mu+1} = D^\mu$ implies $D^{\mu+2} = D^{\mu+1}$. As long as we have strict inclusion in (7.53) the dimension of the distributions D^μ decreases with at least 1 in each step of the algorithm, from which we may conclude that the algorithm will terminate in at most n steps.

(iv) Assume $D \subset \ker dh$ is involutive, has constant dimension as well as $D \cap G$ and G and satisfies $[f,D] \subset D + G$, $[g_i,D] \subset D + G$, $i \in \underline{m}$. Obviously we have $D \subset \ker dh = D^1$. Now assume $D \subset D^\mu$, then

$$\begin{aligned} D &= \ker dh \cap \{X \in V(M) \mid [f,X] \in D + G,\ [g_i,X] \in D + G,\ i \in \underline{m}\} \\ &\subset \ker dh \cap \{X \in V(M) \mid [f,X] \in D^\mu + G,\ [g_i,X] \in D^\mu + G,\ i \in \underline{m}\} \\ &= D^{\mu+1}. \end{aligned}$$

Therefore $D \subset D^\mu$ for all μ, and so $D \subset D^n = D^*$. □

Note that the algorithm (7.53) under the Assumption 7.15 precisely produces the maximal distribution in ker dh meeting the requirements of Theorem 7.5, and thus in order to find a local solution to the DDP we only need to verify the hypothesis (7.52) of Theorem 7.14 for it.

As we will see later the algorithm (7.53) is very much inspired by a corresponding algorithm for computing the maximal controlled invariant subspace for a linear system. For computational reasons we also give a dual version of it, which in some cases is somewhat easier to handle. With the smooth distribution G we define, see Chapter 2, the co-distribution ann G which annihilates G, i.e. for $x \in M$

$$\text{ann } G(x) = \{\omega(x) \mid \omega \text{ is smooth one-form on } M \text{ with } \omega(X) = 0 \text{ for all } X \in G\}. \tag{7.57}$$

Consider the algorithm

$$\begin{cases} P^0 = 0 \\ P^1 = \text{span}\{dh_1,\dots,dh_p\} \\ P^{\mu+1} = P^\mu + L_f(P^\mu \cap \text{ann } G) + \sum_{i=1}^{m} L_{g_i}(P^\mu \cap \text{ann } G) \end{cases} \tag{7.58}$$

In analogy with the Assumption 7.15 for the algorithm (7.53) we assume

Assumption 7.17 *The co-distribution* ann G *and the co-distributions* P^μ *and* $P^\mu \cap$ ann G, $\mu \geq 0$, *have constant dimension on* M *(or equivalently* P^μ + ann G *has constant dimension).*

Under the constant dimension hypothesis the algorithms (7.53) and 7.58) are dual. Precisely:

Proposition 7.18 *Consider the algorithms (7.53) and (7.58) under the Assumption 7.15 respectively 7.17. Then*

$$D^{\mu} = \ker P^{\mu}, \qquad \mu \geq 0, \tag{7.59}$$

or, equivalently

$$\operatorname{ann} D^{\mu} = P^{\mu}, \qquad \mu \geq 0. \tag{7.60}$$

Proof The claim is obviously true for $\mu = 0$ and $\mu = 1$. Let us show the assertion for $\mu = 2$. The proof for arbitrary μ is completely analogous. Let $X \in \ker P^2$, then we have to prove $X \in D^2$. We have, see (7.58), the following three properties for X: (i) $X \in \ker P^1$, (ii) $X \in \ker L_f(P^1 \cap \operatorname{ann} G)$ and (iii) $X \in \ker L_{g_i}(P^1 \cap \operatorname{ann} G)$, $i \in \underline{m}$. So $(L_f\omega)(X) = (L_{g_i}\omega)(X) = 0$ for all $\omega \in P^1 \cap \operatorname{ann} G$ and $i \in \underline{m}$. Now, using the properties of Lie-derivatives for one-forms, see equation (2.169), we have

$$(L_f\omega)(X) = L_f\big(\omega(X)\big) - \omega([f,X])$$

and similarly

$$(L_{g_i}\omega)(X) = L_{g_i}\big(\omega(X)\big) - \omega([g_i,X]), \qquad i \in \underline{m}.$$

As $\omega(X) = 0$ we obtain $\omega([f,X]) = \omega([g_i,X]) = 0$ for $i \in \underline{m}$. Therefore $[f,X],[g_i,X] \in \ker(P^1 \cap \operatorname{ann} G)$, $i \in \underline{m}$. Now under the constant dimension assumptions $\ker(P^1 \cap \operatorname{ann} G) = \ker P^1 + \ker(\operatorname{ann} G) = D^1 + G$, and thus we may conclude $[f,X] \in D^1 + G$ and $[g_i,X] \in D^1 + G$, $i \in \underline{m}$, where $X \in \ker P^1 = D^1$. This shows $\ker P^2 \subset D^2$. In a similar way one shows that $D^2 \subset \ker P^2$. □

From the above proposition we conclude that under the Assumption 7.17 the maximal locally controlled invariant distribution is also given as

$$D^* = \ker P^* = \ker P^n. \tag{7.61}$$

Note that in particular P^* is an involutive codistribution that contains $\operatorname{span}\{dh_1,\dots,dh_p\}$ and for which we have $L_f(P^* \cap G) \subset P^*$ and $L_{g_i}(P^* \cap G) \subset P^*$, $i \in \underline{m}$. Moreover by the duality between D^* and P^*, P^* is the minimal codistribution having these properties. Observe that the Assumption 7.17 about constant dimensions is not really needed for having convergence of the sequence of codistributions $\{P^{\mu}\}_{\mu\geq 0}$ in (7.58) to a limiting codistribution P^*, yielding D^* as $\ker P^*$. If we return to the main result on the solution of the local DDP, i.e. Theorem 7.14, we see that in order to solve this problem, we need to do three things. First we compute D^* via the algorithm (7.53) or the dual algorithm (7.58) and suppose the Assumption 7.15 (or Assumption 7.17) holds. Then, one has to check if the condition (7.52) is fulfilled. If not, Problem 7.13 is not solvable; if (7.52) is true then one solves for the desired (local) regular feedback by using Theorem 7.5. Like we have seen this involves the solution of a set of partial differential equations. However, we will now show that this is *not* necessary. In fact we will give an effective way of determining the codistributions P^μ, $\mu \geq 0$, from the algorithm (7.58) provided Assumption 7.17 holds, and at the same time we obtain a local feedback which renders the limiting codistribution P^* invariant. Because $D^* = \ker P^*$, see Proposition 7.18, this feedback makes D^* invariant.

Algorithm 7.19 (Computing P^μ, $\mu \geq 0$, locally, provided Assumption 7.17 holds).

Step 0 Suppose the dimension of $P^1 = \text{span}\{dh_1,\ldots,dh_p\}$ equals p_1. Then after a possible permutation on the outputs we have around x_0 $P^1 = \text{span}\{dh_1,\ldots,dh_{p_1}\}$.

Step 1 Define the $p_1 \times m$-matrix $A_1(x)$ and the p_1-vector $B_1(x)$ via

$$A_1(x) = \left(L_{g_i} h_j(x)\right)_{\substack{j=1,\ldots,p_1 \\ i=1,\ldots,m}} \tag{7.62a}$$

$$B_1(x) = \left(L_f h_j(x)\right)_{j=1,\ldots,p_1} \tag{7.62b}$$

Because $P^1 \cap \text{ann}\, G$ has constant dimension, the matrix $A_1(x)$ has constant rank, say r_1. After a possible permutation on the outputs we may assume that the first r_1 rows of $A_1(x)$ are linearly independent. Then (see Exercise 2.4) we may select an m-vector $\alpha_1(x)$ and a nonsingular $m \times m$-matrix $\beta_1(x)$ such that

$$A_1(x)\alpha_1(x) + B_1(x) = \begin{pmatrix} 0 \\ \varphi_1(x) \end{pmatrix} \updownarrow r_1 \tag{7.63a}$$

$$A_1(x)\beta_1(x) = \left(\begin{array}{c|c} I_{r_1} & 0 \\ \hline \psi_1(x) & 0 \end{array}\right) \updownarrow r_1 \tag{7.63b}$$

(the first block column having width r_1)

where $\varphi_1(x)$ is a $(p_1 - r_1)$-vector and $\psi_1(x)$ a $(p_1 - r_1) \times r_1$ matrix. Denote the differentials of the entries of φ_1 and ψ_1 as $d\varphi_1$ and $d\psi_1$. Then we have

$$P^2 = \text{span}\{dh_1,\ldots,dh_{p_1}, d\varphi_1, d\psi_1\}. \tag{7.64}$$

Before proving (7.64) we continue the computation of the P^{μ}'s. By assumption P^2 has fixed dimension, say p_2, and we may set $P^2 = \text{span}\{dh_1, \ldots, dh_{p_1}, \ldots, dh_{p_2}\}$ for well chosen differentials of the entries of φ_1 and ψ_1.

Step 2 Repeat step 1 with the functions $h_1, \ldots, h_{p_2}$. This yields a matrix $A_2(x)$ of rank r_2, a vector $B_2(x)$ and new feedback functions $\alpha_2(x)$ and $\beta_2(x)$ such that equations of the form (7.63a,b) hold. The differentials of the entries of the matrices $\varphi_2(x)$ and $\psi_2(x)$ appearing in the modified equations (7.63a,b) enable us to compute P^3 analogously to (7.64) as $P^3 = \text{span}\{dh_1, \ldots, dh_{p_2}, d\varphi_2, d\psi_2\}$.

In a completely similar way the next steps are executed. Clearly, see Propositions 7.16 and 7.18, we are done in at most n steps (more precisely this will be in at most $n-p_1+1$ steps). So going through the above steps enables us to compute the P^{μ}'s. Moreover, one straightforwardly shows that the inductively defined feedback $u = \alpha^*(x) + \beta^*(x)v$ makes P^* and thus D^* invariant. Here α^* and β^* are the matrices determined in the last step. It remains to prove (7.64).

Proof of (7.64) Define the locally defined regular static state feedback $u = \alpha_1(x) + \beta_1(x)v$ and write the modified system as $\dot{x} = \tilde{f}(x) + \sum_{i=1}^{m} \tilde{g}_i(x)v_i$. It is a straightforward exercise to show that the algorithm (7.58) produces the same list of codistributions when a regular feedback is applied. So in particular in a neighborhood of x_0 we have $P^2 = P^1 + L_{\tilde{f}}(P^1 \cap \text{ann}\, G) + \sum_{i=1}^{m} L_{\tilde{g}_i}(P^1 \cap \text{ann}\, G)$. Inspection of (7.63b) yields that

$$dh_i(x) \notin \text{ann}\, G(x), \qquad \text{for } i = 1, \ldots, r_1, \tag{7.65a}$$

as well as

$$dh_i(x) - \sum_{k=1}^{r_1} \big(\psi_1(x)\big)_{ik} dh_k \in \text{ann}\, G(x), \qquad \text{for } i = r_1+1, \ldots, p_1. \tag{7.65b}$$

So the one-forms (7.65b) exactly span the codistribution $P^1 \cap \text{ann}\, G$. Therefore P^2 consists of the one-forms in P^1, i.e. $dh_1, \ldots, dh_{p_1}$, plus the one forms $L_{\tilde{f}}\big(dh_i - \sum_{k=1}^{r_1} (\psi_1)_{ik} dh_k\big)$ and $L_{\tilde{g}_j}\big(dh_i - \sum_{k=1}^{r_1} (\psi_1)_{ik} dh_k\big)$, $i = r_1+1, \ldots, p_1$, $j \in \underline{m}$. Now

$$L_{\tilde{f}}\left(dh_i - \sum_{k=1}^{r_1}(\psi_1)_{ik}dh_k\right) = L_{\tilde{f}}dh_i - \sum_{k=1}^{r_1}\left(L_{\tilde{f}}(\psi_1)_{ik}dh_k + (\psi_1)_{ik}L_{\tilde{f}}dh_k\right)$$

$$= L_{\tilde{f}}dh_i - \sum_{k=1}^{r_1}L_{\tilde{f}}(\psi_1)_{ik}dh_k\,,$$

and similarly

$$L_{\tilde{g}_j}(dh_i - \sum_{k=1}^{r_1}(\psi_1)_{ik}dh_k) = L_{\tilde{g}_j}dh_i - \sum_{k=1}^{r_1}L_{\tilde{g}_j}(\psi_1)_{ik}dh_k$$

for $i = r_1+1,\ldots,p_1$, $j \in \underline{m}$. This because

$$L_{\tilde{f}}dh_k = dL_{\tilde{f}}h_k = d(\text{zero function}) = 0,\ k = 1,\ldots,r_1,$$

respectively

$$L_{\tilde{g}_j}dh_k = d(L_{\tilde{g}_j}h_k) = d(\text{zero function or } 1) = 0,$$
$$k = 1,\ldots,r_1,\ j \in \underline{m}.$$

Therefore, we find

$$P^2 = \text{span}\{dh_1,\ldots,dh_{p_1}\} + \text{span}\{L_{\tilde{f}}dh_k \mid k = r_1+1,\ldots,p_1\} +$$
$$\text{span}\{L_{\tilde{g}_j}dh_k \mid k = r_1+1,\ldots,p_1,\ j \in \underline{m}\}.$$

As $L_{\tilde{f}}dh_k = dL_{\tilde{f}}h_k$ we find, see (7.62a,b)

$$P^2 = \text{span}\{dh_1,\ldots,dh_{p_1},d\varphi_1,d\psi_1\}. \tag{7.64}$$

□

Although the above computations are generally quite complicated, there is a large class of systems for which these computations are not involved that much. This is in particular true, as we will see later, for single input single-output systems and for the static state feedback input-output decouplable systems that will be treated in Chapter 8.

Let us next investigate how the local DDP works out for a linear system, and afterwards treat some typical nonlinear examples.

Example 7.20 Consider as in Example 7.9 the linear system

$$\begin{aligned}\dot{x} &= Ax + Bu + Ed,\\ y &= Cx.\end{aligned} \tag{7.49}$$

Let x_0 be an arbitrary point in $\mathbb{R}^n$ and let us try to solve the DDP in a neighborhood of x_0. Note that in contrast with Example 7.9 we do not restrict ourselves a priori to regular linear static state feedbacks. Because we want to apply Theorem 7.14 we first have to determine the maximal locally controlled invariant distribution D^* contained in the *flat* distribution ker C. For the system (7.49) it is relatively easy to apply the algorithm (7.58) or Algorithm 7.19.

Let $(b_1,\ldots,b_m)$ and $(c_1,\ldots,c_p)$ denote the columns and rows of the matrices B and C. We may interprete the b_i's as constant vectorfields on $\mathbb{R}^n$ and the c_j's as constant one-forms on $\mathbb{R}^n$. According to the Algorithm 7.19 we find $P^0 = 0$ and $P^1 = \operatorname{span}\{c_1,\ldots,c_p\}$, which is a constant dimensional codistribution. Before computing P^2 we observe that $P^1 \cap \operatorname{ann}(\operatorname{span}\{b_1,\ldots,b_m\})$ equals the codistribution $\operatorname{ann}(\ker C + \operatorname{span}\{b_1,\ldots,b_m\})$, which is again a constant dimensional codistribution generated by a set of constant one-forms (in the x-coordinates). Denote these constant one-forms as row vectors $\tilde{c}_1,\ldots,\tilde{c}_k$, and note that $\operatorname{span}\{\tilde{c}_1,\ldots,\tilde{c}_k\} \subset \operatorname{span}\{c_1,\ldots,c_p\}$. Then

$$P^2 = \operatorname{span}\{c_1,\ldots,c_p\} + L_{Ax}\operatorname{span}\{\tilde{c}_1,\ldots,\tilde{c}_k\} + \sum_{i=1}^{m} L_{b_i}\operatorname{span}\{\tilde{c}_1,\ldots,\tilde{c}_k\}. \tag{7.66}$$

In order to compute the last two terms of the right-hand side of (7.66) we have to determine $L_{Ax}\tilde{c}_\ell$ and $L_{b_i}\tilde{c}_\ell$. Using (2.167) we obtain

$$L_{Ax}\tilde{c}_\ell = \tilde{c}_\ell A, \qquad \ell \in \underline{k} \tag{7.67a}$$

$$L_{b_i}\tilde{c}_\ell = 0, \qquad \ell \in \underline{k},\ i \in \underline{m}. \tag{7.67b}$$

Let $\omega(x) = \sum_{\ell=1}^{k} \omega_\ell(x)\tilde{c}_\ell$ be an arbitrary one-form in $\operatorname{span}\{\tilde{c}_1,\ldots,\tilde{c}_k\}$. For an arbitrary vectorfield $X(x)$ we have, see (2.167),

$$L_X\omega(x) = \sum_{\ell=1}^{k} \left(L_X(\omega_\ell(x))\tilde{c}_\ell + \omega_\ell(x)\tilde{c}_\ell\cdot\frac{\partial X}{\partial x}(x)\right. \tag{7.68}$$

Therefore, using (7.67a,b) and (7.68), and the fact that $\operatorname{span}\{\tilde{c}_1,\ldots,\tilde{c}_k\} \subset \operatorname{span}\{c_1,\ldots,c_p\}$, we find

$$P^2 = \operatorname{span}\{c_1,\ldots,c_p\} + \operatorname{span}\{\tilde{c}_1A,\ldots,\tilde{c}_kA\}. \tag{7.69}$$

Thus P^2 is again a codistribution generated by a set of constant one-forms and is therefore of constant dimension. Note that P^2 does not depend explicitly on the input vectorfields $b_1,\ldots,b_m$. The corresponding distribution $D^2 = \ker P^2$ is given as the *flat* distribution generated by the linear subspace $V^2 = \ker C \cap A^{-1}(\ker C + \mathrm{span}\{b_1,\ldots,b_m\})$. (Here $A^{-1}W$ is defined as the linear subspace $\{z \in \mathbb{R}^n \mid Az \in W\}$.)

The next steps in the Algorithm 7.19 proceed in a similar way. Using Proposition 7.18 we obtain the distributions D^μ. A straightforward computation as above shows that the distributions D^μ are *flat* distributions which are generated by the linear subspaces

$$\begin{cases} V^0 = \mathbb{R}^n \\ V^{\mu+1} = \ker C \cap A^{-1}(V^\mu + \mathrm{span}\{b_1,\ldots,b_m\}) \end{cases} \tag{7.70}$$

The algorithm (7.70) is exactly the linear algorithm for computing the maximal controlled invariant subspace of the system (7.49) in the kernel of C. So the maximal locally controlled invariant distribution D^* of the system (7.49) in the distribution $\ker C$ equals the flat distribution corresponding to V^*, the maximal controlled invariant subspace of (7.49) in the linear subspace $\ker C$. Obviously D^* and $D^* \cap \mathrm{span}\{b_1,\ldots,b_m\}$ are constant dimensional. The next step in solving the local DDP for (7.49) is to test (7.52), i.e.

$$\mathrm{span}\{e_1,\ldots,e_\ell\} \subset D^*, \tag{7.71}$$

where $e_1,\ldots,e_\ell$ are the columns of the matrix E. Observe that (7.71) is an inclusion between distributions, which parallels the subspace inclusion

$$\mathrm{span}\{e_1,\ldots,e_\ell\} \subset V^*. \tag{7.72}$$

Equation (7.72) expresses the standard necessary and sufficient condition for the linear DDP. Now, when (7.71) is fulfilled, we know by Theorem 7.14 that around x_0 a solution of the local DDP exists. To find an actual solution one may resort on Theorem 7.5 or on the computations in the Algorithm 7.19. However, as (7.71) and (7.72) are equivalent, things are much easier in this case. Namely, take a regular linear static state feedback $u = Fx + I_m v$, which solves the linear DDP. Thus the matrix F is determined such that $(A+BF)V^* \subset V^*$. Then this same feedback of course also solves the local (nonlinear) DDP. So we gain nothing in trying to solve the DDP for the system (7.49) by allowing for nonlinear feedbacks! Another by-product of the equivalence of (7.71) and (7.72) is that we indeed find a feedback defined on the whole state space, which was not guaranteed by Theorem 7.14 (or Theorem 7.5). □

Next we discuss the D^*-algorithm for a single-input single-output nonlinear system. In this case it is straightforward to develop an explicit formula for the maximal locally controlled invariant distribution.

Theorem 7.21 *Consider the single-input single-output nonlinear system on M*

$$\begin{aligned} \dot{x} &= f(x) + g(x)u, \\ y &= h(x). \end{aligned} \tag{7.73}$$

Let ρ be the smallest nonnegative integer such that the function $L_g L_f^\rho h$ is not identically zero. Assume that $\rho < \infty$ and that

$$L_g L_f^\rho h(x) \neq 0, \qquad \text{for all } x \in M. \tag{7.74}$$

Then

$$D^* = \ker(\operatorname{span}\{dh, dL_f h, \ldots, dL_f^\rho h\}). \tag{7.75}$$

Proof We compute the P^μ's, $\mu \geq 0$, by using the expression (7.64). Computing the 1×1-matrices $A_1(x)$ and $B_1(x)$ from (7.62a,b) yields

$$\begin{cases} A_1(x) = L_g h(x), \\ B_1(x) = L_f h(x). \end{cases} \tag{7.76}$$

In case $\rho > 0$, $A_1(x) = 0$ for all x and we may choose $\alpha_1(x)$ and $\beta_1(x)$ satisfying (7.63a,b) as $\alpha_1(x) = 0$ and $\beta_1(x) = 1$ for all x and so $\varphi_1(x) = L_f h(x)$, which yields

$$P^2 = \operatorname{span}\{dh, dL_f h\}. \tag{7.77}$$

In case $\rho = 0$, the function $A_1(x)$ coincides with the nonvanishing function given in (7.74). A solution of (7.63a,b) in this case is given by $\alpha_1(x) = -\big(L_g h(x)\big)^{-1} L_f h(x)$ and $\beta_1(x) = \big(L_g h(x)\big)^{-1}$, and no functions $\varphi_1(x)$ and $\psi_1(x)$ appear on the right-hand side of (7.63a,b). So

$$P^2 = P^1 = \operatorname{span}\{dh\}, \tag{7.78}$$

and thus, see Propositions 7.18 and 7.16,

$$P^* = \operatorname{span}\{dh\}, \tag{7.79}$$

which is precisely (7.75) for $\rho = 0$. For $\rho > 0$, one iterates the above computations starting from (7.77), until $P^{\rho+1}$ is reached. Clearly

$$P^* = P^{\rho+1} = \operatorname{span}\{dh, dL_f h, \ldots, dL_f^\rho h\} \tag{7.80}$$

from which (7.75) readily follows. Using Algorithm 7.19, the feedback

$u = \alpha^*(x) + \beta^*(x)v$, with $\alpha^*(x) = -\left(L_g L_f^\rho h(x)\right)^{-1} L_f^{\rho+1} h(x)$ and $\beta^*(x) = \left(L_g L_f^\rho h(x)\right)^{-1}$, leaves D^* invariant. □

So far we have developed the theory on the local DDP in the regular case, i.e. we have assumed throughout that the distributions D^*, $D^* \cap G$ and G are constant dimensional. The following example illustrates that in some circumstances this is not needed. Moreover it shows a method how one can heuristically obtain a decoupling control law which not necessarily leaves the maximal locally controlled invariant distribution D^* invariant.

Example 7.22 In Example 1.2 we have seen that the equations for a gas jet controlled spacecraft are given by (see (1.14))

$$\begin{cases} \dot{R} = -RS(\omega) \\ J\dot{\omega} = S(\omega)J\omega + \sum_{i=1}^{m} b_i u_i \end{cases} \tag{7.81}$$

where the orthogonal matrix $R(t)$ denotes the position of the spacecraft with respect to a fixed set of orthonormal axes, $\omega = (\omega_1, \omega_2, \omega_3)^T$ is the angular velocity with respect to the axes. $S(\omega)$ is a skew-symmetric matrix

$$S(\omega) = \begin{pmatrix} 0 & \omega_3 & -\omega_2 \\ -\omega_3 & 0 & \omega_1 \\ \omega_2 & -\omega_1 & 0 \end{pmatrix}. \tag{7.82}$$

The positive definite matrix J, the *inertia matrix*, will be assumed to be diagonal

$$J = \begin{pmatrix} a_1 & 0 & 0 \\ 0 & a_2 & 0 \\ 0 & 0 & a_3 \end{pmatrix}, \qquad a_i > 0,\ i = 1,2,3, \tag{7.83}$$

which means that the eigenvectors of J, the *principal axes*, coincide with the columns of the matrix R. We assume that there are 3 controls on the system, one of them being unknown (a disturbance), acting as torques around the principal axes. Therefore we henceforth consider the system

$$\begin{aligned} &\dot{R}^T = S(\omega)R^T, \\ &\begin{pmatrix} a_1\dot{\omega}_1 \\ a_2\dot{\omega}_2 \\ a_3\dot{\omega}_3 \end{pmatrix} = \begin{pmatrix} 0 & \omega_3 & -\omega_2 \\ -\omega_3 & 0 & \omega_1 \\ \omega_2 & -\omega_1 & 0 \end{pmatrix} \begin{pmatrix} a_1\omega_1 \\ a_2\omega_2 \\ a_3\omega_3 \end{pmatrix} + \begin{pmatrix} 1 \\ 0 \\ 0 \end{pmatrix} u_1 + \begin{pmatrix} 0 \\ 1 \\ 0 \end{pmatrix} u_2 + \begin{pmatrix} 0 \\ 0 \\ 1 \end{pmatrix} d. \end{aligned} \tag{7.84}$$

where the first equation of (7.84) follows from (7.81) and the fact that $R^T R = I_3$, so $\frac{d}{dt}(R^T) = -R^T \dot{R} R^T$, see also Example 3.5. Together with the dynamics (7.84) we consider the output function

$$y = \text{last row of } R^T = \text{last column of } R. \tag{7.85}$$

Let us write

$$R^T = \begin{pmatrix} r_1 & s_1 & t_1 \\ r_2 & s_2 & t_2 \\ r_3 & s_3 & t_3 \end{pmatrix}, \tag{7.86}$$

then

$$y = \begin{pmatrix} y_1 \\ y_2 \\ y_3 \end{pmatrix} = \begin{pmatrix} r_3 \\ s_3 \\ t_3 \end{pmatrix}. \tag{7.87}$$

Note that $r_3^2 + s_3^2 + t_3^2 = 1$ and so the output map (7.87) has rank 2; given y_1 and y_2 the third output is except for a + or - sign completely specified. We want to solve the (local) Disturbance Decoupling Problem for the system (7.84,87). We solve the problem first by considering only the first column of the matrix R^T and the first output $y_1 = r_3$. That is consider the derived system

$$\frac{d}{dt}\begin{pmatrix} r_1 \\ r_2 \\ r_3 \\ \omega_1 \\ \omega_2 \\ \omega_3 \end{pmatrix} = \begin{pmatrix} \omega_3 r_2 - \omega_2 r_3 \\ -\omega_3 r_1 + \omega_1 r_3 \\ \omega_2 r_1 - \omega_1 r_2 \\ b_1 \omega_2 \omega_3 \\ b_2 \omega_1 \omega_3 \\ b_3 \omega_1 \omega_2 \end{pmatrix} + \begin{pmatrix} 0 \\ 0 \\ 0 \\ a_1^{-1} \\ 0 \\ 0 \end{pmatrix} u_1 + \begin{pmatrix} 0 \\ 0 \\ 0 \\ 0 \\ a_2^{-1} \\ 0 \end{pmatrix} u_2 + \begin{pmatrix} 0 \\ 0 \\ 0 \\ 0 \\ 0 \\ a_3^{-1} \end{pmatrix} d, \tag{7.88a}$$

$$y_1 = r_3, \tag{7.88b}$$

where $b_1 = a_1^{-1}(a_2 - a_3)$, $b_2 = a_2^{-1}(a_3 - a_1)$ and $b_3 = a_3^{-1}(a_1 - a_2)$. According to Proposition 7.8 and Theorem 7.14 we need to find a controlled invariant distribution D which contains the disturbance vectorfield $(0,0,0,0,0,a_3^{-1})^T$ and which is contained in the distribution ker dr_3. In what follows we search for an involutive distribution D which is contained in ker dr_3 and which satisfies (7.15a,b), but is not required to have constant dimension. Nevertheless - compare Theorem 7.5 where constant dimension of D is needed - we show that the distribution D is controlled invariant. Let

$$X_1(r,\omega) = (0,0,0,0,0,1)^T. \tag{7.89a}$$

Clearly we need to have $X_1 \in D$. Computing the Lie bracket of the drift vectorfield f in (7.88a) with X_1 yields the vectorfield

$$X_2(r,\omega) = (r_2,-r_1,0,\omega_2,-\omega_1,0)^T. \tag{7.89b}$$

As this vectorfield does not belong to the distribution spanned by the input-vectorfields g_1, g_2 and X_1 we observe that the one-dimensional distribution span$\{X_1\}$ does not satisfy (7.15a). However letting

$$D = \text{span}\{X_1,X_2\}, \tag{7.90}$$

it is rather easily seen that this distribution is involutive and fulfills the conditions (7.15a,b), but is *not* of constant dimension. One may only verify that D is not the maximal locally controlled invariant distribution in ker dr_3. Nevertheless we will show that there exists a feedback $u = \alpha(r,\omega) + \beta(r,\omega)v$ which makes the distribution D invariant (note that Theorem 7.5 only applies around points where D is constant dimensional). A straightforward computation shows that

$$\beta(\omega) = \begin{pmatrix} a_1\omega_2 & a_1\omega_1 \\ -a_2\omega_1 & a_2\omega_2 \end{pmatrix} \tag{7.91}$$

produces new input vectorfields

$$\begin{cases} \tilde{g}_1(r,\omega) = (0,0,0,\omega_2,-\omega_1,0)^T \\ \tilde{g}_2(r,\omega) = (0,0,0,\omega_1,\omega_2,0)^T \end{cases} \tag{7.92}$$

that leave D invariant, though $\beta(\omega)$ is *singular* at points (r,ω) where $\omega_1 = \omega_2 = 0$.

Next we will determine a 2×1-function $\alpha(r,\omega) = \big(\alpha_1(r,\omega),\alpha_2(r,\omega)\big)^T$ such that $\tilde{f}(r,\omega) = f(r,\omega) + \tilde{g}_1(r,\omega)\alpha_1(r,\omega) + \tilde{g}_2(r,\omega)\alpha_2(r,\omega)$ leaves D invariant. This yields the following set of partial differential equations for $\alpha(r,\omega)$

$$\begin{cases} \omega_2X_1\big(\alpha_1(r,\omega)\big) + \omega_1X_1\big(\alpha_2(r,\omega)\big) = -(b_1-1)\omega_2 \\ -\omega_1X_1\big(\alpha_1(r,\omega)\big) + \omega_2X_1\big(\alpha_2(r,\omega)\big) = -(b_2+1)\omega_1 \end{cases} \tag{7.93a}$$

and

$$\begin{cases} \omega_2X_2\big(\alpha_1(r,\omega)\big) + \omega_1X_2\big(\alpha_2(r,\omega)\big) = (b_1+b_2)\omega_1\omega_3 \\ -\omega_1X_2\big(\alpha_1(r,\omega)\big) + \omega_2X_2\big(\alpha_2(r,\omega)\big) = -(b_1+b_2)\omega_2\omega_3 \end{cases}. \tag{7.93b}$$

Similar as in the proof of Theorem 7.5 we find the (non-unique) solution

$$\begin{cases} \alpha_1(r,\omega) = \big((1-b_1)\omega_2^2\omega_3 + (1+b_2)\omega_1^2\omega_3\big).(\omega_1^2+\omega_2^2)^{-1} \\ \alpha_2(r,\omega) = -(b_1+b_2)\omega_1\omega_2\omega_3)(\omega_1^2+\omega_2^2)^{-1} \end{cases}. \tag{7.94}$$

Or, with respect to the original vectorfields g_1 and g_2 we have, using (7.92), $\tilde{f}(r,\omega) = f(r,\omega) + g_1\bar{\alpha}_1(r,\omega) + g_2\bar{\alpha}_2(r,\omega)$, with

$$\begin{cases} \bar{\alpha}_1(r,\omega) = a_1(1-b_1)\omega_2\omega_3 = (a_1-a_2+a_3)\omega_2\omega_3 \\ \bar{\alpha}_2(r,\omega) = -a_2(1+b_2)\omega_1\omega_3 = (+a_1-a_2-a_3)\omega_1\omega_3 \end{cases} \tag{7.95}$$

from which it follows that the everywhere defined static state feedback

$$\begin{bmatrix} u_1 \\ u_2 \end{bmatrix} = \begin{bmatrix} (a_1-a_2+a_3)\omega_2\omega_3 \\ (a_1-a_2-a_3)\omega_1\omega_3 \end{bmatrix} + \begin{bmatrix} a_1\omega_2 & a_1\omega_1 \\ -a_2\omega_1 & a_2\omega_2 \end{bmatrix} \begin{bmatrix} v_1 \\ v_2 \end{bmatrix} \tag{7.96}$$

leaves the distribution D invariant. As noted before, the feedback (7.96) is not regular at points where $\omega_1 = \omega_2 = 0$.

So far we have solved the DDP for the derived system (7.88a,b). We now consider the problem for the complete spacecraft model (7.84) with outputs (7.87). The solution is delivered via the following coup de grâce. Instead of considering the first column $r = (r_1,r_2,r_3)^T$ in (7.88a,b) we could also have used the other two columns $s = (s_1,s_2,s_3)^T$ and $t = (t_1,t_2,t_3)^T$ of R^T with outputs $y_2 = s_3$ respectively $y_3 = t_3$. Posing for these systems the same Disturbance Decoupling Problem gives the same feedback (7.96) as a possible solution, because (7.96) is only depending upon $(\omega_1,\omega_2,\omega_3)^T$ and not on r (or s and t)! Therefore, (7.96) is a decoupling feedback for the system (7.84) which decouples the last row of R^T from the disturbances. The system in decoupled form reads as

$$\dot{R}^T = S(\omega)R^T$$

$$\frac{d}{dt}\begin{bmatrix} \omega_1 \\ \omega_2 \\ \omega_3 \end{bmatrix} = \begin{bmatrix} \omega_2\omega_3 \\ -\omega_1\omega_3 \\ a_3^{-1}(a_1-a_2)\omega_1\omega_2 \end{bmatrix} + \begin{bmatrix} a_1^{-1}\omega_2 \\ -a_2^{-1}\omega_1 \\ 0 \end{bmatrix} v_1 + \begin{bmatrix} a_1^{-1}\omega_1 \\ a_2^{-1}\omega_2 \\ 0 \end{bmatrix} v_2 + \begin{bmatrix} 0 \\ 0 \\ a_3^{-1} \end{bmatrix} d \tag{7.97}$$

□

Although the preceding example does not completely match with the general theory developed so far (as the matrix $\beta(x)$ is not invertible everywhere), it does if we restrict ourselves to an open dense submanifold of the state space M. Such difficulties are common in the treatment of nonlinear control problems and cannot be avoided a priori.

In the formulation of the (local) Disturbance Decoupling Problem we have required that a (locally defined) *regular* static state feedback exists that solves the problem, i.e. in the decoupling feedback

$$u = \alpha(x) + \beta(x)v \tag{7.3}$$

we impose the condition that the $m\times m$-matrix $\beta(x)$ is nonsingular on its domain of definition. This requirement guarantees that we keep as much control on the system as before (but the outputs are isolated from the disturbances), and so we can use the new controls v for other design purposes. Of course, when no further design objectives are imposed, we could be content with a solution of the form (7.3) where the matrix $\beta(x)$ is not necessarily nonsingular. The most extreme situation appears when we have no further access to the system, i.e. $\beta(x) = 0$, and thus one tries to solve the (local) DDP via strict static state feedback $u = \alpha(x)$. At present no complete solution of the (local) DDP is known when allowing for strict static state feedback.

It is clear that in practice one would add further objectives to the (local) DDP. The most logical additional requirement would be that of asymptotic stability of the disturbance decoupled system when setting $v \equiv 0$. We will return to this aspect in Chapter 10.

So far, we used regular static state feedback in order to achieve disturbance decoupling. It is natural - see also the discussion at the end of Chapter 5 - to study the (local) Disturbance Decoupling Problem using different control schemes, such as regular static output feedback or dynamic state or output feedback. For instance in allowing for regular static output feedback one would like to find a control law

$$u = \alpha(z) + \beta(z)v, \tag{7.98}$$

where

$$z = k(x) \tag{7.99}$$

denotes another set of measurements made on the system. For a linear system $\dot{x} = Ax + Bu + Ed$, $y = Cx$, $z = Kx$, this amounts to finding a subspace $\mathcal{V}$, satisfying $\text{Im } E \subset \mathcal{V} \subset \ker C$, which is controlled invariant *and conditioned invariant*; i.e. $A(\mathcal{V} \cap \ker K) \subset \mathcal{V}$. A somewhat similar requirement, involving controlled invariance and conditioned invariance for a distribution D, appears in the nonlinear situation. We will not pursue this further here (see the references at the end of this chapter).

There is a slight modification of the DDP involving regular static state feedback that can be solved completely analogous to the DDP. This is the so-called Modified Disturbance Decoupling Problem.

Problem 7.23 Modified Disturbance Decoupling Problem (MDDP) *Consider the nonlinear dynamics (7.43) with outputs (7.44). Under which conditions can we find a regular static state feedback*

$$u = \alpha(x) + \beta(x)v + \gamma(x)d \tag{7.100}$$

with α and β as in (7.3) and $\gamma(x)$ an $m\times\ell$-matrix, such that in the closed-loop dynamics the disturbances do not influence the outputs?

The difference between the DDP and the MDDP is that in (7.100) we allow for feeding forward the disturbances d in the control law. Obviously this requires knowledge of the disturbances d. Similar to Theorem 7.14 we give the local solution of the MDDP, i.e. for each arbitrary point x_0 in M the control law (7.100) is defined on a neighborhood of x_0.

Theorem 7.24 *Consider the nonlinear system (7.43,44). Suppose that the distributions D^*, $D^* \cap G$ and G are constant dimensional. Then the local Modified Disturbance Decoupling Problem is solvable if and only if*

$$\mathrm{span}\{e_1,\ldots,e_\ell\} \subset D^* + G. \tag{7.101}$$

As the proof of this result parallels that of Theorem 7.14 we will leave it for the reader.

The notion of controlled invariance for affine nonlinear systems (7.1) can also be extended to general nonlinear systems, locally described as $\dot{x} = f(x,u)$. A discussion of this, together with a study of the Disturbance Decoupling Problem for such systems, will be given in Chapter 13.

Notes and References

In linear system theory the notion of controlled invariant subspaces dates back to the end of the sixties, see [BM], [WM]. A modern account of the use of controlled and conditioned invariant subspaces and their use in linear synthesis problems is given in [Wo]. The nonlinear generalization of the notion of controlled invariance together with their applicability in various nonlinear synthesis problems has been initiated by Hirschorn in [Hi] and Isidori et al. in [IKGM1], see also [MW] and [Is1], [Is2]. The characterization of a controlled invariant distribution as given in Theorem 7.5 can be found in [Hi], [IKGM2], [Nij]. The proof given here is a modification of the one given in [Nij]. A relaxation on the constant dimension assumptions of Theorem 7.5 is discussed in [CT]. The algorithm (7.58) for computing the maximal locally controlled invariant distribution has been given in [IKGM1] and its dual (7.53) comes from [Nij]. The Algorithm 7.19 is due to Krener [Kr2]. The difference between locally and globally controlled invariance has been studied in [Kr1] and [BK]. Example 7.22 has been taken from [NvdS3]. Other examples can be found in

[Cl], [GBBI], [MG]. The modified disturbance decoupling problem, Theorem 7.24, has been treated in [MG] and the nonlinear version of (C,A,B)-invariance is discussed in [IKGM1] and [NvdS2]. Controlled invariance for general nonlinear systems is studied in [NvdS1]. Another approach in studying the disturbance decoupling problem based on the so-called generating series of a system can be found in [Cl].

[BM] G. Basile, G. Marro, "Controlled and conditioned invariant subspaces in linear systems theory", J. Optimiz. Th. Applic. 3, pp. 306-315, 1969.

[BK] C.I. Byrnes, A.J. Krener, "On the existence of globally (f,g)-invariant distributions", in **Differential Geometric Control Theory**, (eds. R.W. Brockett, R.S. Millman, H.J. Sussmann), Birkhäuser, Boston, pp. 209-225, 1983.

[dBI] M.D. di Benedetto, A. Isidori, "The matching of nonlinear models via dynamic state feedback", SIAM J. Contr. Optimiz. 24, pp. 1063-1075, 1986.

[Cl] D. Claude, "Decoupling of nonlinear systems", Syst. Contr. Lett. 1, pp. 242-248, 1982.

[CT] D. Cheng, T.J. Tarn, "New results on (f,g)-invariance", Syst. Contr. Lett. 12, pp. 319-326, 1989.

[GBBI] J.P. Gauthier, G. Bornard, S. Bacha, M. Idir, "Réjet des perturbations pour un modèle non linéaire de colonne à distiller", in **Outils et Modèles Mathématiques pour l'Automatique, l'Analyse de Systèmes et le Traitement du Signal**, vol. III (ed. I.D. Landau), Editions du CNRS, Paris, pp. 459-573 1983.

[Hi] R.M. Hirschorn, "(A,B)-invariant distributions and disturbance decoupling of nonlinear systems", SIAM J. Contr. Optimiz. 19, pp. 1-19, 1981.

[IKGM1] A. Isidori, A.J. Krener, C. Gori-Giorgi, S. Monaco, "Nonlinear decoupling via feedback: a differential geometric approach", IEEE Trans. Aut. Contr. AC-26, pp. 331-345, 1981.

[IKGM2] A. Isidori, A.J. Krener, C. Gori-Giorgi, S. Monaco, "Locally (f,g)-invariant distributions", Syst. Contr. Lett. 1, pp. 12-15, 1981.

[Is1] A. Isidori, "Sur la théorie structurelle et la problème de la réjection des perturbations dans les systèmes non linéaires", in **Outils et Modèles Mathématiques pour l'Automatique, l'Analyse de Systèmes et le Traitement du Signal**, Vol. I (ed. I.D. Landau) Editions du CNRS, Paris, pp. 245-294, 1981.

[Is2] A. Isidori, **Nonlinear Control Systems: an Introduction**. Lect. Notes Contr. Inf. Sci. 72, Springer, Berlin, 1985.

[Kr1] A.J. Krener, "(f,g)-invariant distributions, connections and Pontryagin classes", Proceedings 20th IEEE Conf. Decision Control, San Diego, pp. 1322- 1325, 1981.

[Kr2] A.J. Krener, "(Ad f,g), (ad f,g) and locally (ad f,g) invariant and controllability distributions", SIAM J. Contr. Optimiz. 23, pp. 523-549, 1985.

[MG] C.H. Moog and G. Glumineau, "Le problème du réjet de perturbations measurables dans les systèmes non linéaires-applications à l'amarage en un seul point des grands pétroliers", in **Outils et Modèles Mathématiques pour l'Automatique, l'Analyse de Systèmes et le Traitement du Signal**,

Vol III (ed. I.D. Landau), Editions du CNRS, Paris, pp. 689-698, 1983.

[MW] S.H. Mikhail, W.M. Wonham, "Local decomposability and the disturbance decoupling problem in nonlinear autonomous systems", Allerton Conf. Comm. Contr. Comp. 16, pp. 664-669, 1978.

[Nij] H. Nijmeijer, "Controlled invariance for affine control systems" Int. J. Contr. 34, pp. 824-833, 1981.

[NvdS1] H. Nijmeijer, A.J. van der Schaft, "Controlled invariance for nonlinear systems", IEEE Trans. Aut. Contr. AC-27, pp. 904-914, 1982.

[NvdS2] H. Nijmeijer, A.J. van der Schaft, "Controlled invariance by static output feedback", Syst. Contr. Lett. 2, pp. 39-47, 1982.

[NvdS3] H. Nijmeijer, A.J. van der Schaft, "Controlled invariance for nonlinear systems: two worked examples", IEEE Trans. Aut. Contr. AC-29, pp. 361-364, 1984.

[WM] W.M. Wonham, A.S. Morse "Decoupling and pole assignment in linear multivariable systems: a geometric approach", SIAM J. Contr. Optimiz. 8, pp. 1-18, 1970.

[Wo] W.M. Wonham, **Linear multivariable control: a geometric approach**, Springer, Berlin, 1979.

Exercises

7.1 Prove that the Algorithm 7.19 is invariant under regular static state feedback.

7.2 Prove Theorem 7.24.

7.3 Compute the maximal locally controlled invariant distribution D^* for the system (7.88a,b) of Example 7.22.

7.4 Consider a smooth single output nonlinear system on a manifold M, $\dot{x} = f(x) + \sum_{i=1}^{m} g_i(x)u_i + \sum_{i=1}^{\ell} e_i(x)d_i$, $y = h(x)$. With this system we can associate two systems, namely, Σ_u: $\dot{x} = f(x) + \sum_{i=1}^{m} g_i(x)u_i$, $y = h(x)$ and Σ_d: $\dot{x} = f(x) + \sum_{i=1}^{\ell} e_i(x)d_i$, $y = h(x)$. Let ρ, respectively σ be the smallest integer such that $\big(L_{g_1}L_f^{\rho}h(x),\dots,L_{g_m}L_f^{\rho}h(x)\big) \neq (0,\dots,0)$, $\big(L_{e_1}L_f^{\sigma}h(x),\dots,L_{e_\ell}L_f^{\sigma}h(x)\big) \neq (0,\dots,0)$. Assume that these inequalities hold for all $x \in M$.

(a) Compute D_u^*, respectively D_d^*, for Σ_u, respectively Σ_d.

(b) Show that the Disturbance Decoupling Problem for the original system is solvable if and only if $D_d^* \subset D_u^*$.

(c) Show that the condition found under **(b)** is equivalent to $\rho < \sigma$.

7.5 Consider the single-input single-output nonlinear system with disturbance d, Σ: $\dot{x} = f(x) + g(x)u + e(x)d$, $y = h(x)$, around a point x_0 for which $f(x_0) = 0$ and $h(x_0) = 0$. Let Σ_ℓ: $\dot{\bar{x}} = A\bar{x} + b\bar{u} + e\bar{d}$, $\bar{y} = c\bar{x}$ be the linearization of Σ around x_0 and $u = 0$. Let ρ and σ be the integers as defined in Exercise 7.4 and assume $L_gL_f^{\rho}h(x_0) \neq 0$ and

$L_e L_f^{\sigma} h(x_0) \neq 0$. Prove that the Local Disturbance Decoupling Problem for Σ_ℓ is solvable.

7.6 Consider a particle of unit mass moving on the surface of a cylinder according to a potential force given by the potential function V

$$\dot{q}_1 = p_1 \qquad \dot{q}_2 = p_2$$

$$\dot{p}_1 = -\frac{\partial V}{\partial q_1}(q_1, q_2) + u \qquad \dot{p}_2 = -\frac{\partial V}{\partial q_2}(q_1, q_2) + d$$

where $(q_1, q_2, p_1, p_2) \in T^*(S^1 \times \mathbb{R})$, u and d represent the control and disturbance respectively. Let the output be given as $y = q_2$.

(a) Show that the Disturbance Decoupling Problem is solvable.

(b) Let $z = q_1$ be the measurements on the system. Show that if the potential function V can be written as $V(q_1, q_2) = f(q_1) + g(q_2)q_1 + h(q_2)$ for smooth functions f, g and h, then there exists a regular feedback depending on z only, which solves the Disturbance Decoupling Problem.

7.7 Consider on $\mathbb{R}^5$ the system $\dot{x}_1 = x_2 u_1$, $\dot{x}_2 = x_5$, $\dot{x}_3 = x_4 u_1 + x_2$, $\dot{x}_4 = u_2$, $\dot{x}_5 = x_1 u_1 + d$, $y_1 = x_1$, $y_2 = x_3$.

(a) Show that $D^* = 0$ and conclude that the Disturbance Decoupling Problem is not solvable for this system.

(b) Introduce the dynamic compensator $\dot{z} = v_1$, $u_1 = z$, $u_2 = v_2$ and show that for the precompensated system the Disturbance Decoupling Problem is locally solvable around any point $(x_1, \ldots, x_5, z)$ with $x_5 z \neq 0$.

7.8 Let D_1 and D_2 be distributions satisfying the requirements of Theorem 7.5.

(a) Show by means of a counterexample that $D_1 \cap D_2$ is not necessarily locally controlled invariant.

(b) Assume $D_1 \subset D_2$. Prove that around any point x_0 there locally exists a regular state feedback which makes D_1 and D_2 *simultaneously* invariant.

7.9 Prove Theorem 7.24.

7.10 Consider the single-input single-output nonlinear system P: $\dot{x} = f(x) + g(x)u$, $y = h(x)$ (plant), and M: $\dot{x}_m = f_m(x_m) + g_m(x_m)u_m$, $y_m = h_m(x_m)$ (model). The local nonlinear Model Matching Problem can be formulated as follows ([dBI]). Given initial points x_0 and x_{m0}, find a precompensator Q of the form $\dot{x}_c = a(x, x_c) + b(x, x_c)u_m$, $u = c(x, x_c) + d(x, x_c)u_m$, for the system P and a mapping F: $(x, x_m) \mapsto F(x, x_m) = x_c$ such that $y^{P\circ Q}(x, F(x, x_m), t) - y_m(x_m, t)$ is independent of u_m, for all t and all (x, x_m) in a neighborhood of (x_0, x_{m0}). Here $y^{P\circ Q}$ denotes the output of the precompensated system

$P \circ Q$. The solution of this local Model Matching Problem can be obtained as follows. Define the augmented system $\Sigma_a: \dot{x}_a = f_a(x_a) + g_a(x_a)u + p_a(x_a)u_m$, $y_a = h_a(x_a)$, where $f_a(x_a) = (f^T(x), f_m^T(x_m))^T$, $g_a(x_a) = (g^T(x), 0)^T$, $p_a(x_a) = (0, g_m^T(x_m))^T$ and $h_a(x_a) = h(x) - h_m(x_m)$. Prove that the local Model Matching Problem is solvable if and only if $\text{span}\{p_a\} \subset D_a^* + \text{span}\{g_a\}$, where D_a^* is the maximal locally controlled invariant distribution of the system Σ_a contained in $\ker dh_a$. Hint: Relate the problem with the Modified Disturbance Decoupling Problem. See for the multivariable case [dBI].

7.11 Let D_1 and D_2 be two distributions satisfying the requirements of Theorem 7.5. Assume $D_1 \cap D_2 = 0$ and $D_1 + D_2$ is an involutive distribution. Prove that locally D_1 and D_2 can simultaneously be made invariant by applying a regular static state feedback.

8
The Input-Output Decoupling Problem

In this and the next chapter we discuss various versions of the input-output decoupling problem for nonlinear systems. As a typical aspect of input-output decoupling is the invariance of an output on a subset of the inputs, we have to make, like in Chapter 4, some distinction between analytic and smooth systems. In this chapter we first present a general definition of an input-output decoupled system. Next we give an approach to the static state feedback input-output decoupling problem which is most suited to square *analytic* systems. A geometric treatment of the static state feedback input-output decoupling problem, applying to *smooth* systems, will be given in Chapter 9. This last treatment will also allow us to give a solution to the *block* input-output decoupling problem. In Section 8.2 we will treat, for square analytic systems, the *dynamic* state feedback input-output decoupling problem.

8.1. Static State Feedback Input-Output Decoupling for Analytic Systems

Consider the smooth affine nonlinear control system

$$\dot{x} = f(x) + \sum_{i=1}^{m} g_i(x)u_i , \tag{8.1}$$

with outputs

$$y = h(x), \tag{8.2}$$

where $x = (x_1, \ldots, x_n)$ are local coordinates for a smooth manifold M, $f, g_1, \ldots, g_m$ are smooth vectorfields on M and $h = (h_1, \ldots, h_p) : M \to \mathbb{R}^p$ is a smooth mapping. Roughly stated the input-output decoupling problem is as follows. Suppose the outputs (8.2) are partitioned into m different blocks, then the goal is to find - if possible - a feedback law for the system (8.1) such that each of the m output blocks is controlled by one and only one of the newly defined inputs. Depending on the way of output block partitioning and the type of feedback we allow for, we will systematically treat several versions of the input-output decoupling problem.

We start our discussion with assuming that each output block is one-dimensional, so we have

$$p = m, \tag{8.3}$$

i.e. the number of scalar outputs y_i equals the number of scalar controls u_i. A system (8.1,2) satisfying (8.3) will be called a *square system*. We say that the square system (8.1,2) is *input-output decoupled* if after a possible relabeling of the inputs $u_1,\ldots,u_m$, the i-th input u_i only influences the i-th output y_i and does not affect the other outputs y_j, $j \neq i$. More precisely, see Definition 4.11,

Definition 8.1 *The nonlinear system (8.1-3) is called input-output decoupled if, after a possible relabeling of the inputs, the following two properties hold.*
(i) For each $i \in \underline{m}$ the output y_i is invariant under the inputs u_j, $j \neq i$.
(ii) The output y_i is not invariant with respect to the input u_i, $i \in \underline{m}$.

Using Proposition 4.11, we immediately obtain as a necessary condition for input-output decoupling that, cf. Definition 8.1 (i),

$$L_{g_j}L_{X_1}\cdots L_{X_k}h_i(x) = 0,\ \forall k \geq 0,\ X_1,\ldots,X_k \in \{f,g_1,\ldots,g_m\},\ x \in M. \tag{8.4}$$

Next we discuss part (ii) of Definition 8.1. To avoid complications such as demonstrated in Example 4.16 we assume throughout that the system (8.1,2) is *analytic*. In that case, the effect of the control u_i on the output y_i is determined by the functions

$$L_{g_i}L_{X_1}\cdots L_{X_k}h_i(x) = 0,\ k \geq 0,\ X_1,\ldots,X_k \in \{f,g_1,\ldots,g_m\},\ x \in M. \tag{8.5}$$

Consider now the subset of functions of (8.5) given by

$$L_{g_i}L_f^k h_i(x),\ k \geq 0,\ x \in M. \tag{8.6}$$

Clearly when all the functions in (8.6) are identically zero, so

$$L_{g_i}L_f^k h_i \equiv 0,\ k \geq 0, \tag{8.7}$$

then also all the functions given in (8.5) are identically zero, and in no way the input u_i is going to interact with the output y_i, cf. Proposition 4.17. Therefore we assume (8.7) is not true, and we define for $i \in \underline{m}$ the finite nonnegative integers $\rho_1,\ldots,\rho_m$ as the minimal integers for which the function $L_{g_i}L_f^k h_i$ is not identically zero. Thus ρ_i is determined as

$$\begin{cases} L_{g_i}L_f^k h_i \equiv 0, & k = 0,1,\ldots,\rho_i-1, \\ L_{g_i}L_f^{\rho_i} h_i(x) \neq 0 & \text{for some } x \in M. \end{cases} \tag{8.8}$$

Using (8.4) and (8.8) we have for the (ρ_i+1)-th time derivative of y_i

$$y_i^{(\rho_i+1)} = L_f^{\rho_i+1}h_i(x) + L_{g_i}L_f^{\rho_i}h_i(x)u_i, \quad i \in \underline{m} \tag{8.9}$$

and so at the open subset of M

$$M_0 = \{x \in M | L_{g_i}L_f^{\rho_i}h_i(x) \neq 0, \ i \in \underline{m}\} \tag{8.10}$$

the inputs u_i *instantaneously* do influence the output y_i. We are now able to give a formal definition of an input-output decoupled system.

Definition 8.2 *The square analytic system (8.1,2) is said to be strongly input-output decoupled if (8.4) holds and if there exist finite nonnegative integers $\rho_1,\ldots,\rho_m$ as defined in (8.8) such that the subset M_0 given by (8.10) coincides with M.*

We have given here a *global* definition of strong input-output decoupling. We localize it as follows.

Definition 8.3 *Let $x_0 \in M$. The square analytic system (8.1,2) is said to be locally strongly input-output decoupled around x_0 if there exists a neighborhood V of x_0 such that (8.4) holds on V and if there exist finite integers $\rho_1,\ldots,\rho_m$ as defined in (8.8) with M replaced by V such that the subset M_0 given in (8.10) contains V.*

In the Definitions 8.2 and 8.3 we require a strong form of (instantaneous) input-output decoupling. In particular the requirement that the subset M_0 coincides with M is in some cases not entirely natural. The following examples illustrate the difficulties that arise when no assumptions on M_0 are made.

Example 8.4 Consider on $\mathbb{R}^3$ the square analytic system

$$\begin{aligned} \dot{x}_1 &= x_2^3, & y_1 &= x_1, \\ \dot{x}_2 &= u_1, & y_2 &= x_3. \\ \dot{x}_3 &= u_2, \end{aligned} \tag{8.11}$$

It is straightforward to check that (8.4) holds for the system (8.11). Next we compute the functions appearing in (8.8). In the present situation we have $L_{g_1}h_1 = 0$, $L_{g_1}L_fh_1(x) = 3x_2^2$, so $\rho_1 = 1$, and $L_{g_2}h_2(x) = 1$, so $\rho_2 = 0$. The subset M_0 is given as

$$\{x \in \mathbb{R}^3 \mid L_{g_1}L_fh_1(x) \neq 0,\ L_{g_2}h_2(x) \neq 0\} = \{x \in \mathbb{R}^3 \mid x_2 \neq 0\}.$$

Therefore the system (8.11) is *not* strongly input-output decoupled according to Definition 8.2 though it is locally around each point (x_{01},x_{02},x_{03}) with $x_{02} \neq 0$. However, the system (8.11) is globally input-output decoupled in the sense of Definition 8.1. One computes that $L_{g_1}L_f^3h_1(x) = 6 \neq 0$, which shows that the output y_1 is affected by the control u_1, no matter how the initial state (x_{01},x_{02},x_{03}) was chosen (cf. Definition 4.11). □

Example 8.5 Consider on $\mathbb{R}^2$ the square analytic system

$$\begin{aligned} \dot{x}_1 &= x_1u_1, & y_1 &= x_1, \\ \dot{x}_2 &= u_2, & y_2 &= x_2. \end{aligned} \tag{8.12}$$

Again this system (8.12) satisfies the condition (8.4). Furthermore we have $L_{g_1}h_1(x) = x_1$, $L_{g_2}h_2(x) = 1$ yielding $\rho_1 = 0$, and $\rho_2 = 0$. Because $L_{g_1}h_1(x) = 0$ for $(x_1,x_2) = (0,x_2)$ the subset M_0 of (8.10) does not coincide with $M = \mathbb{R}^2$. The system (8.12) is not strongly input-output decoupled but it is locally strongly input-output decoupled around each point (x_{01},x_{02}) provided $x_{01} \neq 0$. In case $x_{01} = 0$ we will have $y_1(t) = x_1(t) = 0$ for all t, no matter which input $u_1(t)$ we choose, which indeed shows that the output y_1 is for such initial states unaffected by the control u_1. □

In Example 8.4 as well as in Example 8.5 the subset M_0 is open and dense in M, but as has been shown this is by itself not sufficient for (global) input-output decoupling. We will see later that by using invariant distributions the distinction between the two examples can be clarified (see Chapter 9). Note that for an analytic system that satisfies Definition 8.1, the subset M_0 is always open and dense in M and thus the system is locally strongly input-output decoupled about each point $x_0 \in M_0$.

If the square system (8.1,2) is not input-output decoupled (or locally input-output decoupled) we may try to alter the system's dynamics by adding control loops such that it becomes input-output decoupled. As a first attempt one may try to achieve this by adding regular static state feedback. Recall that regular static state feedback has been defined as

$$u = \alpha(x) + \beta(x)v \tag{8.13}$$

where $\alpha\colon M \to \mathbb{R}^m$, $\beta\colon M \to \mathbb{R}^{m\times m}$ are analytic mappings with $\beta(x)$ nonsingular for all x and $v = (v_1, \ldots, v_m)$ represents a new control. By applying (8.13) to the dynamics (8.1) we obtain the modified dynamics

$$\dot{x} = \tilde{f}(x) + \sum_{i=1}^{m} \tilde{g}_i(x) v_i, \tag{8.14}$$

where

$$\tilde{f}(x) = f(x) + \sum_{j=1}^{m} g_j(x)\alpha_j(x), \tag{8.15a}$$

$$\tilde{g}_i(x) = \sum_{j=1}^{m} g_j(x)\beta_{ji}(x), \qquad i \in \underline{m}. \tag{8.15b}$$

We now formulate

Problem 8.6 (Regular static state feedback strong input-output decoupling problem) *Consider the square analytic system (8.1,2). Under which conditions does there exist a regular static state feedback (8.13) such that the feedback modified dynamics (8.14) with the outputs (8.2) is strongly input-output decoupled?*

Before we are able to give a (local) solution to Problem 8.6 we need a few more things. Consider the dynamics (8.1) with the outputs (8.2). For $j \in \underline{p}$ we have $y_j(t) = h_j(x(t))$ and so

$$\dot{y}_j = \frac{d}{dt}\big(y_j(t)\big) = L_f h_j(x) + \sum_{i=1}^{m} L_{g_i} h_j(x) u_i . \tag{8.16}$$

Consider the function

$$L_g h_j(x) := \big(L_{g_1} h_j(x), \ldots, L_{g_m} h_j(x)\big). \tag{8.17}$$

This m-valued vector is either identical zero for all $x \in M$, or there exist points in M where it is different from the zero vector. In the last case we define the *characteristic number* ρ_j of the j-th output to be zero. In case (8.17) vanishes for all x in M we differentiate (8.16) once more to obtain

$$\ddot{y}_j = \frac{d}{dt} L_f h_j(x) = L_f^2 h_j(x) + \sum_{i=1}^{m} L_{g_i} L_f h_j(x) u_i . \tag{8.18}$$

Now consider the function

$$L_g L_f h_j(x) := \big(L_{g_1} L_f h_j(x), \ldots, L_{g_m} L_f h_j(x)\big). \tag{8.19}$$

Whenever this m-valued function $L_g L_f h_j(x)$ is nonzero for some x we define $\rho_j = 1$ otherwise we repeat the above procedure.

Definition 8.7 *The characteristic numbers $\rho_1,\ldots,\rho_p$ of the analytic system (8.1,2) are the smallest nonnegative integers such that for $j \in \underline{p}$*

$$L_g L_f^k h_j(x) = \left(L_{g_1} L_f^k h_j(x),\ldots,L_{g_m} L_f^k h_j(x)\right) = 0,$$

$$k = 0,\ldots,\rho_j - 1, \ \forall x \in M, \tag{8.20a}$$

$$L_g L_f^{\rho_j} h_j(x) = \left(L_{g_1} L_f^{\rho_j} h_j(x),\ldots,L_{g_m} L_f^{\rho_j} h_j(x)\right) \neq 0$$

$$\text{for some } x \in M, \tag{8.20b}$$

and when

$$L_g L_f^k h_j(x) = \left(L_{g_1} L_f^k h_j(x),..,L_{g_m} L_f^k h_j(x)\right) = 0 \text{ for all } k \geq 0 \text{ and } x \in M \tag{8.20c}$$

we set

$$\rho_j = \infty. \tag{8.20d}$$

Sometimes the integers $\rho_1,\ldots,\rho_p$ are also referred to as the *relative orders* or *indices* of the system (8.1,2) and they represent the "inherent number of integrations" between the inputs and the output y_j, $j \in \underline{p}$. Thus, see (8.16) and (8.18), the (ρ_j+1)-th time derivative of the output y_j may depend upon the inputs u provided we are at a point x where (8.20b) holds. Note that the integers introduced in (8.8) are exactly the characteristic numbers of a (locally) decoupled system. This because in a decoupled system $L_{g_i} L_f^k h_j = 0$ for all $i \neq j$ and $k \geq 0$, see also equation (8.4). From this observation it should not be too surprising that the characteristic numbers play a key role in the input-output decoupling problem. The following proposition shows that the characteristic numbers are invariant under regular static state feedbacks.

Proposition 8.8 *Let $\rho_1,\ldots,\rho_p$ be the characteristic numbers of the analytic system (8.1,2) and let (8.13) be a regular static state feedback applied to (8.1). Let $\tilde{\rho}_1,\ldots,\tilde{\rho}_p$ be the characteristic numbers of the feedback modified system (8.14). Then*

$$\tilde{\rho}_j = \rho_j, \qquad j \in \underline{p}, \tag{8.21}$$

$$L_{\tilde{f}}^k h_j(x) = L_f^k h_j(x), \ k = 0,\ldots,\rho_j, \ x \in M, \ j \in \underline{p}. \tag{8.22}$$

If $\rho_j < \infty$ then

$$\left(L_{\tilde{g}_1} L_{\tilde{f}}^{\rho_j} h_j(x),..,L_{\tilde{g}_m} L_{\tilde{f}}^{\rho_j} h_j(x)\right) = \left(L_{g_1} L_f^{\rho_j} h_j(x),..,L_{g_m} L_f^{\rho_j} h_j(x)\right)\beta(x),$$

$$j \in \underline{p}, \ x \in M. \tag{8.23}$$

Proof Let $u = \alpha(x) + \beta(x)v$ be a regular static state feedback and recall the defining equations (8.15a,b) of the modified dynamics. Clearly (8.22) is true for $k = 0$. Assuming (8.22) holds for some k with $0 \le k < \rho_j$, we have that

$$L_{\tilde{f}}^{k+1}h_j(x) = L_{\tilde{f}}L_{\tilde{f}}^{k}h_j(x) = L_{\tilde{f}}L_{f}^{k}h_j(x) = L_{f}^{k+1}h_j(x) + \sum_{i=1}^{m}\alpha_i(x)L_{g_i}L_f^k h_j(x) =$$

$$= L_f^{k+1}h_j(x),$$

where the last equality follows from (8.20a). Having established (8.22) we prove (8.23). Using (8.15b) and (8.22) we have

$$\left(L_{\tilde{g}_1}L_{\tilde{f}}^{\rho_j}h_j(x),\ldots,L_{\tilde{g}_m}L_{\tilde{f}}^{\rho_j}h_j(x)\right) = \left(L_{\tilde{g}_1}L_{f}^{\rho_j}h_j(x),\ldots,L_{\tilde{g}_m}L_{f}^{\rho_j}h_j(x)\right) =$$

$$\left(\sum_{k=1}^{m}\beta_{k1}(x)L_{g_k}L_f^{\rho_j}h_j(x),\ldots,\sum_{k=1}^{m}\beta_{km}(x)L_{g_k}L_f^{\rho_j}h_j(x)\right) =$$

$$\left(L_{g_1}L_f^{\rho_j}h_j(x),\ldots,L_{g_m}L_f^{\rho_j}h_j(x)\right)\beta(x).$$

From (8.23) we now immediately conclude that (8.21) holds true. As $\beta(x)$ is nonsingular for all x in M the right-hand side is a nonzero vector for some $\bar{x}$ in M (see (8.20b)) when ρ_j is finite, and thus the left-hand side of (8.23) is a nonvanishing vector at $\bar{x}$. On the other hand from (8.20a) and (8.22) we immediately deduce that $\left(L_{\tilde{g}_1}L_{\tilde{f}}^{k}h_j(x),\ldots,L_{\tilde{g}_m}L_{\tilde{f}}^{k}h_j(x)\right) = 0$, $k = 0,\ldots,\rho_j-1$, for all x in M. The fact that (8.21) is true when $\rho_j = \infty$ immediately follows from (8.22). □

We are now prepared to solve the strong input-output decoupling problem via regular static state feedback. Given a square system (8.1,2) with finite characteristic numbers $\rho_1,\ldots,\rho_m$, we introduce the $m \times m$-matrix $A(x)$ as

$$A(x) = \begin{bmatrix} L_{g_1}L_f^{\rho_1}h_1(x) & \cdots\cdots & L_{g_m}L_f^{\rho_1}h_1(x) \\ \vdots & & \vdots \\ L_{g_1}L_f^{\rho_m}h_m(x) & \cdots\cdots & L_{g_m}L_f^{\rho_m}h_m(x) \end{bmatrix}. \tag{8.24}$$

We then have

Theorem 8.9 *Consider the square analytic system (8.1,2) with finite characteristic numbers $\rho_1,\ldots,\rho_m$. The regular static state feedback strong input-output decoupling problem is solvable if and only if*

$$\operatorname{rank} A(x) = m, \qquad \textit{for all } x \in M. \tag{8.25}$$

Proof Suppose first that the regular static state feedback strong input-output decoupling problem is solvable. So there exists a feedback (8.13) such that the modified system (8.14,2) is strongly input-output decoupled. In particular, combining (8.4) and (8.8) we have that for $j \in \underline{m}$

$$\left(L_{\tilde{g}_1} L_{\tilde{f}}^{\rho_j} h_j(x), \ldots, L_{\tilde{g}_m} L_{\tilde{f}}^{\rho_j} h_j(x)\right) = \left(0, \ldots, 0, L_{\tilde{g}_j} L_{\tilde{f}}^{\rho_j} h_j(x), 0, \ldots, 0\right) \tag{8.26}$$

while by Definition 8.2

$$L_{\tilde{g}_j} L_{\tilde{f}}^{\rho_j} h_j(x) \neq 0, \qquad \text{for all } x \in M,\ j \in \underline{m}. \tag{8.27}$$

So for the modified system (8.14,2) we compute according to (8.24)

$$\tilde{A}(x) = \begin{pmatrix} L_{\tilde{g}_1} L_{\tilde{f}}^{\rho_1} h_1(x) & \cdots\cdots & L_{\tilde{g}_m} L_{\tilde{f}}^{\rho_1} h_1(x) \\ \vdots & & \vdots \\ L_{\tilde{g}_1} L_{\tilde{f}}^{\rho_m} h_m(x) & \cdots\cdots & L_{\tilde{g}_m} L_{\tilde{f}}^{\rho_m} h_m(x) \end{pmatrix} = \begin{pmatrix} L_{\tilde{g}_1} L_{\tilde{f}}^{\rho_1} h_1(x) & & 0 \\ & \ddots & \\ 0 & & L_{\tilde{g}_m} L_{\tilde{f}}^{\rho_1} h_m(x) \end{pmatrix}, \tag{8.28}$$

which is by (8.27) a nonsingular $m \times m$-matrix for all x in M. Now using Proposition 8.8, we obtain from (8.23)

$$\tilde{A}(x) = A(x)\beta(x). \tag{8.29}$$

Since $\beta(x)$ is nonsingular for all $x \in M$ we conclude that the matrix $A(x)$ has rank m for all x in M.

Next we assume (8.25) is satisfied and we have to construct a regular static state feedback $u = \alpha(x) + \beta(x)v$ that achieves input-output decoupling. Recall that by Definition 8.7 of the characteristic numbers we have the following set of equations ($y_i^{(\rho_i+1)}$ denoting the (ρ_i+1)-th time derivative of the i-th output function):

$$\begin{pmatrix} y_1^{(\rho_1+1)} \\ \vdots \\ y_m^{(\rho_m+1)} \end{pmatrix} = \begin{pmatrix} L_f^{\rho_1+1} h_1(x) \\ \vdots \\ L_f^{\rho_m+1} h_m(x) \end{pmatrix} + \begin{pmatrix} L_{g_1} L_f^{\rho_1} h_1(x) & \cdots\cdots & L_{g_m} L_f^{\rho_1} h_1(x) \\ \vdots & & \vdots \\ L_{g_1} L_f^{\rho_m} h_m(x) & \cdots\cdots & L_{g_m} L_f^{\rho_m} h_m(x) \end{pmatrix} \begin{pmatrix} u_1 \\ \vdots \\ u_m \end{pmatrix}, \tag{8.30}$$

which by (8.24) yields

$$\begin{pmatrix} y_1^{(\rho_1+1)} \\ \vdots \\ y_m^{(\rho_m+1)} \end{pmatrix} = \begin{pmatrix} L_f^{\rho_1+1} h_1(x) \\ \vdots \\ L_f^{\rho_m+1} h_m(x) \end{pmatrix} + A(x)u. \tag{8.31}$$

As the matrix $A(x)$ is nonsingular for all x in M we may define the regular static state feedback

$$u = \begin{bmatrix} u_1 \\ \vdots \\ u_m \end{bmatrix} = -\big(A(x)\big)^{-1} \begin{bmatrix} L_f^{\rho_1+1} h_1(x) \\ \vdots \\ L_f^{\rho_m+1} h_m(x) \end{bmatrix} + \big(A(x)\big)^{-1} v. \tag{8.32}$$

Application of this feedback to the system (8.1) yields

$$\begin{bmatrix} y_1^{(\rho_1+1)} \\ \vdots \\ y_m^{(\rho_m+1)} \end{bmatrix} = \begin{bmatrix} v_1 \\ \vdots \\ v_m \end{bmatrix}, \tag{8.33}$$

which obviously shows that the modified system is strongly input-output decoupled. □

Remark It immediately follows from the proof that Theorem 8.9 also holds true for *smooth* systems, if we define strong input-output decoupling for smooth systems in the same way as we did for analytic systems (Definition 8.2). Moreover the rank condition (8.25) implies that the smooth system can be input-output decoupled in the sense of Definition 8.1, as can be readily verified (see Exercise 8.1 and also Exercise 9.2).

At this point let us see what Theorem 8.9 amounts to for linear systems.

Example 8.10 Consider the m-input m-output linear system

$$\begin{aligned} \dot{x} &= Ax + Bu \\ y &= Cx \end{aligned} \tag{8.34}$$

where $x \in \mathbb{R}^n$ and A, B and C are matrices of appropriate size. The system (8.34) is input-output decoupled when the corresponding transfer matrix $C(sI-A)^{-1}B$ is a diagonal invertible matrix. Let us next see what the $m\times m$-matrix $A(x)$ of (8.24) for this system is. Denote the i-th row of C as c_i, and similarly the i-th column of B as b_i, $i \in \underline{m}$. For the linear system (8.34) a function of the form $L_{g_i} L_f^k h_j(x)$ takes the form $L_{b_i} L_{Ax}^k c_j x = c_j A^k b_i$, which is a constant for all $i, j \in \underline{m}$, $k \geq 0$. Therefore, see Definition 8.7, the characteristic numbers ρ_i, are defined as the minimal nonnegative integer for which the row-vector $c_i A^{\rho_i} B \neq 0$, while $\rho_i = \infty$ if $c_i A^k B = 0$ for all $k \geq 0$. From this we conclude that the $m\times m$-matrix $A(x)$ of (8.24) reduces to the *constant* $m\times m$-matrix

$$\begin{bmatrix} c_1 A^{\rho_1} \\ \vdots \\ c_m A^{\rho_m} \end{bmatrix} B. \tag{8.35}$$

From Theorem 8.9 we conclude that the system (8.34) is (strongly) input-output decouplable via a regular static state feedback $u = \alpha(x) + \beta(x)v$ whenever the matrix given in (8.35) is nonsingular. If (8.35) is nonsingular we obtain as a decoupling feedback, see (8.32),

$$u = -\left(\begin{bmatrix} c_1 A^{\rho_1} \\ \vdots \\ c_m A^{\rho_m} \end{bmatrix} B\right)^{-1} \begin{bmatrix} c_1 A^{\rho_1+1} \\ \vdots \\ c_m A^{\rho_m+1} \end{bmatrix} x + \left(\begin{bmatrix} c_1 A^{\rho_1} \\ \vdots \\ c_m A^{\rho_m} \end{bmatrix} B\right)^{-1} v \tag{8.36}$$

which is a regular linear feedback. Of course this is not surprising as the nonsingularity of the matrix given in (8.35) is the necessary and sufficient condition for the linear system to be input-output decouplable by regular *linear* feedback $u = Fx + Gv$. So we may conclude that the linear system is decouplable via a feedback $u = \alpha(x) + \beta(x)v$ if and only if it is via a linear feedback. □

The condition (8.25) needed for the strong input-output decoupling by regular static state feedback has an interesting consequence about the functions $L_f^k h_i$, $i \in \underline{m}$, $k = 0, \ldots, \rho_i$. Define the mapping $S\colon M \to \mathbb{R}^\rho$, $\rho := \sum_{i=1}^{m} (\rho_i + 1)$, as

$$S(x) = \left(h_1(x), L_f h_1(x), \ldots, L_f^{\rho_1} h_1(x), h_2(x), \ldots, \ldots, h_m(x), \ldots, L_f^{\rho_m} h_m(x)\right). \tag{8.37}$$

Then we have

Proposition 8.11 *Consider the smooth square system (8.1,2) and suppose (8.25) holds. Then*

$$\operatorname{rank} S(x) = \rho = \sum_{i=1}^{m} (\rho_i + 1), \qquad \textit{for all } x \in M. \tag{8.38}$$

Proof Assume (8.38) is untrue at some point x_0. So the one-forms

$$dh_1(x_0), dL_f h_1(x_0), \ldots, dL_f^{\rho_1} h_1(x_0), \ldots, dh_m(x_0), \ldots, dL_f^{\rho_m} h_m(x_0) \tag{8.39}$$

are linearly dependent. This is equivalent to the existence of real numbers c_{ik}, $i \in \underline{m}$, $k = 0, \ldots, \rho_i$, such that

$$\sum_{i=1}^{m}\sum_{k=0}^{\rho_i} c_{ik}\, dL_f^k h_i(x_0) = 0. \tag{8.40}$$

Using the definition of the characteristic numbers, we find

$$L_{g_j}\Big(\sum_{i=1}^{m}\sum_{k=0}^{\rho_i} c_{ik} L_f^k h_i(x)\Big) = \sum_{i=1}^{m} c_{i\rho_i} a_{ij}(x), \tag{8.41}$$

where $a_{ij}(x)$ is the (i,j)-entry of the matrix $A(x)$. From (8.40) it follows that this expression should vanish at the point x_0. However this would imply that the rows of the matrix $A(x)$ at x_0 are not linearly independent, so we conclude $c_{1\rho_1} = \ldots = c_{m\rho_m} = 0$, and (8.40) reduces to

$$\sum_{i=1}^{m}\sum_{k=0}^{\rho_i-1} c_{ik}\, dL_f^k h_i(x_0) = 0. \tag{8.42}$$

Then

$$L_{[f,g_j]}\Big(\sum_{i=1}^{m}\sum_{k=0}^{\rho_i-1} c_{ik} L_f^k h_i(x)\Big) = L_f L_{g_i}\Big(\sum_{i=1}^{m}\sum_{k=0}^{\rho_i-1} c_{ik} L_f^k h_i(x)\Big) -$$

$$L_{g_i} L_f\Big(\sum_{i=1}^{m}\sum_{k=0}^{\rho_i-1} c_{ik} L_f^k h_i(x)\Big) = L_{g_i}\Big(\sum_{i=1}^{m}\sum_{k=0}^{\rho_i-1} c_{ik} L_f^{k+1} h_i(x)\Big), \tag{8.43}$$

where we have used Definition 8.7 (see 8.20a). By the same reasoning as before we conclude that the last expression in (8.43) equals $\sum_{i=1}^{m} c_{i\rho_i-1} a_{ij}(x)$. Now (8.40) and the independence of the rows of $A(x)$ at x_0 yield that $c_{1\rho_1-1} = \ldots = c_{m\rho_m-1} = 0$. A repetition of this argument shows that all c_{ik}'s in (8.40) equal zero and thus (8.38) holds true. □

From Proposition 8.11 it follows that the decoupled system (8.1,32,2) admits a local "normal form". Namely, define for $i \in \underline{m}$

$$z^i = (z_{i1}, \ldots, z_{i\rho_i+1}) = \big(h_i(x), \ldots, L_f^{\rho_i} h_i(x)\big), \tag{8.44}$$

and let $\bar{z}$ be $(n-\rho)$ supplementary coordinate functions such that $(\bar{z}, z^1, \ldots, z^m) = S(x)$ forms a local coordinate system. Then with respect to these new coordinates the decoupled system (8.1,32,2) reads as

$$\begin{aligned} \dot{z}^i &= A_i z^i + b_i v_i, && i \in \underline{m}, \\ \dot{\bar{z}} &= \bar{f}(\bar{z}, z^1, \ldots, z^m) + \sum_{i=1}^{m} \bar{g}_i(\bar{z}, z^1, \ldots, z^m) v_i, && \\ y_i &= z_{i1}, && i \in \underline{m}, \end{aligned} \tag{8.45}$$

where the pairs (A_i, b_i), $i \in \underline{m}$, are in Brunovsky canonical form (6.50).

Notice also that the "normal form" (8.45) holds equally well for *smooth* systems satisfying the rank condition (8.25).

Remark 8.12 The regular static state feedback (8.32), suggested in the proof of Theorem 8.9 for achieving input-output decoupling, is not the only solution of the regular static state feedback strong input-output decoupling problem. For instance, see (8.45), for the controls v_i in (8.32) we may introduce additional feedbacks of the form

$$v_i = \tilde{\alpha}_i(y_i, \dot{y}_i, \ldots, y_i^{(\rho_i)}) + \tilde{\beta}_i(y_i, \ldots, y_i^{(\rho_i)})\tilde{v}_i,$$

$$\text{with } \tilde{\beta}_i(y_i, \ldots, y_i^{(\rho_i)}) \neq 0, \quad i \in \underline{m},$$

which keep the system in a decoupled form. In general there exist even more decoupling feedbacks than those suggested above. What might be surprising is that (8.32) makes the decoupled input-output behavior *linear*, see (8.45). Of course the complete feedback modified dynamics (8.1,32) is still nonlinear, except in case $\sum_{i=1}^{m} (\rho_i+1) = n$, see the "normal form" (8.45). □

We emphasize that we require in Theorem 8.9 that the characteristic numbers of the square system (8.1,2) are finite. If this is not the case, say $\rho_i = \infty$ for some i, then the output y_i is not influenced at all by the controls $u_1, \ldots, u_m$, and this will not be altered by applying a regular static state feedback, cf. Proposition 8.8.

In the situation where (8.25) is not met for all $x \in M$ we obtain a result on the local regular static state feedback strong input-output decoupling problem; i.e. given a point x_0, under what conditions does there exist a regular feedback (8.13) defined in a neighborhood V of x_0 such that the system is strongly input-output decoupled (see Definition 8.3)?

Theorem 8.13 *Consider the square system (8.1,2) with finite characteristic numbers $\rho_1, \ldots, \rho_m$. The regular static state feedback strong input-output decoupling problem is locally solvable around a point x_0 in M (i.e. a decoupling feedback is only defined in a neighborhood V of x_0) if and only if*

$$\text{rank } A(x_0) = m. \tag{8.46}$$

Proof The essential observation is that when (8.46) holds then rank $A(x) = m$ for all x in a neighborhood V of x_0. By replacing M by V the proof follows from Theorem 8.9. □

For obvious reasons we will henceforth refer to the matrix $A(x)$, defined in (8.24), as the *decoupling matrix* of a square system (8.1,2).

Example 8.14 Consider the robot-arm configuration introduced in Example 1.1. The dynamics are given as (see equation (1.6))

$$\begin{aligned} \dot{x}^1 &= x^2, \\ \dot{x}^2 &= -M(x^1)^{-1}\left(C(x^1,x^2) + k(x^1)\right) + M(x^1)^{-1}u, \end{aligned} \tag{8.47}$$

where $x^1 = (x_1,x_2) = (\theta_1,\theta_2)$, $x^2 = (x_3,x_4) = (\dot{\theta}_1,\dot{\theta}_2)$, $u = (u_1,u_2)$ and the matrices $M(x^1)$, $C(x^1,x^2)$ and $k(x^1)$ are defined as in (1.7a,b,c). As outputs for the system (8.47) we take the Cartesian coordinates of the endpoint, i.e. see (1.9),

$$\begin{aligned} y_1 &= h_1(x^1,x^2) = \ell_1\sin x_1 + \ell_2\sin(x_1+x_2), \\ y_2 &= h_2(x^1,x^2) = \ell_1\cos x_1 + \ell_2\cos(x_1+x_2). \end{aligned} \tag{8.48}$$

A direct computation yields that the characteristic numbers of the square system (8.47,48) equal $\rho_1 = \rho_2 = 1$, and the decoupling matrix of the system takes the form

$$A(x^1,x^2) = \begin{bmatrix} \ell_1\cos x_1 + \ell_2\cos(x_1+x_2) & \ell_2\cos(x_1+x_2) \\ -\ell_1\sin x_1 - \ell_2\sin(x_1+x_2) & -\ell_2\sin(x_1+x_2) \end{bmatrix} M(x^1)^{-1}. \tag{8.49}$$

Note that the rank of $A(x^1,x^2)$ is depending on x^1. We have

$$\operatorname{rank} A(x^1,x^2) = 2 \tag{8.50}$$

for all (x^1,x^2) in the set

$$\{(x^1,x^2) \mid \ell_1\ell_2\sin x_2 \neq 0\}. \tag{8.51}$$

So we conclude from Theorem 8.12 that the robot-arm is locally strongly input-output decouplable via a regular static state feedback around each point satisfying (8.51). Note that the points we have to exclude are given as $x_2 = 0$ or $x_2 = \pi$ and it is physically clear that at those points we can not have input-output decoupling. For instance $x_2 = 0$ corresponds to a stretched position of the robot-arm and we may not allow that the endpoint (y_1,y_2) reaches outside the working space of the robot arm:

$$\{(y_1,y_2)\,|\,y_1^2 + y_2^2 \le (\ell_1+\ell_2)^2\}.$$

Around each point (x_0^1,x_0^2) satisfying (8.51) we may use the control law (8.32) proposed in Theorem 8.9. As we have in the present situation that $\sum_{i=1}^{2}(\rho_i+1) = 4$ equals the dimension of the state space, we may introduce the coordinate transformation (8.44) around (x_0^1,x_0^2), i.e.

$$\begin{bmatrix} z_{11} \\ z_{21} \\ z_{12} \\ z_{22} \end{bmatrix} = \begin{bmatrix} \ell_1 \sin x_1 + \ell_2 \sin(x_1+x_2) \\ \ell_1 \cos x_1 + \ell_2 \cos(x_1+x_2) \\ x_3 \ell_1 \cos x_1 + (x_3+x_4)\ell_2 \cos(x_1+x_2) \\ -x_3 \ell_1 \sin x_1 - (x_3+x_4)\ell_2 \sin(x_1+x_2) \end{bmatrix}, \tag{8.52}$$

yielding the modified dynamics

$$\begin{aligned} \dot z_{11} &= z_{12} \\ \dot z_{12} &= v_1 \\ \dot z_{21} &= z_{22} \\ \dot z_{22} &= v_2 \end{aligned} \tag{8.53a}$$

with the transformed outputs

$$y_1 = z_1 \quad \text{and} \quad y_2 = z_2. \tag{8.53b}$$

□

8.2 Dynamic State Feedback Input-Output Decoupling

So far we have presented a rather complete description of square analytic systems that can be (locally) strongly input-output decoupled via a regular static state feedback. The essential requirement for decoupling turns out to be the nonsingularity of the decoupling matrix. For smooth nonlinear systems the nonsingularity of the decoupling matrix is again a sufficient condition for the solvability of the input-output decoupling problem (see Exercise 8.1). So square systems having a decoupling matrix of rank smaller than m on the whole state space are certainly not input-output decouplable via a regular static state feedback. This suggests to look at more general control loops for the analytic nonlinear dynamics (8.1) in order to achieve (locally) input-output decoupling. In particular we are going to study if we can achieve input-output decoupling by allowing for dynamic state feedback. Recall, see Definition 5.27, that dynamic state feedback is defined as a relation

$$\begin{cases} \dot z = \gamma(z,x) + \delta(z,x)v \\ u = \alpha(z,x) + \beta(z,x)v \end{cases} \tag{8.54}$$

where $z = (z_1, \ldots, z_q) \in \mathbb{R}^q$, $\gamma: \mathbb{R}^q \times M \to \mathbb{R}^q$, $\delta: \mathbb{R}^q \times M \to \mathbb{R}^{q\times m}$, $\alpha: \mathbb{R}^q \times M \to \mathbb{R}^m$ and $\beta: \mathbb{R}^q \times M \to \mathbb{R}^{m\times m}$ are smooth mappings, and $v = (v_1, \ldots, v_m)$ represents a new input. To see that dynamic state feedback may be of use in the input-output decoupling problem we consider the following example.

Example 8.15 (see Example 5.28) Consider on $\mathbb{R}^3$ the square analytic system

$$\begin{cases} \dot{x}_1 = u_1 \\ \dot{x}_2 = e^{x_1}u_1 + x_3 \\ \dot{x}_3 = u_2 \end{cases} \tag{8.55a}$$

$$\begin{cases} y_1 = x_1 \\ y_2 = x_2 . \end{cases} \tag{8.55b}$$

For the system (8.55) we compute $\rho_1 = \rho_2 = 0$, and the decoupling matrix of the system is

$$A(x) = \begin{pmatrix} 1 & 0 \\ e^{x_1} & 0 \end{pmatrix}, \tag{8.56}$$

which has rank 1 everywhere. So (8.55) is not locally input-output decouplable via a regular static-state feedback. Now let us add to the system (8.55) the dynamic state feedback

$$\begin{cases} \dot{z} = v_1 \\ \begin{pmatrix} u_1 \\ u_2 \end{pmatrix} = \begin{pmatrix} z \\ v_2 \end{pmatrix}, \end{cases} \tag{8.57}$$

we obtain as dynamics

$$\begin{cases} \dot{x}_1 = z \\ \dot{x}_2 = e^{x_1}z + x_3 \\ \dot{x}_3 = v_2 \\ \dot{z} = v_1 . \end{cases} \tag{8.58}$$

Again we compute the characteristic numbers of the modified system (8.58,55b). We find $\tilde{\rho}_1 = \tilde{\rho}_2 = 1$ and the decoupling matrix of this system has the form

$$\tilde{A}(x,z) = \begin{pmatrix} 1 & 0 \\ e^{x_1} & 1 \end{pmatrix}, \tag{8.59}$$

which has rank 2 at each point $(x,z) \in \mathbb{R}^4$. So the modified system is according to Theorem 8.9 input-output decouplable via a regular static state feedback of the form $v = \tilde{\alpha}(x,z) + \beta(x,z)\tilde{v}$. Actually a straightforward calculation shows that the regular feedback

$$\begin{bmatrix} v_1 \\ v_2 \end{bmatrix} = \begin{bmatrix} 0 \\ -e^{x_1}z^2 \end{bmatrix} + \begin{bmatrix} 1 & 0 \\ -e^{x_1} & 1 \end{bmatrix} \begin{bmatrix} \tilde{v}_1 \\ \tilde{v}_2 \end{bmatrix} \tag{8.60}$$

decouples the system (8.58). In fact the dynamic state feedback consisting of the cascade of (8.57) and (8.60) equals the one given in Example 5.28. □

Motivated by the foregoing we formulate the following problem.

Problem 8.16 (Dynamic state feedback strong input-output decoupling problem) *Consider the square analytic system (8.1,2). Under which conditions does there exist a dynamic state feedback (8.54) such that the feedback modified dynamics (8.1,54) with outputs (8.2) is strongly input-output decoupled?*

Before going to address Problem 8.16 for nonlinear systems we briefly discuss it for linear systems.

Example 8.17 Consider the square linear system

$$\begin{cases} \dot{x} = Ax + Bu \\ y = Cx \end{cases} \tag{8.61}$$

with $x \in \mathbb{R}^n$, $u, y \in \mathbb{R}^m$ and A, B and C matrices of appropriate size. The *linear* dynamic state feedback input-output decoupling may be formulated as follows. When does there exist a linear dynamic state feedback

$$\begin{cases} \dot{z} = Pz + Qx + Rv \\ u = Kz + Lx + Mv \end{cases} \tag{8.62}$$

such that the closed-loop system (8.61,62) is input-output decoupled. Because a linear system is input-output decoupled if and only if its transfer matrix is an invertible diagonal matrix, one may check that the necessary and sufficient conditions for the solvability of the linear dynamic state feedback input-output decoupling problem is that the transfer matrix $H(s)$ of the system (8.61),

$$H(s) = C(sI_n - A)^{-1}B \tag{8.63}$$

is nonsingular, i.e. the determinant of $H(s)$ is a nontrivial function in s. For the proof of this we refer to the references given at the end of the chapter. □

We now return to the nonlinear Problem 8.16. In what follows we will address this problem in a local fashion. That is, we will give conditions which assure the existence of a precompensator of the form (8.54) such that the overall system (8.1,54) with outputs (8.2) is input-output decoupled around a point (x_0, z_0) in $M \times \mathbb{R}^q$. In order to obtain these conditions the following algorithm is essential.

Algorithm 8.18 **(Dynamic Extension Algorithm)** Consider the analytic square system (8.1,2).

Step 1 Compute the characteristic numbers $\rho_1^1, \ldots, \rho_m^1$ for the system (8.1,2). Then by Definition 8.7 we obtain the following vector equation

$$\begin{bmatrix} y_1^{(\rho_1+1)} \\ \vdots \\ y_m^{(\rho_m+1)} \end{bmatrix} = E^1(x) + D^1(x)u \tag{8.64}$$

for an analytic m-vector $E^1(x)$ and an analytic $m\times m$-matrix $D^1(x)$. In fact, the exact structure of these matrices $E^1(x)$ and $D^1(x)$ is as given in equation (8.30), and in particular the matrix $D^1(x)$ coincides with the decoupling matrix $A(x)$ given in (8.24). Let

$$r_1(x) = \text{rank } D^1(x). \tag{8.65}$$

By the analyticity of the system (8.1,2) $r_1(x)$ is constant on an open and dense submanifold M_1 of M, say $r_1(x) = r_1$ for x in M_1. Assume we work on a neighborhood contained in M_1. Reorder the output functions $h_1, \ldots, h_m$ such that the first r_1 rows of the matrix D^1 are linearly independent. Note that this may require that we shrink the neighborhood in M_1. Write the output map h as $h = (h^1, \bar{h}^1)$, where $h^1 = (h_1, \ldots, h_{r_1})$ and $\bar{h}^1 = (h_{r_1+1}, \ldots, h_m)$. Denote in the corresponding way y as $(y^1, \bar{y}^1)$ and $\rho^1 = \rho_1^1, \ldots, \rho_{r_1}^1)$ and $\bar{\rho}^1 = (\rho_{r_1+1}^1, \ldots, \rho_m^1)$. We can choose (see Exercise 2.4) an analytic m-vector $\alpha^1(x)$ and an analytic invertible $m\times m$-matrix $\beta^1(x)$ on a (possibly smaller) neighborhood in M_1, such that after applying the state feedback

$$u = \alpha^1(x) + \beta^1(x) \begin{bmatrix} v^1 \\ \bar{v}^1 \end{bmatrix}, \tag{8.66}$$

with $v^1 = (v_1, \ldots, v_{r_1})$ and $\bar{v}^1 = (v_{r_1}, \ldots, v_m)$ the corresponding partitioning of the new input v, we arrive at

$$\begin{bmatrix} (y^1)^{(\rho^1+1)} \\ (\bar{y}^1)^{(\bar{\rho}^1+1)} \end{bmatrix} = \begin{bmatrix} 0 \\ \lambda^1(x) \end{bmatrix} + \begin{bmatrix} I_{r_1} & 0 \\ \mu^1(x) & 0 \end{bmatrix} \begin{bmatrix} v^1 \\ \bar{v}^1 \end{bmatrix}. \tag{8.67}$$

In (8.67) $\lambda^1(x)$ is an analytic $(m-r_1)$-vector and $\mu^1(x)$ an analytic $(m-r_1)\times r_1$-matrix and I_{r_1} the $r_1\times r_1$-identity matrix. Define the modified vectorfields

$$f^1(x) = f(x) + \sum_{j=1}^{m} g_j(x)\alpha_j^1(x), \tag{8.68a}$$

$$g_i^1(x) = \sum_{j=1}^{m} g_j(x)\beta_{ji}^1(x), \quad i \in \underline{m}, \tag{8.68b}$$

and consider the modified dynamics

$$\dot{x} = f^1(x) + \sum_{i=1}^{m} g_i^1(x)u_i . \tag{8.69}$$

Note that in order to simplify notation we have renamed in (8.69) the new controls as $u_1,\dots,u_m$. What has been done so far is nothing else than applying static state feedback to achieve decoupling of the first r_1 input-output channels. In particular when $r_1 = m$ we have repeated the local static state feedback strong input-output decoupling as given in Theorem 8.13.

Step 2 In this step we are only concerned about the outputs $\bar{y}^1 = \bar{h}^1(x)$ and we want to examine their dependency on the inputs $\bar{u}^1 = (u_{r_1+1},\dots,u_m)$. (The inputs $u^1 = (u_1,\dots,u_{r_1})$ one by one control the outputs $y^1 = (y_1,\dots,y_{r_1})$, see (8.67).) In order to do so, we differentiate these outputs with respect to (8.69) to see when $\bar{u}^1$ appears for the first time. Let for $i = r_1+1,\dots,m$, ρ_i^2 be the smallest integer such that the (ρ_i^2+1)-th time derivative of y_i explicitly depends on $\bar{u}^1$. Observe that such a time derivative possibly also depends on the components of u^1 and their time derivatives. Thus ρ_i^2 is the characteristic number of the i-th output channel of the system (8.69) with respect to the inputs $(u_{r_1+1},\dots,u_m)$, where the inputs $(u_1,\dots,u_{r_1})$ and their time derivatives are viewed as parameters. We have

$$\begin{bmatrix} y_{r_1+1}^{(\rho_{r_1+1}^2+1)} \\ \vdots \\ y_m^{(\rho_m^2+1)} \end{bmatrix} = E^2(x,\tilde{U}^1) + D^2(x,\tilde{U}^1)\bar{u}^1, \tag{8.70}$$

for an analytic $(m-r_1)$-vector $E^2(x,\tilde{U}^1)$ and an analytic $(m-r_1)\times(m-r_1)$-matrix $D^2(x,\tilde{U}^1)$, where $\tilde{U}^1$ consists of all components of u^1 and their time-derivatives $u_i^{(j)}$, $i = 1,\ldots,r_1$, $j \geq 0$, which occur in (8.70). We note that the highest derivative of u^1 appearing in $\tilde{U}^1$, and thus in (8.70), is at most of order $n-1$. To see this we simply consider u^1 and their time derivatives in (8.70) as parameters and we observe that for the parametrized system each of the characteristic numbers ρ_i^2 is smaller than $n-1$ (cf. Proposition 8.11 with the single output $y_i = h_i(x)$), and so components of u^1 are at most $n-1$ times differentiated. (Actually, it immediately follows that $\max\{\rho_{r_1}^2 - \rho_{r_1}^1,\ldots,\rho_m^2 - \rho_m^1\}$ is a better upper bound of the time derivatives of u^1 appearing in $\tilde{U}^1$.) So $\mu_1 := \dim \tilde{U}^1 \leq nr_1$, when interpreting u^1 and the time derivatives $(u^1)^{(j)}$ as independent variables. Let

$$r_2(x,\tilde{U}^1) = \operatorname{rank} D^2(x,\tilde{U}^1), \tag{8.71}$$

then $r_2(\cdot,\cdot)$ is constant on an open and dense submanifold M_2 of $M \times \mathbb{R}^{\mu_1}$, say $r_2(x,\tilde{U}^1) = r_2$ for $(x,\tilde{U}^1) \in M_2$. Let

$$q_2 = r_1 + r_2. \tag{8.72}$$

Assume we are working on an open neighborhood in M_2 which is such that step 1 has been performed on the projection of M_2 into M_1. Reorder the output functions $\bar{h}^1$ such that the first r_2 rows of the matrix $D^2(x,\tilde{U}^1)$ are linearly independent on this neighborhood and write $h^2 = (h_{r_1+1},\ldots,h_{q_2})$, $\bar{h}^2 = (h_{q_2+1},\ldots,h_m)$. Accordingly we write $\bar{y}^1 = (y^2,\bar{y}^2)$ and $\rho^2 = (\rho^2_{r_1+1},\ldots,\rho^2_{q_2})$, $\bar{\rho}^2 = (\rho^2_{q_2+1},\ldots,\rho^2_m)$. Then we can choose an analytic $(m-r_1)$-vector $\alpha^2(x,\tilde{U}^1)$ and an invertible analytic $(m-r_1)\times(m-r_1)$-matrix $\beta^2(x,\tilde{U}^1)$ on the above neighborhood such that after applying the control law

$$\bar{u}^1 = \alpha^2(x,\tilde{U}^1) + \beta^2(x,\tilde{U}^1)\begin{pmatrix} v^2 \\ \bar{v}^2 \end{pmatrix} \tag{8.73}$$

we obtain

$$\begin{bmatrix} (y^2)^{(\rho^2+1)} \\ (\bar{y}^2)^{(\bar{\rho}^2+1)} \end{bmatrix} = \begin{bmatrix} 0 \\ \lambda^2(x,\tilde{U}^1) \end{bmatrix} + \begin{bmatrix} I_{r_2} & 0 \\ \mu^2(x,\tilde{U}^1) & 0 \end{bmatrix}\begin{bmatrix} v^2 \\ \bar{v}^2 \end{bmatrix}. \tag{8.74}$$

In (8.73) v^2 is a r_2-control vector and $\bar{v}^2$ a $(m-q_2)$-control vector and in (8.74) $\lambda^2(x,\tilde{U}^1)$ is an analytic $(m-q_2)$-vector and $\mu^2(x,\tilde{U}^1)$ an analytic $(m-q_2)\times r_2$-matrix. Define the modified parametrized vectorfields

$$f^2(x,\tilde{U}^1) = f^1(x) + \sum_{j=r_1+1}^{m} g_j^1(x)\alpha_j^2(x,\tilde{U}^1), \tag{8.75a}$$

$$g_i^2(x,\tilde{U}^1) = g_i^1(x,\tilde{U}) \quad , \ i = 1,\ldots,r_1, \tag{8.75b}$$

$$g_i^2(x,U\ ) = \sum_{j=r_1+1}^{m} g_j^1(x)\beta_{ji}^2(x,\tilde{U}^1), \ i = r_1+1,\ldots,m, \tag{8.75c}$$

and consider the dynamics

$$\dot{x} = f^2(x,\tilde{U}^1) + \sum_{i=1}^{m} g_i^2(x,\tilde{U}^1)u_i\ . \tag{8.76}$$

In (8.76) we have renamed the controls $(u^1,v^2,\bar{v}^2)$ as u, and we have a new splitting for the newly defined control variables u. Namely $u = (U^2,\bar{u}^2)$, where $U^2 = (u_1,\ldots,u_{q_2})$ and $\bar{u}^2 = (u_{q_2+1},\ldots,u_m)$, the controls $u_1,\ldots,u_{r_1}$ and their time derivatives occuring in $\tilde{U}^1$ will be considered as parameters. Alternatively, as we will see later in the proof of Theorem 8.19 we can interpret them as additional state variables and new controls for a suitably defined precompensated system of (8.1). Note that for the modified system (8.75) the first $q_2 = r_1+r_2$ input-output channels are decoupled, see (8.67) and (8.74).

From the foregoing reasoning the general step is easily established.

Step ℓ+1 Assume we have defined a sequence of integers $r_1,\ldots,r_\ell$ and let

$$q_\ell = \sum_{i=1}^{\ell} r_i\,. \tag{8.77}$$

We have a block partitioning of h as $h = (h^1,\ldots,h^\ell,\bar{h}^\ell)$ and the parametrized dynamics

$$\dot{x} = f^\ell(x,\tilde{U}^{\ell-1}) + \sum_{i=1}^{m} g_i^\ell(x,\tilde{U}^{\ell-1})u_i \tag{8.78}$$

where the controls $(u_1,\ldots,u_{q_\ell}) =: U^\ell$ one by one influence the first q_ℓ output channels. Similarly to the second step we examine now the dependency of the remaining outputs $\bar{y}^\ell = (\bar{h}^\ell(x)) = (h_{q_\ell+1}(x),\ldots,h_m(x))$ on the remaining inputs $\bar{u}^\ell = (u_{q_\ell+1},\ldots,u_m)$. So we differentiate these outputs with respect to (8.78) until $\bar{u}^\ell$ appears. Let for $i > q_\ell$, $\rho_i^{\ell+1}$ be the smallest integer such that the $(\rho_i^{\ell+1}+1)$-th time derivative of y_i explicitly depends on $\bar{u}^\ell$. Analogously to (8.70) we obtain the equation

$$\begin{bmatrix} y_{q_{\ell+1}}^{(\rho^{\ell+1}_{q_{\ell+1}}+1)} \\ \vdots \\ y_m^{(\rho^{\ell+1}_m+1)} \end{bmatrix} = E^{\ell+1}(x,\tilde{U}^\ell) + D^{\ell+1}(x,\tilde{U}^\ell)\bar{u}^\ell, \tag{8.79}$$

for an analytic $(m-q_\ell)$-vector $E^{\ell+1}(x,\tilde{U}^\ell)$ and an analytic $(m-q_\ell)\times(m-q_\ell)$-matrix $D^{\ell+1}(x,\tilde{U}^\ell)$, where $\tilde{U}^\ell$ consists of $U^\ell = (u_1,\ldots,u_{q_\ell})$ together with suitable time derivatives of the components $u_1,\ldots,u_{q_\ell}$, which appear when differentiating the outputs $\bar{y}^\ell$ with respect to (8.78). By convention $\tilde{U}^\ell \supset \tilde{U}^{\ell-1}$. Note that, as in step 2, we have that $\mu_\ell := \dim \tilde{U}^\ell$ satisfies $\mu_\ell \leq (n-1)q_\ell$. Let

$$r_{\ell+1}(x,\tilde{U}^\ell) = \operatorname{rank} D^{\ell+1}(x,\tilde{U}^\ell), \tag{8.80}$$

then $r_{\ell+1}(.,.)$ is constant on an open and dense submanifold $M_{\ell+1} = M \times \mathbb{R}^{\mu_\ell}$, say $r_{\ell+1}(x,\tilde{U}^\ell) = r_{\ell+1}$, for $(x,\tilde{U}^\ell) \in M_{\ell+1}$. Let

$$q_{\ell+1} = q_\ell + r_{\ell+1} = \sum_{i=1}^{\ell+1} r_i . \tag{8.81}$$

Assume we are working in a neighborhood in $M_{\ell+1}$, for which the projection of it on M is contained in the projection of M_ℓ on M and which is such that after a relabeling of the outputs $\bar{y}^\ell$ the first $r_{\ell+1}$ rows of $D^{\ell+1}$ are linearly independent on this neighborhood. Write $h^{\ell+1} = (h_{q_\ell+1},\ldots,h_{q_{\ell+1}})$, $\bar{h}^{\ell+1} = (h_{q_{\ell+1}},\ldots,h_m)$, then $\bar{h}^\ell = (h^{\ell+1},\bar{h}^{\ell+1})$ and partition the vector $\bar{y}^\ell$ and $\bar{\rho}^{\ell+1}$ accordingly, i.e. $\bar{y}^\ell = (y^{\ell+1},\bar{y}^{\ell+1})$ and $(\rho^{\ell+1},\bar{\rho}^{\ell+1}) = ((\rho^{\ell+1}_{q_\ell+1},\ldots,\rho^{\ell+1}_{q_{\ell+1}}),(\rho^{\ell+1}_{q_{\ell+1}+1},\ldots,\rho^{\ell+1}_m))$. Choose an analytic $(m-q_\ell)$-vector $\alpha^{\ell+1}(x,\tilde{U}^\ell)$ and an invertible $(m-q_\ell)\times(m-q_\ell)$-matrix $\beta^{\ell+1}(x,\tilde{U}^\ell)$ on the forementioned neighborhood such that by applying the control law

$$\bar{u}^\ell = \alpha^{\ell+1}(x,\tilde{U}^\ell) + \beta^{\ell+1}(x,\tilde{U}^\ell)\begin{bmatrix} v^{\ell+1} \\ \bar{v}^{\ell+1} \end{bmatrix} \tag{8.82}$$

we arrive at

$$\begin{bmatrix} (y^{\ell+1})^{(\rho^{\ell+1})} \\ (\bar{y}^{\ell+1})^{(\bar{\rho}^{\ell+1})} \end{bmatrix} = \begin{bmatrix} 0 \\ \lambda^{\ell+1}(x,\tilde{U}^\ell) \end{bmatrix} + \begin{bmatrix} I_{r_{\ell+1}} & 0 \\ \mu^{\ell+1}(x,\tilde{U}) & 0 \end{bmatrix} \begin{bmatrix} v^{\ell+1} \\ \bar{v}^{\ell+1} \end{bmatrix}. \tag{8.83}$$

In (8.82) $v^{\ell+1}$ is an $r_{\ell+1}$-vector and $\bar{v}^{\ell+1}$ a $(m-q_{\ell+1})$-control vector and the matrices appearing in the right-hand side of (8.83) are analytic in x and $\tilde{U}^\ell$. In fact we have applied a feedback parametrized by $\tilde{U}^\ell$ on the inputs $\bar{u}^\ell$, without changing the inputs u^ℓ. Define the modified parametrized vectorfields

$$f^{\ell+1}(x,\tilde{U}^\ell) = f^\ell(x,\tilde{U}^{\ell-1}) + \sum_{j=q_\ell+1}^{m} g_j^\ell(x,\tilde{U}^{\ell-1})\alpha_j^{\ell+1}(x,\tilde{U}^\ell), \tag{8.84a}$$

$$g_i^{\ell+1}(x,\tilde{U}^\ell) = g_i^\ell(x,\tilde{U}^{\ell-1}) \ , \ i = 1,\dots,q_\ell, \tag{8.84b}$$

$$g_i^{\ell+1}(x,\tilde{U}^\ell) = \sum_{j=q_\ell+1}^{m} g_j^\ell(x,\tilde{U}^{\ell-1})\ \beta_{ji}^{\ell+1}(x,\tilde{U}^\ell), \ i = q_\ell+1,\dots,m, \tag{8.84c}$$

and consider the dynamics

$$\dot{x} = f^{\ell+1}(x,\tilde{U}^\ell) + \sum_{i=1}^{m} g_i^{\ell+1}(x,\tilde{U}^\ell)u_i\,, \tag{8.85}$$

where we have again renamed the controls $(u^\ell,v^{\ell+1},\bar{v}^{\ell+1})$ as u. Notice that the sequence of q_i's as defined in (8.81) is increasing and bounded by m, i.e.

$$q_1 = r_1 \le q_2 \ \dots \le q_\ell \le q_{\ell+1} \le \dots \le m. \tag{8.86}$$

As $q_\ell = q_{\ell+1}$ for some ℓ implies $r_{\ell+1} = 0$, and therefore the remaining outputs $\bar{y}^\ell$ are independent from the remaining inputs $\bar{u}^\ell$, we conclude that the sequence (8.86) stabilizes. That is, there exists a finite integer k such that

$$0 < q_1 < q_2 < \dots < q_{k-1} < q_k = q_{k+1} = \dots \le m. \tag{8.87}$$

This limiting value will be denoted as q^*, so q^* is defined as

$$q^* = \min\{q_\ell \mid q_\ell = q_{\ell+1}\}. \tag{8.88}$$

The integer q^* is well defined on an open and dense submanifold of $\tilde{M}_1 \cap \tilde{M}_2 \cap \dots \cap \tilde{M}_k$, where $\tilde{M}_\ell$ is the projection of the manifold M_ℓ, obtained in the ℓ-th step, on the manifold M. Therefore q^* is well defined on an open and dense submanifold of the state space manifold M. Henceforth q^* will be referred to as the rank *of the square analytic system (8.1,2)*.

□

Remark One can readily verify that for a linear system the above defined rank coincides with the rank of the corresponding transfer matrix, see Example 8.17.

We are now able to give a solution of the local dynamic state feedback input-output decoupling problem.

Theorem 8.19 *Consider the square analytic system (8.1,2). The following two conditions are equivalent.*

(i) The dynamic state feedback strong input-output decoupling problem is locally solvable on an open and dense submanifold of M.

(ii) The rank *of the system (8.1,2) satisfies*

$$q^* = m. \tag{8.89}$$

Proof (ii) ⇒ (i). Suppose (8.89) holds on an open and dense submanifold of M. That is, the successive application of the Algorithm 8.18 yields - see (8.67,74,83) - a parametrized input-output decoupled system on an open and dense submanifold of $M \times \mathbb{R}^{\mu_*}$, where μ_* equals μ_k and k is such that the sequence (8.87) stabilizes. It remains to be shown that this input-output decoupling can be achieved by applying a dynamic state feedback of the form (8.54). To see this we carefully study the steps in the algorithm. At step 1 we apply the regular static state feedback (8.66) yielding input-output decoupling of the first $q_1 = r_1$ input-output channels. Clearly the feedback (8.66) is of the form (8.54). In the second step we proceed as follows. Let ν_1 denote the highest time derivative of the inputs u^1 appearing in $\tilde{U}^1$. Note that $(u^2,\bar{u}^1)$ denote the new input of the feedback modified system (8.69). Introduce the precompensator

$$\dot{z}^1_{ij} = z^1_{ij+1} \quad , \qquad 1 \le j < \nu_1 \quad , \qquad i = 1,\ldots,q_1, \tag{8.90a}$$

$$\dot{z}^1_{i\nu_1} = w_i \qquad\qquad i = 1,\ldots,q_1. \tag{8.90b}$$

Construct the composition of (8.66) with (8.90a,b) via

$$u_i = z^1_{i1} \quad , \qquad\qquad i = 1,\ldots,q_1, \tag{8.91a}$$

$$u_i = \alpha^2_i(x,\tilde{U}^1) + \sum_{j=q_1+1}^{m} \beta^2_{ij}(x,\tilde{U}^1)w_j, \qquad i = q_1+1,\ldots,m. \tag{8.91b}$$

Because ν_1 is the highest time derivative of the inputs $u^1 = (u_1,\ldots,u_{q_1})$ it follows from (8.90a) that all the time derivatives $u_i^{(j)}$ appearing in $\tilde{U}^1$ can be expressed as z^1_{ij}. Therefore the relation between the inputs $(u_1,\ldots,u_m)$ and the newly defined inputs $(w_1,\ldots,w_m)$ in (8.90b,91b) is precisely in the form of a dynamic state feedback (8.54). Notice that the two feedbacks (8.66) together with (8.90,91) indeed form a dynamic state feedback of the form (8.54) with as new inputs $(w_1,\ldots,w_m)$. From (8.67) and (8.74) we see that we have obtained input-output decoupling between the first $r_1+r_2 = q_2$ input-output channels. Observe that, see (8.90), the characteristic numbers of the first $r_1 = q_1$ outputs are increased with the number ν_1. In the third step we proceed in a completely similar way. We will describe here the general $(\ell+1)$-th step. As in the algorithm denote the inputs of the system after applying the feedback (8.66) and $(\ell-1)$ feedbacks of the form (8.90,91) again by $u = (u_1,\dots,u_m)$ instead of $(w_1,\dots,w_m)$. Note that the composition of these ℓ feedbacks is of the desired form (8.54). Let ν_ℓ be the highest time-derivative of the inputs u^ℓ appearing in $\tilde{U}^\ell$. Introduce the precompensator

$$\dot{z}^\ell_{ij} = z^\ell_{ij+1} \quad , \quad 1 \le j < \nu_j, \quad i = 1,\dots,q_\ell, \tag{8.92a}$$

$$\dot{z}^\ell_{i\nu_\ell} = w_i \quad , \quad i = 1,\dots,q_\ell. \tag{8.92b}$$

Compose the first ℓ feedbacks with (8.92a,b) via the linking maps

$$u_i = z^\ell_{i1} \quad , \quad i = 1,\dots,q_\ell, \tag{8.93a}$$

$$u_i = \alpha_i^{\ell+1}(x,\tilde{U}^\ell) + \sum_{j=q_\ell+1}^{m} \beta_{ij}^{\ell+1}(x,\tilde{U}^\ell)w_j, \; i = q_\ell+1,\dots,m. \tag{8.93b}$$

From the definition of $\tilde{U}^\ell$ and the preceding ℓ feedbacks it follows that all time derivatives $u_i^{(j)}$ appearing in $\tilde{U}^\ell$ can be expressed in z^s_{ij}, $s = 1,\dots,\ell$, $i = 1,\dots,q_s$, $j = 1,\dots,\nu_s$. Observe that the composition of the previous ℓ feedbacks with (8.91,92) is again a dynamic state feedback. Now, because (8.89) holds and $q^* = q_k$ for k sufficiently large, we see that after applying a sequence of k feedbacks of the above type the input-output behavior indeed consists of $q^* = m$ decoupled input-output channels. We emphasize that the control laws of the form (8.92,93) are only valid if the matrices $D^{\ell+1}(x,\tilde{U}^\ell)$ have constant rank. As this is the case on an open and dense submanifold by the analyticity of the system, the above procedure yields the existence of a dynamic state feedback (8.54) around an open and dense set of initial points $(x_0, z^1_{ij}(0),\dots,z^k_{ij}(0))$ which achieves strong input-output decoupling. Therefore we have shown that the dynamic state feedback strong input-output decoupling problem is locally solvable around an open and dense submanifold of initial states x_0 in M.

(i) ⇒ (ii) Assume there exists a compensator

$$\begin{cases} \dot{z} = \gamma(z,x) + \delta(z,x)v \\ u = \alpha(z,x) + \beta(z,x)v \end{cases} \tag{8.54}$$

with $z \in \mathbb{R}^q$, which achieves strong input-output decoupling of the overall system (8.1,54,2) around a point $(x_0, z_0) \in M \times \mathbb{R}^q$. By Definition 8.3 it follows that the precompensated system has finite characteristic numbers $\sigma_1, \ldots, \sigma_m$ defined in a neighborhood of (x_0, z_0), and by Theorems 8.13 and 8.9 we may equally well assume that the input-output behavior is locally given as

$$y_i^{(\sigma_i)} = v_i \quad , \quad i \in \underline{m}. \tag{8.94}$$

Observe that (8.94) implies the following local reproducibility property. Given an arbitrary set of analytic functions $\varphi_i(t)$, $i \in \underline{m}$, one is able to find controls $\bar{v}_i(t)$, $i \in \underline{m}$, such that the system (8.1,54) feeded with these controls produces as output $\bar{y}(t) = (\bar{y}_1(t), \ldots, \bar{y}_m(t))$ on a possible small time interval, such that

$$y_i^{(\bar{\sigma}_i)} = \varphi_i(t) \quad , \quad i \in \underline{m}, \tag{8.95}$$

for any fixed set of $\bar{\sigma}_i$ with $\bar{\sigma} \geq \sigma_i$, $i \in \underline{m}$. Therefore the original system (8.1,2) possesses the same reproducibility property, i.e. there exist controls $\bar{u}_i(t)$, $i \in \underline{m}$, such that (8.95) holds for small t, when these controls are applied. This follows from computing the controls $\bar{u}_i(t)$ from (8.54) with inputs $\bar{v}_1(t), \ldots, \bar{v}_m(t)$. To prove that (8.89) is necessary for input-output decoupling we show that if $q^* < m$ then the system (8.1,2) does not possess the above reproducibility property. In order to see this we follow the decoupling procedure based on the algorithm, which according to the proof (ii) $\Rightarrow$ (i) yields as input-output behavior

$$y_i^{(\bar{\bar{\sigma}}_i)} = w_i \quad , \quad i = 1, \ldots, q^*, \tag{8.96}$$

for suitable chosen $\bar{\bar{\sigma}}_i$, $i = 1, \ldots, q^*$, and the outputs y_i, $i = q^*+1, \ldots, m$ do not depend upon the remaining inputs w_i, $i = q^*+1, \ldots, m$. Therefore the forementioned reproducibility property is violated. □

Remark 8.20 The above algorithm can also be applied to smooth systems as long as the constant rank hypothesis of the matrices $D^\ell(x, \tilde{U}^{\ell-1})$ are met. In the same manner as in Theorem 8.19 it follows in this case that, when the number q^* defined in (8.89) equals m on a neighborhood of a point x_0, then the Problem 8.16 is locally solvable about x_0. Notice that the rank of an analytic system is an intrinsic number associated with the system, and which is independent of the particular feedbacks chosen in Algorithm 9.18. Clearly, this follows from the implicit characterization of q^* given in the second part of the proof of Theorem 8.19, as being the number of "reproducible outputs".

Example 8.21 A simplified model of a voltage fed induction motor can be described as

$$\frac{d}{dt}\begin{pmatrix} x_1 \\ x_2 \\ x_3 \\ x_4 \end{pmatrix} = \begin{pmatrix} -(\alpha+\beta) & -\omega & BL_s^{-1} & \omega\sigma^{-1}L_s^{-1} \\ \omega & -(\alpha+\beta) & -\omega\sigma^{-1}L_s^{-1} & \beta L_s^{-1} \\ -\alpha\sigma L_s & 0 & 0 & 0 \\ 0 & -\alpha\sigma L_s & 0 & 0 \end{pmatrix}\begin{pmatrix} x_1 \\ x_2 \\ x_3 \\ x_4 \end{pmatrix} +$$

$$+ \begin{pmatrix} \sigma^{-1}L_s^{-1} & 0 \\ 0 & \sigma^{-1}L_s^{-1} \\ 1 & 0 \\ 0 & 1 \end{pmatrix}\begin{pmatrix} \bar{u}_1 \\ \bar{u}_2 \end{pmatrix}, \tag{8.97}$$

where x_1 and x_2 are the components of the stator current and x_3 and x_4 are the corresponding flux components all measured with respect to a fixed reference frame, $\alpha = R_s\sigma^{-1}L_s^{-1}$, $\beta = R_r\sigma^{-1}L_s^{-1}$ and $\sigma = 1-M^2L_s^{-1}L_r^{-1}$. Here the parameters R_s and R_r denote the stator and rotor resistances, L_s and L_r the stator respectively rotor self-inductance and M is mutual inductance. The mechanical speed ω of the motor is assumed to be constant. In order to have a voltage-frequency control scheme for the induction motor the voltage input vector $(\bar{u}_1,\bar{u}_2)$ is expressed as

$$\begin{pmatrix} \bar{u}_1 \\ \bar{u}_2 \end{pmatrix} = \begin{pmatrix} V\cos\theta \\ V\sin\theta \end{pmatrix}, \quad \text{with } \theta = \int_0^t \omega_a(\tau)d\tau, \tag{8.98}$$

where θ is the angular position, V the amplitude and ω_a the voltage supply frequency, which shows that, more realistically, we should take (V,ω_a) as the input of the induction motor. This can be achieved by adding to (8.96) the state variable $x_5 = \theta$, which yields as dynamics

$$\frac{d}{dt}\begin{pmatrix} x_1 \\ x_2 \\ x_3 \\ x_4 \\ x_5 \end{pmatrix} = \begin{pmatrix} -(\alpha+\beta) & -\omega & \beta L_s^{-1} & \omega\sigma^{-1}L_s^{-1} & 0 \\ \omega & -(\alpha+\beta) & -\omega\sigma^{-1}L_s^{-1} & \beta L_s^{-1} & 0 \\ -\alpha\sigma L_s & 0 & 0 & 0 & 0 \\ 0 & -\alpha\sigma L_s & 0 & 0 & 0 \\ 0 & 0 & 0 & 0 & 0 \end{pmatrix}\begin{pmatrix} x_1 \\ x_2 \\ x_3 \\ x_4 \\ x_5 \end{pmatrix} +$$

$$+ \begin{pmatrix} \sigma^{-1}L_s^{-1}\cos x_5 & 0 \\ \sigma^{-1}L_s^{-1}\sin x_5 & 0 \\ \cos x_5 & 0 \\ \sin x_5 & 0 \\ 0 & 1 \end{pmatrix}\begin{pmatrix} u_1 \\ u_2 \end{pmatrix}, \tag{8.99}$$

where $(u_1,u_2) = (V,\omega_a)$. Together with the dynamics (8.99) we consider the stator flux and stator torque as outputs. So we have

$$\begin{cases} y_1 = h_1(x) = x_3^2 + x_4^2 \\ y_2 = h_2(x) = x_2 x_3 - x_1 x_4 . \end{cases} \tag{8.100}$$

Let us investigate if the square analytic system (8.99,100) is statically or dynamically input output decouplable. It is straightforward to verify that the characteristic numbers ρ_1^1 and ρ_2^1 of (8.99,100) both equal zero, and the first step of Algorithm 8.18 yields

$$\begin{bmatrix} \dot{y}_1 \\ \dot{y}_2 \end{bmatrix} = \begin{bmatrix} 2x_3\dot{x}_3 + 2x_4\dot{x}_4 \\ x_2\dot{x}_3 + x_3\dot{x}_2 - x_1\dot{x}_4 - x_4\dot{x}_1 \end{bmatrix} =$$

$$= \begin{bmatrix} -2\alpha\sigma L_s x_1 x_3 - 2\alpha\sigma L_s x_2 x_4 \\ \omega x_1 x_3 + (\alpha+\beta)x_1 x_4 - (\alpha+\beta)x_2 x_3 + \omega x_2 x_4 - \omega\sigma^{-1}L_s^{-1}x_3^2 - \omega\sigma^{-1}L_s^{-1}x_4^2 \end{bmatrix} +$$

$$\begin{bmatrix} 2x_3 \cos x_5 + 2x_4 \sin x_5 & 0 \\ (x_1+\sigma^{-1}L_s^{-1}x_3)\sin x_5 + (x_2 - \sigma^{-1}L_s^{-1}x_4)\cos x_5 & 0 \end{bmatrix} \begin{bmatrix} u_1 \\ u_2 \end{bmatrix}, \tag{8.101}$$

which will be abbreviated as

$$\begin{bmatrix} \dot{y}_1 \\ \dot{y}_2 \end{bmatrix} = \begin{bmatrix} e_1^1(x) \\ e_2^1(x) \end{bmatrix} + \begin{bmatrix} d_{11}^1(x) & 0 \\ d_{21}^1(x) & 0 \end{bmatrix} \begin{bmatrix} u_1 \\ u_2 \end{bmatrix}. \tag{8.102}$$

Because the 2×2 matrix $D^1(x) = (d_{ij}^1(x))$ has at most rank 1 we conclude that the system (8.99,100) can not be locally decoupled by applying a regular static state feedback, cf. Theorem 8.13. On the other hand, around initial points where either $d_{11}^1(x)$ or $d_{21}^1(x)$ is nonvanishing we find that the rank of the matrix $D^1(x) = r_1 = 1$. To complete step 1 of Algorithm 8.18 we will assume that $d_{11}^1(x) \neq 0$, i.e. we are working in a neighborhood of an initial point x_0 with $d_{11}^1(x_0) \neq 0$. Define around x_0 a regular static state feedback via

$$\begin{cases} u_1 = -(d_{11}^1(x))^{-1}e_1^1(x) + (d_{11}^1(x))^{-1}\tilde{u}_1 \\ u_2 = \tilde{u}_2 \end{cases} \tag{8.103}$$

where $(\tilde{u}_1, \tilde{u}_2)$ denotes the new control. Applying (8.103) around x_0 to the system yields as input-output behavior

$$\begin{cases} \dot{y}_1 = \tilde{u}_1 \\ \dot{y}_2 = e_2^1(x) - d_{21}^1(x)(d_{11}^1(x))^{-1}e_1^1(x) + d_{21}^1(x)(d_{11}^1(x))^{-1}\tilde{u}_1 . \end{cases} \tag{8.104}$$

In the second step we consider the modified dynamics (8.99,103) and verify when the output y_2 is going to depend upon $\tilde{u}_2$. After a straightforward but tedious computation we find

$$\ddot{y}_2 = e^2(x,u_1,u_1) + \left(-d_{21}^1(x)e_1^1(x)\frac{\partial}{\partial x_5}\left(d_{11}^1(x)\right)^{-1} + d_{21}^1(x)\frac{\partial}{\partial x_5}\left(d_{11}^1(x)\right)^{-1}\tilde{u}_1\right)\tilde{u}_2 \tag{8.105}$$

for a certain function $e^2(x,\tilde{u}_1,\dot{\tilde{u}}_1)$. According to the second step of the algorithm we conclude that $\rho_2^2 = 1$ and the "second decoupling matrix" is given as the 1×1-matrix

$$d^2(x,\tilde{u}_1) = -d_{21}^1(x)e_1^1(x)\frac{\partial}{\partial x_5}\left(d_{11}^1(x)\right)^{-1} + d_{21}^1(x)\frac{\partial}{\partial x_5}\left(d_{11}^1(x)\right)^{-1}\tilde{u}_1, \tag{8.106}$$

which is nonvanishing for an open and dense set of points $(x,\tilde{u}_1) \in \mathbb{R}^5 \times \mathbb{R}$. Thus $r_2 = 1$ and $q_2 = r_1+r_2 = 2$. Therefore the model of the induction motor is dynamically decouplable around points where both $d_{11}^1(x)$ and $d^2(x,\tilde{u})$ are nonvanishing. In order to find a decoupling control law we follow the procedure of Theorem 8.19. First we note that the highest time derivative of $\tilde{u}_1$ in the expression (8.105) is $\dot{\tilde{u}}_1$, thus $\nu_1 = 1$, and we define a first order integrator for $\tilde{u}_1$ via

$$\begin{cases} \dot{z}_1 = v_1 \\ \tilde{u}_1 = z_1 \end{cases} \tag{8.107a}$$

and the second input $\tilde{u}_2$ is modified as

$$\tilde{u}_2 = -\left(d^2(x,z_1)\right)^{-1}e_2(x,z_1,v_1) + \left(d^2(x,z_1)\right)^{-1}v_2. \tag{8.107b}$$

Thus a dynamic feedback which decouples the system (8.99,100) locally is given as, see (8.103) and (8.107)

$$\begin{cases} \dot{z}_1 = v_1 \\ \begin{pmatrix} u_1 \\ u_2 \end{pmatrix} = \begin{pmatrix} -\left(d_{11}^1(x)\right)^{-1}e_1^1(x) + \left(d_{11}^1(x)\right)^{-1}z_1 \\ -\left(d^2(x_1z_1)\right)^{-1}e_2(x,z_1,v_1) \end{pmatrix} + \begin{pmatrix} 0 \\ \left(d^2(x,z_1)\right)^{-1} \end{pmatrix} v_2. \end{cases} \tag{8.108}$$

Notice that applying the control law (8.108) locally yields as input-output behavior for the induction motor

$$\begin{cases} \ddot{y}_1 = v_1 \\ \ddot{y}_2 = v_2 \end{cases} \tag{8.109}$$

provided that $d_{11}^1(x)$ and $d^2(x,z_1)$ are nonvanishing. □

Notes and References

The input-output decoupling problem has received a lot of attention over the last two or three decades. The here presented solutions have been given for linear systems in [FW] - using static state feedback - and in [Wa], where dynamic state feedbacks have been allowed. Example 8.10 in fact summarizes the result of [FW], whereas the remark following Algorithm 8.18 refers to [Wa]. Further aspects of the lienar dynamic decoupling problem may be found in [Cr]. A geometric theory on the input-output decoupling problem for not necessarily square linear systems is given in [MW1], [WM], [Wo]. A survey of results has been given in [MW2]. In [Po] the decoupling problem for time-varying square linear systems is treated and probably forms the first approach to generalizing the problem in a nonlinear context. The result on static feedback input-output decoupling as is given in Theorem 8.9 is based on [SR], [Fr], [Si], see also [Is], [IKGM]. The result of Proposition 8.11 is borrowed from [Is]. A different approach to the nonlinear decoupling problem is presented in [Cl]. A nonlinear differential geometric treatment of the problem for smooth not necessarily square systems will be discussed in the next chapter. The dynamic feedback input-output decoupling problem for square nonlinear systems has been studied first in [DM]. The Algorithm 8.18 given here has been taken from [NR1], [NR2]. The limiting value q^* appearing in Algorithm 8.18 is called the rank of the system in analogy with terminology introduced in [Fl1], [Fl2] where a differential algebraic approach for nonlinear systems is given, and is in agreement with the linear terminology. In fact the rank as introduced here equals the rank as given in [Fl1], [Fl2], as has recently been demonstrated in [DGM]. The Example 8.20 on dynamic decoupling comes from [LU].

[Cl] D. Claude, "Decoupling of nonlinear systems", Syst. Contr. Lett. 1, pp. 242-248, 1982.

[Cr] M. Cremer, "A precompensator of minimal order for decoupling a linear multivariable system", Int. J. Contr. 14, pp. 1089-1103, 1971.

[DGM] M. Di Benedetto, J.W. Grizzle, C.H. Moog, "Rank invariants of nonlinear systems", SIAM J. Contr. Optimiz. 27, 1989.

[DM] J. Descusse, C.H. Moog, "Decoupling with dynamic compensation for strong invertible affine nonlinear systems", Int. J. Contr. 42, pp. 1387-1398, 1985.

[FW] P.L. Falb, W.A. Wolovich, "Decoupling in the design and synthesis of multivariable control systems", IEEE Trans. Aut. Contr. AC-12, pp. 651-659, 1967.

[Fl1] M. Fliess, "A note on the invertibility of nonlinear input-output differential systems", Syst. Contr. Lett. 8, pp. 147-151, 1986.

[Fl2] M. Fliess, "Vers une nouvelle théorie du bouclage dynamique sur la sortie des systèmes nonlinéaires", in **Analysis and Optimization of Systems**, (eds. A. Bensoussan, J.L. Lions), Lect. Notes Contr. Inf. Sci. 83, Springer, Berlin, pp. 293-299, 1986.

[Fr] E. Freund, "The structure of decoupled nonlinear systems", Int. J. Contr. 21, pp. 443-450, 1975.

[GN] L.C.J.M. Gras, H. Nijmeijer, "Decoupling in nonlinear systems: from linearity to nonlinearity", IEE Proceedings 136 Pt.D., pp. 53-62, 1989.

[HvdS] H.J.C. Huijberts, A.J. van der Schaft, "Input-output decoupling with stability for Hamiltonian systems", preprint 1987, to appear in Math. Control, Signals, Systems, 1989.

[Is] A. Isidori, **Nonlinear control systems: an introduction**, Lecture Notes Contr. Inf. Sci. 72, Springer, Berlin, 1985.

[IKGM] A. Isidori, A.J. Krener, C. Gori-Giorgi, S. Monaco, "Nonlinear decoupling via feedback: a differential geometric approach", IEEE Trans. Aut. Contr. AC-26, pp. 331-345, 1981.

[LU] A. de Luca, G. Ulivi, "Dynamic decoupling in voltage frequency controlled induction motors", in **Analysis and Optimization of Systems** (eds. A. Bensoussan, J.L. Lions), Lect. Notes Contr. Inf. Sci. 111, pp. 127-137, 1988.

[MBE] R. Marino, W.M. Boothby, D.L. Elliott, "Geometric properties of linearizable control systems", Math. Systems Theory 18, pp. 97-123, 1985.

[MW1] A.S. Morse, W.M. Wonham, "Decoupling and pole assignment by dynamic compensation", SIAM J. Contr. Optimiz. 8, pp. 317-337, 1970.

[MW2] A.S. Morse, W.M. Wonham, "Status of noninteracting control", IEEE Trans. Aut. Contr. AC-16, pp. 568-580, 1971.

[NR1] H. Nijmeijer, W. Respondek, "Decoupling via dynamic compensation for nonlinear control systems", Proc. 25th. CDC, Athens, pp. 192-197, 1986.

[NR2] H. Nijmeijer, W. Respondek, "Dynamic input-output decoupling of nonlinear control systems", IEEE Trans. Aut. Contr., AC-33, pp. 1065-1070, 1988

[Po] W.A. Porter, "Decoupling of and inverses for time-varying linear systems", IEEE Trans. Aut. Contr. AC-14, pp. 378-380, 1969.

[SR] S.N. Singh, W.J. Rugh, "Decoupling in a class of nonlinear systems by state variable feedback", J. Dynamic Systems, Measurements Contr., pp. 323-329, 1972.

[Si] P.K. Sinha, "State feedback decoupling of nonlinear systems", IEEE Trans. Aut. Contr. AC-22, pp. 487-489, 1977.

[Wa] S.H. Wang, "Design of precompensator for decoupling problem", Electronics Letters 6, pp. 739-741, 1970.

[Wo] W.M. Wonham, **Linear multivariable control: a geometric approach**, Springer, Berlin, 1979.

[WM] W.M. Wonham, A.S. Morse, "Decoupling a pole assignment in linear multivariable systems: a geometric appraoch", SIAM J. Contr. Optimiz. 8, psop. 1-18, 1970.

Exercises

8.1 Consider the square *smooth* system (8.1,2) and let $\rho_1,\dots,\rho_m$ be the characteristic numbers defined as in Definition 8.7.

(a) Assume the decoupling matrix $A(x)$ (see (8.24)) is nonsingular at a point $x_0 \in M$. Prove that the regular static state feedback input-output decoupling is locally solvable around x_0 for the system (8.1,2).

(b) Assume the system satisfies (8.4) and the subset M_0 of (8.10) coincides with M. Show that the system is not necessarily input-output decoupled (in the sense of Definition 8.1). Hint: see Example 4.16.

8.2 ([GN]) Consider the square analytic system (8.1,2) about an equilibrium point x_0 of the vectorfield f. Let $\dot{\bar{x}} = A\bar{x} + B\bar{u}$, $\bar{y}_i = c_i\bar{x}$, $i \in \underline{m}$, be the linearization of the system (8.1,2) around x_0.

(a) Prove that if Problem 8.6 is locally solvable about x_0 for the system (8.1,2) then the (linear) input-output decoupling problem is solvable for the linearized system.

(b) Show that the converse of **(a)** is true in case (8.1,2) and its linearization have the same characteristic numbers.

8.3 ([MBE], [HvdS]) Consider an input-output decoupled system of the form (8.1,32,2). This decoupled system admits a local "normal form" (8.45). Assume the distribution $G(x) = \text{span}\{g_1(x),\dots,g_m(x)\}$ is involutive. Prove that there exist local coordinates $z = (\hat{z},z^1,\dots,z^m)$ with z^i as defined in (8.44) such that $\dot{\hat{z}} = \hat{f}(\hat{z},z^1,\dots,z^m)$, i.e. the time-derivative of $\hat{z}$ is not implicitly depending upon the inputs $v_1,\dots,v_m$. Conversely show that when in the "normal form" (8.45) the equations for $\bar{z}$ are independent of the new inputs $v_1,\dots,v_m$ then the distribution G is involutive.

8.4 Consider the square analytic system (8.1,2) with characteristic numbers $\rho_1,\dots,\rho_m$. Assume that rank $A(x_0) = m$. Let D^* be the largest locally controlled invariant distribution in ker dh (see Chapter 7). Show that $D^*(x) = \bigcap_{i=1}^{m} \bigcap_{j=0}^{\rho_i} \ker dL_f^j h_i(x)$ on a neighborhood of x_0.

8.5 Consider as in Chapter 7 an analytic system with disturbances Σ: $\dot{x} = f(x) + \sum_{i=1}^{m} g_i(x)u_i + \sum_{i=1}^{\ell} e_i(x)d_i$, $y_i = h_i(x)$, $i \in \underline{m}$. Let Σ_0 denote the same system with disturbances $d_1 = \dots = d_\ell = 0$. Suppose the strong input-output decoupling problem for Σ_0 is solvable and the local disturbance decoupling problem for Σ is solvable. Prove that locally about every point x_0 there exists a regular state feedback $u = \alpha(x) + \beta(x)v$ such that the closed loop system is input-output decoupled as well as disturbance decoupled.

8.6 Consider the square smooth system (8.1,2) about an equilibrium point x_0 of f. Assume the decoupling matrix $A(x)$ is nonsingular at x_0. Show that the system is feedback linearizable about x_0 (see Definition

6.1) if $\sum_{i=1}^{m} (\rho_i+1) = n$.

8.7 Consider the square analytic system (8.1,2). Define the compensator $\dot{z}_i = v_i$, $u_i = z_i$, $i \in \underline{m}$. Prove that the regular static state feedback input-output decoupling problem for the system (8.1,2) is solvable if and only if it is solvable for the precompensated system.

8.8 Consider the square analytic system (8.1,2) and an analytic compensator of the form (8.54) and assume $\beta(z,x) = 0$ for all (z,x). Assume that for each $x \in M$ the compensator has rank m. Prove that the rank of the system (8.1,2) equals the rank of the precompensated system (8.1,54,2).

8.9 Consider the analytic system (8.1,2) and assume that the number of inputs m is larger than the number of outputs p.

(a) Define the $p \times m$-decoupling matrix $A(x)$ as in (8.24). Prove that there exists locally about x_0 a regular static state feedback $u = \alpha(x) + \beta(x)$ such that the inputs $v_{p+1}, \ldots, v_m$ do not influence the outputs $y_1, \ldots, y_p$ while the closed loop system with inputs $v_1, \ldots, v_p$ is strongly input-output decoupled, if and only if rank $A(x_0) = m$.

(b) Formulate in a similar way the dynamic strong input-output decoupling problem and show that it is locally solvable on an open and dense subset of M if the rank of the system (8.1,2) equals p.

8.10 (see Exercise 7.7) Consider on $\mathbb{R}^5$ the system $\dot{x}_1 = x_2 u_1$, $\dot{x}_2 = x_5$, $\dot{x}_3 = x_4 u_1 + x_2$, $\dot{x}_4 = u_2$, $\dot{x}_5 = x_1 u_1 + d$, $y_1 = x_1$, $y_2 = x_2$. Show that there exists a dynamic compensator for this system which locally achieves strong input-output decoupling as well as disturbance decoupling, provided that $x_5 u \neq 0$.

9
The Input-Output Decoupling Problem: Geometric Considerations

In the previous chapter we have given an analytic approach to the input-output decoupling problem for analytic systems. This resulted in a rather complete description of *square* analytic systems (i.e. systems with an equal number of scalar controls and outputs) that are decouplable either by static or dynamic state feedback. The study presented in Chapter 8 is based on purely analytic considerations such as the determination of the decoupling matrix and computation of its rank. On the other hand, as we will see, for the static state feedback input-output decoupling problem a much more geometric approach, involving controlled invariant distributions, is possible. The key is that for analytic systems the noninteracting condition for an input u_i not to affect an output y_j, see (8.4), is equivalent to the geometric condition derived in Chapter 4 (see Proposition 4.20). Furthermore, also for *smooth* systems such a geometric characterization (Proposition 4.23) will prove to be a good starting point in the study of the input-output decoupling problem. It turns out that the differential geometric approach based on Proposition 4.23 will also enable us to treat the *block* input-output decoupling problem (for smooth systems).

9.1 The Block Input-Output Decoupling Problem for Smooth Nonlinear Systems

Consider a smooth affine nonlinear system

$$\dot{x} = f(x) + \sum_{i=1}^{m} g_i(x)u_i \tag{9.1}$$

together with the partitioned output blocks

$$\begin{cases} y_1 = h_1(x) \\ \quad\vdots \\ y_p = h_p(x) \end{cases} \tag{9.2}$$

where $x = (x_1, \dots, x_n)$ are local coordinates for a smooth manifold M, f, $g_1, \dots, g_m$ are smooth vectorfields on M and $h_i : M \to \mathbb{R}^{p_i}$, $p_i > 0$, $i \in \underline{p}$, are smooth mappings. We assume throughout that the number of output blocks equals the number of scalar controls, thus

$$p = m \ . \tag{9.3}$$

Clearly when all p_i's are identically one, then we are dealing with the square systems of Chapter 8, but when one of the p_i's is larger than one the analysis given so far is no longer adequate. The differential geometric formulation of the static state feedback input-output decoupling problem also differs from the one given in the previous chapter in that we allow for smooth (not necessarily analytic) systems. Throughout we will make the following assumption.

Assumption 9.1 *The system (9.1) satisfies the strong accessibility rank condition at each point* $x_0 \in M$.

Therefore (see Definition 3.19 and Theorem 3.21) the distribution C_0 corresponding to the smallest subalgebra of $V(M)$ which contains the vectorfields $g_1, \ldots, g_m$ and which is invariant under f — so $[f,X] \in C_0$ for all $X \in C_0$ — has dimension n at each point x_0 in M.

In accordance with Definition 8.1 we will say that the system (9.1-3) is input-output decoupled if after a possible relabeling of the inputs $(u_1, \ldots, u_m)$, the i-th control u_i does not influence the outputs y_j, $j \neq i$. Denoting $h_i = (h_{i1}, \ldots, h_{ip_i})$, $i \in \underline{m}$, this yields the necessary condition that (see Chapter 4, Proposition 4.14)

$$L_{g_j} L_{X_1} \cdots L_{X_k} h_{i\ell}(x) = 0 \quad \forall k \geq 0,\ X_1, \ldots, X_k \in \{f, g_1, \ldots, g_m\}\ , \qquad \ell \in \underline{p_i}\ ,\ x \in M. \tag{9.4}$$

We note that at a first glance no condition on the interaction of the i-th input on the i-th output block has been given. However, a system satisfying Assumption 9.1 which fulfils (9.4) *automatically* has the property that the i-th control affects the i-th output block. To see this we introduce the following objects. In the algebra $\mathcal{C}_0$ we define $\mathcal{C}_{0i}$, $i \in \underline{m}$, as the smallest subalgebra which contains the vectorfield g_i and which is such that $[f,X] \in \mathcal{C}_{0i}$, $[g_j, X] \in \mathcal{C}_{0i}$, $j \in \underline{m}$, for all $X \in \mathcal{C}_{0i}$. Furthermore, we let C_{0i} be the corresponding distribution

$$C_{0i}(x) = \text{span}\ \{X(x) \mid X \text{ vectorfield in } \mathcal{C}_{0i}\},\ i \in \underline{m}. \tag{9.5}$$

By definition the distributions C_0 and C_{0i}, $i \in \underline{m}$, are related via

$$C_0(x) = C_{01}(x) + \ldots + C_{0m}(x) \quad , \quad x \in M. \tag{9.6}$$

Now consider the distributions

$$\ker dh_i = \bigcap_{\ell=1}^{p_i} \ker dh_{i\ell} \quad , \qquad i \in \underline{m}. \tag{9.7}$$

From the noninteraction condition (9.4) it follows that

$$C_{0j} \subset \ker dh_i \ , \ \forall\ i,j \in \underline{m},\ j \neq i,\ x \in M\ , \tag{9.8}$$

and so when Assumption 9.1 holds we conclude that for $i \in \underline{m}$

$$\operatorname{Im} h_{i*} = h_{i*}(TM) = h_{i*}(C_0) = h_{i*}(C_{01} + \ldots + C_{01}) = h_{i*}(C_{0i}) \tag{9.9}$$

where $h_{i*} : TM \to T\mathbb{R}^{p_i}$ is the tangent mapping of h_i. This immediately shows that the i-th control *does* influence the i-th output block. Namely if we introduce in analogy with the set $R^V(x_0,T)$ given in (3.21) the subset of reachable points in the i-th output space $(= \mathbb{R}^{p_i})$, $i \in \underline{m}$, for a neighborhood V of x_0 as

$$h_i\big(R^V(x_0,T)\big) = \{\bar{y}_i \in \mathbb{R}^{p_i} \mid \text{There exists a piecewise constant input } u = (u_1,\ldots,u_m)^T : [0,T] \to \mathbb{R}^m \text{ such that the evolution of (9.1) for } x(0) = x_0 \text{ satisfies } x(t) \in V,\ 0 \le t \le T, \text{ and } h_i(x(T)) = \bar{y}_i.\}\ , \tag{9.10}$$

then it follows from (9.9) that $h_i(R^V(x_0,T))$ has nonempty interior in Im h_i for any neighborhood V of x_0 and any $T > 0$. This is what we call *output controllability*. Moreover by manipulating only the i-th input function u_i we can steer in an open set of Im h_i, no matter how the other inputs u_j, $j \neq i$, are chosen (see again (9.9)). For a square system (a system with m scalar outputs) the noninteraction condition (9.4) together with the Assumption 9.1 almost, but not entirely, implies that the system is strongly input-output decoupled in the sense of Definition 8.2, as illiustrated by the next example

Example 9.2 (= Example 8.4) Consider the square system

$$\begin{cases} \dot{x}_1 = x_2^3 \\ \dot{x}_2 = u_1 \\ \dot{x}_3 = u_2 \\ y_1 = x_1 \\ y_2 = x_3 \end{cases} \tag{9.11}$$

Clearly the system (9.11) satisfies the noninteracting condition (9.4). Moreover one straightforwardly checks that (9.11) satisfies the strong accessibility assumption at each point $(x_1,x_2,x_3) \in \mathbb{R}^3$. Computing the distributions C_{01} and C_{02} we obtain

$$C_{01} = \text{span}\{\frac{\partial}{\partial x_1},\frac{\partial}{\partial x_2}\} \quad , \quad C_{02} = \text{span}\{\frac{\partial}{\partial x_3}\} \ , \tag{9.12}$$

and so at each point in $\mathbb{R}^3$ we have

$$h_{1*}(C_{01}) = \text{Im}\ h_{i*} \quad , \quad h_{2*}(C_{02}) = \text{Im}\ h_{i*} \ . \tag{9.13}$$

and we may conclude that the i-th control instantaneously affects the i-th output. However, see Example 8.3, the system (9.11) does not globally meet the requirements of Definition 8.2. □

The following example illustrates the output controllability condition for block outputs.

Example 9.3 (see Example 9.2) Consider the strongly accessible system

$$\begin{aligned} \dot{x}_1 &= x_2^3 \ , & y_1 &= (x_1,x_2)^T \ , \\ \dot{x}_2 &= u_1 \ , & y_2 &= x_3 \ . \\ \dot{x}_3 &= u_2 \ , \end{aligned} \tag{9.14}$$

As in the previous example the noninteracting condition (9.4) for the system (9.14) is immediate. Using (9.12) we see that the output controllability conditions are met. In particular one can use the control u_1 to steer the output $y_1^0 = (x_1^0,x_2^0)$ into an open set of points $y_1^1 = (x_1^1,x_2^1)$ in $\mathbb{R}^2$. □

In what follows we will discuss the possibility of (locally) achieving the noninteraction condition (9.4) for a system (9.1-3) which satisfies Assumption 9.1, by applying regular static state feedback. As in Chapter 8 this means that we search for a regular feedback

$$u = \alpha(x) + \beta(x)v \tag{9.15}$$

with $\alpha\colon M \to \mathbb{R}^m$, $\beta\colon M \to \mathbb{R}^{m\times m}$ smooth mappings and with $\beta(x)$ nonsingular for all x and $v = (v_1,\ldots,v_m)$ representing the new controls, such that in the feedback modified dynamics

$$\dot{x} = \tilde{f}(x) + \sum_{i=1}^{m} \tilde{g}_i(x)v_i \tag{9.16}$$

with

$$\tilde{f}(x) = f(x) + \sum_{j=1}^{m} g_j(x)\alpha_j(x) \ , \tag{9.17a}$$

$$\tilde{g}_i(x) = \sum_{j=1}^{m} g_j(x)\beta_{ji}(x), \ i \in \underline{m}, \tag{9.17b}$$

we have that the i-th new input v_i does not affect the outputs y_j, $j \neq i$. Though in this formulation of the input-output decoupling problem only the noninteracting conditions are required, we emphasize the role of the standing Assumption 9.1. First we derive

Lemma 9.4 *The system (9.1) satisfies the strong accessibility rank condition at each point x_0 in M if and only if the feedback modified system (9.16,17a,b) satisfies the strong accessibility rank condition at each point x_0 in M.*

Proof The result follows in a direct manner from the representation of vectorfields in C_0 as is given in Proposition 3.14. □

The above result implies that whenever we can achieve noninteraction by applying a regular static state feedback to a strongly accessible nonlinear system, then the resulting feedback modified system possesses the output controllability property, and thus the new input v_i may be used to steer the output y_i in some neighborhood with nonempty interior.

As a motivation to the local solution of the nonlinear input-output decoupling problem we first recall the following result for linear systems. (The proof may be found in the literature cited at the end of this chapter.)

Theorem 9.5 *Consider a controllable linear system with m inputs*

$$\begin{aligned} \dot{x} &= Ax + Bu \\ y_i &= C_i x \qquad , \quad i \in \underline{m}. \end{aligned} \tag{9.18}$$

There exists a regular linear state feedback

$$u = Fx + Gv \quad , \quad \det G \neq 0 \ , \tag{9.19}$$

which achieves noninteracting between v_i and y_j, $j \neq i$, if and only if

$$\operatorname{Im} B = \operatorname{Im} B \cap \mathcal{V}_1^* + \dots + \operatorname{Im} B \cap \mathcal{V}_m^* \ , \tag{9.20}$$

where $\mathcal{V}_i^*$ is the maximal controlled invariant subspace of the system contained in the subspace $\bigcap_{j \neq i} \ker C_j$, $i \in \underline{m}$.

The essence of condition (9.20) may be explained as follows. In the decoupling problem one searches for a feedback (9.19) which renders the noninteracting between v_i and y_j, $j \neq i$. In case such a feedback exists we observe that the subspaces

$$\mathcal{R}_i = \text{Im}\big((BG)_i \;\vdots\; (A+BF)(BG)_i \;\vdots \ldots \vdots\; (A+BF)^{n-1}(BG)_i\big), \quad i \in \underline{m}, \tag{9.21}$$

satisfy

$$\mathcal{R}_i \subset \bigcap_{j \neq i} \ker C_j \;. \tag{9.22}$$

Clearly each of the subspaces $\mathcal{R}_i$ defined in (9.21) are controlled invariant (actually the $\mathcal{R}_i$'s are controllability subspaces for the system (9.18)), so they satisfy

$$A\mathcal{R}_i \subset \mathcal{R}_i + \text{Im}\, B \quad , \quad i \in \underline{m}, \tag{9.23}$$

and also (9.22). The subspaces $\mathcal{R}_i$ which satisfy (9.22) and (9.23) are also *compatible*, i.e. there exists a feedback matrix F which makes the $\mathcal{R}_i$'s *simultaneously* invariant:

$$(A+BF)\mathcal{R}_i \subset \mathcal{R}_i \quad , \quad i \in \underline{m}. \tag{9.24}$$

Furthermore we have that $\text{Im}(BG)_i = \text{Im}\, B \cap \mathcal{V}_i^*$, $i \in \underline{m}$, which implies (9.20). Conversely condition (9.20) implies the existence of subspaces $\mathcal{R}_i$, $i \in \underline{m}$, satisfying (9.22), (9.23) and (9.24). So there exists a compatible family of subspaces $\mathcal{R}_1, \ldots, \mathcal{R}_m$ which determine the feedback described in (9.19). In fact the matrix F is chosen as in (9.24) and the matrix G is computed via $\text{Im}(BG)_i = \text{Im}\, B \cap \mathcal{V}_i^*$, $i \in \underline{m}$.

Next we return to the nonlinear decoupling problem. Define for $i \in \underline{m}$ the distribution

$$D_i^* = D^*(f,g; \bigcap_{j \neq i} \ker dh_j), \tag{9.25}$$

as the maximal locally controlled invariant distribution of (9.1) contained in $\bigcap_{j \neq i} \ker dh_j$. We make the following assumption.

Assumption 9.6 *(i) The distributions* $G := \text{span}\{g_1, \ldots, g_m\}$, D_i^* *and* $D_i^* \cap G$, $i \in \underline{m}$, *have constant dimension.*

(ii) The distributions V_i := *involutive closure of* $\sum_{j \neq i} D_j^*$, $i \in \underline{m}$, *have constant dimension.* (9.26)

(iii) The output maps $h_i : M \longrightarrow \mathbb{R}^{p_i}$, $i \in \underline{m}$, *are non-trivial, i.e.* $\ker dh_i(x) \neq 0$ *for all* $x \in M$.

Theorem 9.7 *Consider the system (9.1-3) satisfying Assumptions 9.1 and 9.6. Then the static state feedback input-output decoupling problem is locally solvable if and only if*

$$G = D_1^* \cap G + \ldots + D_m^* \cap G \,. \tag{9.27}$$

Proof ("only if" part) Assume $u = \alpha(x) + \beta(x)v$ achieves locally around a point x_0 the input-output decoupling property. Clearly, it follows that for any $i \in \underline{m}$ the distribution

$$C_i := \text{involutive closure } \left(\text{span}\{ad_{\tilde{f}}^k \tilde{g}_i, \, ad_{\tilde{g}_j}^k \tilde{g}_i, \, k \geq 0, \, j \in \underline{m}\}\right) \tag{9.28}$$

is invariant under $\tilde{f}$ and $\tilde{g}_1, \ldots, \tilde{g}_m$, while

$$C_i \subset \ker dh_j \quad , \; j \neq i \,. \tag{9.29}$$

As $\tilde{g}_i \in C_i$, it follows that

$$G = \text{span}\{\tilde{g}_1, \ldots, \tilde{g}_m\} \subset G \cap C_1 + \ldots + G \cap C_m \,, \tag{9.30}$$

and because the distributions C_i are controlled invariant we have $C_i \subset D_i^*$. Together with (9.30) this yields the desired conclusion (9.27). □

Before proving the converse we need a few intermediate results.

Lemma 9.8 *Consider the system (9.1-3) satisfying Assumptions 9.1 and 9.6. Suppose (9.27) holds true. Then the codistributions* ann $V_1, \ldots,$ ann V_m *are linearly independent at each point* $x \in M$, *i.e.* ann $V_i(x) \cap \left(\text{ann } V_1(x) + \ldots + \text{ann } V_{i-1}(x) + \text{ann } V_{i+1}(x) + \ldots + \text{ann } V_m(x)\right) = 0$.

Proof Let $\omega_1 \in \text{ann } V_1$ and suppose $\omega_1(x) = \sum_{j=2}^{m} \varphi_j(x)\omega_j(x)$ for smooth functions $\varphi_2, \ldots, \varphi_m$ and $\omega_j \in \text{ann } V_j$, $j = 2, \ldots, m$. We have $\omega_1 \in \text{ann } V_1 \subset \text{ann}(D_2^* + \ldots + D_m^*)$, as well as $\omega_1 = \sum_{j=2}^{m} \varphi_j\omega_j \in \text{ann } D_1^*$ because $\omega_j \in \text{ann } V_j \subset \text{ann}(D_1^* + \ldots + D_{j-1}^* + D_{j+1}^* + \ldots + D_m^*)$, $j \neq 1$. This implies that $\omega_1 \in \text{ann } D_1^* \cap \text{ann}(D_2^* + \ldots + D_m^*)$, i.e.

$$\omega_1 \in \text{ann}(D_1^* + D_2^* + \ldots + D_m^*) \,. \tag{9.31}$$

Let $D := D_1^* + D_2^* + \ldots + D_m^*$, then $D \supset G$ and thus the local controlled invariance of the D_i^*'s implies that $[f, D] \subset D$ and $[g_j, D] \subset D$, $j \in \underline{m}$. So D is an invariant distribution for the system (9.1) which contains G. The strong accessibility assumption implies that $\dim D = \dim M$ and thus $\text{ann } D = 0$. From equation (9.31) we then obtain $\omega_1 = 0$ implying that the codistributions ann V_1 and ann $V_2 + \ldots +$ ann V_m are linearly independent. The same argument shows that ann V_i and ann $V_1 + \ldots +$ ann V_{i-1} + ann V_{i+1} $+ \ldots +$ ann V_m are linearly independent for any $i \in \underline{m}$. □

Corollary 9.9 *Under the conditions of Lemma 9.8 there exists around each point* $x_0 \in M$ *coordinates* $x = (x_1,..,x_n)$ *such that* $\text{ann } V_i = \text{span}\{dx^i\}$, $i \in \underline{m}$, *where* $x = (x^0, x^1,..,x^m)$ *is a block partitioning of the x-coordinates.*

Proof By induction. The codistribution $\text{ann } V_1$ is constant dimensional and involutive so there exist coordinates $x = (x_1,...,x_n)$ such that $\text{ann } V_1 = \text{span}\{dx^1\}$. Denoting the remaining components of x as $\bar{x}^1$ we observe that $\text{ann } V_2$ locally can be written as $\text{span}\{d\varphi^2\}$ for $\dim(\text{ann } V_2)$ functions φ^2 (here we use the involutivity of $\text{ann } V_2$). As $\text{ann } V_1 \cap \text{ann } V_2 = 0$ it follows that the rank of $\dfrac{d\varphi^2}{d\bar{x}^1}$ equals $\dim(\text{ann } V_2)$ and so the functions φ^2 may be chosen as a new set of partial local coordinates x^2 ($\dim x^2 = \dim(\text{ann } V_2)$). Repeating the above argument yields the desired set of local coordinates. □

The above corollary describes the structure of the distributions V_i, $i \in \underline{m}$. In the sequel we will describe the distributions D_i^*, $i \in \underline{m}$, under the assumption that (9.27) holds.

Lemma 9.10 *Consider the system (9.1-3) satisfying Assumptions 9.1 and 9.6. Then the condition (9.27) implies that*

$$\dim G \cap D_i^* = 1, \quad i \in \underline{m} \ . \tag{9.32}$$

Proof We first show that for each $i \in \underline{m}$

$$G \cap D_i^* \not\subset \sum_{j\neq i} D_j^* \ . \tag{9.33}$$

Indeed, suppose (9.33) does not hold. Then $G \subset \sum_{j\neq i} D_j^* \subset$ involutive closure of $\sum_{j\neq i} D_j^*$. Note that this last distribution is locally controlled invariant (being the involutive sum of locally controlled invariant distributions) and contains G and therefore has dimension n by Assumption 9.1. However by the definition of the distributions D_i^* it follows that $\sum_{j\neq i} D_j^* \subset$ involutive closure of $\sum_{j\neq i} D_j^* \subset \ker dh_i$, implying that the map h_i is a trivial map, herewith contradicting the assumptions. Now, let $\gamma_1 = \dim(G \cap D_1^*)$ and $\gamma_k = \dim(\sum_{i=1}^{k} G \cap D_i^*) - \dim(\sum_{i=1}^{k-1} G \cap D_i^*)$, $k = 2,...,m$. Obviously $\gamma_k \geq 0$ for all k, and in exactly the same way as we established (9.33) it even follows that $\gamma_k \geq 1$ for $k \in \underline{m}$. On the other hand it follows from (9.27) that $\sum_{k=1}^{m} \gamma_k = \dim(\sum_{i=1}^{m} G \cap D_i^*) = m$. So the γ_k's form a set of m integers ≥ 1 which add up to m. Therefore $\gamma_k = 1$, $k \in \underline{m}$, establishing (9.32). □

Corollary 9.11 *Under the same conditions as in Lemma 9.10 it follows that*

$$G \cap D^* = 0 \ , \tag{9.34}$$

where D^* *is the maximal locally controlled invariant distribution contained in* $\bigcap_{i=1}^{m} \ker dh_i$.

Proof This follows directly from (9.32) and the fact that $D^* \subset D_i^*$ for all $i \in \underline{m}$. □

Lemma 9.12 *Consider the system (9.1-3) and let* D_1 *and* D_2 *be two involutive distributions satisfying*

$$[f,D_i] \subset D_i + G \quad , \qquad i = 1,2 \ , \tag{9.35a}$$

$$[g_j,D_i] \subset D_i + G \quad , \qquad i = 1,2, \ j \in m \ . \tag{9.35b}$$

Assume that

$$G = G \cap D_1 + G \cap D_2 , \tag{9.36}$$

then their intersection $D_1 \cap D_2$ *also satisfies (9.35a,b).*

Proof Let $X \in D_1 \cap D_2$. Then (9.35a) implies $[f,X] = Y_1 + b_1 = Y_2 + b_2$, where $Y_i \in D_i$ and $b_1, b_2 \in G$. This implies $Y_1 - Y_2 = b_2 - b_1 \in G$ and so, using (9.36), we obtain that $Y_1 - Y_2 = b_1' + b_2'$ for some vectorfields $b_1' \in G \cap D_1$ and $b_2' \in G \cap D_2$. Clearly $Y_1 - b_1' = Y_2 + b_2' \in D_1 \cap D_2$ and so we find $[f,X] = Y_1 + b_1 = Y_1 - b_1' + b_1' + b_1 \in D_1 \cap D_2 + G$. A similar reasoning applies to the vectorfields g_j, $j \in \underline{m}$. □

Before giving a canonical characterization of the distributions D_i^* we need one further result. Let

$$D_0 := \ker(\text{ann } V_1 + \ldots + \text{ann } V_m) \ . \tag{9.37}$$

Lemma 9.13 *Consider the system (9.1-3) satisfying the Assumptions 9.1 and 9.6. Then the condition (9.27) implies that*

$$(i) \quad D_0 = \sum_{i=1}^{m} D_i^* \cap V_i \ , \tag{9.38}$$

(ii) D_0 *is locally controlled invariant.* (9.39)

Proof We will establish (9.38) using the local coordinates $x = (x^0, x^1, \ldots, x^m)$ of Corollary 9.9, i.e.

$$\text{ann } V_i = \text{span}\{dx^i\} , \quad i \in \underline{m} . \tag{9.40}$$

Obviously (9.37) and (9.40) imply that

$$D_0 = \text{span}\{\frac{\partial}{\partial x^0}\} . \tag{9.41}$$

Let $X \in D_0$. Since $D_1^* + \ldots + D_m^* = TM$, see Lemma 9.8, we can write $X = X_1 + \ldots + X_m$ with $X_i \in D_i^*$, $i \in \underline{m}$. Then $dx^j(X) = dx^j(X_1) + \ldots + dx^j(X_m) = dx^j(X_j)$, by definition of the distribution V_i. However $X \in D_0$ implies $dx^j(X) = 0$, so $dx^j(X_j) = 0$, $j \in \underline{m}$. Therefore $X_j \in V_j$, i.e. $X \in \sum_{i=1}^{m} D_i^* \cap V_i$. Conversely, for $X \in \sum_{i=1}^{m} D_i^* \cap V_i$, there exist $X_i \in D_i^* \cap V_i$, $i \in \underline{m}$, such that $X = X_1 + \ldots + X_m$. Clearly $dx^j(X_i) = 0$ for $i, j \in \underline{m}$, yielding $X \in D_0$.

In order to establish (ii) we note from (i) that the distribution $\sum_{i=1}^{m} D_i^* \cap V_i$ is involutive and constant dimensional. Moreover each of the distributions D_i^* and V_i, $i \in \underline{m}$, satisfy Lemma 9.12 and therefore each $D_i^* \cap V_i$, $i \in \underline{m}$, is locally controlled invariant. Now D_0, being involutive, is locally controlled invariant as it is the sum of locally controlled invariant distributions. □

We are now prepared for giving a canonical characterization of the distributions D_i^*, provided (9.27) holds true.

Lemma 9.14 *Consider the system (9.1-3) satisfying Assumptions 9.1 and 9.6. Then the condition (9.27) implies that for* $i \in \underline{m}$

$$D_i^* = \ker(\text{ann } V_1 + \ldots + \text{ann } V_{i-1} + \text{ann } V_{i+1} + \ldots + \text{ann } V_m) . \tag{9.42}$$

Proof We establish (9.42) again using the local coordinates $x = (x^0, x^1, \ldots x^m)$ of Corollary 9.9. In these coordinates (9.42) comes down to showing that

$$D_i^* = \text{span}\{\frac{\partial}{\partial x^0}, \frac{\partial}{\partial x^i}\} , \quad i \in \underline{m} . \tag{9.43}$$

As $D_i^* \subset V_j, j \neq i$, it immediately follows that

$$D_i^* \subset \operatorname{span}\{\frac{\partial}{\partial x^0},\frac{\partial}{\partial x^i}\} \quad , \; i \in \underline{m}. \tag{9.44}$$

To show the converse inclusion we first note that the distribution $D_0 = \operatorname{span}\{\frac{\partial}{\partial x^0}\}$ is locally controlled invariant, cf. Lemma 9.13. Moreover, from (9.38) it immediately follows that D_0 is contained in the maximal controlled invariant distribution D^* in $\bigcap_{j\in\underline{m}} \ker dh_j$. As $D^* \subset D_i^*$, $i \in \underline{m}$, we obtain that

$$\operatorname{span}\{\frac{\partial}{\partial x^0}\} = D_0 \subset D_i \quad , \; i \in \underline{m} \;. \tag{9.45}$$

Next let $X \in \operatorname{span}\{\frac{\partial}{\partial x^i}\}$ and write $X = X_1 +\ldots+ X_m$ with $X_j \in D_j^*$, $j \in \underline{m}$. Then $dx_j(X) = dx_j(X_1) +\ldots+ dx_j(X_m) = dx_j(X_j)$. Now for $j \neq i$ we have that $dx_j(X_j) = 0$, thereby yielding $X_j \in D_0$, $j \neq i$. Therefore we obtain that $X = X_1 +\ldots+ X_m \in D_0 + D_i^* \subset D_i^*$. This completes the converse inclusion of (9.44). □

We are now able to prove the remaining part of Theorem 9.7.

Proof of Theorem 9.7 ("if" part) In order to produce locally a feedback which achieves input-output decoupling we proceed as follows. We first apply a locally defined feedback which renders the distribution D_0 invariant, see Lemma 9.13. Using the local coordinates of Corollary 9.9 and writing $\tilde{x} = (x^1,\ldots,x^m)$ and $x = (x^0,\tilde{x})$ we obtain the following decomposition

$$\begin{cases} \dot{x}^0 = \tilde{f}_1(x^0,\tilde{x}) + \sum_{i=1}^{m} \tilde{g}_{1i}(x^0,\tilde{x})u_i \\ \dot{\tilde{x}} = \tilde{f}_2(\tilde{x}) \quad + \sum_{i=1}^{m} \tilde{g}_{2i}(\tilde{x})u_i \end{cases} \tag{9.46a}$$

$$y_i = h_i(x_i) \quad , \quad i \in \underline{m} \;, \tag{9.46b}$$

where $u = (u_1,\ldots,u_m)$ denote the new controls. In the sequel we deal with the dynamics modulo D_0, i.e.

$$\dot{\tilde{x}} = \tilde{f}_2(\tilde{x}) + \sum_{i=1}^{m} \tilde{g}_{2i}(\tilde{x})u_i \quad , \tag{9.47}$$

and we will construct a regular feedback $u = \alpha(\tilde{x}) + \beta(\tilde{x})v$ which achieves input-output decoupling. An easy inspection of (9.47,46b) yields, using the results of the preceeding lemmas, that the maximal locally controlled invariant distribution of (9.47) contained in $\bigcap_{j\neq i} \ker dh_j$ is given as

$$\widetilde{D}_i^* = \mathrm{span}\{\frac{\partial}{\partial x^i}\} \ , \ i \in \underline{m} \ , \tag{9.48}$$

and moreover $\dim(\widetilde{G} \cap \widetilde{D}_i^*) = 1$, $i \in \underline{m}$, where $\widetilde{G}(\tilde{x}) = \mathrm{span}\{\tilde{g}_{21}(\tilde{x}),\ldots,\tilde{g}_{2m}(\tilde{x})\}$. Notice that the system modulo D_0 still satisfies the decoupling condition (9.27). Applying a preliminary feedback involving only a change of the input vectorfields $\tilde{g}_{2i}(\tilde{x})$, $i \in \underline{m}$, we may also assume in the following that $\tilde{g}_{2i}(\tilde{x}) \in \widetilde{D}_i^*$, $i \in \underline{m}$. Using the local controlled invariance of the distributions $\widetilde{D}_i^*$ we find that

$$[\tilde{g}_{2j},\widetilde{D}_i^*] \subset \widetilde{D}_i^* + \widetilde{G} \ , \ j \in m \ , \ i \in m \ , \tag{9.49a}$$

$$[\tilde{f}_2,\widetilde{D}_i^*] \subset \widetilde{D}_i^* + \widetilde{G} \ , \ i \in \underline{m} \ . \tag{9.49b}$$

From (9.49a) we obtain for $j \in \underline{m}$

$$[\tilde{g}_{2j},\widetilde{D}_1^* +\ldots+ \widetilde{D}_{j-1}^* + \widetilde{D}_{j+1}^* +\ldots+ \widetilde{D}_m^*] \subset \widetilde{D}_1^* +\ldots+ \widetilde{D}_{j-1}^* + \widetilde{D}_{j+1}^* +\ldots+ \widetilde{D}_m^* + \mathrm{span}\{\tilde{g}_{2j}\} \ . \tag{9.50}$$

Using Theorem 7.5, we can find a feedback matrix $\beta(\tilde{x})$ such that the modified input vectorfields $\approx{g}_{2j}$ satisfy for $j \in \underline{m}$

$$[\approx{g}_{2j},\widetilde{D}_1^* +\ldots+ \widetilde{D}_{j-1}^* + \widetilde{D}_{j+1}^* +\ldots+ \widetilde{D}_m^*] \subset \widetilde{D}_1^* +\ldots+ \widetilde{D}_{j-1}^* + \widetilde{D}_{j+1}^* +\ldots+ \widetilde{D}_m^* \ , \tag{9.51}$$

which in our local coordinates means that

$$\approx{g}_{2j}(\tilde{x}) = \begin{bmatrix} 0 \\ \vdots \\ 0 \\ \bar{g}_{2j}(x^j) \\ 0 \\ \vdots \\ 0 \end{bmatrix} \in \widetilde{D}_j^* \ , \qquad j \in \underline{m} \ . \tag{9.52}$$

A completely analogous reasoning, again using Theorem 7.5, yields a feedback $\alpha(x)$ such that the modified drift vectorfield $\approx{f}_2(\tilde{x})$ takes the form

$$\approx{f}_2(\tilde{x}) = \begin{bmatrix} \bar{f}_{21}(x^1) \\ \vdots \\ \bar{f}_{2m}(x^m) \end{bmatrix} \tag{9.53}$$

Combining (9.52), (9.53) and (9.46b) we have constructed a decoupled system of the form

$$\begin{cases} \dot{x}^1 = \bar{f}_{21}(x^1) + \bar{g}_{21}(x^1)v_1 \\ \vdots \\ \dot{x}^m = \bar{f}_{2m}(x^m) + \bar{g}_{2m}(x^m)v_m \\ y_i = h_i(x^i) \quad , \; i \in \underline{m} \; . \end{cases} \tag{9.54}$$

Finally, since (9.54) is still strongly accessible (cf. Assumption 9.1), each of the subsystems $\dot{x}^i = \bar{f}_{2i}(x^i) + \bar{g}_{2i}(x^i)v_i$ is strongly accessible, $i \in \underline{m}$, and thus the output controllability of (9.54) is immediate. □

Remark 9.15 We directly obtain for the input-output decoupled system of Theorem 9.7 a local normal form described by, see (9.46,54),

$$\begin{cases} \dot{x}^0 = \bar{f}_1(x^0,x^1,\ldots,x^m) + \sum_{i=1}^{m} \bar{g}_{1i}(x^0,x^1,\ldots,x^m)v_i \\ \dot{x}^1 = \bar{f}_{21}(x^1) + \bar{g}_{21}(x^1)v_1 \\ \vdots \\ \dot{x}^m = \bar{f}_{2m}(x^m) + \bar{g}_{2m}(x^m)v_m \end{cases}$$

$$\begin{cases} y_1 = h_1(x^1) \\ \vdots \\ y_m = h_m(x^m) \end{cases} \tag{9.55}$$

Notice that this extends the normal form of square decouplable systems of Chapter 8 (see (8.45)).

Theorem 9.7 gives a geometric solution to the static state feedback input-output decoupling problem in case the number of output blocks equals the number of inputs (see (9.3)). For a square analytic system with scalar outputs y_i, $i \in \underline{m}$, and satisfying Assumptions 9.1 and 9.7, the geometric decoupling condition (9.27) follows from the analytic decoupling condition (8.25), but, on the other hand (9.27) only implies the rank condition on an open and dense subset of M (see Exercise 9.2). We finally remark that also the more general block-input block-output decoupling problem can be studied geometrically, see for instance Exercise 9.3 or the literature cited at the end of this chapter.

9.2 The Formal Stucture at Infinity and Input-Output Decoupling

In linear system theory there is often a direct way to pass over from a state space problem formulation to a problem description in the *frequency domain*. This is due to the fact that a minimal linear state space system

$$\begin{cases} \dot{x} = Ax + Bu, \quad x(0) = 0 \ , \\ y = Cx \end{cases} \tag{9.56}$$

can be equivalently represented by its transfer matrix $C(sI - A)^{-1}B$. A similar frequency domain description for a nonlinear system

$$\begin{cases} \dot{x} = f(x) + \sum_{i=1}^{m} g_i(x)u_i \ , \quad x(0) = x_0 \ , \\ y = h(x) \end{cases} \tag{9.57}$$

does not exist (see also Chapter 4). Nevertheless, some important structural information, which in the linear case is most conveniently given by the transfer matrix, can be defined for the nonlinear system (9.57) in geometric terms. We will illustrate this with the so called *structure at infinity*. (A linear system (9.56) with transfer matrix $G(s) = C(sI-A)^{-1}B$ is said to have ℓ zeros at infinity of orders $n_1, \ldots, n_\ell$ if the matrix $\tilde{G}(s) = G(1/s)$ has ℓ zeros of orders $n_1, \ldots, n_\ell$ at 0.)

Consider the nonlinear system (9.57) and recall that the maximal locally controlled invariant distribution D^* of (9.57) in ker dh may be computed via (see Chapter 7, (7.53))

$$\begin{cases} D^0 = TM \\ D^{\mu+1} = \ker dh \cap \{X \in V(M) \mid [f,X] \in D^\mu + G, \ [g_i, X] \in D^\mu + G, \ i \in \underline{m}\} \end{cases} \tag{9.58}$$

where $G = \text{span}\{g_1, \ldots, g_m\}$ and the following constant dimensions assumptions hold (see Assumption 7.18).

Assumption 9.16 *For all* $\mu \geq 0$ *the distributions* D^μ *and* $D^\mu \cap G$ *have constant dimension on* M.

Provided Assumption 9.16 holds we have that (see Proposition 7.16)

$$D^*(f,g; \ker dh) = D^n \ . \tag{9.59}$$

From the D^*-algorithm (9.58) the following structural information will be extracted.

Definition 9.17 *Consider the nonlinear system (9.57) for which Assumption 9.16 holds. Let*

$$p^\mu = \dim(G \cap D^{\mu-1}) - \dim(G \cap D^*), \ \mu = 1,2,\ldots \tag{9.60}$$

where $D^* = D^*(f,g;\ \ker dh)$ *and let*

$$n^\mu = \text{number of } p^\nu\text{'s which are greater than or equal to } \mu. \qquad (9.61)$$

Then the system (9.57) is said to have p^1 *formal zeros at infinity of orders* $\{n^\mu\}$.

We observe that for a single-input single-output nonlinear system the formal structure at infinity can be expressed as follows. Let $\rho < \infty$ be the characteristic number of the single-input single-output system (9.57). Then by definition the function $L_g L_f^\rho h(x)$ is nonvanishing in some point $x \in M$. Assume it is nonzero everywhere; then this single-input single-output system has one formal zero at infinity of order $\rho+1$. In order to see this we note that in this case D^μ is given as $D^\mu = \ker(\text{span}\{dh, dL_f h, \ldots, dL_f^{\mu-1}h\})$ and for $k = 0,\ldots,\rho-1$ the function $L_g L_f^k h$ is identically zero, cf. Theorem 7.21. Analogously it follows that a square decouplable system with characteristic indices $\rho_1,\ldots,\rho_m$ has m zeros at infinity of orders $\rho_1+1,\ldots,\rho_m+1$. We stress that the above geometric definition is, for a linear system, indeed equivalent to the more usual definition of structure at infinity of its transfer matrix. In other words, for a linear system the formal structure at infinity of Definition 9.17 coincides with its structure at infinity (see the references).

To show the importance of the formal zeros at infinity for a nonlinear system we will discuss their role in the static state feedback input-output decoupling problem. Consider again the system (9.1-3) and suppose Assumptions 9.1 and 9.6 are met. We introduce the following regularity assumption, which extends Assumption 9.16.

Assumption 9.18 *For each subset* $I \subset \underline{m}$, V_I^*, *the maximal locally controlled invariant distribution in* $\bigcap_{j\in I} \ker dh_j$, *may be computed via the algorithm*

$$\begin{cases} V_I^0 = TM \\ V_I^{\mu+1} = \bigcap_{j\in I} \ker dh_j \cap \{X \in V(M) \mid [f,X] \in G + V_I^\mu,\ [g_i,X] \in G + V_I^\mu, \\ \qquad\qquad i \in \underline{m}\} \end{cases} \qquad (9.62)$$

where for each $\mu \geq 0$ *the distributions* V_I^μ *and* $V_I^\mu \cap G$ *have constant dimension on* M.

In the sequel we will also write

$$D^*_I = V^*_{\underline{m}/I} \ , \ I \subset \underline{m} \tag{9.63}$$

and in this notation $D^*_{\{i\}}$ equals the distribution D^*_i introduced before. Also note the following identities

$$V^*_\emptyset = TM \text{ and } D^*_\emptyset = D^* . \tag{9.64}$$

With the above family of controlled invariant distributions we denote the corresponding lists of orders of zeros at infinity as

$$p^\mu_i = \dim(G \cap V^{\mu-1}_i) - \dim(G \cap V^*_i) \ , \ i \in \underline{m} \ , \ \mu = 1,2,\dots \tag{9.65a}$$

$$p^\mu = \dim(G \cap D^{\mu-1}) - \dim(G \cap D^*) \ , \ \mu = 1,2,\dots \tag{9.65b}$$

$$q^\mu_i = \dim(G \cap D^{\mu-1}_i) - \dim(G \cap D^*_i) \ , \ i \in \underline{m} \ , \ \mu = 1,2,\dots \tag{9.65c}$$

$$p^\mu(I) = \dim(G \cap V^{\mu-1}_I) - \dim(G \cap V^*_I) \ , \ I \subset \underline{m} \ , \ \mu = 1,2,\dots \tag{9.65d}$$

$$q^\mu(I) = \dim(G \cap D^{\mu-1}_I) - \dim(G \cap D^*_I) \ , \ I \subset \underline{m} \ , \ \mu = 1,2,\dots \tag{9.65e}$$

From (9.63-65) the following identities are obvious

$$p^\mu_i = p^\mu(\{i\}) \quad , \quad q^\mu_i = q^\mu(\{i\}) \quad , \quad i \in \underline{m} \ , \ \mu = 1,2,\dots \tag{9.66}$$

$$q^\mu(I) = p^\mu(\underline{m}\backslash I) \quad , \quad I \subset \underline{m} \ , \ \mu = 1,2,\dots \tag{9.67}$$

We will need one further definition. Let $P(\underline{m})$ denote the family of subsets of $\underline{m}$. A function φ: $P(\underline{m}) \longrightarrow \mathbb{N}$ is called a *weight function* if

(i) $\varphi(\emptyset) = 0$, (9.68a)

(ii) $\varphi(I \cup J) = \varphi(I) + \varphi(J) - \varphi(I \cap J)$, for all $I,J \subset \underline{m}$. (9.68b)

Theorem 9.19 *Consider the system (9.1-3) satisfying the Assumptions 9.1, 9.6 and 9.18. Then the following conditions are equivalent*

(i) $G = G \cap D^*_1 + \dots + G \cap D^*_m$. (9.27)

(ii) $p^\mu = \sum_{i \in \underline{m}} p^\mu_i$, $\mu = 1,2,\dots$. (9.69)

(iii) p^μ: $P(\underline{m}) \to \mathbb{N}$ *is a weight function,* $\mu = 1,2,\dots$. (9.70)

(iv) The input-output decoupling problem is locally solvable.

For the proof of this theorem we need some preliminary results.

Lemma 9.20 *Consider the system (9.1-3) satisfying Assumption 9.18. Then if for some* $\mu \geq 0$ *and* $I,J \subset \underline{m}$

$$D^{\mu}_{I\cap J} = D^{\mu}_{I} \cap D^{\mu}_{J} \tag{9.71}$$

and

$$G = G \cap D^{\mu}_{I} + G \cap D^{\mu}_{J} \tag{9.72}$$

then also

$$D^{\mu+1}_{I\cap J} = D^{\mu+1}_{I} \cap D^{\mu+1}_{J} \ . \tag{9.73}$$

Proof $D^{\mu+1}_{I} \cap D^{\mu+1}_{J} = \Big(\bigcap_{j\in \underline{m}\setminus I} \ker dh_j\Big) \cap \Big(\bigcap_{j\in \underline{m}\setminus J} \ker dh_j\Big) \cap$

$$\cap \{X \in V(M) \mid [f,X] \in G + D^{\mu}_{I},\ [g_i,X] \in G + D^{\mu}_{I},\ i \in \underline{m}\} \cap$$

$$\cap \{X \in V(M) \mid [f,X] \in G + D^{\mu}_{J},\ [g_i,X] \in G + D^{\mu}_{J},\ i \in \underline{m}\} =$$

$$= \bigcap_{j\in \underline{m}\setminus I\cap J} \ker dh_j \cap \{X \in V(M) \mid [f,X] \in G \cap D^{\mu}_{I} + G \cap D^{\mu}_{J} + D^{\mu}_{I} \cap D^{\mu}_{J},$$

$$[g_i,X] \in G \cap D^{\mu}_{I} + G \cap D^{\mu}_{J} + D^{\mu}_{I} \cap D^{\mu}_{J},\ i \in \underline{m}\}$$

$$= \bigcap_{j\in \underline{m}\setminus I\cap J} \ker dh_j \cap \{X \in V(M) \mid [f,X] \in G + D^{\mu}_{I\cap J},\ [g_i,X] \in G + D^{\mu}_{I\cap J},\ i \in \underline{m}\}$$

$$= D^{\mu+1}_{I\cap J} \ .$$

□

Lemma 9.21 *Consider the system (9.1-3) satisfying Assumption 9.18. Then (9.27) implies that for all $I,J \subset \underline{m}$ and $\mu \geq 0$*

$$D^{\mu}_{I} \cap D^{\mu}_{J} = D^{\mu}_{I\cap J} \ . \tag{9.71}$$

Proof Choose $I,J \subset \underline{m}$ and let us first assume that $I \cup J = \underline{m}$. Then we have

$$G \supset G \cap D^{*}_{I} + G \cap D^{*}_{J} \supset G \cap \sum_{i\in I} D^{*}_{i} + G \cap \sum_{i\in J} D^{*}_{i} \supset \sum_{i\in \underline{m}} G \cap D^{*}_{i} .$$

So, by (9.27) we obtain $G = G \cap D^{*}_{I} + G \cap D^{*}_{J}$, and thus for all $\mu \geq 0$,

$$G = G \cap D^{\mu}_{I} + G \cap D^{\mu}_{J} \ . \tag{9.72}$$

Induction and Lemma 9.20 then lead to the desired result (9.71). Now, for arbitrary $I,J \subset \underline{m}$ we have

$$D^{\mu}_{I} \cap D^{\mu}_{J} \subset D^{\mu}_{I} \cap D^{\mu}_{J\cup \underline{m}\setminus I} \ , \tag{9.74}$$

and because $I \cup J \cup \underline{m}\setminus I = \underline{m}$, we have that

$$D^{\mu}_{I} \cap D^{\mu}_{J\cup \underline{m}\setminus I} = D^{\mu}_{I\cap(J\cup \underline{m}\setminus I)} = D^{\mu}_{I\cap J} \ . \tag{9.75}$$

On the other hand $D^{\mu}_{I\cap J} \subset D^{\mu}_{I} \cap D^{\mu}_{J}$, which together with (9.74) and (9.75) yields the desired conclusion. □

Lemma 9.22 *Consider the system (9.1-3) satisfying Assumption 9.18. Then*

$$G = \sum_{i\in m} G \cap D^{\mu}_{i} \quad , \quad \mu \geq 0, \tag{9.76}$$

if and only if

$$\forall\, I,J \subset \underline{m} \quad , \quad G \cap D^{\mu}_{I} + G \cap D^{\mu}_{J} = G \cap D^{\mu}_{I\cup J} \,. \tag{9.77}$$

Proof (⇐) $G = G \cap D^{*}_{\underline{m}} = G \cap D^{\mu}_{\underline{m}} = G \cap D^{\mu}_{1} + G \cap D^{\mu}_{\underline{m}\setminus\{1\}} = \sum_{i\in\underline{m}} G \cap D^{\mu}_{i}$.

(⇒) Let $I,J \subset \underline{m}$. Then, for all $\mu \geq 0$, $G \cap D^{\mu}_{I\cup J} = (G \cap D^{\mu}_{I} + G \cap D^{\mu}_{J} + G \cap D^{\mu}_{m\setminus I\cup J}) \cap D^{\mu}_{I\cup J} = G \cap D^{\mu}_{I} + G \cap D^{\mu}_{J} + G \cap D^{\mu}_{I\cup J} \cap D^{\mu}_{m\setminus I\cup J} =$ $G \cap D^{\mu}_{I} + G \cap D^{\mu}_{J} + G \cap D^{\mu}_{\emptyset}$ (by Lemma 9.20) $= G \cap D^{\mu}_{I} + G \cap D^{\mu}_{J} + G \cap D^{\mu} =$ $G \cap D^{\mu}_{I} + G \cap D^{\mu}_{J}$. □

We are now prepared to prove Theorem 9.19.

Proof (of Theorem 9.19)
(i) ⇒ (iii) By Lemmas 9.21 and 9.22 we have for all $I,J \subset \underline{m}$ and $\mu \geq 0$ that

$$\begin{aligned} &G \cap D^{\mu}_{I} + G \cap D^{\mu}_{J} = G \cap D^{\mu}_{I\cup J}\,, \\ &(G \cap D^{\mu}_{I}) \cap (G \cap D^{\mu}_{J}) = G \cap D^{\mu}_{I\cap J}\,, \end{aligned} \tag{9.77}$$

and so it follows using (9.64) that for all $I,J \subset \underline{m}$ and $\mu \geq 0$ that

$$G \cap V^{\mu}_{I} + G \cap V^{\mu}_{J} = G \cap V^{\mu}_{I\cup J}, \tag{9.78}$$

$$(G \cap V^{\mu}_{I}) \cap (G \cap V^{\mu}_{J}) = G \cap V^{\mu}_{I\cup J} \,. \tag{9.79}$$

Therefore, for all $\mu > 0$

$$p^{\mu}(\emptyset) = \dim(G \cap V^{\mu-1}_{\emptyset}) - \dim(G \cap V^{*}_{\emptyset}) = m - m = 0 \,, \tag{9.80}$$

$$p^{\mu}(I \cup J) = \dim(G \cap V^{\mu-1}_{I\cup J}) - \dim(G \cap V^{*}_{I\cup J}). \tag{9.81}$$

Using (9.79) we have

$$\dim(G \cap V^{\mu-1}_{I\cup J}) = \dim\big((G \cap V^{\mu-1}_{I}) \cap (G \cap V^{\mu-1}_{J})\big) =$$

$$\dim(G \cap V^{\mu-1}_{I}) + \dim(G \cap V^{\mu-1}_{J}) - \dim(G \cap V^{\mu-1}_{I} + G \cap V^{\mu-1}_{J}) \,, \tag{9.82}$$

and from (9.78) we obtain

$$\dim(G \cap V_I^{\mu-1} + G \cap V_J^{\mu-1}) = \dim(G \cap V_{I\cap J}^{\mu}). \tag{9.83}$$

Furthermore (9.82) and (9.83) hold true if we replace the superscript μ by $*$, i.e. by taking μ sufficiently large. Combining (9.81-83) leads to

$$p^{\mu}(I \cup J) = \dim(G \cap V_I^{\mu-1}) + \dim(G \cap V_J^{\mu-1}) - \dim(G \cap V_{I\cap J}^{\mu-1}) -$$
$$\dim(G \cap V_I^{*}) - \dim(G \cap V_J^{*}) + \dim(G \cap V_{I\cap J}^{*}) = p^{\mu}(I) + p^{\mu}(J) - p^{\mu}(I \cap J). \tag{9.84}$$

So (9.80) and (9.84) readily yield that p^{μ} is a weight function for $\mu > 0$.
(iii) $\Rightarrow$ (ii) For all $\mu > 0$ we have

$$p^{\mu} = p^{\mu}(\underline{m}) = p^{\mu}(\{1\}) + p^{\mu}(\{2,\ldots,m\}) - p^{\mu}(\phi) =$$
$$= p_1^{\mu} + p^{\mu}(\{2,\ldots,m\}) = \sum_{i\in\underline{m}} p_i^{\mu} . \tag{9.69}$$

(ii) $\Rightarrow$ (i) By induction we show that if (9.69) holds, then for all $\mu \geq 0$

$$G = \sum_{i\in m} G \cap D_i^{\mu} \tag{9.76}$$

as well as

$$D^{\mu} = \bigcap_{i\in m} D_i^{\mu} , \tag{9.85}$$

and

$$D_i^{\mu} = \bigcap_{j\neq i} V_j^{\mu} \quad , \; i \in \underline{m} . \tag{9.86}$$

Clearly the statement is true for $\mu = 0$. Assume (9.76,85,86) hold for a certain $\mu > 0$, then, by repeated application of Lemma 9.20, (9.85) and (9.86) hold true for $\mu + 1$. Furthermore we have for all $i \in \underline{m}$

$$G = G \cap D_i^{\mu} + G \cap V_i^{\mu} . \tag{9.87}$$

Next we compute $\dim(G \cap D_i^{\mu+1} + G \cap V_i^{\mu+1})$.

$$\dim(G \cap D_i^{\mu+1} + G \cap V_i^{\mu+1}) = \dim(G \cap D_i^{\mu+1}) + \dim(G \cap V_i^{\mu+1}) -$$
$$\dim(G \cap D_i^{\mu+1} \cap V_i^{\mu+1}) = \dim(\bigcap_{j\neq i} G \cap V_j^{\mu+1}) + \dim(G \cap V_i^{\mu+1}) -$$
$$\dim(G \cap D^{\mu}) \geq \sum_{j\neq i} \dim(G \cap V_j^{\mu+1}) - (m-2)m + \dim(G \cap V_i^{\mu+1}) -$$
$$\dim(G \cap D^{\mu+1}) = \sum_{j\in\underline{m}} \dim(G \cap V_j^{\mu+1}) - (m-2)m - \dim(G \cap D^{\mu+1}). \tag{9.88}$$

Using (9.69) we obtain the following identities
$m - \dim(G \cap D^*) = \sum_{j \in \underline{m}} (m - \dim(G \cap V_j^*))$, so

$$\sum_{j \in \underline{m}} \dim(G \cap V_j^*) - \dim(G \cap D^*) = (m-1)m \ . \tag{9.89}$$

Moreover,

$$\dim(G \cap D^{\mu+1}) - \dim(G \cap D^*) = \sum_{j \in \underline{m}} \dim(G \cap V_j^{\mu+1}) - \dim(G \cap V_j^*), \tag{9.90}$$

which by (9.89) leads to

$$\sum_{j \in \underline{m}} \dim(G \cap V_j^{\mu+1}) - \dim(G \cap D^{\mu+1}) = (m-1)m. \tag{9.91}$$

So from (9.88) and (9.91) we conclude
$\dim(G \cap D_i^{\mu+1} + G \cap V_i^{\mu+1}) \geq m$, that is ,

$$G \cap D_i^{\mu+1} + G \cap V_i^{\mu+1} = G \ . \tag{9.92}$$

Having established (9.92) for all $i \in \underline{m}$, we see that

$$G = \bigcap_{i \in \underline{m}} \left(G \cap D_i^{\mu+1} + G \cap V_i^{\mu+1}\right) = \sum_{i \in \underline{m}} G \cap D_i^{\mu+1} + G \cap D^{\mu+1},$$

so

$$G = \sum_{i \in \underline{m}} G \cap D_i^{\mu+1} \ , \tag{9.76}$$

and thus (9.27) follows by taking μ sufficiently large. □

It was shown in Corollary 9.11 that given the decoupling condition (9.27), we have $\dim(G \cap D^*) = 0$. Notice that we have not explicitly used this fact in the proof of Theorem 9.19. In fact, as noted before, the theory on input-output decoupling by static state feedback can be extended to the block-input block-output decoupling problem (see also Exercise 9.3) and Theorem 9.19 can be extended to that situation, in which case $\dim(G \cap D^*) = 0$ is no longer necessarily true.

Example 9.23 Consider again the rigid two-link robot manipulator of Example 1.1 (see also Examples 5.20 and 6.8). Its dynamics are given as

$$\frac{d}{dt}\begin{bmatrix} \theta \\ \dot{\theta} \end{bmatrix} = \begin{bmatrix} \dot{\theta} \\ -M(\theta)^{-1} C(\theta,\dot{\theta}) - M(\theta)^{-1}k(\theta) \end{bmatrix} + \begin{bmatrix} 0 \\ M(\theta)^{-1} \end{bmatrix} u \tag{9.93}$$

where $u = (u_1, u_2)$, $\theta = (\theta_1, \theta_2)$, $\dot{\theta} = (\dot{\theta}_1, \dot{\theta}_2)$, and the matrices M, C and k are defined as in Example 1.1. With the dynamics (9.93) we consider the outputs y_1 and y_2 given as

$$y_i = h_i(\theta,\dot{\theta}) = \begin{bmatrix} \theta_i \\ \dot{\theta}_i \end{bmatrix}, \quad i = 1,2 . \tag{9.94}$$

It is easily seen that the dynamics (9.93) is strongly accessible, so Assumption 9.1 is satisfied, see Example 6.8. Next we determine the maximal locally controlled invariant distributions in $\ker dh_i$, $i = 1,2$, using the standard algorithm

$$D_1^* = D_1^1 = \ker dh_1 = \operatorname{span}\{\frac{\partial}{\partial\theta_2},\frac{\partial}{\partial\dot{\theta}_2}\} , \tag{9.95a}$$

$$D_2^* = D_2^1 = \ker dh_2 = \operatorname{span}\{\frac{\partial}{\partial\theta_1},\frac{\partial}{\partial\dot{\theta}_1}\} . \tag{9.95b}$$

Using (9.95) we find indeed

$$G = \operatorname{span}\{\frac{\partial}{\partial\dot{\theta}_1},\frac{\partial}{\partial\dot{\theta}_2}\} = \operatorname{span}\{\frac{\partial}{\partial\dot{\theta}_2}\} + \operatorname{span}\{\frac{\partial}{\partial\dot{\theta}_1}\} = G \cap D_1^* + G \cap D_2^* , \tag{9.96}$$

and so the necessary and sufficient condition (9.27) for input-output decoupling is satisfied. Of course, this is not surprising; an easy inspection of (9.93,94) shows that

$$u = C(\theta,\dot{\theta}) + k(\theta) + M(\theta)\, v \tag{9.97}$$

yields the feedback modified dynamics (see Example 6.8)

$$\frac{d}{dt}\begin{bmatrix} \theta \\ \dot{\theta} \end{bmatrix} = \begin{bmatrix} \dot{\theta} \\ 0 \end{bmatrix} + \begin{bmatrix} 0 \\ I \end{bmatrix} v . \tag{9.98}$$

It is also straightforward to compute the formal structure at infinity for the system (9.93,94). One obtains $p^1 = 2 - 0 = 2$, $p^2 = 0$, and in a similar way $p_1^1 = 2 - 1 = 1$, $p_1^2 = 1 - 1 = 0$, $p_2^1 = 2 - 1 = 1$, $p_2^2 = 1 - 1 = 0$ and thus, indeed, cf. Theorem 9.19, $p^1 = p_1^1 + p_2^1$ and $p^2 = p_1^2 + p_2^2$. □

Notes and References

The geometric theory for the linear static state feedback block decoupling problem has been developed in [BM],[MW],[Wo],[WM]. Various equivalent formulations of the structure at infinity for a linear system have been given in [Ro],[Ha],[Mo]. A geometric characterization of the linear structure at infinity is given in [Ma], and its relevance in the input-output decoupling problem follows from [DLM],[Di]. The differential geometric theory for the nonlinear (block-)input (block)-output decoupling problem has been initiated in [IKGM], and followed up in [Nij3],[NS1],[NS2],[NS4], where the problem under static state feedback is solved. Recently, in [DGM],[MG],[GDM] differential algebraic methods have been developed for solving the dynamic state feedback block decoupling problem. The nonlinear formal zeros at infinity have been introduced for a particular class of nonlinear systems in [Is1],[Is2], whereas the definition as given here comes form [NS2]. Further results on the formal structure at infinity have been reported in [Nij2],[NS3],[Is3]. Theorem 9.7 was first proved in [NS4]. The proof given in [NS4] was based upon the so called "controllability distributions"; the proof given here combines ideas from [Ch] and [NS4],[Nij1], and avoids the introduction of "controllability distributions". The proof of Theorem 9.19 is taken from [NS2]. A study of the input-output decoupling problem with stability has been given in [IG]; see also [HG] for a characterization of decoupling feedbacks.

[BM] G. Basile, G. Marro, "A state space approach to noninteracting controls", Ricerche di Automatica 1, pp. 68-77, 1970.

[Ch] D. Cheng, "Design for noninteracting decomposition of nonlinear systems", IEEE Trans. Aut. Contr. AC-33, pp. 1070-1074, 1988.

[Di] J.M. Dion, "Feedback block decoupling and infinite structure of linear systems", Int. J. Contr. 37, pp. 521-533, 1983.

[DGM] M.D. Di Benedetto, J.W. Grizzle, C.H. Moog, "Rank invariants of nonlinear systems", SIAM J. Contr. Optimiz. 27, 1989.

[DLM] J. Descusse, J.F. Lafay, M. Malabre, "On the structure at infinity of block-decoupable systems: the general case", IEEE Trans. Aut. Contr. AC-28, pp. 1115-1118, 1983.

[GDM] J.W. Grizzle, M.D. Di Benedetto, C.H. Moog, "Computing the differential output rank of a nonlinear system", Proc. 26th IEEE Conf. Decision Control Los Angeles, pp. 142-145, 1987.

[Ha] M.L.J. Hautus, "The formal Laplace transform for smooth linear systems", in **Mathematical Systems Theory**, Lect. Notes Econ. Math. Syst., 131, Springer, Berlin, pp. 29-47, 1976.

[HG] I.J. Ha, E.G. Gilbert, "A complete characterization of decoupling control laws for a genral class of nonlinear systems", IEEE Trans. Automat. Contr., AC-31, pp. 823-830, 1986.

[Is1] A. Isidori, "Nonlinear feedback, structure at infinity and the input-output linearization problem", in **Mathematical Theory of Networks and Systems**, Lect. Notes Contr. Inf. Sci., 58, Springer, Berlin, pp. 473-493, 1983.

[Is2] A. Isidori, "Formal infinite zeros for nonlinear systems", Proc. 22-nd Conf. Decision Control, San Antonio, pp. 647-653, 1983.

[Is3] A. Isidori, "Control of nonlinear systems via dynamic state-feedback", in **Algebraic and Geometric Methods in Nonlinear Control Theory** (eds. M. Fliess, M. Hazewinkel), Reidel, Dordrecht, pp. 121-146, 1986.

[IG] A. Isidori, J.W. Grizzle, "Fixed modes and nonlinear noninteracting control with stability", IEEE Trans. Aut. Contr. AC-33, pp. 907-914, 1988.

[IKGM] A. Isidori, A.J. Krener, C. Gori-Giorgi, S. Monaco, "Nonlinear decoupling via feedback: a differential geometric approach", IEEE Trans. Aut. Contr. AC-26, pp. 331-345, 1981.

[Ma] M. Malabre, "Structure à l'infini des triplets invariants. Application à la poursuite parfaite de modèle", in **Analysis and Optimization of Systems**, (eds. A. Bensoussan & J.L. Lions), Lect. Notes Contr. Inf. Sci. 44, Springer, Berlin, pp. 43-53, 1982.

[MG] C.H. Moog, J.W. Grizzle, "Découplage nonlinéaire vu de l'algèbre linéaire", C.R. Acad. Sci. Paris, t.307, Série I, pp. 497-500, 1988.

[Mo] A.S. Morse, "Structural invariants of linear multivariable systems", SIAM J. Contr. Optimiz. 11, pp. 446-465, 1973.

[MW] A.S. Morse, W.M. Wonham, "Status of noninteracting control", IEEE Trans. Aut. Contr. AC-16, pp. 568-581, 1971.

[Nij1] H. Nijmeijer, "Feedback decomposition of nonlinear control systems", IEEE Trans. Aut. Contr. AC-28, pp. 861-862, 1983.

[Nij2] H. Nijmeijer, "Zeros at infinity for nonlinear systems, what are they and what are they good for?", in **Geometric Theory of Nonlinear Control Systems** (eds. B. Jakubczyk, W. Respondek, K. Tchoń), Scientific Paper of the Institute of Technical Cybernetics of the Technical University of Wroclaw, Poland, no 70, pp. 105-130, 1985.

[Nij3] H. Nijmeijer, "On the input-output decoupling of nonlinear systems" in **Algebraic and Geometric Methods in Nonlinear Control Theory** (eds. M. Fliess, M. Hazewinkel), Reidel, Dordrecht, pp. 101-119, 1986.

[NS1] H. Nijmeijer, J.M. Schumacher, "The regular local noninteracting control problem for nonlinear control systems", Proc. 22-nd IEEE Conf. Decision Control, San Antonio, pp. 388-392, 1983.

[NS2] H. Nijmeijer, J.M. Schumacher, "Zeros at infinity for affine nonlinear control systems", IEEE Trans. Aut. Contr. AC-30, pp. 566-573, 1985.

[NS3] H. Nijmeijer, J.M. Schumacher, "On the inherent integration structure of nonlinear systems", IMA J. Math. Contr. Inf. 2, pp. 87-107, 1985.

[NS4] H. Nijmeijer, J.M. Schumacher, "The regular local noninteracting control problem for nonlinear control systems", SIAM J. Contr. Optimiz. 24, pp. 1232-1245, 1986.

[Ro] H.H. Rosenbrock, **State space and Multivariable Theory**, Wiley, New York, 1970.

[Wo] W.M. Wonham, **Linear Multivariable Control: a Geometric Approach**, (2-nd edition), Springer, Berlin, 1979.

[WM] W.M. Wonham, A.S. Morse, "Decoupling and pole assignment in linear multivariable systems: a geometric approach", SIAM J. Contr. Optimiz. 8, pp. 1-18, 1970.

Exercises

9.1 Show that Assumption 9.1 in Theorem 9.7 may be relaxed as follows. Instead of assuming that the strong accessibility distribution C_0 is n-dimensional we only require that C_0 is constant dimensional and $\operatorname{Im} h_* = h_*(C_0)$. Prove Theorem 9.7 under this weaker assumption.

9.2 Assume the system (9.1-3) is analytic and each output y_i is 1-dimensional. Suppose Assumptions 9.1 and 9.7 are satisfied.

(a) Prove that the analytic decoupling condition (8.46) implies the geometric decoupling condition (9.27). (Hint : Prove that $D_i^* = \ker \operatorname{span}\{dh_j, \ldots, dL_f^{\rho_j} h_j \ , \ j \neq i\}$; see also Theorem 7.21)

(b) Prove that (9.27) implies that (8.46) holds on an open and dense subset of M.

9.3 Consider the nonlinear system (9.1,2) where the number of inputs m is larger than the number of output blocks. Prove under the same assumptions as in Theorem 9.7 that the (block) input-output decoupling is locally solvable if $G = D_1^* \cap G \oplus \ldots \oplus D_p^* \cap G$, where $\oplus$ denotes direct sum. Discuss also the case that $\dim(G \cap D^*) = m - p$, compare Corollary 9.11 (see also [NS4]).

9.4 Consider a system (9.1-3) satisfying all requirements of Theorem 9.7. Recall, see (9.62) and (9.63), the definition of D_I^*, $I \subset \underline{m}$. Prove that for all $I \subset \underline{m}$, $\dim(D_I^* \cap G) = |I|$.

9.5 Consider a system (9.1-3) satisfying all requirements of Theorem 9.7. Moreover, assume that the output map $h \colon M \longrightarrow \mathbb{R}^{p_1} \times \ldots \times \mathbb{R}^{p_m}$ defined as $h(x) = (h_1(x), \ldots, h_m(x)$ is a local diffeomorphism about each point $x \in M$. Prove that $D_i^* = \bigcap_{j \neq i} \ker dh_j$, $i \in \underline{m}$ (see also [Nij2]).

9.6 Consider a nonlinear system (9.57) and assume the system has p^1 formal zeros at infinity of orders $\{n^{\mu}\}$, cf. Definition 9.17. Suppose the system (9.57) is precompensated by the system $\dot{z}_i = v_i$, $u_i = z_i$, $i \in \underline{m}$. Prove that the precompensated system has p^1 formal zeros at infinity of orders $\{n^{\mu+1}\}$.

9.7 For a linear system the number of zeros at infinity is bounded by $\min(m,p)$. Show by means of a counterexample that the number p^1 of formal zeros at infinity of a nonlinear system does not necessarily satisfy $p^1 \leq \min(m,p)$.

9.8 Consider the nonlinear system (9.57) and assume $D^* = 0$.

(a) Assume $m = 1$. Prove that necessarily also $p = 1$. Let $\dot{z} = v$, $u = z$ be a precompensator for this single input single output nonlinear system. Show that the maximal locally controlled invariant distribution of the precompensated system is also identically zero.

(b) Assume $m > 1$ and suppose only the first input is integrated, i.e. we add the precompensator $\dot{z} = v_1$, $u_1 = z$, $u_i = v_i$ $i = 2, \ldots, m$, to the system (9.57). Show by means of a counterexample that the maximal locally controlled invariant distribution of the precompensated system is not identically zero.

9.9 Consider the system (9.1-3), satisfying Assumption 9.1. The Triangular Decoupling Problem consists of finding a regular static state feedback $u = \alpha(x) + \beta(x)v$ such that for the closed loop system the output y_j is not influenced by the controls $v_1,\ldots,v_{j-1}$, $j = 2,\ldots,m$. Let R_k^* be the maximal locally controlled invariant distribution in $\bigcap_{i=m-k}^{m} \ker dh_i$, $k = 0,1,\ldots,m-1$. Prove that under suitable regularity assumptions on the distributions R_k^* the Triangular Decoupling Problem is locally solvable if and only if $\dim(R_k^* \cap G) = k - 1$.

9.10 Consider an arbitrary analytic nonlinear system (9.1,3) with $p = m = 2$. Suppose that $D^* = 0$. Prove that the dynamic input-output decoupling problem is locally solvable on an open and dense subset of M. Show by means of a counterexample that $D^* = 0$ is not a sufficient condition for dynamic input-output decoupling when $m = p > 2$.

9.11 Consider the nonlinear system (9.1,2) about an equilibrium point x_0, i.e. $f(x_0) = 0$. Let Σ_ℓ: $\dot{\bar{x}} = A\bar{x} + B\bar{u}$, $\bar{y} = C\bar{x}$ be the linearization around x_0. Show that in general output controllability of the system (9.1,2) does not imply output controllability of the linearization Σ_ℓ. On the other hand, prove that output controllability of Σ_ℓ implies output controllability of the system (9.1,2) in a neighborhood of x_0 (see also Proposition 3.3). Discuss the input-output decoupling problem for (9.1-3) and Σ_ℓ in light of these results.

10
Local Stability and Stabilization of Nonlinear Systems

In this chapter we will discuss some aspects of local stability and feedback stabilization of nonlinear control systems.

10.1 Local Stability and Local Stabilization via Linearization

We first present some standard definitions and results on the local stability of an autonomous system, i.e. a system without inputs. Consider

$$\dot{x} = f(x), \tag{10.1}$$

where $x = (x_1, \ldots, x_n)$ are local coordinates for a smooth manifold M and f is a smooth vectorfield on M. Let x_0 be an *equilibrium point* of (10.1), i.e.

$$f(x_0) = 0. \tag{10.2}$$

In the sequel we will study the qualitative behavior of the dynamics (10.1) in a neighborhood of the fixed point x_0. The equilibrium point x_0 is said to be *locally stable* if for any neighborhood V of x_0 there exists a neighborhood $\tilde{V}$ of x_0 such that if $\bar{x} \in \tilde{V}$, then the solution $x(t,0,\bar{x})$ belongs to V for all $t \geq 0$. The equilibrium x_0 is *locally asymptotically stable* if x_0 is locally stable and there exists a neighborhood V_0 of x_0 such that all solutions $x(t,0,\bar{x})$ of (10.1) with $\bar{x} \in V_0$, converge to x_0 as $t \to \infty$. In what follows we will study local asymptotic stability. There are two important classical ways to decide about the local asymptotic stability of an equilibrium point x_0. These are the so-called first and second (or direct) method of Lyapunov. In the first method the local stability of x_0 for the system (10.1) is related to the stability of the linearization of (10.1) around the equilibrium point (10.2). So, consider the linear dynamics

$$\dot{\bar{x}} = A\bar{x}, \tag{10.3}$$

with

$$A = \frac{\partial f}{\partial x}(x_0). \tag{10.4}$$

Theorem 10.1 **(First method of Lyapunov)** *The equilibrium point x_0 of the system (10.1) is locally asymptotically stable if the matrix A given in*

(10.4) is asymptotically stable, i.e. the matrix A has all its eigenvalues in the open left half plane. The equilibrium point x_0 is not stable if at least one of the eigenvalues of the matrix A has a positive real part.

Note that it is immediate that the results of theorem 10.1 are not changed under a coordinate transformation $z = S(x)$ around the equilibrium point x_0. Essentially local asymptotic stability and instability can be decided via Theorem 10.1 from the linearized dynamics (10.3) provided that the matrix A given in (10.4) has no eigenvalues with zero real part. An equilibrium point x_0 for which the linearized dynamics has no eigenvalues with zero real part is called a *hyperbolic* equilibrium point.

The second or direct method of Lyapunov for deciding about the (asymptotic) stability of the equilibrium point x_0 involves the introduction of positive definite functions and invariant sets. A smooth function $\mathcal{L}$ defined on some neighborhood V of x_0 is *positive definite* if $\mathcal{L}(x_0) = 0$ and $\mathcal{L}(x) > 0$ for all $x \neq x_0$. A set W in M is an *invariant set* for (10.1) if for all $\bar{x} \in W$ the solutions $x(t,0,\bar{x})$ of (10.1) belong to W for all t.

Theorem 10.2 (Second method of Lyapunov) *Consider the dynamics (10.1) around the equilibrium point (10.2). Let $\mathcal{L}$ be a positive definite function on some neighborhood V_0 of x_0. Then we have*

(i) x_0 is locally stable if

$$L_f\mathcal{L}(x) \leq 0, \qquad \forall x \in V_0. \tag{10.5}$$

(ii) x_0 is locally asymptotically stable if (10.5) holds and the largest invariant set under the dynamics (10.1) contained in the set

$$W = \{x \in V_0 \mid L_f\mathcal{L}(x) = 0\} \tag{10.6}$$

equals $\{x_0\}$; i.e. the only solution $x(t,0,\bar{x})$ starting in $\bar{x} \in W$ which remains in W for all $t \geq 0$, coincides with x_0.

Note that the condition (10.5) expresses that around x_0 the function $\mathcal{L}$ is not increasing along solutions $x(t,0,\bar{x})$ of (10.1). A positive definite function $\mathcal{L}$ satisfying (10.5) is called a *Lyapunov function* for the system (10.1). It follows in particular that x_0 is locally asymptotically stable, when $\mathcal{L}$ is strictly decreasing along all solutions $x(t,0,\bar{x})$, $\bar{x} \in V\backslash\{x_0\}$ because in this case the set W trivially equals $\{x_0\}$.

The main interest of Theorem 10.2 in comparison with Theorem 10.1 lies in the fact that Theorem 10.2 may decide about asymptotic stability in case the linearized dynamics (10.3) has some eigenvalues with real part identically zero. Moreover, although we will not pursue this here, the direct method of Lyapunov may be used in the determination of the domain of attraction of an asymptotic stable equilibrium. On the other hand, the drawback of the second method of Lyapunov for the study of stability of an equilibrium point x_0, is that in general there does not exist a systematic procedure for constructing Lyapunov-functions. An exception is formed by the class of mechanical systems where the total energy serves as a good candidate Lyapunov-function (see also Chapter 12.3).

The following interesting result shows that the converse of Theorem 10.2(ii) is also true (see the references).

Theorem 10.3 *Consider the dynamics (10.1) around the equilibrium point (10.2). Assume the equilibrium point is locally asymptotically stable. Then there exists a Lyapunov-function $\mathcal{L}$ defined on some neighborhood V_0 of x_0 and for which the set W defined in (10.6) equals $\{x_0\}$.*

We emphasize that Theorems 10.1 and 10.2 by themselves only decide about the local nature of the (asymptotic) stability of the equilibrium point x_0. In order to decide about the global character of an asymptotic stable equilibrium more advanced techniques are needed. For this we refer to the literature given at the end of this chapter.

In the sequel we will show how Theorems 10.1 and 10.2 can be exploited in stabilization problems for nonlinear control systems. Consider the control system

$$\dot{x} = f(x,u), \tag{10.7}$$

where $x = (x_1, \dots, x_n)$ are local coordinates for a smooth manifold M, $u = (u_1, \dots, u_m) \in U \subset \mathbb{R}^m$, the input space, and $f(.,u)$ a smooth vectorfield for each $u \in U$. We assume U to be an open part of $\mathbb{R}^m$ and that f depends smoothly on the controls u. Let (x_0, u_0) an equilibrium point of (10.7), so

$$f(x_0, u_0) = 0. \tag{10.8}$$

Our concern is to see if the equilibrium (10.8) is locally asymptotically stable or can be made so by using some suitably chosen control function. In the first case we simply may check if the vectorfield $f(\cdot, u_0)$ satisfies the conditions given in Theorems 10.1 and 10.2. If not, we will see if addition of a strict state feedback $u = \alpha(x)$ to the system (10.7) can improve the stability of the equilibrium (x_0, u_0).

Problem 10.4 (Local feedback stabilization problem) *Under which conditions does there exist a smooth strict static state feedback* $u = \alpha(x)$, $\alpha : M \to U$, *with* $\alpha(x_0) = u_0$, *such that the closed loop system*

$$\dot{x} = f(x, \alpha(x)) \tag{10.9}$$

has x_0 *as a locally asymptotically stable equilibrium?*

A solution of Problem 10.4 can be obtained on the basis of Theorem 10.1 by using the linearization of the system (10.7) around the point (x_0, u_0). That is, we let

$$\dot{\bar{x}} = A\bar{x} + B\bar{u}, \tag{10.10}$$

where

$$A = \frac{\partial f}{\partial x}(x_0, u_0), \qquad B = \frac{\partial f}{\partial u}(x_0, u_0). \tag{10.11}$$

Define $\mathcal{R}$ as the reachable subspace of the linearized system (10.10), see also Chapter 3. So

$$\mathcal{R} = \operatorname{Im}\left(B \vdots AB \vdots \dots \vdots A^{n-1}B\right). \tag{10.12}$$

Clearly the subspace $\mathcal{R}$ is invariant under A, i.e. $A\mathcal{R} \subset \mathcal{R}$, so after a linear change of coordinates (10.10) can be rewritten as

$$\frac{d}{dt}\begin{pmatrix} \bar{x}^1 \\ \bar{x}^2 \end{pmatrix} = \begin{pmatrix} A_{11} & A_{12} \\ 0 & A_{22} \end{pmatrix}\begin{pmatrix} \bar{x}^1 \\ \bar{x}^2 \end{pmatrix} + \begin{pmatrix} B_1 \\ 0 \end{pmatrix}\bar{u} \tag{10.13}$$

where the vectors $(\bar{x}^1, 0)^T$ correspond with vectors lying in $\mathcal{R}$. We then obtain

Theorem 10.5 *The feedback stabilization problem for the system (10.7) admits a local solution around* x_0 *if all eigenvalues af the matrix* A_{22} *appearing in (10.13) are in* $\mathbb{C}^-$, *the open left half plane of* $\mathbb{C}$. *Moreover if one of the eigenvalues of* A_{22} *has a positive real part, then there does not exist a solution to the local feedback stabilization problem.*

Proof Consider the linearized dynamics (10.13) around (x_0, u_0) and assume all eigenvalues of A_{22} belong to $\mathbb{C}^-$. Then a standard result from linear control theory tells us that there is a linear state feedback $\bar{u} = F\bar{x}$ for the system (10.13) which asymptotically stabilizes the origin $\bar{x} = 0$. (Note that we may actually take $\bar{u} = F_1\bar{x}^1$, a feedback only depending on $\bar{x}^1$.) Taking the smooth feedback $u = u_0 + F(x - x_0)$ for the nonlinear system (10.7) we obtain the dynamics

$$\dot{x} = f(x, u_0 + F(x - x_0)), \tag{10.14}$$

of which the linearization around x_0 equals

$$\dot{\bar{x}} = (A + BF)\bar{x}. \tag{10.15}$$

By construction the linear dynamics (10.15) is asymptotically stable and so by Theorem 10.1 we conclude that x_0 is a locally asymptotically stable equilibrium point for (10.15).

Next suppose that at least one of the eigenvalues of the matrix A_{22} in (10.13) has a positive real part. Let $u = \alpha(x)$ be an arbitrary smooth feedback with $\alpha(x_0) = u_0$. Linearizing the dynamics (10.9) around x_0 yields

$$\dot{\bar{x}} = \left(A + B\frac{\partial\alpha}{\partial x}(x_0)\right)\bar{x}, \tag{10.16}$$

which still has the same unstable eigenvalue of the matrix A_{22}. By Theorem 10.1 we may conclude that x_0 is an unstable equilibrium point of (10.9). □

Remark Note that the above theorem yields no definite answer to the feedback stabilization problem when some of the eigenvalues of the matrix A_{22} lie on the imaginary axis (compare Theorem 10.1).

10.2 Local Stabilization using Lyapunov's Direct Method

In the following a stabilization result using Lyapunov's direct method is given. It enables us to improve local stability of an equilibrium point for an affine nonlinear system into local asymptotic stability. Consider the system

$$\dot{x} = f(x) + \sum_{i=1}^{m} g_i(x)u_i \tag{10.17}$$

with

$$f(x_0) = 0. \tag{10.2}$$

In (10.17) $x = (x_1, \dots, x_n)$ are local coordinates around the equilibrium point x_0 on a smooth manifold M and f, $g_1, \dots g_m$ are smooth vectorfields.

Suppose there exists a Lyapunov function $\mathcal{L}$ defined on some neighborhood V_0 of x_0 for the dynamics (10.17) with $u \equiv 0$, so for the system

$$\dot{x} = f(x) \tag{10.1}$$

we have

$$L_f\mathcal{L}(x) \leq 0 \ , \ \forall \ x \in V_0 . \tag{10.5}$$

Then according to Theorem 10.2 the point x_0 is locally stable for the system (10.17) by setting $u = 0$. In what follows we will show that under some additional conditions we are able to produce an asymptotically stabilizing feedback. Consider the smooth feedback $u = \alpha(x)$ with

$$\alpha_i(x) = -L_{g_i}\mathcal{L}(x) \quad , \quad i \in \underline{m} \quad , \quad x \in V_0 , \tag{10.18}$$

yielding the closed loop behavior

$$\dot{x} = f(x) + \sum_{i=1}^{m} g_i(x)\alpha_i(x) . \tag{10.19}$$

Clearly, x_0 is also an equilibrium point for (10.19). At each point $x \in V_0$ we have, using (10.18) and (10.5), that

$$L_f\mathcal{L}(x) + \sum_{i=1}^{m} L_{\alpha_i g_i}\mathcal{L}(x) = L_f\mathcal{L}(x) - \sum_{i=1}^{m} (L_{g_i}\mathcal{L}(x))^2 \leq 0, \tag{10.20}$$

which shows by Theorem 10.2 that x_0 is locally stable for the closed loop dynamics (10.19). In order to study the local asymptotic stability of x_0 for (10.19) we introduce the set

$$\begin{aligned} W &= \{x \in V_0 \mid L_f\mathcal{L}(x) - \sum_{i=1}^{m} (L_{g_i}\mathcal{L}(x))^2 = 0\} \\ &= \{x \in V_0 \mid L_f\mathcal{L}(x) = 0, \ L_{g_i}\mathcal{L}(x) = 0, \ i \in \underline{m}\} . \end{aligned} \tag{10.21}$$

Notice that $x_0 \in W$. Let W_0 be the largest invariant subset of W under the dynamics (10.19). In case that W_0 equals $\{x_0\}$ we conclude from Theorem 10.2 that x_0 is locally an asymptotically stable equilibrium point. Now let $x^\alpha(t,0,\bar{x})$ denote the solution of (10.19) starting at $t = 0$ in $\bar{x} \in V_0$. Observe that any trajectory $x^\alpha(t,0,\bar{x})$ in W_0 is a trajectory of the dynamics (10.1); this because the feedback (10.18) is identically zero for each point in W. Therefore x_0 is locally asymptotically stable for the dynamics (10.19) if the only trajectory of (10.1) contained in W is the trivial solution $x(t) = x_0$, $t \geq 0$. Henceforth we will briefly refer to W_0

as the largest f-invariant subset in W. On the other hand, when the Lyapunov-function $\mathcal{L}$ satisfies

$$d\mathcal{L}(x) \neq 0, \quad \forall x \in V_0 \backslash \{x_0\}, \tag{10.22}$$

also the converse is true. That is, if x_0 is locally asymptotically stable for (10.19), then the only trajectory of (10.1) contained in W is the trivial solution $x(t)$, $t \geq 0$. This holds because along each trajectory of (10.19) belonging to W and which is thus a solution of (10.1), the Lyapunov function is constant; hence the trajectory cannot approach x_0. Summarizing we have obtained.

Lemma 10.6 *Consider the affine control system (10.17) around the equilibrium point (10.2). Suppose there exists a Lyapunov-function $\mathcal{L}$ on a neighborhood V_0 of x_0 such that (10.5) holds true and which satisfies (10.22). The smooth feedback (10.18) locally asymptotically stabilizes the equilibrium point x_0 if and only if the largest f-invariant subset in W (defined in (10.21)) equals $\{x_0\}$.*

The difficulty of applying Lemma 10.6 lies in the fact that we need to know the trajectories of the dynamics when $u \equiv 0$. To avoid the computation of solutions of (10.1) we will establish sufficient conditions in geometric terms. Define the distribution

$$D(x) = \text{span}\{f(x),\ ad_f^k g_i(x),\ i \in \underline{m},\ k \geq 0\},\ x \in V_0. \tag{10.23}$$

Then we have

Lemma 10.7 *Consider the system (10.17) around the point (10.2). Suppose there exists a Lyapunov-function $\mathcal{L}$ on a neighborhood V_0 of the equilibrium point x_0 such that (10.22) and (10.5) hold. Then along any trajectory of the closed loop dynamics (10.19) lying in the set W given in (10.21) the distribution D of (10.23) has dimension strictly smaller than n.*

Proof Consider the function $L_f\mathcal{L}$ on V_0. From (10.5) it follows that this function has a maximum at each point $x \in W$. Therefore

$$dL_f\mathcal{L}(x) = 0, \quad \forall x \in W. \tag{10.24}$$

Consider a trajectory $x^{\alpha}(t,0,\bar{x})$, $t \geq 0$, in W. As noted before, this solution also satisfies (10.1). Define the functions $\varphi_i(t)$ as

$$\varphi_i(t) = L_{g_i}\mathcal{L}(x^\alpha(t,0,\bar{x})), \qquad i \in \underline{m}. \tag{10.25}$$

Obviously these time functions are identically zero and so all time derivatives $(d/dt)^k\varphi_i(t)$ vanish. On the other hand we compute for $k \geq 0$

$$\frac{d^k\varphi_i}{dt^k}(0) = L_{ad^k_f g_i}\mathcal{L}(\bar{x}), \qquad i \in \underline{m}. \tag{10.26}$$

The conclusion now follows from (10.22). □

Lemma 10.7 leads to several sufficient conditions for the local asymptotic stability of x_0 for the closed loop dynamics (10.19). The simplest situation is that where we have

$$\dim D(x_0) = n, \tag{10.27}$$

which implies that on some neighborhood $\tilde{V}_0 \subset V_0$ of x_0

$$\dim D(x) = n. \tag{10.28}$$

Using Proposition 10.7 we conclude that x_0 is locally asymptotically stable for the closed loop system (10.19). Of course, this result also follows from Theorem 10.5 because for the linearization (10.10) of the system (10.17) around x_0, the reachable subspace $\mathcal{R}$ given in (10.12) is n-dimensional as can be seen from (10.27) and using the fact that $f(x_0) = 0$ and $ad^k_f g_i(x_0) = (-1)^k A^k b_i$, $k = 0,1,2,\ldots$ (cf. (3.38)).

Other interesting sufficient conditions based on Proposition 10.7 may be obtained as follows. Define the set

$$\tilde{W} = \{x \in V_0 \,|\, L_{g_i}\mathcal{L}(x) = 0,\ i \in \underline{m}\}. \tag{10.29}$$

Assumption 10.8 *There exist subsets V_1 and V_2 of V_0 with $V_1 \cap V_2 = \emptyset$ and $V_1 \cup V_2 = V_0$ such that*

(i) $x_0 \in V_2$,

(ii) $\dim D(x) = n$, *for all* $x \in V_1$,

(iii) *There exists a neighborhood $\tilde{V}_0$ in V_0 of x_0 such that $\{x_0\}$ is the largest invariant subset of (10.1) in the set $V_2 \cap \tilde{W} \cap \tilde{V}_0$.*

Then we have

Theorem 10.9 *Consider the affine control system (10.17) around the equilibrium point (10.2). Suppose there exists a Lyapunov-function $\mathcal{L}$ on a* *neighborhood* V_0 *of* x_0 *such that (10.5) and (10.22) hold. The smooth feedback (10.18) locally asymptotically stabilizes the equilibrium point* x_0 *if Assumption 10.8 is satisfied.*

Proof Without loss of generality we may suppose that $\tilde{V}_0$ equals V_0. Clearly each nontrivial solution of (10.19) lying in the set W is also contained in $\tilde{W}$. On the other hand Lemma 10.7 implies that such a trajectory is also contained in V_2. The conclusion then follows from Theorem 10.2 by observing that a trajectory of (10.19) in W is also a trajectory of (10.1). □

We note that a simple typical special case of Theorem 10.9 is obtained when

$$\dim D(x) = n, \text{ for all } x \in V_0 \backslash \{x_0\}. \tag{10.30}$$

Example 10.10 Consider the equations for the angular velocities of a rigid body with one external torque (see Example 1.2):

$$I\dot{\omega} = S(\omega)I\omega + bu \tag{10.31}$$

with $\omega = (\omega_1, \omega_2, \omega_3)$,

$$S(\omega) = \begin{pmatrix} 0 & \omega_3 & -\omega_2 \\ -\omega_3 & 0 & \omega_1 \\ \omega_2 & -\omega_1 & 0 \end{pmatrix}, \quad I = \begin{pmatrix} I_1 & 0 & 0 \\ 0 & I_2 & 0 \\ 0 & 0 & I_3 \end{pmatrix}. \tag{10.32}$$

$I_3 > I_2 > I_1 > 0$ denote the principal moments of inertia. Let

$$\begin{cases} I_{23} = (I_2 - I_3) / I_1, \\ I_{31} = (I_3 - I_1) / I_2, \\ I_{12} = (I_1 - I_2) / I_3. \end{cases} \tag{10.33}$$

Then (10.31) may be written as

$$\begin{cases} \dot{\omega}_1 = I_{23}\, \omega_2\, \omega_3 + c_1\, u \\ \dot{\omega}_2 = I_{31}\, \omega_3\, \omega_1 + c_2\, u \\ \dot{\omega}_3 = I_{12}\, \omega_1\, \omega_2 + c_3\, u \end{cases} \tag{10.34}$$

with $c = (c_1, c_2, c_3)^T = I^{-1}b$. Clearly $(\omega_1, \omega_2, \omega_3) = 0$ is an equilibrium point of (10.34) when $u = 0$. An obvious choice for a Lyapunov function for the drift vectorfield in (10.34) is the kinetic energy of the rigid body, i.e.

$$\mathcal{L}(\omega) = \frac{1}{2}(I_1\omega_1^2 + I_2\omega_2^2 + I_3\omega_3^2). \tag{10.35}$$

$\mathcal{L}$ is a smooth positive definite function having a unique minimum in $\omega = 0$. Computing $L_f\mathcal{L}$ yields

$$L_f\mathcal{L}(\omega) = I_1 I_{23}\omega_1\omega_2\omega_3 + I_2 I_{31}\omega_1\omega_2\omega_3 + I_3 I_{12}\omega_1\omega_2\omega_3 = 0, \tag{10.36}$$

which shows that $\omega = 0$ is a stable equilibrium point of (10.34) when $u = 0$. Define the smooth feedback

$$u = -L_c\mathcal{L}(\omega) = -(c_1 I_1\omega_1 + c_2 I_2\omega_2 + c_3 I_3\omega_3). \tag{10.37}$$

In what follows we will see whether or not the feedback (10.37) makes $\omega = 0$ asymptotically stable. Define the distribution

$$D(\omega) = \text{span}\{f(\omega),\ ad_f^k c(\omega),\ k \geq 0\}, \tag{10.38}$$

which after some computations yields

$$D(\omega) = \text{span}\left\{\begin{pmatrix} I_{23}\ \omega_2\omega_3 \\ I_{31}\ \omega_3\omega_1 \\ I_{12}\ \omega_1\omega_2 \end{pmatrix}, \begin{pmatrix} c_1 \\ c_2 \\ c_3 \end{pmatrix}, \begin{pmatrix} I_{23}c_2\omega_3 + I_{23}c_3\omega_2 \\ I_{31}c_1\omega_3 + I_{31}c_3\omega_1 \\ I_{12}c_1\omega_2 + I_{12}c_2\omega_1 \end{pmatrix}, \begin{pmatrix} I_{23}\ c_2c_3 \\ I_{31}\ c_3c_1 \\ I_{12}\ c_1c_2 \end{pmatrix}\right\}. \tag{10.39}$$

In order that we may apply Theorem 10.9 we need to find subsets V_1 and V_2 of $\mathbb{R}^3$ such that Assumption 10.8 holds. Define

$$V_1 = \{\omega \mid \dim D(\omega) = 3\}, \tag{10.40a}$$

$$V_2 = \{\omega \mid \dim D(\omega) < 3\}. \tag{10.40b}$$

Clearly $V_1 \cap V_2 = \emptyset$ and $V_1 \cup V_2 = \mathbb{R}^3$, as well as $0 \in V_2$. Now a straightforward analysis yields that

$$V_2 = V_{21} \cap V_{22}, \tag{10.41}$$

where

$$\begin{aligned} V_{21} = \{\omega \mid\ & I_{23}c_2c_3(I_{31}c_3^2 - I_{12}c_2^2)\omega_1 + I_{31}c_3c_1(I_{12}c_1^2 - I_{23}c_3^2)\omega_2 + \\ & I_{12}c_1c_2(I_{23}c_2^2 - I_{31}c_1^2)\omega_3 = 0\}, \end{aligned} \tag{10.42a}$$

$$\begin{aligned} V_{22} = \{\omega \mid\ & I_{12}c_2c_3(I_{31}c_1^2 - I_{23}c_2^2)\omega_1\omega_2 + c_2I_{31}(I_{23}c_3^2 - I_{12}c_1^2) + \\ & c_1I_{23}(I_{12}c_2^2 - I_{31}c_3^2)\omega_2\omega_3 = 0\}. \end{aligned} \tag{10.42b}$$

Let W be the set defined by

$$W = \{\omega \mid L_c \mathcal{L}(\omega) = 0\}$$

$$= \{\omega \mid I_1 c_1 \omega_1 + I_2 c_2 \omega_2 + I_3 c_3 \omega_3 = 0\}. \tag{10.43}$$

In what follows we will make the following assumption

$$c_1 c_2 c_3 \neq 0. \tag{10.44}$$

So the control axis is not perpendicular to any of the principle axis of the rigid body. To verify the Assumption 10.8 we need to compute the intersection of the sets V_{21}, V_{22} and W. Note that $I_{31}c_3^2 - I_{12}c_1^2 > 0$ and $I_{23}c_2^2 - I_{31}c_1^2 < 0$, which imply that V_{21} is in this case always a two dimensional plane. V_{22} as given in (10.42b) is a cone in $\mathbb{R}^3$, which degenerates into the union of two planes in case that

$$I_{23}c_3^2 - I_{12}c_1^2 = 0. \tag{10.45}$$

Provided (10.45) is *not* satisfied the intersection of the plane V_{21} and the nondegenerate cone V_{22} equals a finite number of lines through $\omega = 0$. Taking the intersection with W we obtain the origin $\omega = 0$ and probably some of these lines in $V_{22} \cap V_{21}$. Let $\tilde{\omega}(.)$ be some trajectory of the system (10.34,37) belonging to one of these lines; obviously $\tilde{\omega}(.)$ is also a trajectory of the vectorfield f. Clearly differentiating the Lyapunov function $\mathcal{L}$ along $\tilde{\omega}(.)$ yields $\frac{d}{dt}\mathcal{L}(\tilde{\omega}(t)) = 0$, $t \geq 0$, so that $\mathcal{L}(\tilde{\omega}(t))$ is constant for all $t \geq 0$. Therefore $\tilde{\omega}(.)$ belongs to an ellipsoid given by $\{\omega \mid \mathcal{L}(\omega) = \text{constant}\}$. Intersection with a line yields that $\tilde{\omega}(.)$ is an equilibrium point of the dynamics (10.34,37). However one immediately checks that given (10.44) the only equilibrium point in W equals $\omega = 0$. So Assumption 10.8 is satisfied if (10.44) holds and $I_{12}c_1^2 - I_{23}c_3^2 \neq 0$, and by Theorem 10.9 we conclude that in this case the feedback (10.37) asymptotically stabilizes $\omega = 0$. Next we investigate the case that (10.45) holds. V_{22} reduces to the union of the planes

$$\{\omega \mid \omega_2 = 0\}, \tag{10.46a}$$

$$\{\omega \mid I_{12}c_3(I_{31}c_1^2 - I_{23}c_2^2)\omega_1 + I_{23}c_1(I_{12}c_2^2 - I_{31}c_3^2)\omega_3 = 0\}. \tag{10.46b}$$

Taking the intersection of (10.46a,b) with the plane W we again obtain a finite number (2) of lines (Notice that we use here (10.44) again.) Similarly as before we conclude that the feedback (10.37) asymptotically stabilizes $\omega = 0$. □

10.3 Local Stabilization via Center Manifold Theory

So far we have discussed the local feedback stabilization problem via procedures based on Lyapunov's first and second method. Another approach to this problem is based on the *Center Manifold* theory. Center manifold theory forms an extension of the first method of Lyapunov in that it provides a way of studying the stability of an autonomous system for which the linearization at the equilibrium has some eigenvalues located on the imaginary axis. In the sequel we first briefly describe this theory and then we explain how it can be exploited in the feedback stabilization problem. Consider again the system (10.1) around the equilibrium point (10.2) and let the $n \times n$-matrix A given by (10.4) denote the linearization of (10.1) at x_0. In order to have local asymptotic stability of (10.1) around the equilibrium point we conclude from Theorem 10.1 that the matrix A should not have eigenvalues in the open right-half plane. Therefore the set of eigenvalues of A, $\sigma(A)$, can be written as the disjoint union

$$\sigma(A) = \sigma_- \cup \sigma_0 , \tag{10.47}$$

where the eigenvalues in σ_- lie in the open left half plane and those in σ_0 lie on the imaginary axis. Let ℓ be the number of eigenvalues (counted with their multiplicity) contained in σ_-. Then there are $n-\ell$ eigenvalues (counted with their multiplicity) in σ_0. Given the splitting of the eigenvalues as above there exists a linear coordinate transformation T such that

$$TAT^{-1} = \begin{pmatrix} A^0 & 0 \\ 0 & A^- \end{pmatrix}, \tag{10.48}$$

where the $(n-\ell, n-\ell)$-matrix A^0 and the (ℓ,ℓ)-matrix A^- have as eigenvalues $\sigma(A^0) = \sigma_0$, respectively $\sigma(A^-) = \sigma_-$. In the transformed coordinates $z = Tx - x_0$ the system (10.1) takes the form

$$\begin{cases} \dot{z}^1 = A^0 z^1 + f^0(z^1, z^2) \\ \dot{z}^2 = A^- z^2 + f^-(z^1, z^2) \end{cases} \tag{10.49}$$

where z^1 and z^2 denote the first $(n-\ell)$, respectively the last ℓ components of z, and f^0 and f^- are smooth appropriate dimensioned vector functions representing the second and higher order terms around the equilibrium $z = 0$. So

$$f^0(0,0) = 0, \tag{10.50a}$$

$$f^{-}(0,0) \;\; = 0, \tag{10.50b}$$

$$df^{0}(0,0) = 0, \tag{10.50c}$$

$$df^{-}(0,0) = 0. \tag{10.50d}$$

We are now prepared to formulate the so called Center Manifold Theorem.

Theorem 10.11 (Center Manifold Theorem) *Consider the local description (10.49) around the equilibrium point $z = 0$ of the autonomous system (10.1). Then for each $k = 2,3,\ldots$ there exists a $\delta_k > 0$ and a C^k-mapping $\varphi\colon \{z^1 \in \mathbb{R}^{n-\ell} \mid \|z^1\| < \delta_k\} \to \mathbb{R}^{\ell}$ with $\varphi(0) = 0$ and $d\varphi(0) = 0$, such that the surface (the* center manifold*)*

$$z^2 = \varphi(z^1), \;\; \|z^1\| < \delta_k, \tag{10.51}$$

is invariant under the dynamics (10.49).

For the proof we refer to the literature cited at the end of this chapter.

Remark 10.12 (i) In general the dynamics (10.49) do not possess a *unique* center manifold, but may have an infinite number of such invariant manifolds.
(ii) The smooth dynamics (10.49) has a C^k center manifold for each (finite) $k = 2,3,\ldots$ However the size of the center manifold (δ_k in (10.51)) depends on k and may shrink with increasing k. Even in case (10.45) is analytic, there does not necessarily exist an analytic center manifold. In what follows we will throughout work with a C^2 center manifold.

The dynamics on the center manifold (10.51) are given as

$$\dot{z}^1 = A^0 z^1 + f^0\big(z^1, \varphi(z^1)\big). \tag{10.52}$$

The following theorem shows that the dynamics (10.52) contain all necessary information about the local asymptotic stability or instability of $z = 0$ for (10.49).

Theorem 10.13 *Let (10.52) represent the dynamics on the center manifold (10.51) of the system (10.49). Then we have: $z^1 = 0$ is locally asymptotically stable, locally stable, or unstable for (10.52) respectively, implies that $(z^1,z^2) = 0$ is locally asymptotically stable, locally stable or unstable for (10.49) respectively.*

Essentially this theorem states that the local asymptotic stability of the n-dimensional system (10.49) can be deduced from the local asymptotic stability of the reduced $(n-\ell)$-dimensional system (10.52) on the center manifold (10.51). Before one can apply this result one needs a way to determine the center manifold (10.51), and in particular the mapping φ. This is in general not possible since it is equivalent to solving (10.49) analytically. Differentiating (10.51) with respect to t and substituting this into (10.49) yields the partial differential equation for φ

$$d\varphi(z^1)\left[A^0z^1 + f^0\left(z^1,\varphi(z^1)\right)\right] = A^-\varphi(z^1) + f^-\left(z^1,\varphi(z^1)\right), \tag{10.53}$$

together with the boundary conditions

$$\varphi(0) = 0, \tag{10.54a}$$

$$d\varphi(0) = 0. \tag{10.54b}$$

Although we may not be able to solve (10.53,54) analytically we can approximate the solution to any degree of accuracy.

Theorem 10.14 *Suppose* $\tilde{\varphi}: \mathbb{R}^{n-\ell} \to \mathbb{R}^{\ell}$ *is a* C^2*-mapping satisfying (10.50) and for some* $q > 1$

$$d\tilde{\varphi}(z)\left[A^0z^1+f^0\left(z^1,\tilde{\varphi}(z^1)\right)\right] - A^-\tilde{\varphi}(z^1) - f^-\left(z^1,\tilde{\varphi}(z^1)\right) = O(\|z^1\|^q) \tag{10.55}$$

for $z^1 \to 0$. *Then*

$$\|\tilde{\varphi}(z^1) - \varphi(z^1)\| = O(\|z^1\|^q) \tag{10.56}$$

for $z^1 \to 0$.

For the proofs of Theorems 10.13 and 10.14 we refer to the literature.

The following example demonstrates the power of center manifold theory.

Example 10.15 Consider on $\mathbb{R}^2$ the system

$$\frac{d}{dt}\begin{pmatrix} x_1 \\ x_2 \end{pmatrix} = \begin{pmatrix} x_1x_2 + x_1^3 + x_1x_2^2 \\ -x_2 - 2x_1^2 + x_1^2x_2 \end{pmatrix} \tag{10.57}$$

around the equilibrium point (0,0). Clearly this system is already in the form (10.49). From Theorem 10.11 we conclude that (10.57) possesses a center manifold of the form

$$x_2 = \varphi(x_1), \tag{10.58}$$

with $\varphi(0) = \varphi'(0) = 0$. Trying an approximate solution $\tilde{\varphi}(x_1)$ of the form $\tilde{\varphi}(x_1) = \tilde{\varphi}_2 x_1^2 + \tilde{\varphi}_3 x_1^3 + \ldots$ we obtain

$$\tilde{\varphi}'(x_1)\left[x_1\tilde{\varphi}(x_1) + x_1^3 + x_1\tilde{\varphi}^2(x_1)\right] + \tilde{\varphi}(x_1) + 2x_1^2 - x_1^2\tilde{\varphi}(x_1) =$$

$$(\tilde{\varphi}_2+2)x_1^2 + \tilde{\varphi}_3 x_1^3 + O(x_1^4). \tag{10.59}$$

Using Theorem 10.14 we see that

$$\varphi(x_1) = -2x_1^2 + O(|x_1|^3). \tag{10.60}$$

Substituting this into the dynamics of the center manifold (10.58) we obtain

$$\dot{x}_1 = -2x_1^3 + x_1^3 + O(x_1^4) = -x_1^3 + O(x_1^4), \tag{10.61}$$

which has $x_1 = 0$ as a local asymptotically stable equilibrium point. Thus from Theorem 10.13 we may conclude that (0,0) is locally asymptotically stable for the system (10.57). □

The computations in Example 10.15 are rather straightforward. In general, however, when the dimension of the center manifold is larger than one, the problem of verifying whether or not the dynamics (10.52) is asymptotically stable is hard.

In the following we investigate how to exploit the center manifold theory in the local feedback stabilization problem. Starting point will be again the control system

$$\dot{x} = f(x,u) \tag{10.7}$$

around the equilibrium point

$$f(x_0,u_0) = 0. \tag{10.8}$$

From Theorem 10.5 we know that the only case of interest arises when the linearization

$$\dot{\bar{x}} = A\bar{x} + B\bar{u} \tag{10.10}$$

around (10.8) - with A and B defined as in (10.11) - has some uncontrollable eigenvalues on the imaginary axis; in all other cases the stability or instability follows from this theorem. This is the case when the reachable subspace $\mathcal{R}$ of (10.10) has a dimension smaller than n, and so after a linear change of coordinates (10.10) can be written as

$$\frac{d}{dt}\begin{pmatrix} \bar{x}^1 \\ \bar{x}^2 \end{pmatrix} = \begin{pmatrix} A_{11} & A_{12} \\ 0 & A_{22} \end{pmatrix}\begin{pmatrix} \bar{x}^1 \\ \bar{x}^2 \end{pmatrix} + \begin{pmatrix} B_1 \\ 0 \end{pmatrix}\bar{u}, \tag{10.13}$$

and the eigenvalues of A_{22} are all located on the imaginary axis. As the pair (A_{11}, B_1) is controllable, we know that there exists a linear feedback $\bar{u} = F_1\bar{x}^1$ such that $A_{11} + B_1F_1$ is an asymptotically stable matrix. This suggests to apply a feedback $u = \alpha(x)$ of the form

$$u = u_0 + F_1(x^1 - x_0^1) + F_2(x_2^2 - x_0^2) + \tilde{\alpha}(x^1 - x_0^1, x^2 - x_0^2) \tag{10.62}$$

to the system (10.7). In (10.62) $x_0 = (x_0^1, x_0^2)$ and $\tilde{\alpha}$ is a smooth feedback function with $\tilde{\alpha}(0,0) = 0$ and $d\tilde{\alpha}(0,0) = 0$. The linearization of (10.7) with the feedback (10.62) equals

$$\frac{d}{dt}\begin{pmatrix} \bar{x}_1 \\ \bar{x}_2 \end{pmatrix} = \begin{pmatrix} A_{11}+B_1F_1 & B_1F_2 \\ 0 & A_{22} \end{pmatrix}\begin{pmatrix} \bar{x}_1 \\ \bar{x}_2 \end{pmatrix}, \tag{10.63}$$

and so we conclude that after a linear change of coordinates (10.7,62) is a system of the form (10.49), which according to Theorem 10.11 has a center manifold

$$z^1 = \varphi^{\alpha}(z^2), \quad \|z^2\| < \delta, \tag{10.64}$$

where $z^1 = x^1 - x_0^1$, $z^2 = x^2 - x_0^2$ and the superscript α denotes the dependency of the center manifold on the feedback chosen. In order that (10.62) locally asymptotically stabilizes the equilibrium x_0 we have to use the freedom in the feedback (10.62) in such a way that the reduced dynamics on the center manifold (10.52) is locally asymptotically stable. Notice that in (10.62) we have a complete freedom in selecting the linear map F_2 and $\tilde{\alpha}$, whereas F_1 is only constrained by the requirement that $A_{11} + B_1F_1$ is asymptotically stable. Clearly, whether or not we can make the reduced dynamics asymptotically stable in this way, depends on the higher order Taylor expansion of the closed loop system dynamics. As stated before, this is the most difficult part of the center manifold theory. To see how the theory works we discuss a smooth control system

$$\dot{x} = f(x) + g(x)u \tag{10.65}$$

on $\mathbb{R}^2$ around the equilibrium point $f(0) = 0$ and assume $g(0) \neq 0$. According to the Flow-Box Theorem, see Theorem 2.26, there exists a coordinate transformation in which $g(x) = \partial/\partial x_2$. In these coordinates (again denoted by $x = (x_1, x_2)$) the system takes the form

$$\begin{cases} \dot{x}_1 = f_1(x_1, x_2), \\ \dot{x}_2 = f_2(x_1, x_2) + u. \end{cases} \tag{10.66}$$

Application of the feedback

$$u = -f_2(x_1, x_2) - x_2 + p_1 x_1 + \tilde{\alpha}(x_1, x_2) \tag{10.67}$$

where $\tilde{\alpha}(0,0) = 0$ and $d\tilde{\alpha}(0,0) = 0$, yields the closed loop dynamics

$$\begin{cases} \dot{x}_1 = f_1(x_1, x_2), \\ \dot{x}_2 = -x_2 + p_1 x_1 + \tilde{\alpha}(x_1, x_2). \end{cases} \tag{10.68}$$

Linearizing (10.68) about (0,0) yields

$$\frac{d}{dt}\begin{bmatrix} \bar{x}_1 \\ \bar{x}_2 \end{bmatrix} = \begin{bmatrix} \frac{\partial f_1}{\partial x_1}(0,0) & \frac{\partial f_1}{\partial x_2}(0,0) \\ p_1 & -1 \end{bmatrix} \begin{bmatrix} \bar{x}_1 \\ \bar{x}_2 \end{bmatrix}. \tag{10.69}$$

First of all we require that

$$\frac{\partial f_1}{\partial x_2}(0,0) = 0, \tag{10.70a}$$

because otherwise the linearization of (10.66) would be controllable and thus (Theorem 10.5) the local feedback stabilization problem is trivially solvable. Also we ask that

$$\frac{\partial f_1}{\partial x_1}(0,0) = 0. \tag{10.70b}$$

In case (10.70b) is violated the system (10.7,67) is locally asymptotically stable if $\frac{\partial f_1}{\partial x_1}(0,0) < 0$ and unstable if $\frac{\partial f_1}{\partial x_1}(0,0) > 0$, no matter how we select the feedback (10.67). Now, assuming all data to be analytic, we may take a Taylor series expansion for the closed loop system (10.7,67). To simplify things we assume that the higher order terms in $\tilde{\alpha}(x_1, x_2)$ in (10.67) only depend on x_1. This yields the following expansions.

$$f_1(x_1, x_2) = a_2 x_1^2 + a_3 x_1^3 + b_1 x_1 x_2 + b_2 x_1 x_2^2 + b_3 x_1^2 x_2 + c_2 x_2^2 + c_3 x_2^3 + .. \tag{10.71a}$$

$$\tilde{\alpha}(x_1, x_2) = p_2 x_1^2 + p_3 x_1^3 + \ldots \tag{10.71b}$$

Writing $\tilde{x}_1 = x_1$ and $\tilde{x}_2 = x_2 - p_1 x_1$ (in order to obtain the Jordan form for the linearized dynamics as in (10.49)), and substituting (10.71a,b) into the dynamics we obtain

$$\begin{cases} \dot{\tilde{x}}_1 = a_2\tilde{x}_1^2 + a_3\tilde{x}_1^3 + b_1\tilde{x}_1\tilde{x}_2 + b_1p_1\tilde{x}_1^2 + \dots \\ \dot{\tilde{x}}_2 = -\tilde{x}_2 + p_2\tilde{x}_1^2 - p_1(a_2\tilde{x}_1^2+a_3\tilde{x}_1^3+b_1\tilde{x}_1\tilde{x}_2+b_1p_1\tilde{x}_1^2+\dots). \end{cases} \tag{10.72}$$

From Theorem 10.11 we know there exists a center manifold

$$\tilde{x}_2 = \varphi(\tilde{x}_1) \tag{10.73}$$

for (10.72). Note that by definition $\varphi(0) = \varphi'(0) = 0$. To get an approximation of φ - see Theorem 10.14 - we let

$$\tilde{\varphi}(\tilde{x}_1) = \varphi_2\tilde{x}_1^2 + \varphi_3\tilde{x}_1^3 \tag{10.74}$$

and substitute this in equation (10.55) yielding

$$\begin{aligned} &(2\varphi_2\tilde{x}_1 + 3\varphi_3\tilde{x}_1^2)(a_2\tilde{x}_1^2 + a_3\tilde{x}_1^3 + b_1\varphi_2\tilde{x}_1^3 + b_1\varphi_3\tilde{x}_1^4 + b_1p_1\tilde{x}_1^2 + \dots) \\ &+ \varphi_2\tilde{x}_1^2 + \varphi_3\tilde{x}_1^3 - p_2\tilde{x}_1^2 + p_1a_2\tilde{x}_1^2 + p_1a_3\tilde{x}_1^3 + b_1p_1^2\tilde{x}_1^2 + b_1\varphi_2\tilde{x}_1^3 + \dots \\ &= O(|x_1|^q). \end{aligned} \tag{10.75}$$

Letting $q = 3$ we find that

$$\varphi_2 = p_2 - p_1a_2 - b_1p_1^2 \tag{10.76}$$

yields a second order approximation of the center manifold. The reduced equation (10.52) then takes the form

$$\dot{\tilde{x}}_1 = (a_2+b_1p_1)\tilde{x}_1^2 + (a_3+b_1p_2-b_1p_1a_2-b_1^2p_1^2)\tilde{x}_1^3 + \dots, \tag{10.77}$$

from which we conclude that a necessary condition for the asymptotic stability is that

$$a_2 + b_1p_1 = 0, \tag{10.78}$$

and given (10.78) the reduced equation (10.77) is locally asymptotically stable if

$$a_3 + b_1p_2 - b_1p_1a_2 - b_1^2p_1^2 < 0. \tag{10.79}$$

Therefore we conclude that (10.77) is locally asymptotically stable if (10.78) and (10.79) hold, cf. Theorem 10.13. Note that these two conditions can be fullfilled by selecting a suitable feedback (10.67,71b) if the coefficient b_1 is nonzero. In terms of the original dynamics (10.66) this is equivalent to the requirement that

$$\frac{\partial^2 f_1}{\partial x_1 \partial x_2}(0,0) \neq 0. \tag{10.80}$$

We conclude our exposition about the use of center manifold techniques in the smooth stabilization problem with the following example.

Example 10.16 Consider the equations for the angular velocities of a rigid body with two external torques aligned with two principal axes (see Example 1.2):

$$I\dot{\omega} = S(\omega)I\omega + b_1 u_1 + b_2 u_2, \tag{10.81}$$

where $\omega = (\omega_1, \omega_2, \omega_3)$ and

$$S(\omega) = \begin{pmatrix} 0 & \omega_3 & -\omega_2 \\ -\omega_3 & 0 & \omega_1 \\ \omega_2 & -\omega_1 & 0 \end{pmatrix}, \quad I = \begin{pmatrix} I_1 & 0 & 0 \\ 0 & I_2 & 0 \\ 0 & 0 & I_3 \end{pmatrix}, \tag{10.28}$$

$$b_1 = (0 \quad 1 \quad 0)^T, \qquad b_2 = (0 \quad 0 \quad 1)^T. \tag{10.82}$$

I_1, I_2 and I_3 denote the principal moments of inertia. We will assume $I_3 > I_2 > I_1 > 0$. The system (10.81) can be rewritten as

$$\begin{cases} \dot{\omega}_1 = I_{23}\omega_2\omega_3 \\ \dot{\omega}_2 = I_{31}\omega_3\omega_1 + c_2 u_1 \\ \dot{\omega}_3 = I_{12}\omega_1\omega_2 + c_3 u_2 \end{cases} \tag{10.83}$$

where

$$\begin{cases} I_{23} = (I_2 - I_3)/I_1 \,, \\ I_{31} = (I_3 - I_1)/I_2 \,, \\ I_{12} = (I_1 - I_2)/I_3 \,, \end{cases} \tag{10.33}$$

and

$$c_2 = I_2^{-1}, \qquad c_3 = I_3^{-1}. \tag{10.84}$$

Consider the feedback law

$$\begin{cases} u_1 = \dfrac{1}{c_2}(-\omega_2 + p_1\omega_1^2 + p_2\omega_1^3) \\ u_2 = \dfrac{1}{c_3}(-\omega_3 + q_1\omega_1^2 + q_2\omega_1^3) \end{cases} \tag{10.85}$$

which yields, as in (10.49), the equations

$$\begin{cases} \dot{\omega}_1 = I_{23}\omega_2\omega_3 \\ \dot{\omega}_2 = -\omega_2 + I_{31}\omega_3\omega_1 + p_1\omega_1^2 + p_2\omega_1^3 \\ \dot{\omega}_3 = -\omega_3 + I_{12}\omega_1\omega_2 + q_1\omega_1^2 + \omega_2\omega_1^3. \end{cases} \tag{10.86}$$

From Theorem 10.11 we deduce that there exists a center manifold described by

$$\begin{cases} \omega_2 = \varphi_1(\omega_1) \\ \omega_3 = \varphi_2(\omega_1) \end{cases} \tag{10.87}$$

with

$$\varphi_1(0) = \varphi_1'(0) = \varphi_2(0) = \varphi_2'(0) = 0. \tag{10.88}$$

We approximate (see Theorem 10.14) the center manifold by $\tilde{\varphi}_1(\omega_1) = \alpha_1\omega_1^2 + \alpha_2\omega_1^3$ and $\tilde{\varphi}_2(\omega_1) = \beta_1\omega_1^2 + \beta_2\omega_1^3$, resulting in the equations

$$\begin{aligned} &(2\alpha_1\omega_1 + 3\alpha_2\omega_1^2)I_{23}(\alpha_1\omega_1^2 + \alpha_2\omega_1^3)(\beta_1\omega_1^2 + \beta_2\omega_1^3) + \alpha_1\omega_1^2 + \alpha_2\omega_1^3 \\ &- I_{31}(\beta_1\omega_1^2 + \beta_2\omega_1^3)\omega_1 - p_1\omega_1^2 - p_2\omega_1^3 = O(\omega_1^4), \end{aligned} \tag{10.89a}$$

$$\begin{aligned} &(2\beta_1\omega_1 + 3\beta_2\omega_1^2)I_{23}(\alpha_1\omega_1^2 + \alpha_2\omega_1^3)(\beta_1\omega_1^2 + \beta_2\omega_1^3) + \beta_1\omega_1^2 + \beta_2\omega_1^3 \\ &- I_{12}\omega_1(\alpha_1\omega_1^2 + \alpha_2\omega_1^3) - q_1\omega_1^2 - q_2\omega_1^3 = O(\omega_1^4). \end{aligned} \tag{10.89b}$$

From (10.89a,b) we obtain

$$\begin{cases} \alpha_1 - p_1 = 0, \\ \alpha_2 - I_{31} - p_2 = 0, \\ \beta_1 - q_1 = 0, \\ \beta_2 - I_{12}\alpha_1 - q_2 = 0. \end{cases} \tag{10.90}$$

So the center manifold (10.87) is approximated by

$$\begin{cases} \tilde{\varphi}_1(\omega_1) = p_1\omega_1^2 + (I_{31}q_1 + p_2)\omega_1^3, \\ \tilde{\varphi}_2(\omega_1) = q_1\omega_1^2 + (I_{12}p_1 + q_2)\omega_1^3. \end{cases} \tag{10.91}$$

Substituting this in the (approximated) dynamics on the center manifold yields

$$\dot{\omega}_1 = I_{23}\big(p_1\omega_1^2 + (I_{31}q_1+p_2)\omega_1^3\big)\big(q_1\omega_1^2 + (I_{12}p_1+q_2)\omega_1^3\big), \tag{10.92}$$

or,

$$\dot{\omega}_1 = I_{23}p_1q_1\omega_1^4 + \big[I_{23}p_1(I_{12}p_1+q_2) + I_{23}q_1(I_{31}q_1+p_2)\big]\omega_1^5 + O(\omega_1^6). \tag{10.93}$$

In order that $\omega_1 = 0$ is a locally asymptotically stable equilibrium of (10.93) we need to have $I_{23}p_1q_1 = 0$, so

$$p_1 = 0, \tag{10.94a}$$

or

$$q_1 = 0. \tag{10.94b}$$

The system (10.93), and therefore the system (10.86), is then locally asymptotically stable if either

$$I_{23}q_1(I_{31}q_1 + p_2) < 0, \tag{10.95a}$$

or

$$I_{23}p_1(I_{12}p_1 + q_2) < 0. \tag{10.95b}$$

□

In the feedback stabilization problem we have concentrated on the local existence of *smooth* stabilizing state feedbacks. This smoothness assumption fits naturally into the context of this chapter. However, one may relax this assumption and allow for instance k-times continuously differentiable feedback functions. Clearly, when α is only continuous but not differentiable, there is an extra problem, since the solutions of $\dot{x} = f(x,\alpha(x))$ need not be uniquely defined for positive time. Moreover, when α is only continuous one can no longer use a result as Theorem 10.5 for testing the stabilizability via this non-smooth feedback and, in fact, the requirement that the linearization of the system should not possess an unstable uncontrollable mode in order to be stabilizable, no longer need to be true for the existence of a C^0 stabilizing state feedback. As an example one can show that for the system $\dot{x}_1 = u$, $\dot{x}_2 = x_2 - x_1^3$, the feedback $u = -x_1 + Ex_2^{1/3} + K(x_2-x_1^3)$ is a stabilizing feedback for certain E and K, although the system is not (locally) stabilizable by a C^1 feedback. We will not pursue the non-smooth stabilization problem here, but instead refer to the relevant literature cited in the references.

Notes and References

The stability theory for autonomous differential equations has a long standing history and is today still far away from its completion. From the many textbooks on the basic results on stability we mention [LL,Ha,HS]. The first and second method of Lyapunov were originally described in [Ly]. Theorem 10.3 on the local existence of a Lyapunov-function for a stable equilibrium can be found in [Mas,Mal,Ha,Wils], see also [Br]. The feedback stabilization problem for nonlinear control system is widely studied in the control literature. Theorem 10.5 can already be found in [LM]. The feedback stabilization problem using a Lyapunov-function as in Lemma 10.6 - Theorem 10.9 - is studied in [JQ,Sl,KT,LA]; we have more or less followed the survey paper [Ba]. Example 10.10 is borrowed from [AS] where in a slightly different way the same result is obtained. A standard reference on center manifold theory is [Ca]. Center manifold theory as a tool in the (smooth) feedback stabilization problem was first studied by Aeyels, see [Ae1,Ae2, Ae3] and [AS]. A survey of this approach is given in [Ba]. The application of center manifold theory for a two dimensional control system follows that of [Ba]. Example 10.16 is essentially due to [AS]. The non-smooth feedback stabilization problem for a two dimensional control system was studied in [Ka]; the example given at the end of this chapter has been taken from this reference. For an approach to feedback stabilization based on the notion of zero dynamics we refer to Chapter 11. A recent survey about the feedback stabilization problem has been given in [So].

[Ae1] D. Aeyels, "Stabilisation of a class of nonlinear systems by a smooth feedback control", Systems Control Lett. 5, pp. 289-294, 1985.

[Ae2] D. Aeyels, "Stabilisation by smooth feedback of the angular velocity of a rigid body", Systems Control Lett. 6, pp. 59-64, 1985.

[Ae3] D. Aeyels, "Local and global stabilizability for nonlinear systems", in **Theory and applications of nonlinear control systems** (eds. C.I. Byrnes, A. Lindquist), North-Holland, Amsterdam, pp. 93-105, 1986.

[AS] D. Aeyels, M. Szafranski, "Comments on the stabilizability of the angular velocity of a rigid body", Systems Control Lett. 10, pp. 35-40, 1988.

[Ba] A. Bacciotti, "The local stabilizability problem for nonlinear systems", IMA J. Math. Contr. Inform. 5, pp. 27-39, 1988.

[Br] R.W. Brockett, "Asymptotic stability and feedback stabilization", in **Differential geometric control theory** (eds. R.W. Brockett, R.S. Millmann, H.J. Sussmann), Birkhauser, Boston, pp. 181-191, 1983.

[Ca] J. Carr, **Applications of centre manifold theory**, Springer, New York, 1981.

[Ha] W. Hahn, **Stability of motion**, Springer, New York, 1967.

[HS] M.W. Hirsch, S. Smale, **Differential equations, dynamical systems and linear algebra**, Academic Press, New York, 1974.

[JQ] V. Jurdjevic, J.P. Quinn, "Controllability and stability", J. Diff. Equat. 28, pp. 381-389, 1978.

[KT] N. Kalouptsidis, J. Tsinias, "Stability improvement of nonlinear systems by feedback", IEEE Trans. Aut. Contr. AC-29, pp. 364-367, 1984.

[Ka] M. Kawski, "Stabilization of nonlinear systems in the plane", Systems Control Lett. 12, pp. 169-175, 1989.

[LA] K.K. Lee, A. Arapostathis, "Remarks on smooth feedback stabilization of nonlinear systems", Systems Control Lett. 10, pp. 41-44, 1988.

[LL] J. LaSalle, S. Lefschetz, **Stability by Lyapunov's direct method with applications**, Academic Press, New York, 1961.
[LM] E.B. Lee, L. Markus, **Foundations of optimal control theory**, John Wiley, New York, 1967.
[Ly] M.A. Lyapunov, "Problème general de la stabilité du mouvement", reprinted in **Annals of Mathematical Studies**, 17, Princeton University Press, Princeton, 1949.
[Mal] I.G. Malkin, "On the question of reversibility of Lyapunov's theorem on asymptotic stability", Prikl. Mat. Meh. 18, pp. 129-138, 1954.
[Mas] I.L. Massera, "Contributions to stability theory", Ann. Math. 64, pp. 182-206, 1956. Erratum in Ann. Math. 68, p. 202, 1958.
[Sl] M. Slemrod, "Stabilization of bilinear control systems with applications to nonconservative problems in elasticity", SIAM J. Contr. Optimiz., 16, pp. 131-141, 1978.
[So] E.D. Sontag, "Feedback stabilization of nonlinear systems", Proceedings MTNS-89, Amsterdam, Birkhäuser, Boston, to appear.
[Will] J.L. Willems, **Stability theory of dynamical systems**, Nelson, London, 1970.
[Wils] F.W. Wilson Jr., "The structure of the level surfaces of a Lyapunov functions", J. Diff. Equat., 3, pp. 323-329, 1967.

Exercises

10.1 Consider on $\mathbb{R}^n$ the smooth system Σ: $\dot{x} = f(x) + g(x)u$ with $f(0) = 0$. Suppose $x = 0$ is locally stable for $\dot{x} = f(x)$ and let V be a Lyapunov-function for $\dot{x} = f(x)$. Define the distribution D via $D(x) = \text{span}\{f(x), ad_f^k g(x),\ k \geq 0\}$. Suppose there exists a neighborhood W of 0 such that $L_f^k L_X V(x) = 0$ for all $X \in D$ and $k \geq 0$ implies that $x = 0$. Prove that the feedback $u = -L_g V(x)$ locally asymptotically stabilizes the origin (see [LA]).

10.2 Consider a smooth single-input single-output system on $\mathbb{R}^n$ of the form $\dot{x} = f(x) + g(x)u$, $y = h(x) = x_n$. Assume that for all $x \in \mathbb{R}^n$ $L_g h(x) \neq 0$. Let $K = \{x \in \mathbb{R}^n \mid h(x) = 0\}$.

(a) Show that there exists a smooth function α: $\mathbb{R}^n \to \mathbb{R}$ such that K is an invariant set for the dynamics $\dot{x} = f(x) + g(x)\alpha(x)$ and show that $\alpha|_K$ is uniquely determined.

(b) Prove that any point $x_0 \in \mathbb{R}^n$ can be steered into K in finite time by a suitably chosen input u.

(c) Suppose $\|L_f h(x)\| \leq c_0 |x_n|$ and $\|L_g h(x)\| \geq \epsilon > 0$. Define $B_K^r = \{x \in \mathbb{R}^n \mid \text{distance}(x,K) \leq r\}$. Prove that there exists a constant $c \neq 0$ such that all points $x_0 \in B_K^r \backslash K$ are steered by either $u = +c$ or $u = -c$ in finite time into K.

(d) Assume $f(0) = 0$ and let Σ_ℓ: $\dot{\bar{x}} = A\bar{x} + b\bar{u}$, $\bar{y} = c\bar{x}$ be the linearization of Σ about $x = 0$. Let $h(s) = c(sI-A)^{-1}b$ be the transfer function of Σ_ℓ with $h(s) = q(s)/p(s)$, $q(s) = q_{n-1}s^{n-1} + \ldots$ $\ldots + q_0$, $p(s) = s^n + p_{n-1}s^{n-1} + \ldots + p_0$ and assume $p(s)$ and $q(s)$

have no common factors. Show that $q_{n-1} = cb \neq 0$ and show that the linearization of the system found under **(a)** is given as $\dot{\bar{x}} = \left(A-(cb)^{-1}bcA\right)\bar{x}$ restricted to the subspace ker c. Prove that the characteristic polynomial of this matrix equals $q(s)$.

(e) Show that when all zeros of $q(s)$ lie in the open left half plane, then there exists a neighborhood N of $x = 0$ such that (i) $y(t) = 0$ for t sufficiently large, and (ii) $x(t,0,x_0,u) \to 0$ for $t \to \infty$, $x_0 \in N$, where the control u is defined to be $\pm c$ outside $K \cap N$ and $u = \alpha(x)$ in $K \cap N$.

10.3 Investigate the center manifold approach for the system (10.66) in case that in the Taylor-series expansion (10.71a) condition (10.80) is not fullfilled.

10.4 Show that in Example 10.10 the assumption (10.44) is essentially needed for the asymptotic stability of the closed loop dynamics (10.34,37).

10.5 Consider a nonlinear system $\dot{x} = f(x,u)$ about an equilibrium point (x_0,u_0). Suppose the local feedback stabilization problem is solvable for this system. Prove that the local feedback stabilization problem is solvable for the extended system $\dot{x} = f(x,u)$, $\dot{u} = w$.

10.6 ([Br]) Consider on $\mathbb{R}^n$ the smooth system $\dot{x} = f(x,u)$ and let $(x,u) = (0,0)$ be an equilibrium point. Prove that a necessary condition for the solvability of the local feedback stabilization problem is that the mapping $(x,u) \mapsto f(x,u)$ is onto on an open set containing $x = 0$ for (x,u) belonging to a neighborhood of $(0,0)$.

10.7 ([Ael]) Show that in Example 10.16 the feedback $u_1 = -c_2^{-1}I_{31}\omega_3\omega_1 - c_2^{-1}\omega_2 + c_2^{-1}\omega_1$, $u_2 = -c_3^{-1}I_{12}\omega_1\omega_2 - c_3^{-1}\omega_3 - c_3^{-1}\omega_1^2$ locally asymptotically stabilizes the origin.

10.8 ([Ael]) Show that for the system $\dot{x}_1 = x_1x_2$, $\dot{x}_2 = u$ there does not exist a linear feedback $u = a_1x_1 + a_2x_2$ rendering the origin asymptotically stable, but the local feedback stabilization problem is solvable via a quadratic feedback function.

10.9 Consider the bilinear system $\dot{x} = Ax + (Bx)u$ on $\mathbb{R}^2$, with $A = \begin{pmatrix} 0 & 2 \\ 2 & 0 \end{pmatrix}$ and $B = \begin{pmatrix} 3 & 5 \\ 5 & 3 \end{pmatrix}$. Show that the matrices A and B can be diagonalized simultaneously. Determine all possible constant feedbacks $u = c$ such that the closed loop system $\dot{x} = (A+Bc)x$ is asymptotically stable.

10.10 Consider on $\mathbb{R}$ the system $\dot{x} = x + u^k$, $k \in \mathbb{N}$. Determine for which values of k there exist a continuous feedback $u = \alpha(x)$, with $\alpha(0) = 0$, such that the closed loop system is asymptotically stable.

11 Controlled Invariant Submanifolds and Nonlinear Zero Dynamics

In Chapter 3.3 we have seen that the notion of an A-invariant subspace $\mathcal{V} \subset \mathbb{R}^n$ for a linear set of differential equations $\dot{x} = Ax$, $x \in \mathbb{R}^n$, can be conveniently generalized to nonlinear differential equations $\dot{x} = f(x)$, $x \in M$, by introducing the notion of an *invariant foliation* or *invariant* (constant dimensional and involutive) *distribution*. Subsequently in Chapter 7 (and also in Chapter 9) it has been shown that an appropriate nonlinear generalization of the concept of a *controlled invariant subspace*, at least for applications such as disturbance decoupling and input-output decoupling, is that of a *controlled invariant distribution*. In the present chapter we will show that the concept of a (controlled) invariant subspace also allows for a *different* nonlinear generalization, namely that of a (controlled) invariant *submanifold*. Furthermore, we will show that this second generalization is the appropriate tool for dealing with problems such as interconnection and inversion of nonlinear systems, and for defining the nonlinear analog of the concept of *transmission zeros* of a linear system.

11.1 Locally Controlled Invariant Submanifolds

Consider a linear set of differential equations

$$\dot{x} = Ax, \quad x \in \mathbb{R}^n. \tag{11.1}$$

With any linear subspace $\mathcal{V} \subset \mathbb{R}^n$ we can associate the foliation $F_{\mathcal{V}} = \{x + \mathcal{V} \mid x \in \mathbb{R}^n\}$ of $\mathbb{R}^n$. If $\mathcal{V}$ is A-invariant, i.e. $A\mathcal{V} \subset \mathcal{V}$, then this implies that the foliation $F_{\mathcal{V}}$ is invariant for (11.1). On the other hand $A\mathcal{V} \subset \mathcal{V}$ is also equivalent to the requirement that the solutions of (11.1) for $x(0) \in \mathcal{V}$ *remain* in $\mathcal{V}$ for all $t \geq 0$. While the first interpretation of A-invariance gives rise to the nonlinear generalization of an invariant foliation or invariant distribution, the second interpretation leads to the notion of an *invariant submanifold*. Indeed consider a vectorfield on M, locally represented as

$$\dot{x} = f(x). \tag{11.2}$$

A submanifold $N \subset M$ is called *invariant* for (11.2) if

$$f(x) \in T_x N, \qquad \text{for all } x \in N. \tag{11.3}$$

If N is connected then this immediately implies that the solutions of (11.2) for $x(0)$ in N remain in N for all $t \geq 0$. (In the preceding chapter we already encountered the more general notion of an invariant *subset* of M for (11.2), cf. Theorem 10.2.)

Now let us consider the smooth nonlinear dynamics

$$\dot{x} = f(x) + \sum_{j=1}^{m} g_j(x)u_j, \qquad u = (u_1, \ldots, u_m) \in \mathbb{R}^m, \tag{11.4}$$

where $x = (x_1, \ldots, x_n)$ are local coordinates for some n-dimensional manifold M.

Definition 11.1 *A submanifold $N \subset M$ is (locally) controlled invariant for (11.4) if there exists (locally on N) a strict static state feedback $u = \alpha(x)$, $x \in N$, such that*

$$f(x) + \sum_{j=1}^{m} g_j(x)\alpha_j(x) \in T_x N, \qquad \textit{for all } x \in N, \tag{11.5}$$

i.e., N is invariant for $\dot{x} = f(x) + \sum_{j=1}^{m} g_j(x)\alpha_j(x)$.

We immediately obtain (compare with Theorem 7.5)

Proposition 11.2 *Consider (11.4) and a submanifold $N \subset M$. Denote*

$$G(x) = \operatorname{span}\{g_1(x), \ldots, g_m(x)\}, \qquad x \in M, \tag{11.6}$$

and assume that $\dim(T_x N + G(x))$ is constant for every $x \in N$. Then N is locally controlled invariant for (11.4) if and only if

$$f(x) \in T_x N + G(x), \qquad \textit{for every } x \in N. \tag{11.7}$$

Proof The "only if" direction is trivial. Suppose (11.7) holds. Write $g(x) = \big(g_1(x) \vdots \ldots \vdots g_m(x)\big)$. Locally we may choose coordinates $x = (x^1, x^2)$ such that $N = \{x \mid x^1 = 0\}$. Write accordingly

$$f(x) = \begin{pmatrix} f^1(x^1, x^2) \\ f^2(x^1, x^2) \end{pmatrix}, \quad g(x) = \begin{pmatrix} g^1(x^1, x^2) \\ g^2(x^1, x^2) \end{pmatrix}. \tag{11.8}$$

Then by assumption $g^1(0, x^2)$ has constant rank, while (11.7) is equivalent to $f^1(0, x^2) \in \operatorname{Im} g^1(0, x^2)$. It follows (see Exercise 2.4) that we can locally find an m-vector $\alpha(0, x^2)$, depending smoothly on x^2, such that $g^1(0, x^2)\alpha(0, x^2) + f^1(0, x^2) = 0$. □

Remark If the assumption of constant dimensionality of $T_xN + G(x)$, $x \in N$, is not satisfied, then Proposition 11.2 is not valid anymore, as shown by the following example

$$\begin{aligned} \dot{x}_1 &= x_2 - x_2^2 u, \\ \dot{x}_2 &= k(x_1,x_2), \end{aligned} \qquad (x_1,x_2) \in \mathbb{R}^2. \tag{11.9}$$

Let $N = \{x | x_1 = 0\}$. It is easily seen that (11.7) is satisfied. However there does not exist a smooth feedback $u = \alpha(x)$, defined on a neighborhood of any point $(x_1^o,0)$, which renders N invariant. (Indeed the feedback suggested by (11.9) is $u = 1/x_2$, for $x_2 \neq 0$, which cannot be extended to a smooth feedback around $x_2 = 0$.)

Now let us consider (11.4) together with output equations, i.e.

$$\begin{aligned} \dot{x} &= f(x) + \sum_{j=1}^{m} g_j(x)u_j, & u &= (u_1,\ldots,u_m) \in \mathbb{R}^m, \\ y &= h(x), & y &\in \mathbb{R}^p. \end{aligned} \tag{11.10}$$

A submanifold $N \subset M$ is called *output-nulling* if $N \subset h^{-1}(0)$, i.e., if the output value corresponding to states in N is zero. Recall that in Chapter 7 algorithms have been given to compute, under constant rank assumptions, the maximal locally controlled invariant distribution contained in the distribution ker dh (cf.(7.53), (7.58) and Algorithm 7.19). Similarly, we now want to compute the *maximal locally controlled invariant output-nulling submanifold* for the system (11.10).

Algorithm 11.3 (Constrained dynamics algorithm) Consider the system (11.10), and suppose $h(x_0) = 0$. Denote $G(x)$ as in (11.6). Let $O(x_0)$ be a neighborhood of x_0.

Step 0 Define $N_1 = h^{-1}(0) \cap O(x_0)$.

Step k > 0 Assume that N_k is a submanifold through x_0. Then define

$$N_{k+1} = \{x \in N_k \,|\, f(x) \in T_xN_k + G(x)\}.$$

If we can find $O(x_0)$ such that at every step N_k, $k \geq 0$, is a submanifold through x_0, then x_0 is called a *regular point* for the algorithm.

Let x_0 be a regular point for Algorithm 11.3. Then we obtain a descending sequence of submanifolds

$$N_1 \supset N_2 \supset \ldots \supset N_k \supset N_{k+1} \supset \ldots \tag{11.11}$$

Since $\dim N_{k+1} \le \dim N_k$, $k = 1,2,\ldots$, it follows that there exists a $k^* \le n$ such that $N_{k^*+j} = N_{k^*}$, $j = 1,2,\ldots$. Denote the maximal connected component of N_{k^*} containing x_0 by N^*.

Proposition 11.4 *Suppose x_0 is a regular point for Algorithm 11.3. Then N^* satisfies (11.7). Furthermore for any output-nulling submanifold N satisfying (11.7) there exists some neighborhood $O(x_0)$ of x_0 such that $N \cap O(x_0) \subset N^*$.*

Proof Since on a neighborhood of x_0, $N^* = N_{k^*} = N_{k^*+1}$, it immediately follows from the definition of N_{k^*+1} that N^* satisfies (11.7). Let $N \subset h^{-1}(0)$ satisfy (11.7). By induction to k it follows that $N \cap O_k(x_0) \subset N_k$ for suitable neighborhoods $O_k(x_0)$ for all k. □

Thus N^* is the *maximal* output-nulling submanifold through x_0 with respect to property (11.7). If additionally $\dim (T_x N^* + G(x))$ is constant for $x \in N^*$ then it immediately follows from Proposition 11.2 that N^* is the *maximal locally controlled invariant output-nulling* submanifold around x_0.

Remark 11.5 For a *linear* system $\dot{x} = Ax + Bu$, $y = Cx$, Algorithm 11.3 simply reduces to the algorithm (7.70), since the definition of $V^{\mu+1}$ can be rewritten as

$$V^{\mu+1} = \{x \in V^\mu \mid Ax \in V^\mu + \operatorname{span}\{b_1,\ldots,b_m\}\}$$

We will now give a more constructive version of Algorithm 11.3, which actually is very much related to Algorithm 7.19.

Algorithm 11.6 (Constrained dynamics algorithm) Consider the system (11.10) and suppose that $h(x_0) = 0$ and $f(x_0) = 0$.

Step 0 *Assume* that $h = (h_1,\ldots,h_p)$ has constant rank s_1 in a neighborhood of x_0 in $h^{-1}(0)$. Then locally around x_0 the set $N_1 = h^{-1}(0)$ is an $(n-p_1)$-dimensional submanifold, where $p_1 := s_1$. Permute the outputs in such a way that $h_1,\ldots,h_{p_1}$ are independent around x_0.

Step 1 Define the $p_1 \times m$ matrix $A_1(x)$ and the $p_1 \times 1$ vector $B_1(x)$ as

$$\begin{aligned} A_1(x) &= \left(L_{g_i} h_j(x)\right)_{\substack{j=1,\ldots,p_1 \\ i=1,\ldots,m}} \\ B_1(x) &= \left(L_f h_j(x)\right)_{j=1,\ldots,p_1} \end{aligned} \tag{11.12}$$

Assume that $A_1(x)$ has constant rank r_1 in a neighborhood of x_0 in N_1. After a possible permutation of the output functions we may assume that the first r_1 rows of A_1 are linearly independent. Then by the implicit function theorem (see Exercise 2.4) there exists on a neighborhood of x_0 in N_1 a feedback $u = \alpha_1(x)$, with $\alpha_1(x_0) = 0$, such that

$$A_1(x)\alpha_1(x) + B_1(x) = \begin{bmatrix} 0 \\ \varphi_1(x) \end{bmatrix} \updownarrow r_1 \quad , \ x \in N_1 . \tag{11.13}$$

Assume that φ_1 has constant rank s_2 on a neighborhood of x_0 in N_1. Then locally around x_0, $N_2 = \{x \in N_1 | \varphi_1(x) = 0\}$ is an $(n-p_2)$-dimensional submanifold, with $p_2 := p_1 + s_2$. (Notice that because $h(x_0) = 0$ and $f(x_0) = 0$ we have $\varphi_1(x_0) = 0$.) Permute the entries of φ_1 such that the first s_2 entries are independent on N_1, and denote them as $h_{p_1+1}, \ldots, h_{p_2}$.

Step k > 0 Let N_k be a smooth $(n-p_k)$-dimensional submanifold through x_0, given as $\{x \in N_{k-1} | h_{p_{k-1}+1}(x) = \ldots = h_{p_k}(x) = 0\}$. Define the $p_k \times m$ matrix $A_k(x)$ and the $p_k \times 1$ vector $B_k(x)$ as

$$\begin{aligned} A_k(x) &= \left(L_{g_i} h_j(x)\right)_{\substack{j=1,\ldots,p_k \\ i=1,\ldots,m}} \\ B_k(x) &= \left(L_f h_j(x)\right)_{j=1,\ldots,p_k} \end{aligned} \quad , \quad x \in N_{k-1} . \tag{11.14}$$

Assume that $A_k(x)$ has constant rank r_k in a neighborhood of x_0 in N_k. After a possible permutation of the functions h_j we may assume that the first r_k rows of A_k are linearly independent. Therefore there exists on a neighborhood of x_0 in N_k a feedback $u = \alpha_k(x)$, with $\alpha_k(x_0) = 0$, such that

$$A_k(x)\alpha(x) + B_k(x) = \begin{bmatrix} 0 \\ \varphi_k(x) \end{bmatrix} \updownarrow r_k \quad , \qquad x \in N_k . \tag{11.15}$$

Assume that φ_k has constant rank s_{k+1} on a neighborhood of x_0 in N_k. Then locally around x_0, $N_{k+1} := \{x \in N_k | \varphi_k(x) = 0\}$ is an $(n-p_{k+1})$-dimensional submanifold, where $p_{k+1} := p_k + s_{k+1}$. (Notice that $\varphi_k(x_0) = 0$, since $f(x_0) = 0$ and $\alpha(x_0) = 0$.) Take s_{k+1} entries of φ_k which are independent on N_k and denote them as $h_{p_k+1}, \ldots, h_{p_{k+1}}$.

If at every step of the algorithm the two constant rank assumptions are satisfied then we call x_0 a *regular point* for the algorithm.

Remark 11.7 As already indicated in Algorithm 11.3 it is not necessary to assume that $f(x_0) = 0$. In fact if $f(x_0) \neq 0$, then at the k-th step there may still exist a feedback $u = \alpha_k(x)$ satisfying (11.15) such that $\varphi_k(x_0) = 0$. (However in general $\alpha_k(x_0)$ can not be taken equal to zero in this case.) If this holds for every $k = 1,\ldots,k^*$, and the constant rank assumptions of Algorithm 11.6 are satified then we will still call x_0 a *regular point* for Algorithm 11.6.

Let us now check that the submanifolds N_k as produced by Algorithm 11.6 coincide with those as defined in Algorithm 11.3. Suppose x_0 is a regular point for Algorithm 11.6. Clearly N_1 for Algorithms 11.3 and 11.6 are the same around x_0. Now assume that N_j of Algorithm 11.3 coincides with N_j of Algorithm 11.6 for $j \leq k$. Let $x = (x^1,x^2)$ be local coordinates around x_0 such that $N_k = \{x \mid x^1 = 0\}$. Indeed, let us take $x^1 = (h_1,\ldots,h_{p_k})$. (Strictly speaking the functions $h_{p_j+1},\ldots,h_{p_{j+1}}$ are only defined on N_j, $j \leq k$; however they can be easily extended to independent functions on a whole neighborhood of x_0 in M.) Partition accordingly f and $g = (g_1 \vdots \ldots \vdots g_m)$ as

$$f(x) = \begin{bmatrix} f^1(x^1,x^2) \\ f^2(x^1,x^2) \end{bmatrix}, \quad g(x) = \begin{bmatrix} g^1(x^1,x^2) \\ g^2(x^1,x^2) \end{bmatrix} \tag{11.16}$$

Notice that in these coordinates $A_k = g^1$, while $B_k = f^1$. Therefore (11.15) amounts to

$$f^1(0,x^2) + g^1(0,x^2)\alpha_k(x^2) = \begin{bmatrix} 0 \\ \varphi_k(x^2) \end{bmatrix} \tag{11.17}$$

Now, $f(0,x^2) \in T_xN_k + G(x)$ if and only if $f^1(0,x^2) \in \operatorname{Im} g^1(0,x^2)$. In view of (11.17) this last inclusion holds if and only if $\begin{bmatrix} 0 \\ \varphi_k(x^2) \end{bmatrix} \in \operatorname{Im} g^1(0,x^2)$, and since the first r_k rows of $g^1(0,x^2)$ are independent this is true if and only if $\varphi_k(x^2) = 0$, thereby proving that N_{k+1} of Algorithm 11.3 coincides with N_{k+1} of Algorithm 11.6.

In comparison with Algorithm 11.3, in the k-th step of Algorithm 11.6 the additional assumption is made that the rank of A_k is constant on a neighborhood of x_0 in N_k, thus enabling the construction of the feedbacks $\alpha_k(x)$, $k = 1,\ldots,k^*$, solving (11.15). In particular, at the k^*-th step we obtain by definition of k^* that $\varphi_{k^*} = 0$ on $N_{k^*} =: N^*$ (or equivalently $s_{k^*+1} = 0$), and thus $\alpha_{k^*}(x) =: \alpha^*(x)$ satisfies (see (11.15))

$$A^*(x)\alpha^*(x) + B^*(x) = 0, \quad x \in N^*, \tag{11.18}$$

with $A^* := A_{k^*}$, $B^* := B_{k^*}$, rank $A^* = r^* := r_{k^*}$. Hence $u = \alpha^*(x)$ is a feedback which renders N^* invariant. Thus, if x_0 is a regular point for Algorithm

11.6, then N^* is automatically *locally controlled invariant* (compare with Proposition 11.4 and Proposition 11.2). Finally, because the submanifolds N_k as produced by Algorithm 11.6 equal the *intrinsically* defined submanifolds N_k of Algorithm 11.3, it follows that the N_k of Algorithm 11.6 do *not* depend on the particular choice of α_k satisfying (11.15) or the selection of the independent entries of φ_k.

Remark Notice that Algorithm 11.6 is very close to Algorithm 7.19, used for computing the maximal locally controlled invariant *distribution* contained in ker dh. The main difference is that in Algorithm 11.6 we do not have to compute the matrices $\beta_k(x)$, nor $\psi_k(x)$ (see also Exercises 11.1,2). Additionally, the constant rank assumptions in Algorithm 11.6 need only be made on neighborhoods of x_0 *contained* in the submanifolds N_k.

The indices s_k and r_k defined in Algorithm 11.6 are related as follows.

Lemma 11.8 *Let* x_0 *be a regular point for Algorithm 11.6, yielding indices* s_k, r_k, $k = 1,\dots,k^*$. *Then*

$$s_1 \le p, \tag{11.19a}$$

$$s_2 \le s_1 - r_1, \tag{11.19b}$$

$$s_{k+1} \le s_k - (r_k - r_{k-1}), \quad k = 2,\dots,k^* - 1. \tag{11.19c}$$

Proof Clearly $s_1 \le p$. Since φ_1 in (11.13) is an $(s_1 - r_1)$-vector, (11.19b) immediately follows. Now consider (11.15) for k and k-1, i.e.

$$A_k(x)\alpha_k(x) + B_k(x) = \begin{bmatrix} 0 \\ \varphi_k(x) \end{bmatrix}, \tag{11.20}$$

$$A_{k-1}(x)\alpha_{k-1}(x) + B_{k-1}(x) = \begin{bmatrix} 0 \\ \varphi_{k-1}(x) \end{bmatrix}. \tag{11.21}$$

First, the right-hand side of (11.20) has s_k more entries than the right-hand side of (11.21). Since rank $A_k(x)$-rank $A_{k-1}(x) = r_{k-1} - r_{k-2}$ it follows that the right-hand side of (11.20) contains $r_{k-1} - r_{k-2}$ more zero entries than the right-hand side of (11.21). Furthermore we can choose α_k and α_{k-1} satisfying (11.20), resp. (11.21), in such a way that all the entries of φ_{k-1} also appear as entries of φ_k (since the rows of A_{k-1}, resp. B_{k-1}, are also rows of A_k, resp. B_k). It follows that $\varphi_k(x)$ consists of the entries of $\varphi_{k-1}(x)$ together with $s_k - (r_k - r_{k-1})$ additional entries. Only these additional entries can contribute to the rank of φ_k on N_k, and thus (11.19c) follows. □

By rewriting (11.19c) as $s_{k+1}+r_k \leq s_k+r_{k-1}$ and recalling that $s_{k^*+1} = 0$ and $r_{k^*} =: r^*$, we immediately obtain

Corollary 11.9 *The indices s_k, r_k satisfy*

$$r^* \leq s_{k^*} + r_{k^*-1} \leq \ldots \leq s_{k+1} + r_k \leq s_k + r_{k-1} \leq \ldots \leq s_1 \leq p. \quad (11.22)$$

In particular if $p = m$ and $r^* = m$, then (11.22), or equivalently (11.19), holds with equality.

Remark 11.10 As noticed in the proof of Lemma 11.8 we can choose $\alpha_1, \ldots, \alpha_{k^*}$ in such a way that all the entries of φ_{k-1} also appear as entries of φ_k, $k = 2,\ldots,k^*$. It follows that if $p = m = r^*$, then the entries of φ_k *not* appearing in φ_{k-1} are independent on N_k.

Remark 11.11 The condition $r^* = m$ is equivalent to the existence of a *unique* solution $\alpha^*(x)$ of (11.18).

Remark 11.12 The k-th step of Algorithm 11.6 can be recast into the following different but equivalent form. Instead of constructing a feedback $u = \alpha_k(x)$ on N_k satisfying (11.15) we can construct on a neighborhood of x_0 in N_k a $(p_k-r_k) \times p_k$ matrix $R_k(x)$ of full row rank satisfying (see Exercise 2.4)

$$R_k(x)A_k(x) = 0, \qquad x \in N_k. \quad (11.23)$$

Then locally around x_0, N_{k+1} is alternatively given as

$$N_{k+1} = \{x \in N_k \mid R_k(x)B_k(x) = 0\} \quad (11.24)$$

Indeed, since the first r_k rows of A_k are independent, it follows that R_k is of the form $R_k = (* \vdots T_k)$, with T_k an invertible $(p_k-r_k) \times (p_k-r_k)$ matrix (In fact we can choose R_k such that T_k is the identity matrix.) Premultiplying (11.15) by $R_k(x)$ yields

$$R_k(x)\, B_k(x) = T_k(x)\varphi_k(x) \quad (11.25)$$

and thus $R_k(x)B_k(x) = 0$ for $x \in N_k$ if and only if $\varphi_k(x) = 0$.

Finally let us give the following particular case, where the computation of N^* and α^* becomes especially easy.

Proposition 11.13 *Consider the system (11.10). Assume that the* *characteristic numbers* $\rho_1,\ldots,\rho_p$ *as defined in Definition 8.7 are all finite. Furthermore assume that the* $p \times m$ *matrix (the decoupling matrix)*

$$A(x) = \left(L_{g_j} L_f^{\rho_i} h_i(x)\right)_{\substack{i=1,\ldots,p \\ j=1,\ldots,m}} \tag{11.26}$$

has rank equal to p *on the set*

$$N^* = \{x \mid h_i(x) = \ldots = L_f^{\rho_i} h_i(x) = 0,\ i \in \underline{p}\} \tag{11.27}$$

If N^* *is non-empty (for instance if there exists* x_0 *with* $h(x_0) = 0$ *and* $f(x_0) = 0$*), then the functions* $h_i,\ldots,L_f^{\rho_i} h_i$, $i \in \underline{p}$, *are independent on* N^*, *and thus* N^* *is a smooth submanifold of dimension* $n - \sum_{i=1}^{p} (\rho_i + 1)$. *Furthermore* N^* *equals the maximal controlled invariant output-nulling submanifold for (11.10). The feedbacks* $u = \alpha^*(x)$, $x \in N^*$, *which render* N^* *invariant are given as the solutions of*

$$A(x)\alpha^*(x) + B(x) = 0\ , \qquad x \in N^*, \tag{11.28}$$

where $B(x)$ *is the* p*-vector with* i*-th component* $L_f^{(\rho_i+1)} h_i(x)$. *In fact,* $A(x) = A^*(x)$ *and* $B(x) = B^*(x)$.

Proof (see also Exercise 11.3). Independence of $h_i,\ldots,L_f^{\rho_i} h_i$, $i \in \underline{p}$, follows as in Proposition 8.11. It is immediately seen that $u = \alpha(x)$ satisfying (11.28) renders N^* as defined in (11.27) invariant, and thus N^* is controlled invariant and output-nulling. Clearly, if the outputs $y(t)$ have to be kept zero then also all their time-derivatives have to be zero. Maximality of N^* now follows by definition of ρ_i, $i \in \underline{p}$, since $y_i^{(j)} = L_f^j h_i$, $j=1,\ldots,\rho_i$, $i \in \underline{p}$, implying that the functions $h_i,\ldots,L_f^{\rho_i} h_i$ have to be zero on *any* controlled invariant output-nulling submanifold. For the last equalities we refer to Exercise 11.3. □

Remark 11.14 Notice that N^* as obtained in Proposition 11.13 is *globally* defined, not just in a neighborhood of a particular point x_0. Furthermore, if rank $A(x) = p$ everywhere, then the maximal controlled invariant *distribution* D^* contained in ker dh is given as ker span $\{dh_i,\ldots,dL_f^{\rho_i} h_i,\ i \in \underline{p}\}$ (see Exercise 8.4 and Theorem 7.21), and thus N^* given in (11.27) is an *integral manifold* of D^*.

11.2 Constrained Dynamics and Zero Dynamics

In the previous section we have given algorithms to compute the maximal locally controlled invariant output-nulling submanifold N^* for (11.10) around a point x_0 satisfying $f(x_0) = 0$, $h(x_0) = 0$. The resulting dynamics on N^* is given as

$$\dot{x} = f(x) + \sum_{j=1}^{m} g_j(x)\alpha_j^*(x) \quad , \quad x \in N^*, \tag{11.29}$$

where $\alpha^*(x) = (\alpha_1^*(x),\ldots,\alpha_m^*(x))$ is any solution of (11.18). Now let $\bar{\alpha}(x)$ be one particular solution of (11.18). Locally around x_0 we can find an $m \times (m-r^*)$ matrix $\beta(x)$ of full column rank such that $A^*(x)\beta(x) = 0$ (see Exercise 2.4). Then the full solution set of (11.18) is given as

$$\alpha^*(x) = \bar{\alpha}(x) + \beta(x)v \quad , \qquad v \in \mathbb{R}^{m-r^*}, \tag{11.30}$$

and thus locally around x_0 the resulting dynamics on N^* is also given as

$$\dot{x} = \tilde{f}(x) + \sum_{j=1}^{\tilde{m}} \tilde{g}_j(x)v_j \quad , \qquad x \in N^*, \tag{11.31}$$

where $\tilde{f}(x) := f(x) + \sum_{j=1}^{m} g_j(x)\bar{\alpha}_j(x)$, $\tilde{g}_j(x) := \sum_{i=1}^{m} g_i(x)\beta_{ij}(x)$, $j \in \underline{\tilde{m}}$, $\tilde{m} := m - r^*$, with inputs $v_1,\ldots,v_{\tilde{m}}$. Alternatively (11.31) is obtained by applying to (11.10) the *degenerate feedback* $u = \bar{\alpha}(x) + \beta(x)v$, $v \in \mathbb{R}^{\tilde{m}}$, given by (11.30), and then *restricting* the closed-loop system to N^*. We will call (11.31) the *constrained* (or *clamped*) *dynamics* for the system (11.10), i.e., all motions of (11.10) *compatible* with the *constraints* $h(x) = 0$.

Next let us consider for the constrained dynamics (11.31) the distribution C_0 (Definition 3.19), characterizing local strong accessibility with respect to the controls $v_1,\ldots,v_{\tilde{m}}$. By Proposition 3.47 C_0 is the smallest distribution on N^* that is invariant for (11.31) and contains the vectorfield $\tilde{g}_1,\ldots,\tilde{g}_{\tilde{m}}$. *Assume* that C_0 has constant dimension on a neighborhood of x_0. Then by Theorem 3.49 we can find local coordinates $\bar{x} = (\bar{x}^1,\bar{x}^2)$ for N^* around x_0 such that (11.31) takes the form

$$\begin{aligned} \dot{\bar{x}}^1 &= \tilde{f}^1(\bar{x}^1,\bar{x}^2) + \sum_{j=1}^{\tilde{m}} \tilde{g}_j^1(\bar{x}^1,\bar{x}^2)v_j, \\ \dot{\bar{x}}^2 &= \tilde{f}^2(\bar{x}^2). \end{aligned} \tag{11.32}$$

Definition 11.15 *The dynamics* $\dot{\bar{x}}^2 = \tilde{f}^2(\bar{x}^2)$ *are called the* zero dynamics *of the system (11.10) around* x_0.

Remark For a controllable and observable *linear* system $\dot{x} = Ax + Bu$ the eigenvalues of the zero dynamics (which in this case are linear) coincide precisely with the *transmission zeros* of the transfer matrix $C(Is-A)^{-1}B$ (see the references cited at the end of this chapter).

Notice that the zero dynamics do not depend on the particular choice of $\bar{\alpha}(x)$ satisfying (11.18), and so are intrinsically defined. Furthermore if $r^* = m$ then (see Remark 11.11) the zero dynamics equal the constrained dynamics.

Finally, let us consider the special case considered in Proposition 11.13, with the additional requirements that $p = m$ and that rank $A(x) = m$ *everywhere* (and not only on N^* and therefore on a neighborhood of N^*), i.e. the case of an input-output decouplable square system (see Theorem 8.9). Then the functions

$$z_{ij} = L_f^{j-1}h_i(x), \; j = 1,\ldots,\rho_i + 1, \qquad i \in \underline{m}, \tag{11.33}$$

are everywhere independent (Proposition 8.11), and as explained in Chapter 8, the decoupling feedback $u = A^{-1}(x)(-B(x) + v)$, with $A(x)$ and $B(x)$ as in (11.28), transforms the system (11.10) into the "normal form" (cf. (8.45))

$$\begin{cases} \dot{z}^1 = A_1 z^1 + b_1 v_1 & , \; y_1 = z_{11}, \\ \quad \vdots & \quad \vdots \\ \dot{z}^m = A_m z^m + b_m v_m & , \; y_m = z_{m1}, \\ \dot{\bar{z}} = \bar{f}(\bar{z}, z^1,\ldots,z^m) + \sum_{j=1}^{m} \bar{g}_j(\bar{z}, z^1,\ldots,z^m)v_j, \end{cases} \tag{11.34}$$

with $z^i := (z_{i1},\ldots,z_{i(\rho_i+1)})$, $i \in \underline{m}$, and $\bar{z}$ being additional coordinates, satisfying $\bar{z}(x_0) = 0$. Furthermore the matrix pairs (A_i, b_i), $i \in \underline{m}$, are in Brunovsky canonical form (see (6.50)). It immediately follows that in this case the zero dynamics are equal to the constrained dynamics and are given as

$$\dot{\bar{z}} = \bar{f}(\bar{z}, 0,\ldots,0), \tag{11.35}$$

and that $\bar{z}$ can be regarded as local coordinates for N^* around x_0.

From linear control theory it is well-known that the stability, respectively instability, of the zero dynamics is very important for design purposes, and we may expect this to be true in the nonlinear case as well. We immediately obtain

Theorem 11.16 *Consider a square system (11.10), with* $f(x_0) = 0$ *and* $h(x_0) = 0$, *and rank* $A(x_0) = m$. *Assume that its zero dynamics are locally asymptotically stable around* $x_0 \in N^*$. *Then there exists a decoupling regular static state feedback* $u = \alpha(x) + \beta(x)\bar{v}$, *such that the closed loop system for* $\bar{v} = 0$ *is locally asymptotically stable around* x_0.

Proof First apply the decoupling feedback law $u = A^{-1}(x)(-B(x)+v)$, such that locally around x_0 the closed-loop system takes the form (11.34). Then apply additional linear feedback

$$v_i = k_i z^i + \bar{v}_i, \qquad i \in \underline{m}, \tag{11.36}$$

such that the matrices $\bar{A}_i := A_i + b_i k_i$, $i \in \underline{m}$, are all asymptotically stable. Notice that the closed-loop system with inputs $\bar{v}_1,\ldots,\bar{v}_m$ is still input-output decoupled. Setting $\bar{v}_i = 0$, $i \in \underline{m}$, we obtain the system

$$\begin{cases} \dot{z}^1 = \bar{A}_1 z^1 \\ \quad\vdots \\ \dot{z}^m = \bar{A}_m z^m \\ \dot{\bar{z}} = \bar{f}(\bar{z},z^1,\ldots,z^m) + \sum_{j=1}^{m} \bar{g}_j(\bar{z},z^1,\ldots,z^m)k_j z^i. \end{cases} \tag{11.37}$$

Clearly, the eigenvalues of the linearization of (11.37) in $\bar{z} = 0, z^i = 0$, $i \in \underline{m}$, are the eigenvalues of $\bar{A}_1,\ldots,\bar{A}_m$ together with the eigenvalues of the linearization of the zero dynamics (11.35) in $\bar{z} = 0$. Accordingly to (10.49) we can write (11.35) into the form (with $z = (s^1,s^2)$)

$$\begin{aligned} \dot{s}^1 &= A^o s^1 + f^o(s^1,s^2) \\ \dot{s}^2 &= A^- s^2 + f^-(s^1,s^2) \end{aligned} \tag{11.38}$$

where the eigenvalues of A^o are on the imaginary axis and the eigenvalues of A^- are in the open left half plane, and f^o and f^- only contain second- and higher-order terms. Thus by the Center Manifold Theorem (Theorem 10.11) there exists around x_0 an invariant submanifold $s^2 = \varphi(s^1)$ of N^* with dynamics

$$\dot{s}^1 = A^o s^1 + f^o(s^1,\varphi(s^1)). \tag{11.39}$$

Since N^* is an invariant submanifold for the whole dynamics (11.37) it folllows that the submanifold $s^2 = \varphi(s^1)$ of N^* is *also* a center manifold for (11.37). Since the dynamics (11.35) on N^* are assumed to be locally asymptotically stable about x_0 it follows that (11.39) is locally asymptotically stable about x_0, and thus by an application of

Theorem 10.13, x_0 is a locally asymptotically stable equilibrium for the whole dynamics (11.37). □

Remark The above sufficient condition for input-output decoupling with local asymptotic stability is far from being necessary, cf. the references at the end of this chapter. The reason is that decoupling feedback laws do not necessarily have to make N^* invariant.

Theorem 11.16 immediately suggests an alternative approach to the problem of local *feedback stabilization* of a nonlinear system

$$\dot{x} = f(x) + \sum_{j=1}^{m} g_j(x)u_j, \qquad f(x_0) = 0, \tag{11.40}$$

as treated in the preceding chapter. Indeed suppose that we can find m (dummy) output functions

$$y_i = h_i(x), \qquad i \in \underline{m}, \tag{11.41}$$

such that the system (11.40),(11.41) has nonsingular decoupling matrix and its zero-dynamics are locally asymptotically stable. Then by an application of Theorem 11.16 the system can be made locally asymptotically stable by feedback. This idea is illustrated in the next example.

Example 11.17 Consider the equations for the angular velocities of a rigid body with two external torques aligned with two of the principal axes (see Example 10.16)

$$\begin{aligned} \dot{\omega}_1 &= I_{23}\omega_2\omega_3 \\ \dot{\omega}_2 &= I_{31}\omega_3\omega_1 + c_2u_1 \\ \dot{\omega}_3 &= I_{12}\omega_1\omega_2 + c_3u_2 \end{aligned} \tag{11.42}$$

By a preliminary feedback the system can be transformed into

$$\begin{aligned} \dot{\omega}_1 &= I_{23}\omega_2\omega_3 \\ \dot{\omega}_2 &= v_1 \\ \dot{\omega}_3 &= v_2 \end{aligned} \tag{11.43}$$

Now consider the dummy output functions

$$\begin{aligned} y_1 &= \omega_2 + \omega_1 \\ y_2 &= \omega_3 + \omega_1^2 \end{aligned} \tag{11.44}$$

In the coordinates y_1, y_2, ω_1 the system takes the form

$$\begin{aligned}
\dot{y}_1 &= I_{23}\omega_2\omega_3 + v_1 \\
\dot{y}_2 &= 2I_{23}\omega_1\omega_2\omega_3 + v_2 \\
\dot{\omega}_1 &= I_{23}(y_1-\omega_1)(y_2-\omega_1^2)
\end{aligned} \tag{11.45}$$

Clearly the zero dynamics of (11.45) is given as

$$\dot{\omega}_1 = I_{23}\omega_1^3, \tag{11.46}$$

which is asymptotically stable by the fact that $I_{23} = (I_2-I_3)/I_1 < 0$. Thus for example the feedback

$$\begin{aligned}
v_1 &= -I_{23}\omega_2\omega_3 - k_1y_1, \qquad & k_1 > 0, \\
v_2 &= -2I_{23}\omega_1\omega_2\omega_3 - k_2y_2, \qquad & k_2 > 0,
\end{aligned} \tag{11.47}$$

asymptotically stabilizes (11.45). (Compare with the stabilizing feedback obtained in Example 10.16.) □

Finally, in order to make contact with Theorem 10.5, let us suppose that the linearization of (11.40) in x_0 and $u = 0$ is *not* controllable. As explained in Chapter 10 (cf. Theorem 10.5), this means that there are some eigenvalues of the linearization, called the *uncontrollable modes*, which are invariant under feedback (namely the eigenvalues of A_{22} in (10.13)). Now suppose there exist output equations (11.41) for which the decoupling matrix $A(x)$ has everywhere rank m, such that after feedback the system takes the form (11.34). For simplicity (see Exercise 11.6), assume that we can choose the additional coordinates $\bar{z}$ in such a way that $\bar{g}_j = 0$ in (11.34) $j \in \underline{m}$. (This is the case if and only if the distribution $\mathrm{span}\{g_1(x),\ldots,g_m(x)\}$ is involutive about x_0, see Exercise 8.3.) Clearly the uncontrollable modes of the linearization of (11.40) are the same as the uncontrollable modes of the linearization of (11.34), i.e. of

$$\begin{aligned}
\dot{\xi}^i &= A_i\xi^i + b_iv_i, \qquad i \in \underline{m}, \\
\dot{\bar{\xi}} &= \frac{\partial \bar{f}}{\partial \bar{z}}(0,\ldots,0)\bar{\xi} + \sum_{i=1}^{m} \frac{\partial \bar{f}}{\partial z^i}(0,\ldots,0)\xi^i \, .
\end{aligned} \tag{11.48}$$

Since (A_i, b_i), $i \in \underline{m}$, are controllable pairs it follows that the uncontrollable modes of the linearization are necessarily modes of the *linearized* zero dynamics (cf. (11.35))

$$\dot{\bar{\xi}} = \frac{\partial \bar{f}}{\partial \bar{z}}(0,\ldots,0)\bar{\xi}. \tag{11.49}$$

In view of Theorem 10.5 the problem of local feedback stabilization becomes difficult when (some of) the uncontrollable modes of the linearization are on the imaginary axis. We conclude that these particular modes also appear as eigenvalues of the linearization of the zero dynamics.

11.3 Interconnection of Systems and Inverse Systems

In many instances systems are at first instance given as the *interconnection* of a number of (small and relatively simple) subsystems. Even if these subsystems have explicit state space models, the system resulting from the interconnection is described *a priori* by a mixed set of differential and algebraic relations, and a major problem is the transformation of these relations into a set of *explicit* differential and algebraic relations, i.e. a state space model (which is usually more convenient for control and simulation purposes). In the present section we will show how such a transformation may be naturally interpreted as the computation of some kind of constrained dynamics.

Consider k affine nonlinear systems

$$\dot{x}^i = f^i(x^i) + \sum_{j=1}^{m_i} g_j^i(x^i)u_j^i, \qquad i \in \underline{k}, \tag{11.50}$$

where $x^i = (x_1^i,\ldots,x_{n_i}^i)$ are local coordinates for an n_i-dimensional manifold M^i, $i \in \underline{k}$. Suppose these systems are *interconnected* by an interconnection constraint of the form

$$\varphi(x^1,x^2,\ldots,x^k) = 0, \tag{11.51}$$

where $\varphi\colon M^1 \times M^2 \times \ldots \times M^k \to \mathbb{R}^c$ is a smooth mapping. In order to construct the system resulting from this interconnection we consider the product system

$$\begin{aligned} \dot{x}^1 &= f^1(x^1) + \sum_{j=1}^{m_1} g_j^1(x^1)u_j^1 \\ &\vdots \\ \dot{x}^k &= f^k(x^k) + \sum_{j=1}^{m_k} g_j^k(x^k)u_j^k \end{aligned} \tag{11.52a}$$

with state space $M^1 \times \ldots \times M^k$, input space $\mathbb{R}^{m_1} \times \ldots \times \mathbb{R}^{m_k}$ and (dummy) output equations

$$y_j = \varphi_j(x^1,\ldots,x^k), \qquad j \in \underline{c}. \tag{11.52b}$$

The system resulting from the interconnection is precisely given as the constrained dynamics for (11.52), which can be computed using one of the algorithms given in the first section of this chapter.

Remark If the interconnection constraint φ in (11.51) depends on any of the inputs $u_1,\ldots,u_k$, then the output map (11.52b) will depend on the inputs and we have to take recourse to the computation of the constrained dynamics for a general nonlinear system, see Chapter 13.

Example 11.18 Consider two rigid two-link robot manipulators (see Example 1.1), mounted on a same platform, at a fixed distance d.

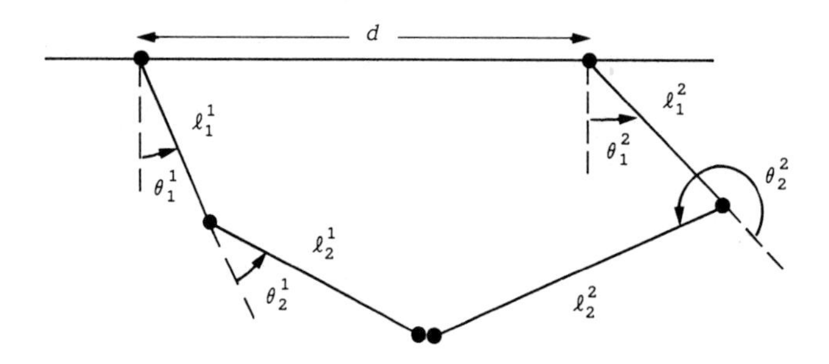

Fig. 11.1. Co-operating robot manipulators.

The equations of motion of both robot manipulators are described in Example 1.1. Now suppose that both manipulators have to *co-operate;* in particular, suppose that the endpoints of both manipulators have to be at the same place at every time t (for example for simultaneously grasping an object). Also suppose that the endpoints of both manipulators merely have to touch each other, in such a way that no reaction forces are present (as an idealization of the situation that the object that has to be grasped by both manipulators is easily damaged). Then we have the interconnection constraint equations (cf. (1.9))

$$\ell_1^1 \sin\theta_2^1 + \ell_2^1 \sin(\theta_1^1+\theta_2^1) = \ell_1^2 \sin\theta_1^2 + \ell_2^2 \sin(\theta_1^2+\theta_2^2) + d,$$

$$\ell_1^1 \cos\theta_1^1 + \ell_2^1 \cos(\theta_1^1+\theta_2^1) = \ell_1^2 \cos\theta_1^2 + \ell_2^2 \cos(\theta_1^2+\theta_2^2). \tag{11.53}$$

It is easily seen that the product system of the two robot manipulators with the two output mappings corresponding to the constraints (11.53) satisfies the assumptions of Proposition 11.13, with $\rho_1 = \rho_2 = 1$. Indeed N^* is non-empty and is given as the set of all points $(\theta_1^1, \theta_2^1, \dot\theta_1^1, \dot\theta_2^1, \theta_1^2, \theta_2^2, \dot\theta_1^2, \dot\theta_2^2)$ satisfying equations (11.53) and their first-order time-derivatives

$$
\begin{aligned}
&\ell_1^1\, \dot\theta_1^1 \cos\theta_1^1 + \ell_2^1\, (\dot\theta_1^1 + \dot\theta_2^1) \cos(\theta_1^1 + \theta_2^1) = \\
&\qquad \ell_1^2\, \dot\theta_1^2 \cos\theta_1^2 + \ell_2^2(\dot\theta_1^2 + \dot\theta_2^2) \cos(\theta_1^2 + \theta_2^2) \\
&\ell_1^1\, \dot\theta_1^1 \sin\theta_1^1 + \ell_2^1\, (\dot\theta_1^1 + \dot\theta_2^1) \sin(\theta_1^1 + \theta_2^1) = \\
&\qquad \ell_1^2\, \dot\theta_1^2 \sin\theta_1^2 + \ell_2^2(\dot\theta_1^2 + \dot\theta_2^2) \sin(\theta_1^2 + \theta_2^2)
\end{aligned}
\tag{11.54}
$$

and the required feedback is computed as in (11.27). □

Let us apply this same methodology to the problem of *inversion* of a nonlinear control system. That is, we want to reconstruct the input functions u on the basis of the knowledge of the output functions for $t \geq 0$ and the initial state of the system. (This is usually called the problem of *left-inversion*; the dual problem of finding input functions such that the resulting output functions for a fixed initial state are equal to some desired functions of time is called the problem of *right-inversion*, and is touched upon in Chapter 8, cf. the proof of Theorem 8.19.)

Let us again consider the system

$$
\begin{aligned}
\dot x &= f(x) + \sum_{j=1}^{m} g_j(x) u_j, \qquad x(0) = x_0, \\
\bar y_i &= h_i(x), \qquad i \in \underline{p},
\end{aligned}
\tag{11.55}
$$

where $x = (x_1, \ldots, x_n)$ are local coordinates for the state space M. In addition, consider an auxiliary system consisting of p parallel n-fold integrators

$$
\begin{aligned}
w^{(n)} &= v, \\
& \qquad\qquad w \in \mathbb{R}^p,\ v \in \mathbb{R}^p,\ y \in \mathbb{R}^p, \\
y &= w,
\end{aligned}
\tag{11.56}
$$

with state $(w, \dot w, \ldots, w^{(n-1)}) \in \mathbb{R}^{np}$, input v and output y, which is initialized at some point $(y_0, \dot y_0, \ldots, y_0^{(n-1)}) \in \mathbb{R}^{np}$. Let us *interconnect* (11.55) to (11.56) by the interconnection constraint

$$S_0(x,y) := h(x) - y = 0. \tag{11.57}$$

Now apply Algorithm 11.6 to the product of systems (11.55) and (11.56) with output equations $S_0(x,y)$, and assume that $(x_0, y_0, \dot{y}_0, \ldots y_0^{(n-1)})$ is a regular point (in the sense of Remark 11.7). Then we obtain a descending sequence of submanifolds $N_1 \supset N_2 \supset \ldots \supset N_{k^*} = N^*$ of $M \times \mathbb{R}^{np}$ around $(x_0, y_0, \ldots y_0^{(n-1)})$, of the form

$$N_{k+1} = \{(x,y,\ldots,y^{(n-1)}) \in N_k \,|\, S_k(x,y,\ldots,y^{(n-1)}) = 0\}, \tag{11.58}$$

where S_k has s_{k+1} components. By the special form of (11.56) and (11.57) we have the following extra information concerning the mappings S_k. At the first step of the algorithm the matrix A_1 takes the form

$$A_1(x) = \left[\begin{array}{cc} (L_{g_i} h_j(x))_{\substack{j=1,\ldots,p_1 \\ i=1,\ldots,m}} & 0_{p_1 \times p} \end{array} \right]. \tag{11.59}$$

Moreover, since the components of $S_0(x,y) = h(x) - y$ are independent on $M \times \mathbb{R}^{np}$, we have $s_1 = p_1 = p$. It follows that $r_1 = \operatorname{rank}(L_{g_i} h_j(x))_{\substack{j=1,\ldots,p \\ i=1,\ldots,m}}$. Furthermore the matrix B_1 is given as

$$B_1(x,y,\dot{y}) = (L_f h_j(x) - \dot{y}_j)_{j=1,\ldots,p}. \tag{11.60}$$

Therefore S_1 has $s_2 = p - r_1$ entries, and may be taken of the form

$$S_1(x,y,\dot{y}) = K_1(x,y) + G_1(x,y)\dot{y}, \tag{11.61}$$

where $\operatorname{rank} G_1(x,y) = s_2$ in a neighborhood of (x_0, y_0). (This becomes especially clear if we use the procedure given in Remark 11.12.) In general we obtain that the mappings S_k can be taken of the form

$$S_k(x,y,\dot{y},\ldots,y^{(k)}) = K_k(x,y,\ldots,y^{(k-1)}) + G_k(x,y,\ldots,y^{(k-1)})y^{(k)}, \tag{11.62}$$

where $\operatorname{rank} G_k(x,y,\ldots,y^{(k-1)}) = s_{k+1}$ around $(x_0, y_0, \ldots, y_0^{(k-1)})$. Furthermore (cf. (11.22))

$$s_k = s_{k-1} - (r_{k-1} - r_{k-2}), \qquad k = 3,\ldots,n, \tag{11.63}$$

and for $k < n$ the last p columns of the matrices $A_k(x,y,\ldots,y^{(k-1)})$ (corresponding to dependencies on the inputs w) are identically zero. Finally, in view of (11.62) and the special form of the dynamics (11.56), the equation

$$A_n(x,y,\dot{y},\ldots,y^{(n-1)})\, \alpha(x,y,\ldots,y^{(n)}) + B_n(x,y,\ldots,y^{(n)}) = 0 \tag{11.64}$$

can be always solved for α on N_n, implying that $k^* \le n$. Using (11.63) we see that (11.22) holds with equality everywhere, and thus $r^* = p$.

Let us now define one additional integer, namely

$$\rho^* := \operatorname{rank} \tilde{A}_{k^*}, \tag{11.65}$$

where $\tilde{A}_{k^*}$ denotes the first m columns of A_{k^*}. For $k^* < n$ it immediately follows that $\rho^* = r^*$ (since the last p columns of A_{k^*} for $k^* < n$ are zero). However if $k^* = n$, we only obtain $\rho^* \le r^*$.

Now assume that $p = m$ and $\rho^* = m$. Regarding the algorithm as above as the constrained dynamics algorithm for the system (11.55), *parametrized* by $y, \dot{y}, \ldots, y^{(n-1)}$, it follows from Corollary 11.9 (see Remark 11.10) that

$$\operatorname{rank} \frac{\partial S_k}{\partial x} (x, y, \ldots, y^{(k)}) = s_{k+1} \text{ on } N_k, \qquad k = 1,2,..,k^*-1. \tag{11.66}$$

Thus from the $p_{k^*} = s_1 + \ldots + s_{k^*}$ equations defining N^*, i.e.

$$\begin{aligned} S_0(x,y) &= 0, \\ S_1(x,y,\dot{y}) &= 0, \\ &\vdots \\ S_{k^*-1}(x,y,\dot{y},\ldots,y^{(k^*-1)}) &= 0, \end{aligned} \tag{11.67}$$

we can locally solve for p_{k^*} components of the state x, as functions of $y, \dot{y}, \ldots, y^{(k^*-1)}$ and of the remaining $n-p_{k^*}$ state components, which we denote by $z \in \mathbb{R}^{n-p_{k^*}}$. Furthermore since $\rho^* = m$, the equations

$$\tilde{A}_{k^*} u + B_{k^*} = 0 \tag{11.68}$$

have a unique solution u as function of x and $y, \dot{y}, \ldots, y^{(k^*)}$. Combining this, we obtain equations of the form

$$\dot{z} = F(z, y, \dot{y}, \ldots, y^{(k^*-1)}), \qquad z(0) = z_0, \tag{11.69a}$$

$$u = G(z, y, \dot{y}, \ldots, y^{(k^*)}). \tag{11.69b}$$

Equations (11.69a) together with the auxiliary system (11.56) describe the dynamics on the constrained dynamics submanifold $N^* \subset M \times \mathbb{R}^{np}$. The system (11.69) is called an *inverse system*, since for every output function $y(t)$, $t \ge 0$ small, with $(y(0), \dot{y}(0), .., y^{(n-1)}(0))$ close to $(y_0, \dot{y}_0, \ldots, y_0^{(n-1)})$, it reconstructs the unique input function $u(t)$, $t \ge 0$ small, yielding this particular output function for initial state $x(0) = x_0$ if we set the components of z_0 to be equal to the corresponding components of x_0. Summarizing we have obtained

Proposition 11.19 *Consider the system (11.55) with $p = m$. Suppose $(x_0, y_0, \ldots, y_0^{(n-1)})$ is a regular point for Algorithm 11.6 applied to the product system (11.55),(11.56), with output equation (11.57). Assume that $\rho^* = m$, where ρ^* is defined as in (11.65). Then there exists an inverse system (11.69) for (11.55), with z_0 determined by x_0, which reconstructs in a unique manner the input function from the observed output function.*

Example 11.20 Consider the system on $\mathbb{R}^4$

$$\begin{aligned} \dot{x}_1 &= u_1 , & y_1 &= x_1 , & \dot{x}_3 &= x_4 , \\ \dot{x}_2 &= x_3 + x_4 u_1 , & y_2 &= x_2 , & \dot{x}_4 &= u_2 . \end{aligned} \tag{11.70}$$

Thus $S_0(x,y) = \begin{pmatrix} x_1 - y_1 \\ x_2 - y_2 \end{pmatrix}$, and so $A_1(x) = \begin{pmatrix} 1 & 0 \\ x_4 & 0 \end{pmatrix}$, $B_1(x,y) = \begin{pmatrix} -\dot{y}_1 \\ x_3 - \dot{y}_2 \end{pmatrix}$. Take $S_1(x,y,\dot{y}) = x_4\dot{y}_1 + x_3 - \dot{y}_2$, then $A_2(x,y) = \begin{pmatrix} 1 & 0 \\ x_4 & 0 \\ 0 & \dot{y}_1 \end{pmatrix}$ has rank 2 for $\dot{y}_1 \neq 0$, while $k^* = 2$. Solving for u_1 and u_2, one obtains

$$\begin{aligned} u_1 &= \dot{y}_1 \\ u_2 &= (- x_4 - x_4\ddot{y}_1 + \ddot{y}_2)/\dot{y}_1 . \end{aligned} \tag{11.71a}$$

From the equations $S_0 = 0$ and $S_1 = 0$ we can solve for x_1, x_2 and x_3, thereby leaving the equation

$$\dot{x}_4 = u_2 = (-x_4 - x_4\ddot{y}_1 + \ddot{y}_2)/\dot{y}_1, \tag{11.71b}$$

and thus (11.71a,b) constitutes the inverse system for $\dot{y}_1 \neq 0$. □

In analogy with Proposition 11.13 we now state an important particular case, where the computation of the inverse system becomes especially easy. Suppose that $p = m$ and that the decoupling matrix $A(x)$ of (11.55) has rank m everywhere. Then by Proposition 8.11 the functions $z_{ij} = L_f^{j-1}h_i$, $j = 1,\ldots,\rho_i+1$, $i \in \underline{m}$, are everywhere independent, and we can take local coordinates $(z^1,\ldots,z^m,\bar{z})$, with $z^i = (z_{i1},\ldots,z_{i(\rho_i+1)})$. By definition of ρ_i, $i \in \underline{m}$, we have

$$z_{ij} = y_i^{(j-1)} , \qquad j = 1,\ldots,\rho_i + 1, \ i \in \underline{m}, \tag{11.72}$$

while

$$\begin{bmatrix} y_1^{(\rho_1+1)} \\ \vdots \\ y_m^{(\rho_m+1)} \end{bmatrix} = A(x)u + B(x), \tag{11.73}$$

where the m-vector $B(x)$ has i-th component equal to $L_f^{(\rho_i+1)}h_i(x)$. It follows that u is uniquely given as

$$u = A^{-1}(z^1,..,z^m,\bar{z}) \left[- B(z^1,..,z^m,\bar{z}) + \begin{bmatrix} y_1^{(\rho_1+1)} \\ \vdots \\ y_m^{(\rho_m+1)} \end{bmatrix} \right], \tag{11.74a}$$

where we have expressed $A(x)$ and $B(x)$ in the new coordinates $(z^1,..,z^m,\bar{z})$. Furthermore, the dynamics of the $\bar{z}$-coordinates are given as (see (11.34))

$$\dot{\bar{z}} = \bar{f}(\bar{z},z^1,...,z^m) + \sum_{j=1}^{m} \bar{g}_j(\bar{z},z^1,..,z^m)y_j^{(\rho_j+1)} \tag{11.74b}$$

with $\bar{f}$ and $\bar{g}_j$, $j \in \underline{m}$, as in (11.34). Substituting in (11.74a),(11.74b) for z^i the vector $(y_i,..,y_i^{(\rho_i)})$, $i \in \underline{m}$, it immediately follows that (11.74) is an inverse system with state $\bar{z}$. In fact, since $\bar{z}$ are local coordinates for N^* (see (11.34)) it is concluded that in this case the inverse system is *globally* defined with state space N^*, and coincides for $y_i = .. = y_i^{(\rho_i)} = 0$, $i \in \underline{m}$, with the *zero dynamics*.

Remark 11.21 Let us finally indicate the connections of the algorithm given above for computing an inverse system with the dynamic extension algorithm as dealt with in Chapter 8 (Algorithm 8.18). In particular let us show that ρ^* as defined in (11.65) coincides with q^*, the rank of the system, as defined in the dynamic extension algorithm (cf.(8.88)). The key observation is as follows. Consider the first step of the dynamic extension algorithm. We obtain (cf. (8.67))

$$(y^1)^{(\rho^1+1)} = v^1 =: u^1. \tag{11.75}$$

Therefore we can replace in the feedback defined in Step 2 of Algorithm 8.18

$$\bar{u}^1 = \alpha^2(x,\tilde{U}^1) + \beta^2(x,\tilde{U}^1) \begin{pmatrix} v^2 \\ \bar{v}^2 \end{pmatrix}, \tag{11.76}$$

the variables $\tilde{U}^1$ (i.e., u^1 and its time-derivatives) by $(y^1)^{(\rho^1+1)}$ and its time-derivatives. The same holds at the next steps, i.e. the dependency of the feedbacks in Algorithm 8.18 on $\tilde{U}^\ell$ can be regarded as dependency on time-derivatives of the output functions. Therefore these feedbacks reduce to the feedbacks as used in the above algorithm for constructing an inverse system (see also Exercise 11.11 and the references). Let now x_0 be in the open and dense submanifold where q^* is well-defined. Then it readily follows that

$$q^* := \text{rank } D^{\ell^*}(x,\widetilde{U}^{\ell^*}) = \rho^* \tag{11.77}$$

(cf. 8.78), where ℓ^* is the least integer such that $q_{\ell^*} = q_{\ell^*+1} (= q^*)$.

Notes and References

The notion of zero dynamics was first identified, in the single-input single-output case, in [BI1] and [Ma], in particular in connection with high-gain feedback. (The relation between the maximal controlled invariant *distribution* and linear zero dynamics was previously stressed in [KI], [IKGM].) Applications of the notion of zero dynamics to feedback stabilization and adaptive control were already obtained in [BI2,3]. The general definition of constrained dynamics, and zero dynamics for $r^* = m$, was given in [IM]; see also [vdS1] for Hamiltonian constrained dynamics. The version of Algorithm 11.6 as given in Remark 11.12 was formulated in [IM]; it is a generalization of Hirschorn's algorithm [Hi], which in turn is based on the linear structure algorithm [Si1]. The equivalence between Algorithm 11.6 and its version in Remark 11.12 has been shown in [vdS4,5], see also [BI4]. Proposition 11.13 for $p < m$ can be found in [vdS2]. Algorithm 11.3, which is a coordinate-free version of Algorithm 11.6, was first formulated in [BI4]. The definition of zero dynamics for $r^* < m$ is taken from [vdS3,5]. The idea of exploiting zero dynamics for feedback stabilization was further elaborated in [BI4,5,6]; Example 11.17 is taken from [BI6]. The observation that the uncontrollable modes of the lienarization are contained in the linearization of the zero-dynamics is due to [BI4]. The algorithm described in equations (11.57)-(11.68) is a version of Singh's algorithm [Si]; see [IM] for a clear exposition of this algorithm, and [dBGM] for a somewhat alternative formulation. The present treatment based on interconnections was inspired by [vdS4]. The definition of inverse system is taken from [Hi], [IM]. Example 11.20 is taken from [IM]. The equivalence of Singh's algorithm with the dynamic extension algorithm has been shown in [dBGM].

[BI1] C. Byrnes, A. Isidori, "A frequency domain philosophy for nonlinar systems, with applications to stabilization and adaptive control", Proc. 23rd IEEE Conf. Decision Control, Las Vegas pp. 1569-1573, 1984.

[BI2] C. Byrnes, A. Isidori, "Asymptotic expansions, root-loci and the global stability of nonlinear feedback systems", in **Algebraic and Geometric Methods in Nonlinear Control Theory** (eds. M. Fliess, M. Hazewinkel), Reidel, Dordrecht, pp. 159-179, 1986.

[BI3] C. Byrnes, A. Isidori, "Global feedback stabilization of nonlinear systems", Proc. 24th Conf. Decision Control, Ft. Lauderdale, pp. 1031-1037, 1985.

[BI4] C. Byrnes, A. Isidori, "Local stabilization of minimum-phase nonlinear systems", Systems Control Lett. 11, pp. 9-17, 1988.

[BI5] C. Byrnes, A. Isidori, "Attitude stabilization of rigid spacecraft", Automatica.

[BI6] C. Byrnes, A. Isidori, "New results and examples in nonlinear feedback stabilization", Systems Control Lett. 12, pp. 437-442, 1989.

[dB] M.D. di Benedetto, "A condition for the solvability of the nonlinear model matching problem", in **New Trends in Nonlinear Theory** (eds. J. Descusse, M. Fliess, A. Isidori, D. Leborgne), Lect. Notes Contr. Inf. Sci. 122, pp. 102-115, Springer, Berlin, 1989.

[dBGM] M.D. di Benedetto, J.W. Grizzle, C.H. Moog, "Rank invariants of nonlinear systems", SIAM J. Contr. Optimiz. 27, pp. 658-672, 1989.

[DP] B. D'Andrea, L. Praly, "About finite nonlinear zeros for decouplable systems", Systems Control Lett. 10, pp. 103-109, 1988.

[Hi] R.M. Hirschorn, "Invertibility of multivariable nonlinear control systems", IEEE Trans. Automat. Contr. AC-24, pp. 855-865, 1979.

[IKGM] A. Isidori, A.J. Krener, C. Gori-Giorgi, S. Monaco, "Nonlinear decoupling via feedback; a differential geometric approach", IEEE Trans. Autom. Contr. AC-21, pp. 331-345, 1981.

[IM] A. Isidori, C.H. Moog, "On the nonlinear equivalent of the notion of transmission zeros", in **Modelling and Adaptive Control** (eds. C.I. Byrnes, A. Kurzhanski), Lect. Notes Contr. Inf. Sci., 105, , pp. 146-158, 1988.

[KI] A.J. Krener, A. Isidori, "Nonlinear zero distributions", Proc. 19th IEEE Conf. Decision Control, Albuquerque, pp. 665-668, 1980.

[Ma] R. Marino, "High-gain feedback in nonlinear control systems", Int. J. Contr. 42, pp. 1369-1385, 1985.

[Mo] C.H. Moog, "Nonlinear decoupling and structure at infinity", Math. Control Systems, 1, pp. 257-268, 1988.

[Nij] H. Nijmeijer, "Right-invertibility for a class of nonlinear control systems: a geometric approach", Systems Control Lett. 2, pp. 125-132, 1986.

[Si] S.N. Singh, "A modified algorithm for invertibility in nonlinear systems", IEEE Trans. Automat. Contr. AC-26, pp. 595-598, 1981.

[Sil] L.M. Silverman, "Inversion of multivariable linear systems", IEEE Trans. Automat. Contr., AC-14, pp. 270-276, 1969.

[vdS1] A.J. van der Schaft, "On feedback control of Hamiltonian systems" in **Theory and Applications of Nonlinear Control Systems** (eds. C.I. Byrnes, A. Lindquist), North-Holland, Amsterdam, pp. 273-290, 1986.

[vdS2] A.J. van der Schaft, "On realization of nonlinear systems described by higher-order differential equations", Math. Systems Theory 19, pp. 239-275, 1987.

[vdS3] A.J. van der Schaft, "Realizations of higher-order nonlinear differential equations", Proc. 25th IEEE Conf. Decision Control, Athens, pp. 1569-1573, 1986.

[vdS4] A.J. van der Schaft, "On clamped dynamics of nonlinear systems", Memo nr. 634, University of Twente, 1987.

[vdS5] A.J. van der Schaft, "On clamped dynamics of nonlinear systems" in **Analysis and Control of Nonlinear Systems** (eds. C.I. Byrnes,

C.F. Martin, R.E. Saeks), North-Holland, Amsterdam, pp. 499-506, 1988.

Exercises

11.1 Consider the system (11.10) with $f(x_0) = 0$ and $h(x_0) = 0$, and suppose x_0 is a regular point for Algorithm 11.6. Define the distribution $\Delta^* = \ker \operatorname{span}\{dh_1, \ldots, dh_{p_{k^*}}\}$

(a) Show that Δ^* is an invariant distribution for the vectorfield $f + \sum_{j=1}^{m} g_j \alpha_j^*$, where α^* is any solution of (11.18).

(b) Is Δ^* a controlled invariant distribution for (11.10)?

11.2 Consider the system (11.10) with $f(x_0) = 0$, $h(x_0) = 0$, and suppose that x_0 is a regular point for Algorithm 7.19 as well as for Algorithm 11.6, yielding respectively the distribution D^*, and the submanifold N^*. Show that $\dim N^* \geq \dim D^*$.

11.3 Assume that the conditions of Proposition 11.13 are satisfied. Show that Algorithm 11.6 yields N^* as given in (11.27).

11.4 Let D be a distribution which is invariant for $\dot{x} = f(x)$, i.e. $[f,D] \subset D$. Suppose $f(x_0) \in D(x_0)$ for some x_0. Let N be an integral manifold of D through x_0. Prove that N is an invariant manifold for $\dot{x} = f(x)$.

11.5 [DP] Assume that the conditions of Proposition 11.13 are satisfied, and that $f(x_0) = 0$, $h(x_0) = 0$. Show that the linearization of the zero dynamics at x_0 equals the (linear) zero dynamics for the system (11.10) linearized at x_0.

11.6 [BI4] Show that in general (i.e. also if $\operatorname{span}\{g_1(x), \ldots, g_m(x)\}$ is *not* involutive) the uncontrollable modes of the linearization of (11.34) are modes of the linearized zero dynamics (11.49).

11.7 (see Example 11.20) Show that the constrained dynamics for (11.70) is trivial; i.e. $N^* = \{0\}$.

11.8 (*Output tracking*) Consider the system (11.10) with initial state $x(0) = x_0$. Assume that the conditions of Proposition 11.13 are satisfied. Suppose one wants to track a desired output trajectory $y_0(t)$, $t \geq 0$.

(a) Show that this is possible using the control (compare with (11.74a))

$$(1) \quad u = A^{-1}(z^1, \ldots, z^m, \bar{z})\left(-B(z^1, \ldots, z^m, \bar{z}) + \begin{pmatrix} y_{d1}^{(\rho_1+1)} \\ \vdots \\ y_{dm}^{(\rho_m+1)} \end{pmatrix}\right)$$

if

(2) $y_{dj}^{(k)}(0) = L_f^k h_j(x_0),\ k = 1,\dots,p_j, \quad j \in \underline{m}.$

(Here z^i, $i \in \underline{m}$, and $\bar{z}$ are as in (11.34)).

(b) Consider instead of (1) the control law

$$u = A^{-1}(z^1,\dots,z^m,\bar{z})\Big(-B(z^1,\dots,z^m,\bar{z}) +$$

$$(3) \qquad + \begin{pmatrix} y_{d1}^{(\rho_1+1)} + a_{1\rho_1}(y_{d1}-y_1)^{(\rho_1)} + \dots + a_{10}(y_{d1}-y_1) \\ \vdots \\ y_{dm}^{(\rho_m+1)} + a_{m\rho_1}(y_{dm}-y_m)^{(\rho_m)} + \dots + a_{m0}(y_{dm}-y_m) \end{pmatrix}\Big)$$

where the polynomials $a^i(s) = s^{\rho_i+1} + a_{i\rho_i}s^{\rho_i} + \dots + a_{i0}$, $i \in \underline{m}$, are all Hurwitz. Show that the following higher-order differential equations are satisfied for $e_i = y_{di} - y_i$, $i \in \underline{m}$,

$$e_i^{(\rho_i+1)}(t) + a_{i\rho_i}e_i^{(\rho_i)}(t) + \dots + a_{i0}e_i(t) = 0, \qquad i \in \underline{m},$$

and thus that $e_i(t) \to 0$, for $t \to \infty$, $i \in \underline{m}$. Therefore, if (2) is not satisfied we obtain *asymptotic* output tracking.

11.9 [dB] Consider the nonlinear model matching problem (MMP) that was defined in Exercise 7.10. If we instead require the existence of an initial state x_{c0} of Q such that $y^{P\circ Q}(x_0,x_{c0},t) - y_m(x_{m0},t) = 0$ for all t, we refer to the problem as the local strong model matching problem (SMMP). Assume the constrained dynamics manifold N_a^* of the augmented system Σ_a (cf. Exercise 7.10) exists.

(a) Show that the local SMMP is solvable from $x_{a0} := (x_0,x_{m0})$ if and only if $x_{a0} \in N_a^*$.

(b) Show that the local SMMP is solvable from any $x_{a0} \in N_a^*$ if and only if the local MMP is solvable around any $\tilde{x}_{a0}$.

(c) Prove (a) and (b) for a square multivariable plant and model. For (b) assume that the plant has nonsingular decoupling matrix.

11.10 [Mo] Consider the algorithm for computing an inverse system, i.e. (11.55)-(11.69). Define

$$p_1' := \rho^*$$

$$p_k' := \rho^* - r_{k-1}, \qquad k \geq 2,$$

and

$$n_i' := \text{number of } p_k's \text{ that are greater than or equal to } i,\ i \geq 1.$$

This defines an alternative formal structure at infinity; prove that for a system with nonsingular decoupling matrix it coincides with the formal structure at infinity defined in Chapter 9.

11.11 ([dBGM], [Nij]) Prove (11.77). Furthermore, show that

$$\rho^* = \frac{\partial(\dot{y},\ddot{y},\dots,y^{(n)})}{\partial(u,\dot{u},\dots,u^{(n-1)})},$$

where the right-hand side denotes the Jacobian matrix with respect to $u, \dot{u}, \ldots, u^{(n-1)}$ of the vector with block components $\dot{y}_1, \ldots, y^{(n)}$, inductively defined by (with $(y^{(0)} = h(x))$

$$y^{(k+1)} = \frac{\partial y^{(k)}}{\partial x} (f(x) + \sum_{j=1}^{m} g_j(x) u_j) + \sum_{i=0}^{k-1} \frac{\partial y^{(k)}}{\partial u^{(i)}} u^{(i+1)}, \ k > 1.$$

12
Mechanical Nonlinear Control Systems

In the present chapter we focus on a *special* subclass of nonlinear control systems, which can be called *mechanical* nonlinear control systems. Roughly speaking these are control systems whose dynamics can be described by the *Euler-Lagrangian* or *Hamiltonian* equations of motion. It is well-known that a large class of physical systems admits, at least partially, a representation by these equations, which lie at the heart of the theoretical framework of physics.

Let us consider a mechanical system with n degrees of freedom, locally represented by n generalized configuration (position) coordinates $q = (q_1,\dots,q_n)$. In classical mechanics the following equations of motion are derived

$$\frac{d}{dt}\left(\frac{\partial T}{\partial \dot{q}_i}\right) - \frac{\partial T}{\partial q_i} = F_i , \qquad i \in \underline{n}. \tag{12.1}$$

Here $T(q,\dot{q})$, with $\dot{q} = (\dot{q}_1,\dots,\dot{q}_n)$ the generalized velocities, denotes the total *kinetic energy* (strictly speaking kinetic *co-energy*) of the system, while F_i are the *forces* acting on the system. Usually the forces F_i are decomposed into a part which are called *conservative* forces, i.e., forces that are derivable from a *potential energy*, and a remaining part F_i^e, $i \in \underline{n}$, consisting of *dissipative* and generalized *external* forces:

$$F_i = -\frac{\partial V}{\partial q_i}(q) + F_i^e , \qquad i \in \underline{n}, \tag{12.2}$$

with $V(q)$ being the potential energy function. Defining the *Lagrangian* function $L_0(q,\dot{q})$ as $T(q,\dot{q}) - V(q)$, one arrives at the celebrated *Euler-Lagrange* equations

$$\frac{d}{dt}\left(\frac{\partial L_0}{\partial \dot{q}_i}\right) - \frac{\partial L_0}{\partial q_i} = F_i^e , \qquad i \in \underline{n}. \tag{12.3}$$

From (12.3) a *control system* is obtained by disregarding dissipative forces and interpreting the external forces F_i^e in (12.3) as *input* or *control* variables u_i. More generally, if only some degrees of freedom can be directly controlled, then one obtains the control system

$$\frac{d}{dt}\left(\frac{\partial L_0}{\partial \dot{q}_i}\right) - \frac{\partial L_0}{\partial q_i} = \begin{cases} u_i & , \ i = 1,\dots,m, \\ 0 & , \ i = m+1,\dots,n, \end{cases} \tag{12.4}$$

with $u_1,\dots,u_m$ being the controls. Notice that (12.4) is not yet in the standard state space form $\dot{x} = f(x) + \sum_{j=1}^{m} g_j(x)u_j$; indeed (12.4) is a set of implicit second-order differential equations. However for mechanical systems the kinetic energy $T(q,\dot{q})$ is of the form

$$T(q,\dot{q}) = \frac{1}{2}\dot{q}^T M(q)\dot{q} \tag{12.5}$$

for some positive-definite matrix $M(q)$. Thus (12.4) takes in obvious vector notation the form

$$M(q)\ddot{q} + C(q,\dot{q}) + k(q) = Bu \tag{12.6}$$

with $k_i(q) = \frac{\partial V}{\partial q_i}(q)$, $i \in \underline{n}$, and B the $n \times m$ matrix

$$B = \begin{pmatrix} I_m \\ 0 \end{pmatrix}, \tag{12.7}$$

yielding the 2n-dimensional standard state space system

$$\frac{d}{dt}\begin{pmatrix} q \\ \dot{q} \end{pmatrix} = \begin{pmatrix} \dot{q} \\ -M^{-1}(q)\left(C(q,\dot{q}) + k(q)\right) \end{pmatrix} + \begin{pmatrix} 0 \\ M^{-1}(q)B \end{pmatrix}u \tag{12.8}$$

Example 12.1 Consider the rigid frictionless two-link robot manipulator treated in Example 1.1, with generalized configuration coordinates $q_1 = \theta_1$, $q_2 = \theta_2$ (relative angles), and as controls u_1 and u_2 the torques at the joints. Alternatively one can consider the same system with only one torque present (at the upper or lower joint). □

Equations (12.8) constitute a class of mechanical control systems which is often encountered in applications. However control or input variables in mechanical systems do not necessarily have to appear as external forces as in (12.4) or (12.8), as is illustrated by the following simple example.

Example 12.2 Consider a linear mass-spring system attached to a moving frame

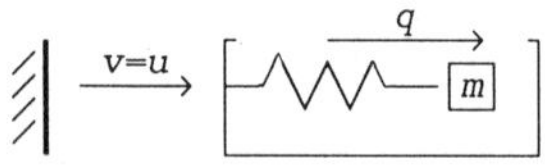

Fig. 12.1. Moving linear mass-spring system.

where the input u is the velocity of the frame. In this case the kinetic energy $\frac{1}{2}m(\dot{q} + u)^2$ depends directly on u, and so does the Lagrangian function $L(q,\dot{q},u) = \frac{1}{2}m(\dot{q} + u)^2 - \frac{1}{2}kq^2$, yielding the equations of motion (cf. (12.3) with $F_i^e = 0$)

$$\frac{d}{dt}\left(\frac{\partial L}{\partial \dot{q}}(q,\dot{q},u)\right) - \frac{\partial L}{\partial q}(q,\dot{q},u) = m(\ddot{q} + \dot{u}) - kq = 0. \qquad \square$$

In general, for a mechanical system of n degrees of freedom with a Lagrangian $L(q,\dot{q},u)$ depending directly on u we obtain, in the absence of other external forces, the equations of motion

$$\frac{d}{dt}\left[\frac{\partial L}{\partial \dot{q}_i}(q,\dot{q},u)\right] - \frac{\partial L}{\partial q_i}(q,\dot{q},u) = 0, \qquad i \in \underline{n}. \tag{12.9}$$

Notice that (12.4) can be regarded as a *special* case of (12.9) by taking in (12.9) the Lagrangian

$$L(q,\dot{q},u) = L_0(q,\dot{q}) + \sum_{j=1}^{m} q_j u_j . \tag{12.10}$$

We will call (12.9) a *Lagrangian control system.*

Let us now pass on to the *Hamiltonian* formulation. For the Lagrangian control system (12.9) we define the generalized momenta

$$p_i = \frac{\partial L}{\partial \dot{q}_i}(q,\dot{q},u), \qquad i \in \underline{n}. \tag{12.11}$$

In general the $n \times n$ matrix with (i,j)-th element $\dfrac{\partial^2 L}{\partial \dot{q}_i \partial \dot{q}_j}$ will be non-singular everywhere (for example if L is as defined in (12.10) with $L_0 = T - V$), implying that $p = (p_1,\ldots,p_n)$ are independent functions. One now defines the Hamiltonian function $H(q,p,u)$ as the Legendre transform of $L(q,\dot{q},u)$, i.e.

$$H(q,p,u) = \sum_{i=1}^{n} p_i \dot{q}_i - L(q,\dot{q},u), \tag{12.12}$$

where $\dot{q}$ and p are related by (12.11). Since by (12.11) the partial derivatives of the right-hand side of (12.12) with respect to $\dot{q}_i$ are all zero, one immediately concludes that H indeed does not depend on $\dot{q}$. It is well-known that with (12.11) and (12.12) the Euler-Lagrange equations (12.9) transform into the *Hamiltonian* equations of motion

$$\dot{q}_i = \frac{\partial H}{\partial p_i}(q,p,u), \tag{12.13a}$$

$$i \in \underline{n}.$$

$$\dot{p}_i = -\frac{\partial H}{\partial q_i}(q,p,u), \tag{12.13b}$$

Indeed (12.13b) follows from substituting (12.11) into (12.9), while (12.13a) follows from (12.12). We call (12.13) a *Hamiltonian control system*. A main advantage of (12.13) in comparison with (12.9) is that (12.13) immediately constitutes a control system in standard state space form, with *state* variables (q,p) (in physics usually called the *phase* variables). Moreover, as we will also see later on, the variables q and p are conjugate variables and the Hamiltonian $H(q,p,u)$ can be directly related to the *energy* of the system. In particular, if $L(q,\dot{q},u)$ is given as in (12.10) then it immediately follows that

$$H(q,p,u) = H_0(q,p) - \sum_{j=1}^{m} q_j u_j \tag{12.14}$$

with $H_0(q,p)$ the Legendre transform of $L_0(q,\dot{q})$. If L_0 is given as $T(q,\dot{q}) - V(q)$ with T as in (12.5) then it follows that (since $p = M(q)\dot{q}$)

$$H_0(q,p) = \frac{1}{2} p^T M^{-1}(q) p + V(q), \tag{12.15}$$

which is the *internal energy* of the system.

Example 12.3 Consider again the two-link rigid robot manipulator from Example 1.1. Denote $\theta_1 = q_1$, $\theta_2 = q_2$. The total kinetic energy is given as $T(q,\dot{q}) = \frac{1}{2}\dot{q}^T M(q)\dot{q}$, with $M(q)$ as in (1.7a). Take for simplicity $m_1 = m_2 = 1$ and $\ell_1 = \ell_2 = 1$. Then the generalized momenta are given as

$$p_1 = \frac{\partial T}{\partial \dot{q}_1} = (3 + 2\cos q_2)\dot{q}_1 + (1 + \cos q_2)\dot{q}_2,$$

$$p_2 = \frac{\partial T}{\partial \dot{q}_2} = (1 + \cos q_2)\dot{q}_1 + \dot{q}_2,$$

while the Hamiltonian $H(q,p,u)$ is given as (see also Example 3.40)

$$\frac{1}{2} p^T M^{-1}(q) p + V(q) - q_1 u_1 - q_2 u_2 =$$

$$(1 + \sin^2 q_2)^{-1}\left(\tfrac{1}{2}p_1^2 - (1 + \cos q_2)p_1 p_2 + \tfrac{1}{2}(3 + 2\cos q_2)p_2^2\right)$$

$$- 2g\cos q_1 - g\cos(q_1 + q_2) - q_1 u_1 - q_2 u_2 .$$

□

Example 12.4 Consider the system of Example 12.2 with $L = \frac{1}{2}m(\dot{q} + u)^2 - \frac{1}{2}kq^2$. We obtain $p = m(\dot{q} + u)$ and $H(q,p,u) = H_0(q,p) - up$, with $H_0(q,p) = \frac{1}{2m}p^2 + \frac{1}{2}kq^2$ being the internal energy. □

Example 12.5 Consider k point masses m_i, with positions $q^i \in \mathbb{R}^3$, $i \in \underline{k}$, in their own gravitational field corresponding to the potential energy $V(q^1,\ldots,q^k) = \sum_{i<j} m_i m_j / \|q^i - q^j\|$. Suppose the *positions* of the first ℓ masses

($\ell \leq k$) can be *controlled*. We obtain a Hamiltonian control system

$$\dot{q}_j^i = \frac{\partial H}{\partial p_j^i}, \quad \dot{p}_j^i = -\frac{\partial H}{\partial q_j^i}, \quad i = \ell+1,\ldots,k, \; j = 1,2,3,$$

with $H(q^{\ell+1},..,q^k,p^{\ell+1},..,p^k,u^1,..,u^\ell) = V(u^1,..,u^\ell,q^{\ell+1},..,q^k) + \sum_{i=\ell+1}^{k} \frac{1}{2m_i} \|p^i\|^2$. □

In this chapter we mainly confine ourselves to *affine* Hamiltonian control systems, i.e., Hamiltonian control systems (12.13) with a Hamiltonian of the form

$$H(q,p,u) = H_0(q,p) - \sum_{j=1}^{m} H_j(q,p)u_j, \tag{12.16}$$

where $H_0(q,p)$ is the *internal* Hamiltonian (energy) and $H_j(q,p)$, $j \in \underline{m}$, are the *interaction* or *coupling* Hamiltonians.

We now come to the definition of the *natural outputs* of a Hamiltonian control system (12.13), which we define as

$$y_j = -\frac{\partial H}{\partial u_j}(q,p,u), \qquad j \in \underline{m}. \tag{12.13c}$$

In particular for an affine Hamiltonian (12.16) we have $y_j = H_j(q,p)$, $j \in \underline{m}$, and we obtain the *affine Hamiltonian input-output system*, or briefly *Hamiltonian system*

$$\begin{aligned}
\dot{q}_i &= \frac{\partial H_0}{\partial p_i}(q,p) - \sum_{j=1}^{m} \frac{\partial H_j}{\partial p_i}(q,p)u_j, \\
\dot{p}_i &= -\frac{\partial H_0}{\partial q_i}(q,p) + \sum_{j=1}^{m} \frac{\partial H_j}{\partial q_i}(q,p)u_j,
\end{aligned} \qquad i \in \underline{n}, \tag{12.17}$$

$$y_j = H_j(q,p), \qquad j \in \underline{m}.$$

There are several reasons for adopting this definition of natural outputs. First in this way *duality* between inputs and outputs is induced. For instance, if $u_1,\ldots,u_m$ are generalized external forces, then $y_1,..,y_m$ will be the corresponding generalized configuration coordinates. (This is usually called the case of *collocated* sensors and actuators.) Secondly a strong type of symmetry or *reciprocity* between the inputs and the natural outputs results, as will become clear especially in Section 12.2. Thirdly, differentiating the internal energy $H_0(q,p)$ along the system (12.17) we obtain the *energy balance*

$$\frac{dH_0}{dt} = \sum_{i=1}^{n}\left(\frac{\partial H_0}{\partial q_i}\dot{q}_i + \frac{\partial H_0}{\partial p_i}\dot{p}_i\right) =$$

$$\sum_{i=1}^{n}\left(\frac{\partial H_0}{\partial q_i}\frac{\partial H_0}{\partial p_i} - \frac{\partial H_0}{\partial p_i}\frac{\partial H_0}{\partial q_i}\right) + \sum_{j=1}^{m}\sum_{i=1}^{n}\left(-\frac{\partial H_0}{\partial q_i}\frac{\partial H_j}{\partial p_i} + \frac{\partial H_0}{\partial p_i}\frac{\partial H_j}{\partial q_i}\right)u_j =$$

$$= 0 + \sum_{j=1}^{m}\sum_{i=1}^{n}\left[\frac{\partial H_j}{\partial q_i}\left(\frac{\partial H_0}{\partial p_i} - \sum_{k=1}^{m}u_k\frac{\partial H_k}{\partial p_i}\right) + \frac{\partial H_j}{\partial p_i}\left(-\frac{\partial H_0}{\partial q_i} + \sum_{k=1}^{m}u_k\frac{\partial H_k}{\partial q_i}\right)\right]u_j$$

$$= \sum_{j=1}^{m}\frac{dH_j}{dt}u_j = \sum_{j=1}^{m}\dot{y}_j u_j \,, \tag{12.18}$$

expressing that the increase of internal energy equals the total external work performed on the system (the time-integral of $\sum_{j=1}^{m}\dot{y}_j u_j$).

Example 12.3 (continued) The natural outputs are the relative angles $q_1 = \theta_1$, $q_2 = \theta_2$. If, on the other hand, the controls would be the horizontal and vertical forces at the endpoint of the manipulator, then the natural outputs are the Cartesian coordinates of the end-point, cf. (1.9). □

Example 12.4 (continued) The natural output is the momentum p. Notice that in this case $\dot{y} = p$ is the *reaction force* of the moving frame, so that $u\dot{y}$ is again the instantaneous external work. □

Example 12.5 (continued) Natural outputs are $y^i = -\dfrac{\partial V}{\partial u^i}$, $i \in \underline{\ell}$, i.e. the forces as experienced by the first ℓ controlled point masses. This is an example of a *non-affine* Hamiltonian system. □

The adopted definition of a Lagrangian or Hamiltonian control system basically only covers internally *conservative* mechanical systems. In particular it follows from (12.18) that for $u = 0$ the internal Hamiltonian $H_0(q,p)$ is a *conserved quantity*, expressing the conservation of internal energy. On the other hand, in practice mechanical systems always *do* possess inherent damping (which, however, is often difficult to quantify). Nevertheless the conservative idealization (disregarding dissipative forces) is usually a natural starting point for analysis, as well as for control purposes. In fact the actual presence of unmodelled damping will in many cases only *improve* the characteristics of the controlled system, see also Section 12.3.

Dissipation of energy *can* be included from the very beginning by adding to the Euler-Lagrange equations (12.3) a dissipation term in the following way

$$\frac{d}{dt}\left(\frac{\partial L_0}{\partial \dot{q}_i}\right) - \frac{\partial L}{\partial q_i} + \frac{\partial R}{\partial \dot{q}_i} = F_i^e, \qquad i \in \underline{n}, \tag{12.19}$$

where $R(\dot{q})$ is called a *Rayleigh's dissipation* function. This will be further elaborated in Section 12.3.

12.1 Definition of a Hamiltonian Control System

We now want to give a *global*, and coordinate-free, definition of an (affine) Hamiltonian control system (12.17). First we have to give an intrinsic definition of the fact that the state space for an affine Hamiltonian system has local coordinates $q_1,\dots,q_n,p_1,\dots,p_n$, where the configuration coordinates $q_1,\dots,q_n$ are in some sense *conjugate*, or *dual*, to the momentum coordinates $p_1,\dots,p_n$.

Definition 12.6 *Let M be an manifold and let $C^\infty(M)$ be the smooth real functions on M. A Poisson structure on M is a bilinear map from $C^\infty(M) \times C^\infty(M)$ into $C^\infty(M)$, called the Poisson bracket and denoted as*

$$(F,G) \mapsto \{F,G\}, \qquad F,G \in C^\infty(M), \tag{12.20}$$

which satisfies for any $F,G,H \in C^\infty(M)$ the following properties

$$\{F,G\} = -\{G,F\} \qquad (\textit{skew-symmetry}), \tag{12.21a}$$

$$\{F,\{G,H\}\} + \{G,\{H,F\}\} + \{H,\{F,G\}\} = 0 \quad (\textit{Jacobi identity}), \tag{12.21b}$$

$$\{F,GH\} = \{F,G\}H + G\{F,H\} \quad (\textit{Leibniz' rule}). \tag{12.21c}$$

M together with a Poisson structure is called a Poisson manifold.

Remark It follows from (12.21a,b) and the bilinearity, that $C^\infty(M)$ endowed with the Poisson bracket is a Lie algebra (see Definition 2.28).

Now let us define for given $F \in C^\infty(M,\mathbb{R})$ and arbitrary $x \in M$ the mapping $X_F(x)\colon C^\infty(M) \to \mathbb{R}$ as

$$X_F(x)(G) := \{F,G\}(x) \tag{12.22}$$

Then it follows from (12.21c) that $X_F(x)(GH) = \{F,G\}(x)H(x) + G(x)\{F,H\}(x)$ and thus (see Definition 2.21) $X_F(x) \in T_xM$ for any $x \in M$. Hence for any F we obtain a smooth vectorfield X_F on M satisfying

$$X_F(G) = \{F,G\}. \tag{12.23}$$

X_F is called the *Hamiltonian vectorfield* corresponding to the Hamiltonian function F.

Example 12.7 Let $M = \mathbb{R}^{2n}$ with natural coordinates $(x_1,\ldots,x_{2n}) =: (q_1,\ldots,q_n,p_1,\ldots,p_n)$. Define the Poisson bracket of two functions $F(q,p)$, $G(q,p)$ as

$$\{F,G\}(q,p) = \sum_{i=1}^{n} \left(\frac{\partial F}{\partial p_i}\frac{\partial G}{\partial q_i} - \frac{\partial F}{\partial q_i}\frac{\partial G}{\partial p_i}\right)(q,p) \tag{12.24}$$

(It is easily verified that this bracket satisfies (12.21).) Let $H(q,p) \in C^\infty(M)$, then the Hamiltonian vectorfield X_H is given as

$$X_H(q,p) = \left(\frac{\partial H}{\partial p_1},\ldots,\frac{\partial H}{\partial p_n}, -\frac{\partial H}{\partial q_1},\ldots, -\frac{\partial H}{\partial q_n}\right)^T(q,p). \tag{12.25}$$

Indeed

$$\begin{aligned} (\dot{q}_i =)\ X_H(q_i) &= \{H,q_i\} = \frac{\partial H}{\partial p_i} \\ (\dot{p}_i =)\ X_H(p_i) &= \{H,p_i\} = -\frac{\partial H}{\partial q_i} \end{aligned} \quad , \ i \in \underline{n}. \tag{12.26}$$

The bracket (12.34) is called the *standard* Poisson bracket on $\mathbb{R}^{2n}$. □

Example 12.8 (see also Exercise 12.1) Consider $M = \mathbb{R}^3$ with natural coordinate functions $x = (x_1,x_2,x_3)$. Define for any $F,G \in C^\infty(\mathbb{R}^3)$ the bracket

$$\begin{aligned} \{F,G\}(x) = {} & x_1\left(\frac{\partial F}{\partial x_2}\frac{\partial G}{\partial x_3} - \frac{\partial F}{\partial x_3}\frac{\partial G}{\partial x_2}\right)(x) + x_2\left(\frac{\partial F}{\partial x_3}\frac{\partial G}{\partial x_1} - \frac{\partial F}{\partial x_1}\frac{\partial G}{\partial x_3}\right)(x) \\ & + x_3\left(\frac{\partial F}{\partial x_1}\frac{\partial G}{\partial x_2} - \frac{\partial F}{\partial x_2}\frac{\partial G}{\partial x_1}\right)(x) \end{aligned} \tag{12.27}$$

It can be verified that this bracket satisfies (12.21), and thus defines a Poisson bracket on $\mathbb{R}^3$. Consider the Hamiltonian

$$H(x) = \frac{x_1^2}{2a_1} + \frac{x_2^2}{2a_2} + \frac{x_3^2}{2a_3} \tag{12.28}$$

for certain constants a_1,a_2,a_3. Then the Hamiltonian vectorfield X_H becomes

$$\dot{x}_1 = \{H, x_1\} = \frac{a_2 - a_3}{a_2 a_3} x_2 x_3,$$

$$\dot{x}_2 = \{H, x_2\} = \frac{a_3 - a_1}{a_3 a_1} x_3 x_1, \tag{12.29}$$

$$\dot{x}_3 = \{H, x_3\} = \frac{a_1 - a_2}{a_1 a_2} x_1 x_2.$$

Identifying x_i with $a_i \omega_i$, $i = 1,2,3$, we have obtained the *Euler equations* for the dynamics of the angular velocities of a rigid body; see Example 1.2 and Example 3.24. □

The map $F \mapsto X_F$ is a *linear* map from $C^\infty(M)$ to $V^\infty(M)$, the linear space of smooth vectorfields on M. Actually this map is a *Lie algebra morphism* from $C^\infty(M)$, endowed with the Poisson bracket, to $V^\infty(M)$, endowed with the usual Lie bracket of vectorfields:

Lemma 12.9 *For any $F,G \in C^\infty(M)$ we have*

$$[X_F, X_G] = X_{\{F,G\}}. \tag{12.30}$$

Proof Let $H \in C^\infty(M)$. By (12.21b) and (12.23)

$$[X_F, X_G](H) = X_F(X_G(H)) - X_G(X_F(H)) =$$

$$\{F,\{G,H\}\} - \{G,\{F,H\}\} = \{\{F,G\},H\} = X_{\{F,G\}}(H).$$ □

A diffeomorphism $\varphi\colon M \to M$ for a Poisson manifold M is called a *Poisson automorphism* if

$$\{F\circ\varphi, G\circ\varphi\} = \{F,G\}\circ\varphi, \quad \text{for all } F,G \in C^\infty(M), \tag{12.31}$$

and a vectorfield X on M is called an *infinitesimal Poisson automorphism* if the time t-integrals X^t of X (cf. (2.77)) are Poisson automorphisms for all t for which X^t is defined, i.e. for any F,G

$$\{F\circ X^t, G\circ X^t\} = \{F,G\}\circ X^t, \quad \text{for } t \geq 0. \tag{12.32}$$

Lemma 12.10 *Let M be a Poisson manifold. A vectorfield X on M is an infinitesimal Poisson automorphism if and only if*

$$X(\{F,G\}) = \{X(F),G\} + \{F,X(G)\}, \quad \textit{for all } F,G \in C^\infty(M). \tag{12.33}$$

Proof First note that for any t_0

$$\frac{d}{dt}\{F\circ X^t, G\circ X^t\}\circ X^{-t}\Big|_{t=t_0} = -\left(X(\{F\circ X^{t_0}, G\circ X^{t_0}\})\right)\circ X^{-t_0} +$$

$$\{X(F\circ X^{t_0}), G\circ X^{t_0}\}\circ X^{-t_0} + \{F\circ X^{t_0}, X(G\circ X^{t_0})\}\circ X^{-t_0}. \tag{12.34}$$

Now let (12.32) be satisfied, then $\{F\circ X^t, G\circ X^t\}\circ X^{-t} = \{F,G\}$, and thus the left-hand side of (12.34) vanishes. Taking $t_0 = 0$ in (12.34) we obtain (12.33). Conversely if (12.33) is satisfied then the right-hand side of (12.34) is zero for any t_0, and thus $\{F\circ X^t, G\circ X^t\}\circ X^{-t} = \{F,G\}$, which implies (12.32). □

Notice that for a Hamiltonian vectorfield X_H the Jacobi-identity yields $X_H\{F,G\} = \{H,\{F,G\}\} = \{\{H,F\},G\} + \{F,\{H,G\}\} = \{X_H(F),G\} + \{F,X_H(G)\}$, and thus X_H satisfies (12.33). Hence any Hamiltonian vectorfield is necessarily an infinitesimal Poisson automorphism.

Let M be a Poisson manifold with local coordinates $x_1,\dots,x_r$. Then there exist locally smooth functions $w_{ij}(x)$, $i,j\in \underline{r}$, such that the Poisson bracket is given as

$$\{F,G\}(x) = \sum_{i,j=1}^{r} w_{ij}(x)\frac{\partial F}{\partial x_i}(x)\frac{\partial G}{\partial x_j}(x). \tag{12.35}$$

Indeed, since

$$\{F,G\}(x) = (X_F G)(x) = dG(X_F)(x), \text{ and}$$
$$\{F,G\}(x) = -\{G,F\}(x) = -(X_G F)(x) = -dF(X_G)(x),$$

the Poisson bracket $\{F,G\}(x)$ only depends on $dF(x)$ and $dG(x)$, and thus is of the form (12.35). Furthermore from $\{F,G\} = -\{G,F\}$ it follows that

$$w_{ij}(x) = -w_{ji}(x), \qquad i,j\in \underline{r}. \tag{12.36}$$

Also note that w_{ij} is directly determined by

$$w_{ij}(x) = \{x_i, x_j\}, \qquad i,j\in r, \tag{12.37}$$

and from the Jacobi-identity $\{x_i,\{x_j,x_k\}\} + \{x_j,\{x_k,x_i\}\} + \{x_k,\{x_i,x_j\}\} = 0$ it thus follows that

$$\sum_{\ell=1}^{r}\left(w_{\ell j}\frac{\partial w_{ik}}{\partial x_\ell} + w_{\ell i}\frac{\partial w_{kj}}{\partial x_\ell} + w_{\ell k}\frac{\partial w_{ji}}{\partial x_\ell}\right) = 0, \qquad i,j,k\in \underline{r}. \tag{12.38}$$

Conversely, if some functions w_{ij}, $i,j\in \underline{r}$, satisfy (12.36) and (12.38), then by (12.35) they define a Poisson bracket. We conclude that locally any Poisson bracket is determined by a skew-symmetric matrix

$$W(x) = \big(w_{ij}(x)\big)_{\substack{i=1,\dots,r\\ j=1,\dots,r}} \tag{12.39}$$

with $w_{ij}(x)$ satisfying (12.38). The matrix $W(x)$ is called the *structure matrix*. The *rank* of the Poisson bracket in every $x \in M$ is simply defined as the rank of the structure matrix $W(x)$. Since $W(x)$ is skew-symmetric necessarily its rank is *even*. A Poisson bracket is said to be *non-degenerate* if rank $W(x) = \dim M$ for every $x \in M$. In particular for a non-degenerate Poisson bracket we have for any x

$$\operatorname{rank} W(x) = \dim M = 2n, \text{ for some } n. \tag{12.40}$$

Example 12.7 (continued) In this case

$$W(x) = \begin{bmatrix} 0 & -I_n \\ I_n & 0 \end{bmatrix}, \quad x \in \mathbb{R}^{2n}, \tag{12.41}$$

and thus the standard Poisson bracket is non-degenerate. □

Example 12.8 (continued) Here

$$W(x) = \begin{bmatrix} 0 & x_3 & -x_2 \\ -x_3 & 0 & x_1 \\ x_2 & -x_1 & 0 \end{bmatrix} \tag{12.42}$$

which has rank 2 at any point $(x_1, x_2, x_3) \neq (0,0,0)$. □

The following theorem shows that locally every *non-degenerate* Poisson structure is as the standard Poisson bracket as given in Example 12.7.

Theorem 12.11 *Let M be a $2n$-dimensional manifold with non-degenerate Poisson bracket $\{\,,\}$. Then locally around any $x_0 \in M$ we can find coordinates $(q,p) = (q_1,\dots,q_n,p_1,\dots,p_n)$, called canonical coordinates, such that*

$$\{F,G\}(q,p) = \sum_{i=1}^{n} \Big(\frac{\partial F}{\partial p_i}\frac{\partial G}{\partial q_i} - \frac{\partial F}{\partial q_i}\frac{\partial G}{\partial p_i}\Big)(q,p) \tag{12.43}$$

Remark Equivalently, $q_1,\dots,q_n,p_1,\dots,p_n$ are *canonical coordinates* if and only if

$$\{p_i, q_j\} = \delta_{ij}, \ \{q_i, q_j\} = 0, \ \{p_i, p_j\} = 0, \ i,j \in \underline{n}. \tag{12.44}$$

In the proof of Theorem 12.11 we use

Lemma 12.12 *Let $G_1,\dots,G_k$, $H \in C^\infty(M)$, and $F \in C^\infty(\mathbb{R}^k)$. Then*

$$\{F(G_1,\ldots,G_k),H\}(x) = \sum_{i=1}^{k} \frac{\partial F}{\partial x_i}(G_1(x),\ldots,G_k(x))\{G_i,H\}(x) \tag{12.45}$$

Proof Follows immediately from the coordinate expression (12.35). □

Proof of Theorem 12.11 Take any function q_1 with $dq_1(x_0) \neq 0$. Since the Poisson bracket is non-degenerate it follows that $X_{q_1}(x_0) \neq 0$. Then by the Flow-Box Theorem (Theorem 2.26) there exist coordinates $x'_1,\ldots,x'_{2n}$ around x_0 such that $X_{q_1} = \frac{\partial}{\partial x'_1}$. Denote $p_1 := -x'_1$. It follows that $\{p_1,q_1\} = -\{q_1,p_1\} = -X_{q_1}(p_1) = 1$. Furthermore by Lemma 12.9, $[X_{p_1},X_{q_1}] = X_{\{p_1,q_1\}} = 0$. Hence by Theorem 2.26 we can find independent functions $x_3,\ldots,x_{2n}$ such that

$$X_{q_1}(x_i) = \{q_1,x_i\} = 0 = X_{p_1}(x_i) = \{p_1,x_i\}, \tag{12.46}$$

for $i = 3,\ldots,2n$. Since $-X_{q_1}(p_1) = 1 = X_{p_1}(q_1)$ it follows that $q_1,p_1,x_3,\ldots,x_{2n}$ form a coordinate system around x_0. Then by the Jacobi-identity

$$\{\{x_i,x_j\},q_1\} = 0 = \{\{x_i,x_j\},p_1\}, \tag{12.47}$$

and thus by Lemma 12.12

$$\frac{\partial\{x_i,x_j\}}{\partial p_1}\{p_1,q_1\} = 0 = \frac{\partial\{x_i,x_j\}}{\partial q_1}\{q_1,p_1\}. \tag{12.48}$$

It follows that the functions $\{x_i,x_j\}$, $i = 3,\ldots,2n$ do not depend on q_1 and p_1, so that

$$\{x_i,x_j\} = v_{ij}(x_3,\ldots,x_{2n}) \tag{12.48}$$

for some functions v_{ij}, $i,j = 3,\ldots,2n$, which define a non-degenerate Poisson structure on $\mathbb{R}^{2n-2}$. The theorem now follows by induction to n. □

Recall that for any Poisson bracket the Hamiltonian vectorfields are necessarily infinitesimal Poisson automorphisms. The converse is generally not true as illustrated by

Example 12.13 Consider $\mathbb{R}^3$ with natural coordinates $x = (x_1,x_2,x_3)$ and Poisson bracket

$$\{F,G\}(x) = \left(\frac{\partial F}{\partial x_2}\frac{\partial G}{\partial x_1} - \frac{\partial F}{\partial x_1}\frac{\partial G}{\partial x_2}\right)(x) \tag{12.49}$$

Then $X = \frac{\partial}{\partial x_3}$ is easily seen to satisfy (12.33), and so is an infinitesimal Poisson automorphism. However X is not a Hamiltonian vectorfield since this would require the existence of a function $H(x)$ satisfying $1 = \dot{x}_3 = \{H,x_3\}$. However by (12.49) $\{H,x_3\} = 0$ for any H. □

The following proposition shows that for *non-degenerate* Poisson brackets the converse *does* hold, at least locally.

Proposition 12.14 *Let M be a Poisson manifold with non-degenerate Poisson bracket. Let the vectorfield X on M be an infinitesimal Poisson automorphism. Then locally around any $x_0 \in M$ there exists a function H such that $X = X_H$.*

Proof By Theorem 12.11 we can take canonical coordinates $(q,p) = (q_1,\ldots,q_n,p_1,\ldots,p_n)$ around x_0. Write the vectorfield X in these coordinates as $X = (X_1^q,\ldots,X_n^q,X_1^p,\ldots,X_n^p)^T$. Since X is an infinitesimal Poisson automorphism we obtain for any $i,j \in \underline{n}$

$$0 = X(\delta_{ij}) = X(\{p_i,q_j\}) = \{X(p_i),q_j\} + \{p_i,X(q_j)\} =$$

$$= \{X_i^p,q_j\} + \{p_i,X_j^q\} = \frac{\partial X_i^p}{\partial p_j} + \frac{\partial X_j^q}{\partial q_i}.$$

Similarly

$$0 = X\{q_i,q_j\} = \{X_i^q,q_j\} + \{q_i,X_j^q\} = \frac{\partial X_i^q}{\partial p_j} - \frac{\partial X_j^q}{\partial p_i},$$

$$0 = X\{p_i,p_j\} = \{X_i^p,p_j\} + \{p_i,X_j^p\} = -\frac{\partial X_i^p}{\partial q_j} + \frac{\partial X_j^p}{\partial q_i}.$$

Thus the one-form represented by the row-vector $(-X_1^p,\ldots,-X_n^p,\ X_1^q,\ldots,X_n^q)$ is *closed* (cf. 2.163), and hence locally there exists $H(q,p)$ such that $-X_i^p = \frac{\partial H}{\partial q_i}$, $X_i^q = \frac{\partial H}{\partial p_i}$, $i \in \underline{n}$. □

A Poisson automorphism $\varphi\colon M \to M$ for a non-degenerate Poisson bracket is usually called a *canonical mapping*. Also, a *coordinate transformation* S mapping canonical coordinates $(q,p) = (q_1,\ldots,q_n,p_1,\ldots,p_n)$ for a non-degenerate Poisson bracket into new coordinates $(Q,P) = (Q_1,\ldots,Q_n,P_1,\ldots,P_n)$ is called a *canonical coordinate transformation* if (Q,P) are again canonical coordinates, i.e. $\{P_i,Q_j\} = \delta_{ij}$, $\{Q_i,Q_j\} = 0 = \{Q_i,P_j\}$, $i,j \in \underline{n}$.

For later use, see Section 12.4, we mention that to any non-degenerate Poisson structure another, *dual*, geometric object can be associated. Namely, let $\{,\}$ be a non-degenerate Poisson bracket on M, locally given by the skew-symmetric structure matrix $W(x)$. Define then the bilinear map

$$\omega_x : T_xM \times T_xM \to \mathbb{R} \tag{12.50}$$

by setting

$$\omega_x(X_F(x), X_G(x)) = \{F,G\}(x) \tag{12.51}$$

Although ω_x is in principle now only defined on tangent vectors of the form $X_F(x) \in T_xM$ for $F \in C^\infty(M)$ it follows by linearity and the fact that we can choose $2n$ functions $F_1, \ldots, F_{2n}$ such that $X_{F_1}(x), \ldots, X_{F_{2n}}(x)$ are *independent* that (12.51) defines ω_x completely as a bilinear map (12.50). Since in local coordinates the column vector $X_H(x)$ for $H \in C^\infty(M)$ is given as $\left(dH(x)\omega_x\right)^T$, with $dH(x)$ a row vector, it follows in view of (12.35) that ω_x has the matrix representation

$$[\omega_x] = -\left(W(x)\right)^{-1} \tag{12.52}$$

By letting x vary we obtain a so-called differential two-form ω, which is called a *symplectic form* on M. In canonical coordinates (q,p) ω_x equals the constant matrix $W(x)$ in (12.41). Furthermore note that by (12.51), and $\{F,G\} = -\{G,F\} = -X_G(F) = -dF(G)$, the Hamiltonian vectorfield X_H corresponding to the Hamiltonian H is uniquely determined by the relation

$$\omega_x(X_H(x), Z) = -dH(x)(Z), \text{ for any } Z \in T_xM,\ x \in M. \tag{12.53}$$

The manifold M endowed with the symplectic form ω is called a *symplectic manifold.*

We are now able to give a *coordinate free* definition of a Hamiltonian control system (12.17).

Definition 12.15 *Let M be a manifold with non-degenerate Poisson bracket. Let $H_0, H_1, \ldots, H_m \in C^\infty(M)$. Then*

$$\begin{aligned} \dot{x} &= X_{H_0}(x) - \sum_{j=1}^{m} X_{H_j}(x)u_j, \\ y_j &= H_j(x), \qquad j \in \underline{m}, \end{aligned} \qquad x \in M, \tag{12.54}$$

is an affine Hamiltonian input-output system, *or briefly,* Hamiltonian system.

With the aid of Theorem 12.11 we immediately obtain

Corollary 12.16 *Let (12.54) be a Hamiltonian system on M. Then around any x_0 there exist canonical coordinates $q_1,\dots,q_n,p_1,\dots,p_n$ for M, such that (12.54) takes the form (12.17).*

Remark 12.17 Sometimes it is useful to relax the definition of Hamiltonian system by requiring that H_0 is only *locally* defined, or equivalently (see Proposition 12.14) that the Hamiltonian system is given as

$$\begin{aligned} \dot{x} &= X_0(x) - \sum_{j=1}^{m} X_{H_j}(x)u_j . \\ y_j &= H_j(x), \qquad j \in \underline{m}, \end{aligned} \tag{12.54'}$$

with X_0 an infinitesimal Poisson automorphism.

Remark 12.18 If the Poisson bracket in Definition 12.15 is *degenerate*, then (12.54) will be called a *Poisson system*.

12.2 Controllability and Observability; Local Decompositions

Let us consider a Hamiltonian system (12.54). By Definition 3.29 the observation space $\mathcal{O}$ is the linear space of functions on M spanned by all repeated Lie derivatives

$$L_{X_1}L_{X_2}\cdots L_{X_k}H_j, \qquad j \in \underline{m},\ k = 0,1,2,\dots. \tag{12.55}$$

with X_i, $i \in \underline{k}$, in the set $\{X_H, -X_{H_1},\dots,-X_{H_m}\}$. Now by (12.23) $L_{X_F}H_j = \{F,H_j\}$ for all $F \in C^\infty(M)$, and thus we immediately obtain

Proposition 12.19 *The observation space $\mathcal{O}$ of (12.54) is the linear space of functions on M, containing $H_1,\dots,H_m$ and all repeated Poisson brackets*

$$\{F_1,\{F_2,\{\dots\{F_k,H_j\}\}\dots\}\}, \qquad j \in \underline{p},\ k \in \mathbb{N}, \tag{12.56}$$

with F_i, $i \in \underline{k}$, in the set $\{H_0,H_1,\dots,H_m\}$.

With regard to controllability of (12.54), we note that by Proposition 3.20 every element of the Lie algebra $\mathcal{C}_0$ characterizing local strong accessibility (see Theorem 3.21) is a linear combination of elements

$$[X_k,[X_{k-1},[\ldots,[X_1,-X_{H_j}]]\ldots]],\qquad j\in \underline{m},\ k=0,1,\ldots, \tag{12.57}$$

where X_i, $i \in \underline{k}$, is in the set $\{X_{H_0},-X_{H_1},\ldots,-X_{H_m}\}$. Using Lemma 12.9 and Proposition 12.19 we immediately obtain

Proposition 12.20 *Every element of* $\mathcal{C}_0$ *is a linear combination of Hamiltonian vectorfields* X_F *with* $F \in \mathcal{O}$.

Remark By Lemma 12.9 it also follows that every element of $\mathcal{C}_0$ is a Hamiltonian vectorfield.

Remark Proposition 12.19 en 12.20 also hold for Poisson systems.

Since the Poisson bracket for a Hamiltonian system is non-degenerate it follows that the kernel of the map $F \mapsto X_F$ defined by the Poisson bracket consists precisely of the *constant functions* on M. From Propositions 12.19, 12.20 and Theorem 3.21, Corollary 3.23, Theorem 3.32 and Corollary 3.33 the following is now immediate.

Proposition 12.21 *Suppose* $\dim d\mathcal{O}(x_0) = 2n$ $(= \dim M)$ *for the Hamiltonian system (12.54). Then the system is locally observable at* x_0, *and locally strongly accessible at* x_0. *Conversely, assume that* $\dim d\mathcal{O}(x) =$ *constant. Then (12.54) is locally observable if and only if it is locally strongly accessible.*

Example 12.22 Consider the two-link robot manipulator from Example 12.3 (see also Example 1.1), with for simplicity $m_1 = m_2 = \ell_1 = \ell_2 = 1$. The Hamiltonian $H(q,p,u) = H_0(q,p) - H_1(q,p)u_1 - H_2(q,p)u_2$ is as given in Example 12.3. It is easily seen that the observation space $\mathcal{O}$ corresponding to H_0, H_1, H_2 satisfies $\dim d\mathcal{O}(q,p) = 4$ for all points (q,p), and thus the system is locally observable, as well as strongly accessible.

If we only take one input, say u_2, and the corresponding output $y = q_2$, then things become more complicated. However it can be checked that in this case still $\dim d\mathcal{O}(q,p) = 4$ everywhere, see also Exercise 12.8. □

Let us now return to the characterization of the observation space $\mathcal{O}$ of a Hamiltonian system, as given in Proposition 12.19. Comparing this characterization with the characterization of $\mathcal{C}_0$ as given in Proposition 3.20 and replacing Lie brackets with Poisson brackets it immediately follows from Definition 3.19 and Proposition 3.20 that $\mathcal{O}$ is alternatively given as follows.

Proposition 12.23 *The observation space $\mathcal{O}$ of a Hamiltonian control system is the smallest subalgebra of $C^\infty(M)$ (under the Poisson bracket) which contains $H_1,\dots,H_m$ and satisfies $\{H_0,F\} \in \mathcal{O}$ for all $F \in \mathcal{O}$.*

Let us introduce some additional terminology. A collection $\mathcal{F}$ of smooth functions on a symplectic manifold M will be called a *function space* if

i) $\mathcal{F}$ is a linear subspace over $\mathbb{R}$ of $C^\infty(M)$. (12.58a)

ii) Let $F_1,\dots,F_s \in \mathcal{F}$, and $G:\mathbb{R}^s \to \mathbb{R}$ be a smooth function, then $G(F_1,\dots,F_s) \in \mathcal{F}$. (12.58b)

Furthermore, we call $\mathcal{F}$ a *function group* if also the following holds.

iii) Let $F_1,F_2 \in \mathcal{F}$, then $\{F_1,F_2\} \in \mathcal{F}$. (12.58c)
($\mathcal{F}$ is a subalgebra of $C^\infty(M)$ under the Poissonbracket)

Given some functions F_i, $i \in I$ (index set), on M we denote by

$$\text{span}\,\{F_i\,;\ i \in I\} \tag{12.59}$$

the smallest function space in $C^\infty(M)$ containing F_i, $i \in I$. Recall from Chapter 3 (Definition 3.19) that the subalgebra $\mathcal{C}_0$ defines a *distribution* C_0. In an analogous way the subalgebra $\mathcal{O} \subset C^\infty(M)$ now defines the *function space* $\mathcal{F}_{\mathcal{O}} :=$ span $\mathcal{O}$. In fact, since $\mathcal{O}$ is an algebra it immediately follows that $\mathcal{F}$ is a *function group* (just like C_0 is an *involutive* distribution). Furthermore note that

$$d\mathcal{O}(x) = d\mathcal{F}_{\mathcal{O}}(x), \quad \text{for any } x \in M, \tag{12.60}$$

and in particular dim $d\mathcal{O}(x) = $ dim $d\mathcal{F}_{\mathcal{O}}(x)$ for every $x \in M$.

We now want to give a Kalman decomposition of a Hamiltonian system similarly to the decomposition of a general nonlinear system as obtained in Theorem 3.51, but adapted to the Hamiltonian structure. The following theorem is crucial in doing this. First we introduce for any function space $\mathcal{F}$ its *polar group* $\mathcal{F}^\perp$ as

$$\mathcal{F}^\perp = \{G \in C^\infty(M) \mid \{G,F\} = 0,\ \forall F \in \mathcal{F}\}. \tag{12.61}$$

It immediately follows that $\mathcal{F}^\perp$ satisfies properties (12.58a,b,c) and thus that $\mathcal{F}^\perp$ is also a function group. Indeed, let $G_1,G_2 \in \mathcal{F}^\perp$ and $K:\mathbb{R}^2 \to \mathbb{R}$. Then by the Jacobi-identity

$$\{\{G_1,G_2\},F\} = -\{\{G_2,F\},G_1\} - \{\{F,G_1\},G_2\}\} = 0 \tag{12.62}$$

for any $F \in \mathcal{F}$, and thus $\{G_1,G_2\} \in \mathcal{F}^\perp$. Furthermore by Lemma 12.12

$$\{K(G_1,G_2),F\} = \frac{\partial K}{\partial x_1}(G_1,G_2)\{G_1,F\} + \frac{\partial K}{\partial x_2}(G_1,G_2)\{G_2,F\} = 0 \tag{12.63}$$

for any $F \in \mathcal{F}$, and thus $K(G_1,G_2) \in \mathcal{F}^\perp$.

Theorem 12.24 *Let M be a manifold with non-degenerate Poisson bracket, and let $\mathcal{F}$ be a function group on M. Assume that* $\dim d\mathcal{F}(x)$ *= constant, and that* $\dim d(\mathcal{F} \cap \mathcal{F}^\perp)(x)$ *= constant. Let* $\dim d\mathcal{F} = k$, *and* $\dim d(\mathcal{F} \cap \mathcal{F}^\perp) = r$. *Then locally there exist canonical coordinates $(q_1,\ldots,q_n,p_1,\ldots,p_n)$ for M such that*

$$\mathcal{F} = \text{span}\,\{q_1,\ldots,q_\ell,p_1,\ldots,p_\ell,p_{\ell+1},\ldots,p_{\ell+r}\}, \tag{12.64}$$

with $2\ell + r = k$.

Proof Since $\dim d\mathcal{F} = k$ we can locally find k independent functions $F_1,\ldots,F_k \in \mathcal{F}$ such that $\mathcal{F} = \text{span}\{F_1,\ldots,F_k\}$. Since $\mathcal{F}$ is a function group it follows that

$$\{F_i,F_j\} = w_{ij}(F_1,\ldots,F_k), \qquad i,j \in \underline{k}, \tag{12.65}$$

for some smooth functions $w_{ij} : \mathbb{R}^k \to \mathbb{R}$. It is immediately checked that these functions $w_{ij}(y_1,\ldots,y_k)$ satisfy (12.36) and (12.38), and thus define a Poisson structure $\{\ ,\ \}_{\mathbb{R}^k}$ on $\mathbb{R}^k$. Furthermore by the assumption that $\dim d(\mathcal{F} \cap \mathcal{F}^\perp)(x) = r$ it immediately follows that this Poisson structure has constant rank 2ℓ, where $2\ell = k - r$. Then by an immediate generalization of Theorem 12.11 to degenerate Poisson brackets (see Exercise 12.2), it follows that there exist coordinates $q'_1,\ldots,q'_\ell,p'_1,\ldots$ $\ldots,p'_\ell,p'_{\ell+1},\ldots,p'_{\ell+r}$ on $\mathbb{R}^k$ satisfying

$$\begin{aligned} &\{q'_i,p'_j\}_{\mathbb{R}^k} = \delta_{ij},\ i,j \in \underline{\ell}, \\ &\{q'_i,q'_j\}_{\mathbb{R}^k} = 0 = \{p'_i,p'_s\}_{\mathbb{R}^k},\ i,j \in \underline{\ell},\ s \in \underline{\ell + r}. \end{aligned} \tag{12.66}$$

Now define, with $F = (F_1,\ldots,F_k)^T$,

$$\begin{aligned} q_i &= q'_i \circ F, \qquad i = 1,\ldots,\ell, \\ p_i &= p'_i \circ F, \qquad i = 1,\ldots,\ell + r. \end{aligned} \tag{12.67}$$

In view of (12.66) the functions q_i, $i \in \underline{\ell}$, p_i, $i \in \underline{\ell + r}$, are *partial* canonical coordinates for M. A straightforward generalization of Theorem 12.11 then yields the existence of additional canonical coordinates $q_{\ell+1},\ldots,q_n,p_{\ell+r+1},\ldots,p_n$, and (12.64) results. □

Now let us return to the function group $\mathcal{F}_{\mathcal{O}}$ defined by the observation space $\mathcal{O}$. Recall that in Theorem 3.51 we assumed that the distributions C_0, ker $d\mathcal{O}$ and C_0 + ker $d\mathcal{O}$ all have constant dimension. By Proposition 12.19 and (12.60) it is immediate that if dim $d\mathcal{F}_{\mathcal{O}}(x)$ = constant then C_0 and ker $d\mathcal{O}$ have constant dimension. Furthermore it immediately follows from the definition of $\mathcal{F}_{\mathcal{O}}^{\perp}$ (see (12.61)) that C_0 + ker $d\mathcal{O}$ has constant dimension if and only if dim $d(\mathcal{F}_{\mathcal{O}} \cap \mathcal{F}_{\mathcal{O}}^{\perp})(x)$ = constant. This motivates the assumptions made in the following theorem, which is a generalization of Theorem 3.51.

Theorem 12.25 *Consider a Hamiltonian system (12.54) with observation space* $\mathcal{O}$. *Assume that* dim $d\mathcal{F}_{\mathcal{O}}(x)$ = *constant* = k *and that* dim $d(\mathcal{F}_{\mathcal{O}} \cap \mathcal{F}_{\mathcal{O}}^{\perp})(x)$ = *constant* = r. *Then locally there exist canonical coordinates*

$$(q_1,\dots,q_\ell,q_{\ell+1},\dots,q_{\ell+r},q_{\ell+r+1},\dots,q_n,p_1,\dots,p_\ell,p_{\ell+1},\dots,p_{\ell+r},p_{\ell+r+1},\dots,p_n)$$
$$=: (q^1,q^2,q^3,p^1,p^2,p^3) \tag{12.68}$$

where $k = 2\ell + r$, *such that*

$$\mathcal{F}_{\mathcal{O}} = \text{span}\,(q^1,q^2,p^1), \quad \mathcal{F}_{\mathcal{O}}^{\perp} = \text{span}\,(q^2,q^3,p^3). \tag{12.69}$$

Furthermore we have

$$\{H_0,\mathcal{F}_{\mathcal{O}}\} \subset \mathcal{F}_{\mathcal{O}}, \quad \{H_0,\mathcal{F}_{\mathcal{O}}^{\perp}\} \subset \mathcal{F}_{\mathcal{O}}^{\perp}, \tag{12.70}$$

which implies that H_0 *is locally of the form*

$$H_0(q,p) = H_0^1(q^1,p^1,q^2) + H_0^3(q^3,p^3,q^2) + (p^2)^T f(q^2) \tag{12.71}$$

for some smooth map $f : \mathbb{R}^r \to \mathbb{R}^r$. *Moreover* $H_j \in \mathcal{F}_{\mathcal{O}}$, $j \in \underline{m}$, *only depend on* q^1,q^2,p^1. *In the coordinates (12.68) the system takes the form*

$$\begin{cases} \dot{q}^1 = \dfrac{\partial H_0^1}{\partial p^1}(q^1,p^1,q^2) - \displaystyle\sum_{j=1}^m \dfrac{\partial H_j}{\partial p^1}(q^1,p^1,q^2)u_j, \\ \dot{p}^1 = -\dfrac{\partial H_0^1}{\partial q^1}(q^1,p^1,q^2) + \displaystyle\sum_{j=1}^m \dfrac{\partial H_j}{\partial q^1}(q^1,p^1,q^2)u_j, \end{cases} \tag{12.72a}$$

$$\begin{cases} \dot{q}^2 = f(q^2), & (12.72b) \\ \dot{p}^2 = -\left(\dfrac{\partial f}{\partial q^2}(q^2)\right)^T p^2 & (12.72c) \\ \qquad - \dfrac{\partial H_0^1}{\partial q^1}(q^1,p^1,q^2) - \dfrac{\partial H_0^3}{\partial q^2}(q^3,p^3,q^2) + \displaystyle\sum_{j=1}^m \dfrac{\partial H_j}{\partial q^2}(q^1,p^1,q^2)u_j, \end{cases}$$

$$\begin{cases} \dot{q}^3 = \dfrac{\partial H_0^3}{\partial p^3}(q^3,p^3,q^2), \\ \dot{p}^3 = -\dfrac{\partial H^3}{\partial q^3}(q^3,p^3,q^2), \end{cases} \tag{12.72d}$$

$$y_j = H_j(q^1,p^1,q^2), \qquad j \in \underline{m}. \tag{12.72e}$$

Furthermore, ker $d\mathcal{O}$ = span $\{\frac{\partial}{\partial q^3}, \frac{\partial}{\partial p^2}, \frac{\partial}{\partial p^3}\}$ *and* C_0 = span $\{\frac{\partial}{\partial q^1}, \frac{\partial}{\partial p^1}, \frac{\partial}{\partial p^2}\}$.

Remark As in Theorem 3.51 we can call q^2,p^2,p^3 the *unobservable* part of the system, and q^1,p^1,p^2 the *strongly accessible* part. The sub-system (12.72 a,b,e) is locally observable, while for fixed q_2 the sub-systems (12.72a,c) as well as (12.72a) are locally strongly accessible.

Proof Application of Theorem 12.24 yields (12.69). By definition of $\mathcal{O}$ we have $\{H_0,\mathcal{F}_{\mathcal{O}}\} \subset \mathcal{F}_{\mathcal{O}}$ (cf. Proposition 12.23). By the Jacobi-identity (12.21b) we have for any $F \in \mathcal{F}_{\mathcal{O}}$ and $F^{\perp} \in \mathcal{F}_{\mathcal{O}}$

$$\{\{H_0,F^{\perp}\},F\} = -\{\{F^{\perp},F\},H_0\} - \{\{F,H_0\},F^{\perp}\} = 0, \tag{12.73}$$

since $\{F,H_0\} \in \mathcal{F}_{\mathcal{O}}$. Thus $\{H_0,F^{\perp}\} \in (\mathcal{F}_{\mathcal{O}}^{\perp})^{\perp}$. For a function group $\mathcal{F}$ with dim $d\mathcal{F}(x)$ = constant it easily follows that $(\mathcal{F}_{\mathcal{O}}^{\perp})^{\perp} = \mathcal{F}_{\mathcal{O}}$, and thus (12.70) follows. Writing out (12.70) in the coordinates (12.68) yields

$$\begin{aligned} &\{H_0,q^1\} = \frac{\partial H_0}{\partial p^1} \in \mathcal{F}_{\mathcal{O}}, \quad \{H_0,p^1\} = -\frac{\partial H_0}{\partial q^1} \in \mathcal{F}_{\mathcal{O}}, \\ &\{H_0,q^2\} = \frac{\partial H_0}{\partial p^2} \in \mathcal{F}_{\mathcal{O}} \cap \mathcal{F}_{\mathcal{O}}^{\perp}, \\ &\{H_0,q^3\} = \frac{\partial H_0}{\partial p^3} \in \mathcal{F}_{\mathcal{O}}^{\perp}, \quad \{H_0,p^3\} = -\frac{\partial H_0}{\partial q^3} \in \mathcal{F}_{\mathcal{O}}^{\perp}, \end{aligned} \tag{12.74}$$

and (12.71) and (12.72) follow. □

It follows from the local representation (12.72) that the *input-output behavior* of the Hamiltonian system is the same as the input-output behavior of the lower-dimensional system (12.72a,b,e), which can be regarded as the *lower-dimensional Hamiltonian system*

$$\dot{q}^1 = \frac{\partial H_0^1}{\partial p^1}(q^1,p^1,q^2) - \sum_{j=1}^{m} \frac{\partial H_j}{\partial p^1}(q^1,p^1,q^2)u_j,$$

$$\dot{p}^1 = -\frac{\partial H_0^1}{\partial q^1}(q^1,p^1,q^2) + \sum_{j=1}^{m} \frac{\partial H_j}{\partial q^1}(q^1,p^1,q^2)u_j, \tag{12.75}$$

$$y_j = H_j(q^1,p^1,q^2), \qquad j \in \underline{m},$$

with state (q^1,p^1), driven by the autonomous dynamics (12.72b).

12.3 Stabilization of Hamiltonian Control Systems

For clarity of exposition we will restrict ourselves in this section to an important but particular subclass of Hamiltonian systems, called *simple Hamiltonian systems*. Let Q be an n-dimensional manifold, denoting the configuration space, and let T^*Q be its cotangent bundle, denoting the phase space or state space. On T^*Q there is a naturally defined Poisson bracket, which is defined in local coordinates as follows. Let $q_0 \in Q$ and let $q = (q_1,\ldots,q_n)$ be local coordinates for Q around q_0. Then there exist, see Chapter 2.2.3, natural coordinates $(q_1,\ldots,q_n,p_1,\ldots,p_n)$ for T^*Q. Now let F and G be two smooth functions on T^*Q, then we define their Poisson bracket as

$$\{F,G\}(q,p) = \sum_{i=1}^{n} \left(\frac{\partial F}{\partial p_i}\frac{\partial G}{\partial q_i} - \frac{\partial F}{\partial q_i}\frac{\partial G}{\partial p_i}\right)(q,p). \tag{12.76}$$

In order to check that this Poisson bracket is well-defined, i.e. does not depend on the particular choice of natural coordinates, we let $(\bar{q}_1,\ldots,\bar{q}_n) = \bar{q}$ be another set of local coordinates around q_0. If $\bar{q}$ is linked to q by a coordinate transformation

$$\bar{q} = S(q) \tag{12.77a}$$

(with $\bar{q}$ and q being column vectors), then it follows that $\bar{p} = (\bar{p}_1,\ldots,\bar{p}_n)$ is related to p as (see (2.149))

$$p = \bar{p}\,\frac{\partial S}{\partial q}(q) \tag{12.77b}$$

(with $\bar{p}$ and p being row vectors). It immediately follows from (12.77) that for any $F,G \in C^\infty(T^*Q)$

$$\sum_{i=1}^{n} \left(\frac{\partial \bar{F}}{\partial \bar{p}_i}\frac{\partial \bar{G}}{\partial \bar{q}_i} - \frac{\partial \bar{F}}{\partial \bar{q}_i}\frac{\partial \bar{G}}{\partial \bar{p}_i}\right)(\bar{q},\bar{p}) = \sum_{i=1}^{n} \left(\frac{\partial F}{\partial p_i}\frac{\partial G}{\partial q_i} - \frac{\partial F}{\partial q_i}\frac{\partial G}{\partial p_i}\right)(q,p), \tag{12.78}$$

where $\bar{F}$ and $\bar{G}$ are the functions F and G expressed in the new coordinates $\bar{q}$, $\bar{p}$, and where the variables $(\bar{q},\bar{p})$ in the left hand side are related to the variables (q,p) in the right hand side by (12.77). In particular, it follows that natural coordinates for T^*Q are canonical coordinates for the above Poisson bracket on T^*Q.

Definition 12.26 *A simple Hamiltonian system on T^*Q is a Hamiltonian system (12.54) where $H_0, H_1, \ldots, H_m$ are of the form (in natural coordinates (q,p) for T^*Q)*

$$H_0(q,p) = \frac{1}{2} p^T G(q) p + V(q) \tag{12.79a}$$

with $G(q)$ a positive definite symmetric $n \times n$ matrix for every q, and

$$H_j(q,p) = H_j(q), \qquad j \in \underline{m}. \tag{12.79b}$$

The expression $\frac{1}{2} p^T G(q) p$ is called the kinetic energy and $V(q)$ the potential energy.

Let us consider a simple Hamiltonian system with equilibrium point $x_0 = (q_0, p_0)$. It immediately follows from $\dot{q} = \frac{\partial H_0}{\partial p} = G(q)p$ that necessarily $p_0 = 0$. Furthermore q_0 satisfies $dV(q_0) = 0$. The following theorem is an immediate consequence of Theorem 10.3.

Theorem 12.27 *Let $(q_0, 0)$ be an equilibrium point of the simple Hamiltonian system given by (12.79). Suppose that $V(q) - V(q_0)$ is a positive definite function on some neighborhood of q_0. Then the system for $u = 0$ is stable (but not asymptotically stable).*

Proof We will show that $\mathcal{L}(q,p) := H_0(q,p) - V(q_0)$ is a Lyapunov function for the system with $u = 0$. Indeed (compare with 12.18)

$$\frac{d}{dt} \mathcal{L}(q,p) = \frac{d}{dt} H_0(q,p) = \sum_{j=1}^{n} \left(\frac{\partial H_0}{\partial q_i}\frac{\partial H_0}{\partial p_i} - \frac{\partial H_0}{\partial p_i}\frac{\partial H_0}{\partial q_i}\right)(q,p) = 0. \tag{12.80}$$

Furthermore since $G(q) > 0$, and by assumption $V(q) - V(q_0)$ is positive definite, it follows that $\mathcal{L}(q,p)$ is positive on some neighborhood of $(q_0, 0)$. Since by (12.80) $\frac{d}{dt} \mathcal{L}(q,p) = 0$ it also follows that $(q_0, 0)$ can *not* be an asymptotically stable equilibrium. □

Now we are heading for a specialization of the *stabilization* result using Lyapunov's direct method as given in Chapter 10 (in particular Theorem

10.9). It follows from Theorem 12.27 that if $V(q) - V(q_0)$ is a positive definite function on some neighborhood of a point q_0 with $dV(q_0) = 0$, then $\mathcal{L}(q,p) = H_0(q,p) - V(q_0)$ is a *Lyapunov function* for the system with equilibrium $(q_0,0)$. Furthermore the feedback proposed in (10.18) takes the form

$$\begin{aligned} \alpha_i(q,p) &= -L_{-X_{H_i}}\mathcal{L}(q,p) = L_{X_{H_i}}H_0(q,p) \\ &= X_{H_i}(H_0)(q,p) = -\{H_0,H_i\}(q,p). \end{aligned} \tag{12.81}$$

Furthermore we have by (12.79b)

$$\frac{d}{dt} y_i = \{H_0 - \sum_{j=1}^{m} u_j H_j, H_i\} = \{H_0, H_i\}, \tag{12.82}$$

and thus the feedback is simply given as

$$u_i = \alpha_i(q,p) = -\dot{y}_i, \qquad i \in \underline{m}, \tag{12.83}$$

which physically can be interpreted as adding *damping* to the system. (Notice that $y_i = H_i(q)$ can be regarded as a generalized configuration coordinate, and thus $\dot{y}_i$ as a generalized velocity.) Indeed with this choice of feedback we obtain (see (10.20))

$$\frac{d}{dt}\mathcal{L}(q,p) = \frac{d}{dt}H_0(q,p) = -\sum_{i=1}^{m}(\dot{y}_i)^2, \tag{12.84}$$

which expresses the fact that the rate of decrease of internal energy equals $\sum_{i=1}^{m}(\dot{y}_i)^2$, the dissipation of energy due to damping (Compare with (12.18), where a similar expression has been derived in a more general situation.) We now come to the following specialization of Theorem 10.9. Define the codistribution

$$P(q,p) = \text{span}\{dH_0(q,p),\ d(ad^k_{H_0} H_i)(q,p),\ i \in \underline{m},\ k \geq 0\} \tag{12.85}$$

where we have defined inductively

$$ad^0_{H_0} H_i = H_i,\ ad^k_{H_0} H_i = \{H_0, ad^{k-1}_{H_0} H_i\}, \qquad k = 1,2,\ldots \tag{12.86}$$

Theorem 12.28 *Consider the simple Hamiltonian control system (12.79) on* T^*Q. *Let* q_0 *satisfy* $dV(q_0) = 0$, *and let* $V(q) - V(q_0)$ *be positive definite on a neighborhood* U_0 *of* q_0 *such that* $dV(q) \neq 0$, $q \neq q_0$, $q \in U_0$. *Then there exists some neighborhood* W_0 *of* $(q_0,0)$ *such that* $\mathcal{L}(q,p) = H_0(q,p) - V(q_0)$ *is positive definite on* W_0 *and* $d\mathcal{L}(q,p) \neq 0$ *for all* $(q,p) \in W_0$ *with*

$(q,p) \neq (q_0,0)$. *Assume there exist subsets* W_1 *and* W_2 *of* W_0 *with* $W_1 \cap W_2 = \emptyset$ *and* $W_1 \cup W_2 = W_0$ *such that*

(i) $(q_0,0) \in W_2$,

(ii) $\dim P(q,p) = 2n, \ \forall (q,p) \in W_1$,

(iii) there exists a neighborhood $\widetilde{W}_0 \subset W_0$ *of* $(q_0,0)$ *such that* $\{(q_0,0)\}$ *is the largest invariant subset of the dynamics* $\dot{q} = \frac{\partial H_0}{\partial p}$, $\dot{p} = -\frac{\partial H_0}{\partial q}$ *in the set* $W_1 \cap \widetilde{W}_0 \cap \{(q,p) \in W_0 \mid \dot{y}_i = \{H_0,H_i\}(q,p) = 0, \ i \in \underline{m}\}$

Then the feedback (12.83) locally asymptotically stabilizes the system in $(q_0,0)$.

Proof Follows immediately from Theorem 10.9 by noting that by Lemma 12.19 $\dim P(q,p) = 2n$ if and only if $\dim D(q,p) = 2n$ where (see (10.23))

$$D(q,p) = \operatorname{span}\{X_{H_0}(q,p), \ ad^k_{X_{H_0}} X_{H_i}(q,p), \ i \in \underline{m}, \quad k \geq 0\}. \tag{12.87}$$

□

A typical special case of Theorem 12.28 is obtained when

$$\dim P(q,p) = 2n, \text{ for all } (q,p) \in W_0 \text{ with } q \neq q_0, \tag{12.88}$$

i.e., when $W_2 = \{(q,p) \in W_0 \mid q = q_0\}$. Indeed since

$$\dot{q} = \frac{\partial H_0}{\partial p} = G(q)p, \tag{12.89}$$

and $G(q) > 0$ it immediately follows that $\{(q_0,0)\}$ is the largest invariant subset contained in V_2. Furthermore we mention the following simple

Corollary 12.29 *Let* $H_1(q),\ldots,H_m(q)$ *be independent about* q_0, *and let* $m = n$. *Then* $\dim P(q,p) = 2n$, *for all* (q,p). *Hence if* $V(q) - V(q_0)$ *is positive definite on a neighborhood* U_0 *of* q_0 *such that* $dV(q) \neq 0$ *for all* $q \in U_0$ *with* $q \neq q_0$, *then (12.83) locally asymptotically stabilizes the system in* $(q_0,0)$.

Proof Since $H_1,\ldots,H_n$ are independent we may take local coordinates $\bar{q}_i = H_i$, $i \in n$ for Q. Then in corresponding natural coordinates $(\bar{q},\bar{p})$

$$\{H_0,H_i\} = \{\tfrac{1}{2}\,\bar{p}^T G(\bar{q})\bar{p} + V(\bar{q}),\bar{q}_i\} = \big(G(\bar{q})\bar{p}\big)_i \tag{12.90}$$

Since $G(\bar{q}) > 0$ for all $\bar{q}$ it follows that $\dim \{dH_i, d\{H_0,H_i\}, \ i \in \underline{n}\} = 2n$. □

Remark 12.30 Note that the feedback (12.83) can be alternatively regarded as the addition of a Rayleigh's *dissipation function* (see (12.19))

$R(\dot{q}) = \frac{1}{2}\sum_{i=1}^{m}\dot{y}_i^2$, $y_i = H_i(q)$, $i \in \underline{m}$, to the system equations for $u = 0$.

A main assumption in Theorem 12.28 was the positive definiteness of $V(q)-V(q_0)$. On the other hand, application of the linear proportional output feedback

$$u_j = - k_i y_i + v_i \quad , \quad i \in \underline{m}, \tag{12.91}$$

with v_i the new controls, to the simple Hamiltonian control system (12.79) is easily seen to result in another simple Hamiltonian control system

$$\begin{aligned} \dot{q}_i &= \frac{\partial \bar{H}_0}{\partial p_i}(q,p), \\ & \qquad\qquad\qquad\qquad\qquad i \in \underline{n}, \\ \dot{p}_i &= - \frac{\partial \bar{H}_0}{\partial q_i}(q,p) + \sum_{j=1}^{m} \frac{\partial H_j}{\partial q_i}(q) v_j , \end{aligned} \tag{12.92}$$

where $\bar{H}_0(q,p) := \frac{1}{2} p^T G(q) p + \bar{V}(q)$, and $\bar{V}(q)$ is the *new potential energy*

$$\bar{V}(q) = V(q) + \frac{1}{2} \sum_{i=1}^{m} k_i y_i^2 . \tag{12.93}$$

Hence by a feedback (12.91) we have the additional possibility of *shaping* the potential energy. The following lemmas will give a partial answer to the question when it is possible to shape the potential energy in such a way that it becomes *positive definite*.

Lemma 12.31 *Let Q be an $n \times n$ symmetric matrix and let C be a surjective $m \times n$ matrix. Then there exists an $m \times m$ symmetric matrix H such that $Q + C^T H C > 0$ if and only if Q restricted to* ker *C is positive definite. Furthermore we can take H to be a diagonal matrix.*

Proof The "only if" direction is clear. Let now Q restricted to ker C be positive definite. Let W be an $n \times (n-m)$ matrix whose columns span ker C, and let V be an $n \times m$ matrix whose columns span the orthogonal complement of $Q(\ker C)$. First we prove that the $n \times n$ matrix $(V \vdots W)$ is nonsingular. Indeed let $V\alpha + W\beta = 0$, with $\alpha \in \mathbb{R}^m$ and $\beta \in \mathbb{R}^{n-m}$. Then

$$0 = W^T Q(V\alpha + W\beta) = W^T Q W \beta .$$

Since Q restricted to ker C is positive definite this implies $\beta = 0$ and hence $\alpha = 0$. It is easy to see that

$$(V \vdots W)^T (Q + C^T HC)(V \vdots W) = \begin{bmatrix} V^T QV + V^T C^T HCV & 0 \\ 0 & W^T QW \end{bmatrix}$$

Since rank $V^T C^T HCV$ = rank $(V \vdots W)^T C^T HC(V \vdots W)$ = rank $C^T HC$ it follows that $Q + C^T HC$ can be made positive definite by choosing an appropriate $H = H^T$ (if necessary diagonal). □

Lemma 12.32 *Consider the simple Hamiltonian control system (12.79) with $dV(q_0) = 0$ and $H_j(q_0) = 0$, $j \in \underline{m}$. Assume that the matrix*

$$\left(\frac{\partial^2 V}{\partial q_i \partial q_j}(q_0) \right)_{i,j \in \underline{n}} \tag{12.94}$$

is positive definite when restricted to the subspace

$$\bigcap_{j=1}^{m} \ker dH_j(q_0) \tag{12.95}$$

Then there exists a feedback (12.91) such that $\bar{V}(q) = V(q) + \frac{1}{2} \sum_{j=1}^{m} k_j y_j^2$ is positive definite on a neighborhood U_0 of q_0, and $d\bar{V}(q) \neq 0$ for all $q \in U_0$ with $q \neq q_0$.

Proof Apply Lemma 12.31 to

$$Q = \left(\frac{\partial^2 V}{\partial q_i \partial q_j}(q_0) \right)_{i,j \in \underline{n}}, \quad C = \left(\frac{\partial H_i}{\partial q_j}(q_0) \right)_{i \in \underline{m}, j \in \underline{n}}. \tag{12.96}$$

This yields the existence of a diagonal matrix $H = \text{diag}(k_1, \ldots, k_m)$ such that $Q + C^T HC > 0$. Now consider the function $\bar{V}(q) = V(q) + \frac{1}{2} \sum_{j=1}^{m} k_j y_j^2$. Then

$$\left(\frac{\partial^2 \bar{V}}{\partial q_i \partial q_j}(q_0) \right)_{i,j \in \underline{n}} = Q + C^T HC > 0, \tag{12.97}$$

and thus $\bar{V}$ is as required. □

We conclude that if the potential energy $V(q)$ satisfies the assumptions of Lemma 12.32 then there exists a proportional output feedback (12.91) such that $\bar{V}$ as defined by (12.93) satisfies the assumptions of Theorem 12.28. Hence by Theorem 12.27 the Hamiltonian system for $v_i = 0$ is *stable*. Furthermore if also the remaining assumptions of Theorem 12.28 are met for the system with internal energy $\bar{H}_0 = H_0 + \frac{1}{2} \sum_{j=1}^{m} k_j y_j^2$, then the derivative feedback

$$v_i = -\dot{y}_i, \quad i \in \underline{m}, \tag{12.98}$$

will result in local *asymptotic* stability.

Remark 12.33 If $m = n$ and $H_1, \ldots, H_n$ are independent then the assumptions of Lemma 12.32 are automatically met. Hence by Corollary 12.29 the system can be always made locally asymptotically stable by a feedback of the form

$$u_i = -k_i y_i - \dot{y}_i, \quad i \in \underline{n}. \tag{12.99}$$

It is clear that Theorem 12.28 remains valid if we replace the feedback (12.83) by the more general expression

$$u_i = -c_i \dot{y}_i, \; c_i > 0, \qquad i \in \underline{m}. \tag{12.100}$$

Thus if the assumptions of Lemma 12.32 and Theorem 12.28 (for the system with internal energy $\bar{H}_0$) are met, then every feedback of proportional-derivative (PD) type

$$u_i = -k_i y_i - c_i \dot{y}_i, \; k_i > 0, \; c_i > 0, \quad i \in \underline{m}, \tag{12.101}$$

with k_i sufficiently large will locally asymptotically stabilize the system. Furthermore, the freedom in the choice of the gain parameters k_i, c_i, $i \in \underline{m}$, can be used for ensuring a satisfactory transient behavior (analogously to classical PD control for linear second order-systems).

Remark 12.34 Motivated by the fact that the damping terms $-c_i \dot{y}_i$, $i \in \underline{m}$, in (12.101) correspond to the Rayleigh dissipation function $\frac{1}{2}\sum_{i=1}^{m} c_i \dot{y}_i^2$, and the terms $-k_i y_i$, $i \in \underline{m}$, correspond to the extra potential energy $\frac{1}{2}\sum_{i=1}^{m} k_i y_i^2$ we could even generalize (12.101) to the "nonlinear PD controller"

$$u_i = -\frac{\partial P}{\partial y_i}(y) - \frac{\partial R}{\partial \dot{y}_i}(\dot{y}), \quad i \in \underline{m}, \tag{12.102}$$

corresponding to the addition of a general potential energy term $P(y)$ and Rayleigh dissipation function $R(\dot{y})$.

Example 12.35 (see also Example 12.3). Consider the two-link rigid robot manipulator from Example 1.1, where we take for simplicity $\ell_1 = \ell_2 = 1$, $m_1 = m_2 = 1$. Furthermore we take $u_2 = 0$. Suppose one wishes to make the configuration $q_1 = \pi$, $q_2 = \pi$ asymptotically stable by smooth feedback. First we apply the linear feedback

$$u_1 = -k(q_1 - \pi) + v_1. \tag{12.103}$$

It is easily seen that for $k > 2g$ the potential energy $V + \frac{1}{2}k(q_1 - \pi)^2$ has a unique minimum in (π,π). Since $\dim P(q,p) = 4$ the additional derivative feedback

$$v_1 = -c_1\dot{y}_1, \qquad c_1 > 0, \tag{12.104}$$

will thus result in global (except for the point $q_1 = \pi$, $q_2 = 0$) asymptotic stability. □

12.4 Constrained Hamiltonian Dynamics

In this section we make a closer study of the constrained and zero dynamics, as treated in Chapter 11, in the case of a Hamiltonian system (12.54). We confine ourselves to the case as dealt with in Proposition 11.3, i.e. we assume *throughout* that the $m \times m$ decoupling matrix

$$A(x) = \left[L_{-X_{H_j}} L_{X_{H_0}}^{\rho_i} H_i(x) \right]_{i,j \in \underline{m}} \tag{12.106}$$

has rank equal to m on the set

$$N^* = \{x \in M \mid H_i(x) = \ldots = L_{X_{H_0}}^{\rho_i} H_i(x) = 0, \quad i \in \underline{m}\} \tag{12.107}$$

Then we know from Chapter 11 that the constrained dynamics for the Hamiltonian system (12.54) are given as

$$\dot{x} = X_{H_0}(x) - \sum_{j=1}^{m} X_{H_j}(x)\alpha_j^*(x), \qquad x \in N^*, \tag{12.108}$$

where $\alpha^*(x)$ is the *unique* solution of (11.28). Moreover since $p = m$ the constrained dynamics equals the zero dynamics, cf. (11.35). We will show that because of the Hamiltonian structure the zero dynamics (12.108) has a very special form. First of all we note that the matrix $A(x)$ can be rewritten into the following more convenient way. By (12.23) we have

$$L_{X_{H_0}}^k H_i(x) = ad_{H_0}^k H_i(x), \qquad k = 0,1,\ldots,\rho_i, \qquad i \in \underline{m}, \tag{12.109}$$

where the repeated Poisson bracket $ad_{H_0}^k H_i$ is defined as in (12.86), and ρ_i is the smallest nonnegative integer such that

$$\{H_\ell, ad_{H_0}^{\rho_i} H_i\}(x) \neq 0, \quad \text{for some } \ell \in \underline{m} \text{ and } x \in M. \tag{12.110}$$

Therefore

$$A(x) = -\left(\{H_j, ad_{H_0}^{\rho_i} H_i\}(x)\right)_{i,j\in\underline{m}} = \left(\{ad_{H_0}^{\rho_i} H_i, H_j\}(x)\right)_{i,j\in\underline{m}}. \tag{12.111}$$

In particular it follows that for the computation of $A(x)$ we do not have to go through the equations of motion (12.54); the knowledge of the Hamiltonian $H_0 - \sum_{j=1}^{m} H_j u_j$ suffices.

A submanifold N of a manifold M with symplectic form ω (defined by a non-degenerate Poisson bracket, cf. (12.51)) is called a *symplectic submanifold* if the following holds. *Restrict* the bilinear form ω_x: $T_xM \times T_xM \to \mathbb{R}$, $x \in M$, to a bilinear form $\bar{\omega}_x$: $T_xN \times T_xN \to \mathbb{R}$, $x \in N$, i.e.

$$\bar{\omega}_x(X,Y) := \omega_x(X,Y), \qquad X,Y \in T_xN, \ x \in N. \tag{12.112}$$

Then N is called symplectic if $\bar{\omega}_x$ is a non-degenerate bilinear form for every $x \in N$, i.e. if the rank of a matrix representation of $\bar{\omega}_x$ equals $\dim N$ for every $x \in N$. (In particular N is even-dimensional.)

Theorem 12.36 *Consider the Hamiltonian system (12.54) on M. Assume that N^* is non-empty and that* rank $A(x) = m$ *for every $x \in N^*$. Then N^* is a symplectic submanifold of M. Moreover, denote the functions $H_i(x), ad_{H_0}H_i(x), \ldots, ad_{H_0}^{\rho_i}H_i(x)$, $i \in \underline{m}$ (which are independent on N^* by Proposition 11.3) as*

$$\psi_j(x), \ j = 1,\ldots,s := \sum_{i=1}^{m} (\rho_i + 1), \tag{12.113}$$

then the $s \times s$ skew-symmetric matrix

$$C(x) := \left(\{\psi_i, \psi_j\}(x)\right)_{i,j\in\underline{s}} \tag{12.114}$$

has rank s for every $x \in N^*$.

Proof First note that by the Jacobi-identity we have for any $i,j \in \underline{m}$

$$\{ad_{H_0}^{\rho_i} H_i, H_j\} = -\{ad_{H_0}^{\rho_i - 1} H_i, \{H_0, H_j\}\} + \{H_0, \{ad_{H_0}^{\rho_i - 1} H_i, H_j\}\}. \tag{12.115}$$

By definition of ρ_i, cf. (12.110), the last term is zero, and inductively we obtain

$$\{ad_{H_0}^{\rho_i} H_i, H_j\} = (-1)^k \{ad_{H_0}^{\rho_i - k} H_j, ad_{H_0}^{k} H_j\}, \ k = 0,1,\ldots,\rho_i. \tag{12.116}$$

By permuting the indices $1,\ldots,m$ we may assume that $\rho_1 \geq \rho_2 \geq \ldots \geq \rho_m$. First suppose that $\rho_1 > \rho_2 > \ldots > \rho_m$. Then it follows that

$$\{ad_{H_0}^{\rho_i} H_i, H_j\} = (-1)^{\rho_i} \{H_i, ad_{H_0}^{\rho_i} H_j\} = 0, \qquad \text{for } j < i. \tag{12.117}$$

Hence $A(x)$ (cf. (12.111)) is an upper triangular matrix. Since by assumption $A(x)$ is non-singular for $x \in N^*$, it follows that the diagonal elements $\{ad_{H_0}^{\rho_i} H_i, H_j\}(x)$, $x \in N^*$, $i \in \underline{m}$, are all non-zero. By (12.116) this implies that $\{ad_{H_0}^{\rho_i - k} H_i, ad_{H_0}^{k} H_i\}(x) \neq 0$, for $k = 0,1,\ldots,\rho_i$, $i \in \underline{m}$, $x \in N^*$. Hence for every function ψ_i as in (12.113) there exists another function ψ_j as in (12.113) such that $\{\psi_i, \psi_j\}(x) \neq 0$, $x \in N^*$. Now suppose that $\rho_1 = \rho_2 > \rho_3 > \ldots > \rho_m$. By the same argument $A(x)$ has the form

$$\begin{bmatrix} * & * & \cdots & & * \\ * & * & & & \vdots \\ 0 & 0 & * & & \vdots \\ \vdots & & & \ddots & \\ 0 & \cdots & & 0 & * \end{bmatrix} \tag{12.118}$$

where the 2×2 submatrix $\left(\{ad_{H_0}^{\rho_i} H_i, H_j\}\right)_{i,j=1,2}$ has rank 2. Take a fixed point $x \in N^*$. If $\{ad_{H_0}^{\rho_i} H_i, H_i\}(x) \neq 0$, $i = 1,2$, then we are in the same situation as above. If $\{ad_{H_0}^{\rho_1} H_1, H_1\}(x) = 0$ then necessarily $\{ad_{H_0}^{\rho_1} H_1, H_2\}(x) \neq 0$. But since $\rho_2 = \rho_1 = \rho$ this implies by (12.116) that $\{ad_{H_0}^{\rho-k} H_1, ad_{H_0}^{k} H_2\}(x) \neq 0$ for $k = 0,1,\ldots,\rho$. Hence, again for any function ψ_i as in (12.112) there exists another function ψ_j as in (12.113) such that $\{\psi_i, \psi_j\}(x) \neq 0$. If more integers ρ_i are equal then we proceed in the same way by looking at the corresponding non-singular submatrix of A.

Now take $x \in N^*$ and an arbitrary $X \in T_x N^*$. By definition of N^* we have $X(\psi_j)(x) = 0$ for all ψ_j defined in (12.113). Therefore

$$\omega_x(X, X_{\psi_j}(x)) = X(\psi_j)(x) = 0, \quad j \in \underline{s}, \tag{12.119}$$

and thus the vectors $X_{\psi_j}(x)$, $j \in \underline{s}$, are all elements of

$$(T_x N)^{\perp} := \left\{ Y \in T_x M \,\middle|\, \omega_x(X,Y) = 0, \quad \text{for all } X \in T_x N \right\} \tag{12.120}$$

Since $\dim (T_x N)^{\perp} = \dim T_x M - \dim T_x N$, and $\dim T_x N = \dim M - s$ it follows that the vectors $X_{\psi_j}(x)$, $j \in \underline{s}$, form a basis for $(T_x N)^{\perp}$. Since for any ψ_i there exists ψ_j such that

$$\omega_x(X_{\psi_i}(x), X_{\psi_j}(x)) = \{\psi_i, \psi_j\}(x) \neq 0, \tag{12.121}$$

it follows that for any $X \in (T_x N)^{\perp}$ there exists $Y \in (T_x N)^{\perp}$ such that $\omega_x(X,Y) \neq 0$. Thus $(T_x N)^{\perp} \cap ((T_x N)^{\perp})^{\perp} = (T_x N)^{\perp} \cap T_x N = \{0\}$, implying that for any $\widetilde{X} \in T_x N$ there exist $\widetilde{Y} \in T_x N$ such that $\omega_x(\widetilde{X}, \widetilde{Y}) \neq 0$. Thus N^* is a symplectic submanifold.

Finally suppose that $C(x)$ is *singular* in some $x \in N^*$. Then there are constants $a_1,\ldots,a_s$ such that

$$\sum_{i=1}^{s} a_i\{\psi_i,\psi_j\}(x) = 0, \qquad j \in \underline{s},$$

implying that $\sum_{i=1}^{s} a_i X_{\psi_i}(\psi_j)(x) = 0$, $j \in \underline{s}$, and thus that $\sum_{i=1}^{s} a_i X_{\psi_i}(x) \in T_x N^*$. This would yield $\omega_x(\sum_{i=1}^{s} a_i X_{\psi_i}(x),X) = \sum_{i=1}^{s} a_i \omega_x(X_{\psi_i}(x),X) = \sum_{i=1}^{s} a_i X(\psi_i)(x) = 0$, for all $X \in T_x N^*$ and thus ω_x restricted to $T_x N^*$ is degenerate, which is a contradiction. □

Remark 12.37 Similarly it can be shown that for *arbitrary* independent functions ψ_j, $j \in \underline{s}$, the non-empty submanifold $N := \{x \in M | \psi_j(x) = 0, j \in \underline{s}\}$ is symplectic if and only if the corresponding $C(x)$ is non-singular for all $x \in N$ (see Exercise 12.16).

Example 12.38 Consider a simple Hamiltonian system on $\mathbb{R}^{2n}$ (Definition 12.26) with $H_0(q,p) = \frac{1}{2} p^T G(q)p + V(q)$, and take $H_j(q) = q_j$, $j = 1,\ldots,m \le n$. Then $\rho_i = 1$, $i \in \underline{m}$, and $ad^1_{H_0}H_i = \sum_{j=1}^{n} g_{ij}(q)p_j$, $i \in \underline{m}$, with $g_{ij}(q)$ the (i,j)-th element of $G(q)$. Hence $A(q,p) = G_{11}(q)$, where $G_{11}(q)$ is the $m \times m$ leading submatrix of $G(q)$, and thus $A(q,p)$ is non-singular, implying that $N^* = \{(q,p) \in \mathbb{R}^{2n} | q_1 = \ldots = q_m = 0, \sum_{j=1}^{n} q_{1j}(q)(p_j = \ldots = \sum_{j=1}^{n} g_{mj}(q)p_j = 0\}$ is a symplectic submanifold of $\mathbb{R}^{2n}$. Denoting the functions $q_1,\ldots,q_m$, $\sum_{j=1}^{n} g_{1j}(q)p_j,\ldots,\sum_{j=1}^{n} g_{mj}(q)p_j$ as $\psi_1,\ldots,\psi_{2m}$ it follows that

$$C(q,p) = \begin{bmatrix} 0 & -G_{11}(q) \\ G_{11}(q) & S(q,p) \end{bmatrix}, \tag{12.122}$$

where S is the $m \times m$ matrix with (i,j)-th element

$$s_{ij}(q,p) = \sum_{k,\ell=1}^{n} \left(g_{i\ell} \frac{\partial g_{jk}}{\partial q_\ell} - g_{j\ell} \frac{\partial g_{ik}}{\partial q_\ell} \right)(q)p_k \tag{12.123}$$

□

Note that the non-singularity of $A(x)$ for $x \in N^*$ implies that $A(x)$ is nonsingular on a neighborhood in M of every point $x \in N^*$, and similarly for $C(x)$. In order to simplify the exposition we will henceforth assume that $A(x)$, and thus $C(x)$, is non-singular for *all* $x \in M$.

Now let us consider a symplectic non-empty submanifold N given as

$$N = \{x \in M | \psi_j(x) = 0, \quad j \in \underline{s}\} \tag{12.124}$$

for some arbitrary independent functions ψ_j, $j \in \underline{s}$ (not necessarily as in (12.113)), such that the matrix $C(x)$ as defined in (12.114) is non-singular everywhere. Using the restricted symplectic form $\bar{\omega}$ on N (cf. (12.112)), we can define for any $F \in C^{\infty}(N)$ the vectorfield $\bar{X}_F$ on N by setting (see (12.53))

$$\bar{\omega}_x(\bar{X}_F(x), Z) = - d\bar{F}(x)(Z), \text{ for any } Z \in T_xN, \ x \in N. \tag{12.125}$$

In particular, denoting the *restriction* of $H \in C^{\infty}(M)$ to N as $\bar{H} \in C^{\infty}(N)$ we can define the vectorfield $\bar{X}_{\bar{H}}$ on N. In general $\bar{X}_{\bar{H}}$ will be different from X_H (the Hamiltonian vectorfield on M with Hamiltonian H) restricted to N. In fact

Lemma 12.39 *Let $H \in C^{\infty}(M)$, then $\bar{X}_{\bar{H}} = X_H$ on N if and only if $\{H,\psi_j\} = 0$ on N for $j \in \underline{s}$. Furthermore for any $H \in C^{\infty}(M)$ define*

$$\alpha_i(x) = - \sum_{j=1}^{s} c^{ij}(x) \{H,\psi_j\}(x), \quad x \in M, \tag{12.126}$$

with $\left(c^{ij}(x)\right)_{i,j\in\underline{s}}$, the inverse matrix of $C(x)$, cf. (12.114), and

$$H^a(x) = H(x) - \sum_{i=1}^{s} \psi_i(x)\alpha_i(x), \quad x \in M, \tag{12.127}$$

then $\bar{X}_{\bar{H}} = X_{H^a}$ on N.

Proof The first statement is easily deduced from the fact that $X_H(x) \in T_xN$ for all $x \in N$ if and only if $X_H(\psi_j) = \{H,\psi_j\} = 0$ on N for $j \in \underline{s}$. The second statement follows from the fact that

$$\{H^a,\psi_k\} = \{H,\psi_k\} - \sum_{i=1}^{s} \{\psi_i,\psi_k\}\alpha_i - \sum_{i=1}^{s} \psi_i \{\alpha_i,\psi_k\}. \tag{12.128}$$

The last term on the right-hand side is zero on N. Furthermore the second term equals by definition of α_i

$$\sum_{i,j=1}^{s} c_{ik}c^{ij}\{H,\psi_j\} = - \sum_{i,j=1}^{s} c_{ki}c^{ij}\{H,\psi_j\} = - \{H,\psi_k\}, \tag{12.129}$$

implying that $\{H^a,\psi_k\} = 0$ on N for $j \in \underline{s}$. Since $\bar{H}^a = \bar{H}$ it follows from the first statement that $\bar{X}_{\bar{H}} = X_{H^a}$ on N. □

We will now show that for any $H \in C^{\infty}(M)$ the vectorfield $X_{\bar{H}}$ on N is a

Hamiltonian vectorfield on N. First we note that the restricted symplectic form $\bar{\omega}$ defines a *bracket* on N by setting

$$\{F,G\}_N(x) = \bar{\omega}_x(\bar{X}_F(x),\bar{X}_G(x)), \quad x \in N, \tag{12.130}$$

where $\bar{X}_F$ and $\bar{X}_G$ are defined as in (12.125).

Lemma 12.40 *For any $F,G \in C^\infty(N)$*

$$\{F,G\}_N(x) = \{F,G\}(x) - \sum_{i,j=1}^{s} \{F,\psi_i\}(x)c^{ij}(x)\{\psi_j,G\}(x), \quad x \in N \tag{12.131}$$

where the right-hand side is computed for any smooth extensions of F and G to a neighborhood of x in M. Furthermore $\{\ ,\ \}_N$ is a non-degenerate Poisson bracket on N, called the Dirac bracket, and for any $F \in C^\infty(N)$ the vectorfield $\bar{X}_F$ is the Hamiltonian vectorfield on N with respect to F and the Poisson bracket $\{\ ,\ \}_N$, i.e. for any $G \in C^\infty(N)$

$$\bar{X}_F(G)(x) = \{F,G\}_N(x), \quad x \in N. \tag{12.132}$$

Proof By (12.130) and Lemma 12.39 we have for all $x \in N$

$$\{F,G\}_N(x) = \bar{\omega}_x(\bar{X}_F(x),\bar{X}_G(x)) = \bar{\omega}_x(X_{F^a}(x),X_{G^a}(x)) =$$

$$\omega_x(X_{F^a}(x),X_{G^a}(x)) = \{F^a,G^a\}(x) =$$

$$\{F + \sum_{i,j=1}^{s} c^{ij}\{F,\psi_j\}\psi_i, G + \sum_{k,\ell=1}^{s} c^{k\ell}\{G,\psi_\ell\}\psi_k\}(x) =$$

$$\{F,G\}(x) + \sum_{k,\ell=1}^{s} c^{k\ell}(x)\{G,\psi_\ell\}(x)\{F,\psi_k\}(x) + \sum_{i,j=1}^{s} c^{ij}(x)\{F,\psi_j\}(x)\{\psi_i,G\}(x)$$

$$+ \sum_{i,j,k,\ell=1}^{s} c^{ij}(x)c^{k\ell}(x)c_{ik}(x)\{F,\psi_j\}(x)\{G,\psi_\ell\}(x) =$$

$$\{F,G\}(x) + \sum_{k,\ell=1}^{s} (-c^{k\ell}(x) + c^{\ell k}(x) + c^{k\ell}(x))\{F,\psi_k\}(x)\{\psi_\ell,G\}(x) =$$

$$\{F,G\}(x) - \sum_{i,j=1}^{s} \{F,\psi_i\}(x)c^{ij}(x)\{\psi_j,G\}(x). \tag{12.133}$$

Clearly the bracket $\{,\}_N$ as given in (12.131) satisfies (12.21a) and (12.21c), while the Jacobi-identity (12.21b) follows straightforwardly. Thus $\{\ ,\ \}_N$ is a Poisson bracket. By (12.130) this implies that $\bar{\omega}$ is the symplectic form on N corresponding to $\{\ ,\ \}_N$, and thus by (12.125) $\bar{X}_F$ is the corresponding Hamiltonian vectorfield on N for any $F \in C^\infty(N)$. □

Now let us come back to the constrained dynamics or zero dynamics (12.108) evolving on the symplectic submanifold N^*. First we note that since $H_j(x) = 0$ for $x \in N^*$ the zero dynamics can be rewritten as

$$\dot{x} = X_{H_0}(x) - \sum_{j=1}^{m} X_{H_j}(x)\alpha_j^*(x) - \sum_{j=1}^{m} H_j(x) X_{\alpha_j^*}(x) = X_{H_0 - \sum_{j=1}^{m} H_j\alpha_j^*}(x), \quad x \in N^*. \tag{12.134}$$

Furthermore by the definition of ρ_i (cf. (12.110)) we have for $x \in N^*$

$$\{H_0 - \sum_{j=1}^{m} H_j\alpha_j^*,\ ad_{H_0}^{k} H_i\}(x) = ad_{H_0}^{k+1} H_i(x) = 0,\ k = 0,1,..,\rho_i-1,\ i \in \underline{m} \tag{12.135}$$

while for $x \in N^*$

$$\{H_0 - \sum_{j=1}^{m} H_j\alpha_j^*,\ ad_{H_0}^{\rho_i} H_i\}(x) =$$

$$ad_{H_0}^{\rho_i+1} H_i(x) - \sum_{j=1}^{m} \{H_j, ad_{H_0}^{\rho_i} H_i\}(x)\alpha_j^*(x) = 0, \quad i \in \underline{m}, \tag{12.136}$$

by definition of $\alpha^*(x)$. (In fact (11.28) exactly amounts to (12.136).) Therefore, with ψ_i, $i \in \underline{s}$, as defined in (12.113),

$$\{H_0 - \sum_{j=1}^{m} H_j\alpha_j^*, \psi_i\}(x) = 0, \quad \text{for } x \in N^*,\ i \in \underline{s}, \tag{12.137}$$

and thus by Lemma 12.39 we have

$$X_{H_0 - \sum_{j=1}^{m} H_j\alpha_j^*}(x) = \bar{X}_{\bar{H}_0}, \quad \text{on } N^*, \tag{12.138}$$

since the restrictions of $H_0 - \sum_{j=1}^{m} H_j\alpha_j^*$ and H_0 to N^* are clearly the same. Using Lemma 12.40 we finally conclude

Theorem 12.41 *Consider the Hamiltonian system (12.54) on M. Assume that N^* is non-empty and that* rank $A(x) = m$ *for all $x \in M$. Then the zero dynamics (12.108) is the Hamiltonian vectorfield on N^* with respect to the Dirac bracket (12.131) (with N replaced by N^*, and ψ_j, $j \in \underline{s}$, as given in 12.113) and Hamiltonian $\bar{H}_0$, i.e. the zero dynamics is given as*

$$\dot{\bar{x}} = \{\bar{H}_0, \bar{x}\}_{N^*}, \quad \bar{x} \text{ coordinates for } N^*. \tag{12.139}$$

Remark Clearly the theorem also holds if rank $A(x) = m$ only for $x \in N^*$.

Example 12.42 Consider the Hamiltonian system on $\mathbb{R}^{2n}$ with canonical coordinates (q,p), given as

$$H_0(q,p) = \sum_{i=1}^{n} (\tfrac{1}{2} p_i^2 + \tfrac{1}{2} k_i q_i^2), \quad H_1(q,p) = \sum_{i=1}^{n} q_i^2 - 1. \tag{12.140}$$

i.e. n independent mass-spring systems with unit masses and spring constants k_i, whose configuration evolves for zero output on the unit sphere $\{q \mid \|q\| = 1\}$ in $\mathbb{R}^n$. The zero dynamics takes the following form. Since $\{H_0,H_1\}(q,p) = 2\sum_{i=1}^{n} p_i q_i$ we have $\rho_1 = 1$. Hence $A(q,p) = 4\sum_{i=1}^{n} q_i^2 \neq 0$ for $\|q\| \neq 0$, and $N^* = \{(q,p) \mid \sum_{i=1}^{n} q_i^2 - 1 = 0, \ \sum_{i=1}^{n} p_i q_i = 0\}$ (the tangent bundle to the unit sphere) is a symplectic submanifold of $\mathbb{R}^{2n}$. The feedback $u = \alpha^*(q,p)$ which renders N^* invariant is computed as the solution of $A(q,p)\alpha^*(q,p) + ad^2_{H_0}H_1(q,p) = 4\sum_{i=1}^{n} q_i^2\alpha^*(q,p) + 2\sum_{i=1}^{n} p_i^2 - 2\sum_{i=1}^{n} k_i q_i^2 = 0$ on N^*, yielding (since $\sum_{i=1}^{n} q_i^2 = 1$ on N^*), $\alpha^*(q,p) = -\frac{1}{2}\sum_{i=1}^{n} p_i^2 + \frac{1}{2}\sum_{i=1}^{n} k_i q_i^2$. Thus the zero dynamics is given as $\dot{x} = X_H(x)$, $x \in N^*$, with $H = H_0 - \alpha^* H_1$, resulting in the equations

$$\ddot{q}_i = \dot{p}_i = -k_i q_i + \sum_{j=1}^{n} (k_j q_j^2 - p_j^2) q_i, \quad i \in \underline{n}, \tag{12.141}$$

under the constraint $\sum_{i=1}^{n} q_i^2 = 1$. The vector $2\alpha^*(q_1,\dots,q_n)^T$ yielding the second term on the right-hand side of (12.141) is the normal force required to constrain the motion of the mass-spring systems to the unit sphere. The Dirac bracket $\{\ ,\ \}_{N^*}$ is given as (with $\psi_1 = \sum_{i=1}^{n} q_i^2 - 1$, $\psi_2 = \sum_{i=1}^{n} p_i q_i$)

$$\{F,G\}_{N^*} = \{F,G\} + \tfrac{1}{2}\{F,\psi_1\}\cdot\{\psi_2,G\} - \tfrac{1}{2}\{F,\psi_2\}\cdot\{\psi_1,G\}. \tag{12.142}$$

For $n = 2$ a convenient set of canonical coordinates for $\{\ ,\ \}_{N^*}$ is

$$q := \arctan\frac{q_2}{q_1}, \quad p := q_1 p_2 - q_2 p_1, \tag{12.143}$$

and we can express $\bar{H}_0$ in these coordinates as

$$\bar{H}_0(q,p) = \tfrac{1}{2} p^2 + \tfrac{1}{2} k_1 \cos^2 q + \tfrac{1}{2} k_2 \sin^2 q. \tag{12.144}$$

yielding the following equivalent equations for the zero dynamics

$$\dot{q} = p, \quad \dot{p} = -(k_1 - k_2)\cos q \sin q \tag{12.145}$$

□

Example 12.43 Consider the simple Hamiltonian system as treated in Example 12.38, i.e., $H_0(q,p) = \frac{1}{2}p^TG(q)p + V(q)$, $H_j(q) = q_j$, $j \in \underline{m}$. The Dirac bracket $\{\ ,\ \}_{N^*}$ is given as in (12.131), where c^{ij} is the (i,j)-th element of the inverse matrix of (12.122):

$$C^{-1} = \begin{bmatrix} G_{11}^{-1}SG_{11}^{-1} & G_{11}^{-1} \\ -G_{11}^{-1} & 0 \end{bmatrix}. \tag{12.146}$$

It is easily checked that the canonical coordinate functions $q_{m+1},\dots,q_n$, $p_{m+1},\dots,p_n$ for $\mathbb{R}^{2n}$ when restricted to N^* form a set of *canonical* coordinates for N^* with Poisson bracket $\{,\}_{N^*}$. Denote $(\bar{q},\bar{p}) = (q_{m+1},\dots,q_n, p_{m+1},\dots,p_n)^T$ and $(\tilde{q},\tilde{p}) = (q_1,\dots,q_m,p_1,\dots,p_m)^T$, then $\bar{H}_0(\bar{q},\bar{p})$ is computed as follows. Write

$$G = \begin{bmatrix} G_{11} & G_{12} \\ G_{12}^T & G_{22} \end{bmatrix}, \quad G_{11}\ m \times m \text{ matrix}, \tag{12.147}$$

then the equations $\sum_{j=1}^{n} g_{1j}p_j = \dots = \sum_{j=1}^{n} g_{mj}p_j = 0$ can be rewritten as $G_{11}\tilde{p} + G_{12}\bar{p} = 0$, yielding $\tilde{p} = -G_{11}^{-1}G_{12}\bar{p}$. Expanding $\frac{1}{2}p^TGp = \frac{1}{2}\tilde{p}^TG_{11}\tilde{p} + \tilde{p}^TG_{12}\bar{p} + \frac{1}{2}\bar{p}^TG_{22}\bar{p}$ we thus obtain

$$\bar{H}_0(\bar{q},\bar{p}) = \frac{1}{2}\bar{p}^T\left(G_{22} - G_{12}^TG_{11}^{-1}G_{12}\right)(0,\bar{q})\bar{p} + V(0,\bar{q}), \tag{12.148}$$

where $\left(G_{22} - G_{12}^TG_{11}^{-1}G_{12}\right)(0,\bar{q})$ is again positive definite. The zero dynamics is simply given as

$$\dot{\bar{q}} = \frac{\partial \bar{H}_0}{\partial \bar{p}}, \quad \dot{\bar{p}} = -\frac{\partial \bar{H}_0}{\partial \bar{q}} \tag{12.149}$$

More concretely, consider the simple Hamiltonian system as treated in Example 12.3 (two-link rigid robot manipulator with unit masses and lengths). Consider first the output $y = q_1$ and set $u_2 = 0$. Then $\bar{H}_0(q_2,p_2)$ is given as (substitute $q_1 = 0$ and $p_1 = (1 + \cos q_2)p_2$ in H_0)

$$\bar{H}_0(q_2,p_2) = \frac{1}{2}p_2^2 - g\cos q_2 - 2g, \tag{12.150}$$

resulting in the zero dynamics $\dot{q}_2 = \frac{\partial \bar{H}_0}{\partial p_2}$, $\dot{p}_2 = -\frac{\partial \bar{H}_0}{\partial q_2}$. On the other hand, the zero dynamics for $u_1 = 0$ and holding the output y_2 equal to zero is governed by the Hamiltonian

$$\bar{H}_0(q_1,p_1) = \frac{1}{10}p_1^2 - 3g\cos q_1 \tag{12.151}$$

Physically the zero dynamics are obvious in both cases, as is illustrated by the following figures.

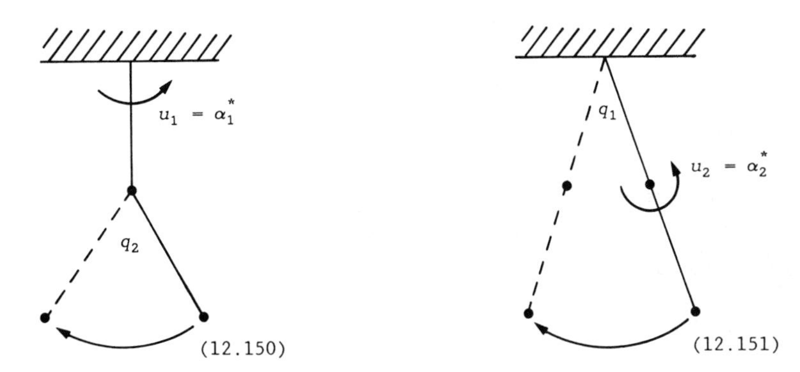

Fig. 12.2. Zero dynamics for the two-link manipulator.

□

Using Theorem 12.41 we obtain the following analog of Theorem 11.16.

Proposition 12.44 *Consider the Hamiltonian system (12.54) on M, with $X_{H_0}(x_0) = 0$ and $H_j(x_0) = 0$, $j \in \underline{m}$, such that* rank $A(x) = m$ *for all* $x \in M$. *Suppose that $\bar{H}_0$ (i.e. the restriction of H_0 to N^*) has a strict local minimum in $x_0 \in N^*$. Then there exists a decoupling regular static state feedback $u = \alpha(x) + \beta(x)\bar{v}$ such that the closed-loop system for $\bar{v} = 0$ is locally stable around x_0.*

Proof First we note that since the zero-dynamics (12.139) is Hamiltonian with Hamiltonian function $\bar{H}_0$, the equilibrium x_0 is locally stable if $\bar{H}_0$ has a strict local minimum in x_0. Then we apply Theorem 11.16, with local asymptotic stability replaced by local stability. □

Remark Since an equilibrium of a Hamiltonian vectorfield is *never* locally asymptotically stable (see Exercise 12.10), but at most stable, it follows that for a Hamiltonian system we can never obtain decoupling with local asymptotic stability if dim $N^* > 0$ and if we take decoupling feedbacks which render N^* invariant.

12.5 Conservation Laws and Reduction of Order

Let us first consider the Hamiltonian system (12.54) for $u = 0$, i.e. the Hamiltonian vectorfield

$$\dot{x} = X_{H_0}(x), \quad x \in M, \tag{12.152}$$

on a $2n$-dimensional symplectic manifold M. A function $F : M \to \mathbb{R}$ is called a *conserved quantity* (or *first integral*) for (12.152) if $X_{H_0}(F) = \{H_0, F\} = 0$. The existence of a conserved quantity F with $dF(x) \neq 0$ clearly *reduces* the solution of (12.152) for $x(0) = x_0$ to the solution of a $(2n-1)$-dimensional level set $F^{-1}(F(x_0))$. The Hamiltonian structure, however, allows us to reduce the order by *two*, in the following sense. Let F be a conserved quantity for (12.152) with $dF(x_0) \neq 0$. Denote $P_1 := F$. By the Flow-Box Theorem (Theorem 2.26) we can find, locally about x_0, a function Q_1 with $dQ_1(x_0) \neq 0$, such that near x_0

$$\{P_1, Q_1\}(x) = X_{P_1}(Q_1)(x) = 1. \tag{12.153}$$

Then, as in the proof of Theorem 12.11, we can find additional coordinate functions $(\bar{Q}, \bar{P}) = (Q_2, \ldots, Q_n, P_2, \ldots, P_n)^T$ such that $Q_1, \ldots, Q_n, P_1, \ldots, P_n$ are *canonical* coordinates about x_0. Since $0 = \{H_0, P_1\} = -\frac{\partial H_0}{\partial Q_1}$ the Hamiltonian H_0 only depends on $P_1, \bar{Q}, \bar{P}$, and hence in such coordinates the differential equations (12.152) take the form

$$\dot{Q}_1 = \frac{\partial H_0}{\partial P_1}(\bar{Q}, \bar{P}, P_1) \tag{12.154a}$$

$$\dot{P}_1 = 0 \tag{12.154b}$$

$$\left\{ \begin{aligned} \dot{\bar{Q}} &= \frac{\partial H_0}{\partial \bar{P}}(\bar{Q}, \bar{P}, P_1) \\ \dot{\bar{P}} &= -\frac{\partial H_0}{\partial \bar{Q}}(\bar{Q}, \bar{P}, P_1) \end{aligned} \right. \tag{12.154c}$$

It is clear that by solving the $(2n-2)$-dimensional set of Hamiltonian equations (12.154c) for $P_1 = P_1(x_0)$ one immediately obtains the solution of the full system (12.154); in fact Q_1 is obtained from $\bar{Q}, \bar{P}$ and the constant P_1 by a simple *integration* of (12.154a).

Remark 12.45 The knowledge of conserved quantities also leads to a sharpening of the conditions of Theorem 12.28. Indeed, let F_i, $i \in \underline{k}$, be conserved quantities for X_{H_0}. Then it can be seen that the condition $\dim P(q,p) = 2n$ for $(q,p) \in V_1$ can be replaced by the (weaker) condition $\dim \big(P(q,p) + \operatorname{span}\{dF_i(q,p) \mid i \in \underline{k}\}\big) = 2n$ for $(q,p) \in V_1$.

Let us now consider the Hamiltonian system (12.54) for non-zero inputs u. Analogously, we call $F : M \to \mathbb{R}$ a *conserved quantity* for (12.54) if

$\{H_0 - \sum_{j=1}^{m} H_j u_j, F\} = 0$, for all u, or equivalently

$$\{H_j, F\} = 0, \quad j = 0,1,\ldots,m. \tag{12.155}$$

Note that the above definition of a conserved quantity is quite restrictive. In fact by applying the Jacobi-identity (12.21b) we obtain from (12.155) that $\{G,F\} = 0$ for *all* $G \in \mathcal{O}$, with $\mathcal{O}$ the *observation space* of the system (12.54) (cf. Proposition 12.19). In particular, if $\dim d\mathcal{O}(x) = \dim M$ for all $x \in M$ (and thus the system is locally observable as well as strongly accessible, cf. Proposition 12.21) then F is necessarily a *constant* function, and thus a *trivial* conserved quantity.

A more general concept is that of a *conservation law*. Let $F: M \to \mathbb{R}$, and let $F^e(y,u)$ be a smooth function defined on $\mathbb{R}^m \times \mathbb{R}^m$, the space of outputs and inputs. The pair (F,F^e) is called a *conservation law* for (12.54) if

$$X_{H_0 - \sum_{j=1}^{m} H_j u_j}(F)(x) = F^e(H_1(x),\ldots,H_m(x),u_1,\ldots,u_m), \tag{12.156}$$

which expresses that the time-derivative of F along the system (12.54) is a function *only* of the outputs and inputs. Because

$$X_{H_0 - \sum_{j=1}^{m} H_j u_j}(F) = \{H_0, F\} - \sum_{j=1}^{m} \{H_j, F\} u_j,$$

it immediately follows that $F^e(y,u)$ is of the form $V(y) + \sum_{j=1}^{m} K_j(y)u_j$, for certain smooth functions K_j, $j \in \underline{m}$, and V on $\mathbb{R}^m$, and thus (12.156) reduces to

$$\begin{aligned} \{H_0, F\}(x) &= V\big(H_1(x),\ldots,H_m(x)\big), \\ \{H_j, F\}(x) &= -\,K_j\big(H_1(x),\ldots,H_m(x)\big), \quad j \in \underline{m}. \end{aligned} \tag{12.157}$$

In particular, if $V = 0$ then F is a conserved quantity for X_{H_0}. Under an extra assumption we will show that V in (12.157) always can be *made* equal to zero by applying a special type of *feedback*.

Proposition 12.46 *Consider a conservation law* (F,F^e), *with* $F^e(y,u) = V(y) + \sum_{j=1}^{m} K_j(y)u_j$, *for the Hamiltonian system (12.54). Let* $x_0 \in M$ *and denote* $y_0 = (H_1(x_0),\ldots,H_m(x_0))^T$. *Assume that the vector* $(K_1(y_0),\ldots,K_m(y_0))^T$ *is non-zero. Then there exists, locally about* x_0, *a regular static state feedback* $u = \alpha(x) + \beta(x)v$ *transforming (12.54) into another Hamiltonian system*

$$\dot{x} = X_{\bar{H}_0}(x) - \sum_{j=1}^{m} X_{\bar{H}_j}(x) v_j,$$

$$\bar{y}_j = \bar{H}_j(x), \qquad j \in \underline{m}, \tag{12.158}$$

where $\bar{H}_0(x) = H_0(x) + S\big(H_1(x),\ldots,H_m(x)\big)$ *for a certain function* $S(y)$, *and* $\bar{H}_j(x) = R_j\big(H_1(x),\ldots,H_m(x)\big)$, $j \in \underline{m}$, *with* $(R_1,\ldots,R_m)^T: \mathbb{R}^m \to \mathbb{R}^m$ *being a coordinate transformation about* y_0, *in such a way that*

$$\{\bar{H}_0, F\} = 0, \quad \{\bar{H}_j, F\} = -\delta_{1j}, \quad j \in \underline{m}, \tag{12.159}$$

i.e., (F, u_1) *is a conservation law for (12.158).*

Proof Define on $Y = \mathbb{R}^m$ the vectorfield $K(y) = \big(K_1(y),\ldots,K_m(y)\big)^T$. By the Flow-Box Theorem (Theorem 2.26) there exists a local coordinate transformation $(\bar{y}_1,\ldots,\bar{y}_m) = \big(R_1(y),\ldots,R_m(y)\big)$ about y_0 such that $K(y) = \dfrac{\partial}{\partial \bar{y}_1}$. Define the preliminary feedback $u = \beta(x)\bar{u}$ as

$$(u_1,\ldots,u_m) = (\bar{u}_1,\ldots,\bar{u}_m)\left(\frac{\partial R_i}{\partial y_j}(y)\right)_{i,j \in \underline{m}} \tag{12.160}$$

with $y = (H_1(x),\ldots,H_m(x))^T$, transforming (12.54) into the Hamiltonian system $\dot{x} = X_{H_0}(x) - \sum_{j=1}^{m} X_{\bar{H}_j}(x)\bar{u}_j$, $\bar{y}_j = \bar{H}_j(x)$, $j \in \underline{m}$, where $\bar{H}_j(x) := R_j\big(H_j(x),\ldots,H_m(x)\big)$, $j \in \underline{m}$. Denote $V(y_1,\ldots,y_m)$ expressed in the new coordinates $\bar{y}_1,\ldots,\bar{y}_m$ by $\bar{V}(\bar{y}_1,\ldots,\bar{y}_m)$. Next define a function $\bar{S}(\bar{y}_1,\ldots,\bar{y}_m)$ such that $\bar{V}(\bar{y}_1,\ldots,\bar{y}_m) = \dfrac{\partial \bar{S}}{\partial \bar{y}_1}(\bar{y}_1,\ldots,\bar{y}_m)$ (this is always possible), and introduce the feedback

$$\bar{u}_i = v_i - \frac{\partial \bar{S}}{\partial \bar{y}_i}(\bar{y}_1,\ldots,\bar{y}_m), \qquad i \in \underline{m}, \tag{12.161}$$

with $\bar{y}_j = \bar{H}_j(x)$, $j \in \underline{m}$. It is immediately checked that $(\bar{y}_1,\ldots,\bar{y}_m,\bar{v}_1,\ldots,\bar{v}_m)$ are canonical coordinates for $T^*\mathbb{R}^m \simeq \mathbb{R}^m \times \mathbb{R}^m$ and that the feedback (12.160) and (12.161) transforms (12.54) into (12.158), where $S(y_1,\ldots,y_m)$ equals $\bar{S}(\bar{y}_1,\ldots,\bar{y}_m)$ expressed in the original coordinates $y_1,\ldots,y_m$. Finally

$$\{\bar{H}_j, F\} = \sum_{i=1}^{m} \frac{\partial R_j}{\partial y_i}\{H_i, F\} = -\sum_{i=1}^{m} \frac{\partial R_j}{\partial y_i} K_i = L_K R_j = -\delta_{1j}, \qquad j \in \underline{m},$$

and thus

$$\{\bar{H}_0, F\} = \{H_0, F\} + \{S \circ (H_1,\ldots,H_m), F\} = V + \sum_{j=1}^{m} \frac{\partial \bar{S}}{\partial \bar{y}_j}\{R_j, F\} = 0,$$

yielding (12.159). □

Remark 12.47 The above proof shows that $\mathbb{R}^m \times \mathbb{R}^m$, the space of outputs and inputs, is most naturally seen as $T^*\mathbb{R}^m$ endowed with its natural Poisson bracket (cf. (12.76)). Furthermore the set-up naturally generalizes to a space of outputs and inputs given as T^*Y, where Y is any m-dimensional output manifold.

Motivated by the preceding proposition let us consider a conservation law (F,F^e) with $F^e(y,u) = u_1$. It follows that F is a conserved quantity for X_{H_0}, and thus we can construct in the same way as before (cf. (12.154)) canonical coordinates $(Q,P) = (Q_1,\ldots,Q_n,P_1,\ldots,P_n)^T$, with $P_1 := F$. Then, as before, since $\frac{\partial H_0}{\partial Q_1} = -\{H_0,P_1\} = 0$ the function H_0 only depends on $(\bar{Q},\bar{P}) := (Q_2,\ldots,Q_n,P_2,\ldots,P_n)^T$ and P_1. Furthermore the condition $\frac{\partial H_j}{\partial Q_1} = -\{H_j,P_1\} = \delta_{1j}$, $j \in \underline{m}$, implies that H_j for $j = 2,\ldots,m$ only depends on $(\bar{Q},\bar{P})$ and P_1, while H_1 is of the form

$$H_1(Q,P) = Q_1 + \tilde{H}_1(\bar{Q},\bar{P},P_1) \tag{12.162}$$

For ease of notation we write $\tilde{H}_j(\bar{Q},\bar{P},P_1) := H_j(\bar{Q},\bar{P},P_1)$, $j = 2,\ldots,m$. Then in the new canonical coordinates (Q,P) the Hamiltonian system (12.54) takes the form (compare with (12.154))

$$\dot{Q}_1 = \frac{\partial H_0}{\partial P_1}(\bar{Q},\bar{P},P_1) - \sum_{j=1}^{m} \frac{\partial \tilde{H}_j}{\partial P_1}(\bar{Q},\bar{P},P_1)u_j \tag{12.163a}$$

$$\dot{P}_1 = u_1 \tag{12.163b}$$

$$\left\{\begin{aligned} \dot{\bar{Q}} &= \frac{\partial H_0}{\partial \bar{P}}(\bar{Q},\bar{P},P_1) - \sum_{j=1}^{m} \frac{\partial \tilde{H}_j}{\partial \bar{P}}(\bar{Q},\bar{P},P_1)u_j \\ \dot{\bar{P}} &= -\frac{\partial H_0}{\partial \bar{Q}}(\bar{Q},\bar{P},P_1) + \sum_{j=1}^{m} \frac{\partial \tilde{H}_j}{\partial \bar{Q}}(\bar{Q},\bar{P},P_1)u_j \end{aligned}\right. \tag{12.163c}$$

$$\left\{\begin{aligned} y_1 &= Q_1 + \tilde{H}_1(\bar{Q},\bar{P},P_1) \\ y_j &= \tilde{H}_j(\bar{Q},\bar{P},P_1), \qquad j = 2,\ldots,m. \end{aligned}\right. \tag{12.163d}$$

Thus we have obtained the $(2n-2)$-dimensional Hamiltonian control system (12.163c), which is driven by the variable $F = P_1 = \int u_1 dt$. Furthermore the remaining variable Q_1 is obtained from $\bar{Q},\bar{P},P_1$ and u by a simple integration of (12.163a). Pictorially we have

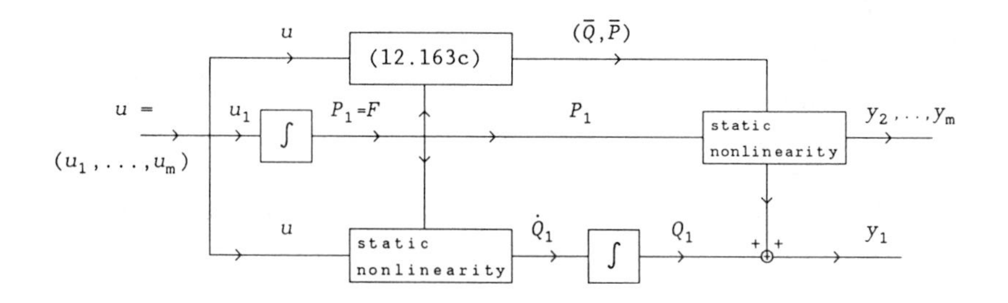

Fig. 12.3. Hamiltonian system with conservation law (F,u_1).

Summarizing we have proven

Theorem 12.48 *Consider a Hamiltonian system (12.54), with conservation law* $(F,F^e) = (F,u_1)$ *satisfying* $dF(x_0) \neq 0$. *Then locally around* x_0 *there exist canonical coordinates* $(Q,P) = (Q_1,\bar{Q},P_1 := F,\bar{P})$ *in which the Hamiltonian system takes the form (12.163).*

Remark 12.49 If F is a conserved quantity for (12.54) (or, equivalently, if (F,F^e) with $F^e = 0$ is a conservation law) then the Hamiltonian system takes the form (12.163) with (12.163b) replaced by $\dot{P}_1 = 0$, and with (12.163d) replaced by $y_j = \tilde{H}_j(\bar{Q},\bar{P},P_1)$, $j \in \underline{m}$.

Example 12.50 Consider the Hamiltonian system on $\mathbb{R}^4$ with canonical coordinates (q_1,q_2,p_1,p_2), given by the internal Hamiltonian

$$H_0(q_1,q_2,p_1,p_2) = \frac{1}{2}p_1^2 + \frac{1}{2}p_2^2 + V(q_2-q_1), \tag{12.164a}$$

and the interaction Hamiltonian

$$H_1(q_1,q_2,p_1,p_2) = q_2. \tag{12.164b}$$

This describes the motion of two particles of unit mass moving on a line, whose interaction comes from a potential $V(r)$ depending on the *distance* r between the two particles (for example, $V(r) = \frac{1}{2}kr^2$ corresponds to a linear spring between the two masses). Moreover the second particle is controlled by an external force u. It is easily checked that (F,F^e) with $F(q,p) = p_1 + p_2$ (the total linear momentum) and $F^e(y,u) = u$ is a conservation law. Introduce new canonical coordinates

$$P_1 = F = p_1 + p_2, \quad Q_1 = q_1, \quad Q_2 = q_2 - q_1, \quad P_2 = p_2. \tag{12.165}$$

In these variables the internal Hamiltonian is given as

$$H_0(Q_2,P_2,P_1) = \frac{1}{2}P_1^2 + P_2^2 - P_1P_2 + V(Q_2), \tag{12.166a}$$

while

$$H_1(Q_1,Q_2,P_1,P_2) = Q_1 + Q_2 \tag{12.166b}$$

and the system takes the form (cf. (12.163))

$$\begin{aligned}
\dot{Q}_1 &= P_1 - P_2, \qquad y = Q_1 + Q_2, \\
\dot{P}_1 &= u, \\
\dot{Q}_2 &= -P_1 + 2P_2, \\
\dot{P}_2 &= -\frac{dV}{dQ_2}(Q_2) + u.
\end{aligned} \tag{12.167}$$

Notice that for almost all potential functions $V(Q_2)$ the observation space $\mathcal{O}$ satisfies $\dim d\mathcal{O}(Q,P) = 4$, in which case there cannot exist non-trivial conserved quantities. On the other hand, if we change H_1 as given in (12.164b) into $H_1(q_1,q_2,p_1,p_2) = q_2 - q_1$ (i.e., an actuator is controlling the distance between the two particles), then $F = p_1 + p_2$ *is* a conserved quantity for the resulting Hamiltonian system. □

Remark 12.51 (*Symmetries*) Let F be a conserved quantity for X_H, i.e. $X_H(F) = 0$. Then clearly $X_F(H) = \{F,H\} = -\{H,F\} = -X_H(F) = 0$. A vectorfield X satisfying $X(H) = 0$ is called a *symmetry* for the Hamiltonian H. We thus see that F is a conserved quantity for X_H *if and only if* X_F is a symmetry for H. (Statements of this type, relating symmetries to conserved quantities, usually go under the heading of *Noether's theorem*.) Moreover $\{F,H\} = 0$ implies (cf. Lemma 12.9) that $[X_F,X_H] = 0$, which in view of Lemma 2.25 is equivalent to the fact that the flows X_F^t, X_H^t satisfy $X_F^t \circ X_H^s = X_H^s \circ X_F^t$ for all s,t for which X_F^t, X_H^s exist. In particular this means that for any t the mapping X_F^t maps solutions of X_H *onto* solutions of X_H; therefore X_F is called a *symmetry* for the *vectorfield* X_H.

This generalizes to *conservation laws* (12.156) as follows. Let (F,F^e) be a conservation law for the Hamiltonian system (12.54). Then we consider the *pair* of vectorfields (X_F,X_{F^e}) with X_F the Hamiltonian vectorfield on M, and X_{F^e} the Hamiltonian vectorfield on $T^*\mathbb{R}^m \simeq \mathbb{R}^m \times \mathbb{R}^m$ (space of outputs and inputs), with respect to its symplectic structure as a cotangent bundle. It can be proved (cf. the references cited at the end of this chapter) that the mappings $(X_F^t,X_{F^e}^t)\colon M \times (\mathbb{R}^m \times \mathbb{R}^m) \to M \times (\mathbb{R}^m \times \mathbb{R}^m)$ map solutions $\big(x(s),y(s),u(s)\big)$ of the Hamiltonian system *onto* other solutions, and thus (X_F, X_{F^e}) can be called a *symmetry* for the Hamiltonian system. Conversely it can be shown that if (X_F, X_{F^e}) is a symmetry, then (F, F^e) is a conservation law.

Notes and References

Classical references to the Euler-Lagrange and Hamilton equations are [Go] and [Wh]. For the modern geometrical treatment, using symplectic geometry, we refer to e.g. [AM], [Ar] and [LM]; in particular in [LM] an extensive treatment of Poisson structures can be found, see also [We].

The definition of a Hamiltonian system with inputs and outputs is due to [Br], and was further developed in [vdS1,vdS3]. The theory of controllability and observability for Hamiltonian systems is taken from [vdS1,vdS4]. For realization theory of Hamiltonian systems we refer to [CI1,CI2], [J1,J2], [vdS1] and [CvdS]. The stabilization of Hamiltonian systems using "PD-controllers", as in 12.3, was first advocated, especially in a robotics context, by [TA], see also [Ko]. The present treatment is largely based on [vdS8], see also [Ma], [TK]. The treatment of constrained or clamped dynamics for Hamiltonian systems as given here was developed in [vdS9,vdS10]. For a study of Hamiltonian systems with additional physical constraints we refer to [MB]. The treatment of Dirac brackets is largely taken from [DLT], see also [MB]. Theorem 12.44 can be found in [HvdS]. The use of conserved quantities for order reduction of Hamiltonian vectorfields and ultimately, for explicitly solving a set of Hamiltonian differential equations is very classical, see e.g. [Wh]. The present treatment was much influenced by [Ol]. For the relation between symmetries and conserved quantities (Noether's theorem) we refer to [AM] and [Ol]. The problem of using multiple conserved quantities or a group of symmetries for order reduction is much more involved. Its geometric treatment is a vast subject, see e.g. [AM], [MW], [MR]. Conservation laws, as well as symmetries, for Hamiltonian systems were dealt with in [vdS1,vdS2,vdS5], where a generalization of Noether's theorem was obtained. The use of symetries for decomposition of systems was emphasized in [GM1,GM2]; in particular in [GM2] the decomposed form (12.164) was extended to an Abelian group of symmetries. Conserved quantities for Hamiltonian and Poisson control systems and their use for reduction purposes were studied e.g. in [Kr], [KM], [Sa] and [SOKM]. For a treatment of general Hamiltonian systems (not affine in the control variables) and applications to optimal control theory, we refer to e.g. [vdS1].

[Ar] V.I. Arnold, **Mathematical Methods of Classical Mechanics**, Springer, Berlin, 1978 (translation of the 1974 Russian edition).

[AM] R.A. Abraham, J.E. Marsden, **Foundations of Mechanics** (2nd edition), Benjamin/Cummings, Reading, Mass., 1978.

[Br] R.W. Brockett, "Control theory and analytical mechanics", in **Geometric Control Theory** (eds. C. Martin, R. Hermann), Vol. VII of Lie groups: History, Frontiers and Applications, Math. Sci. Press., Brookline, pp. 1-46, 1977.

[CI1] P.E. Crouch, M. Irving, "On finite Volterra series which admit Hamiltonian realizations", Math. Systems Theory, 17, pp. 293-318, 1984.

[CI2] P.E. Crouch, M. Irving, "Dynamical realizations of homogeneous Hamiltonian systems", SIAM J. Contr. Optimiz., 24, pp. 374-395, 1986.

[CvdS] P.E. Crouch, A.J. van der Schaft, **Variational and Hamiltonian Control Systems**, Lect. Notes Contr. Inf. Sci., 101, Springer, Berlin, 1987.

[DLT] P. Deift, F. Lund, E. Trubowitz, "Nonlinear wave equations and constrained harmonic motion", Comm. Math. Phys., 74, pp. 141-188, 1980.

[Go] H. Goldstein, **Classical Mechanics**, Addison-Wesley, Reading, Mass., 1950.

[GM1] J.W. Grizzle, S.I. Marcus, "The structure of nonlinear control systems possessing symmetries", IEEE Trans. Autom. Contr. 30, 248-258, 1985.

[GM2] J.W. Grizzle, S.I. Marcus, "A Jacobi-Liouville theorem for Hamiltonian control systems", 23rd IEEE Conf. Decision Control, Las Vegas, pp. 1598-1602, 1984.

[HvdS] H.J.C. Huijberts, A.J. van der Schaft, "Input-output decoupling with stability for Hamiltonian systems", to appear in Math. Control, Signals and Systems, 1989.

[Ja1] B. Jakubczyk, "Poisson structures and relations on vectorfields and their Hamiltonians", Bull. Pol. Ac.: Math., 34, pp. 713-721, 1986.

[Ja2] B. Jakubczyk, "Existence of Hamiltonian realizations of nonlinear causal operators", Bull. Pol. Ac.: Math., 34, pp. 737-747, 1986.

[Ko] D.E. Koditschek, "Natural motion for robot arms", 23rd IEEE Conf. Decision Control, Las Vegas, pp. 733-735, 1984.

[Kr] P.S. Krishnaprasad, "Lie-Poisson structures, dual-spin spacecraft and asymptotic stability", Nonl. Anal. Th. Meth. Appl., 9, pp. 1011-1035, 1985.

[KM] P.S. Krishnaprasad, J.E. Marsden, "Hamiltonian structures and stability for rigid bodies with flexible attachments", Arch. Rat. Mech. Anal., 98, pp. 71-93, 1987.

[LM] P. Libermann, C.M. Marle, **Symplectic geometry and analytical mechanics**, Reidel, Dordrecht, 1987.

[Ma] R. Marino, "Stabilization and feedback equivalence to linear coupled oscillators", Int. J. Control, 39, pp. 487-496, 1984.

[MB] N.H. McClamroch, A.M. Bloch, "Control of constrained Hamiltonian systems and applications to control of constrained robots", **Dynamical Systems Approaches to Nonlinear Problems in Systems and Circuits** (eds. F.M.A. Salam, M.L. Levi), SIAM, pp. 394-403, 1988.

[MR] J.E. Marsden, T. Ratiu, "Reduction of Poisson manifolds", Lett. Math. Phys., pp. 161-169, 1986.

[MW] J.E. Marsden, A. Weinstein, "Reduction of symplectic manifolds with symmetry", Rep. Math. Phys, pp. 121-130, 1974.

[Ol] P.J. Olver, **Applications of Lie groups to differential equations**, Springer, New York, 1986.

[Sa] G. Sanchez de Alvarez, **Geometric Methods of Classical Mechanics applied to Control Theory**, Ph.D. Thesis, Dept. Mathematics, Univ. of California, Berkeley, 1986.

[SOKM] N. Sreenath, Y.G. Oh, P.S. Krishnaprasad, J.E. Marsden, "The dynamics of coupled planar rigid bodies, Part I: Reduction, equilibria & stability", Dynamics and Stability of Systems, 3, pp. 25-49, 1988.

[vdS1] A.J. van der Schaft, **System theoretic descriptions of physical systems**, CWI Tract 3, CWI, Amsterdam, 1984.

[vdS2] A.J. van der Schaft, "Symmetries and conservation laws for Hamiltonian systems with inputs and outputs: A generalization of Noether's theorem", Systems Control Lett., 1, pp. 108-115, 1981.

[vdS3] A.J. van der Schaft, "Hamiltonian dynamics with external forces and observations", Math. Systems Th., 15, pp. 145-168, 1982.

[vdS4] A.J. van der Schaft, "Controllability and observability for affine nonlinear Hamiltonian systems", IEEE Trans. Autom. Contr., AC-27, pp. 490-492, 1982.

[vdS5] A.J. van der Schaft, "Symmetries, conservation laws and time-reversibility for Hamiltonian systems with external forces", J. Math. Phys., 24, pp. 2095-2101, 1983.

[vdS6] A.J. van der Schaft, "Linearization of Hamiltonian and gradient systems", IMA J. Math. Control Information 1, pp. 185-198, 1984.

[vdS7] A.J. van der Schaft, "Controlled invariance for Hamiltonian systems", Math. Systems Th., 18, pp. 257-291, 1985.

[vdS8] A.J. van der Schaft, "Stabilization of Hamiltonian systems", Nonl. An. Th. Meth. Appl., 10, pp. 1021-1035, 1986.

[vdS9] A.J. van der Schaft, "On feedback control of Hamiltonian systems", in **Theory and Applications of Nonlinear Control Systems** (eds. C.I. Byrnes, A. Lindquist), North-Holland, Amsterdam, pp. 273-290, 1986.

[vdS10] A.J. van der Schaft, "Equations of motion for Hamiltonian systems with constraints", J. Phys. A. Math. Gen., 20, pp. 3271-3277, 1987.

[TK] J. Tsinias, N. Kalouptsidis, "On stabilizability of nonlinear systems", 21st IEEE Conf. Decision Control, pp. 712-716, 1982.

[We] A. Weinstein, "The local structure of Poisson manifolds", J. Differential Geom., 18, pp. 523-557, 1983.

[Wh] E.T. Whittaker, **A treatise on the analytical dynamics of particles and rigid bodies**, 4th edition, Cambridge University Press, Cambridge, 1959.

Exercises

12.1 Let V be an r-dimensional Lie algebra (Def. 2.28), with basis $v_1,\dots,v_r$ satisfying the commutation relations

$$[v_i,v_j] = \sum_{k=1}^{r} c_{ij}^k v_k, \qquad i,j \in \underline{r}$$

with c_{ij}^k, $i,j,k \in \underline{r}$, the structure constants (see Exercise 2.15). Consider $M = \mathbb{R}^r$ with natural coordinate functions $x = (x_1,\dots,x_r)$.

(a) Show that

$$\{F,G\}(x) = \sum_{i,j,k=1}^{r} c_{ij}^k x_k \frac{\partial F}{\partial x_i}(x)\frac{\partial G}{\partial x_j}(x), \qquad F,G\colon M \to \mathbb{R},$$

defines a Poisson bracket on M, called the *Lie-Poisson* bracket.

Furthermore, identify M with V^*, the dual of V. Then show that the above Poisson bracket has the coordinate free form

$$\{F,G\}(x) = \langle x,[dF(x),dG(x)]\rangle$$

where $dF(x)$ and $dG(x)$ are regarded as elements of $(V^*)^* = V$, and where $\langle\ ,\ \rangle$ denotes the natural pairing between V and V^*.

(b) Consider the three-dimensional Lie algebra $so(3)$ of the rotation group $SO(3)$. Using the basis

$$v_1 = \begin{pmatrix} 0 & 0 & 0 \\ 0 & 0 & -1 \\ 0 & 1 & 0 \end{pmatrix}, \quad v_2 = \begin{pmatrix} 0 & 0 & 1 \\ 0 & 0 & 0 \\ -1 & 0 & 0 \end{pmatrix}, \quad v_3 = \begin{pmatrix} 0 & -1 & 0 \\ 1 & 0 & 0 \\ 0 & 0 & 0 \end{pmatrix}$$

we have the commutation relations $[v_1,v_2] = v_3$, $[v_2,v_3] = v_1$, $[v_3,v_1] = v_2$. Show that the resulting Poisson bracket on $so^*(3)$ is given as in (12.27).

12.2 Prove the following generalization of Theorem 12.11. Let M be an m-dimensional manifold with Poisson bracket $\{\ ,\ \}$ having constant rank $2n$ ($2n \leq m$). Show that locally around any $x_0 \in M$ we can find coordinates $(q,p,z) = (q_1,\ldots,q_n,p_1,\ldots,p_n,z_1,\ldots,z_\ell)$ for M (so $2n + \ell = m$), satisfying

$$\{p_i,q_j\} = \delta_{ij},\ \{q_i,q_j\} = 0,\ \{p_i,p_j\} = 0, \qquad i,j \in \underline{n}$$

$$\{p_i,z_k\} = 0,\ \{q_i,z_k\} = 0,\ \{z_r,z_k\} = 0, \qquad i \in \underline{n},\ r,k \in \underline{\ell}$$

(see for the non-constant rank case [Ol] or [We]).

12.3 Let $W(x)$ be the structure matrix of a non-degenerate Poisson structure on M with local coordinates $x = (x_1,\ldots,x_{2n})$. Denote the (i,j)-th element of the inverse matrix $W^{-1}(x)$ by $w^{ij}(x)$, $i,j \in \underline{2n}$. Show that the functions $w^{ij}(x)$ satisfy

(i) $\qquad w^{ij}(x) = -w^{ji}(x), \qquad i,j \in \underline{2n}$,

(ii) $\qquad \dfrac{\partial w^{ij}}{\partial x_k}(x) + \dfrac{\partial w^{ki}}{\partial x_j}(x) + \dfrac{\partial w^{jk}}{\partial x_i}(x) = 0, \qquad i,j,k \in \underline{2n}$.

Conversely, show that any matrix $(2n\times2n)$-matrix $S(x)$ whose elements $s_{ij}(x)$ satisfy (i) and (ii) defines a symplectic structure on $\mathbb{R}^{2n}$; i.e. $-(S(x))^{-1}$ is the structure matrix of a Poisson bracket.

12.4 Consider a Poisson control system on a manifold with degenerate Poisson structure. Let C_0 be the distribution as in Definition 3.13. Show that $\dim C_0(x)$ is always less than the rank of the Poisson structure in x, $x \in M$. In particular show that a Poisson control system cannot be locally strongly accessible. What about local observability?

12.5 ([vdS1]) Consider two Hamiltonian systems,

$$\Sigma_\ell^i:\ \dot{x}^i = X_{H_0^i}(x^i) - \sum_{j=1}^m X_{H_j^i}(x^i)u_j,\ x^i \in M^i,\ y_j = H_j^i(x^i),\ i = 1,2.$$

where M^1 and M^2 are symplectic manifolds with Poisson brackets $\{\ ,\ \}_1$, respectively $\{\ ,\ \}_2$. Assume that both observation spaces $\mathcal{O}^i$ satisfy $\dim d\mathcal{O}^i(x^i) = \dim M^i$, for all $x^i \in M^i$, $i = 1,2$.

Suppose that Σ^1 and Σ^2 are equivalent with equivalence mapping $\varphi: M^1 \to M^2$, i.e. if $x^1(t)$ is a solution of Σ^1 for a particular $u(t)$, then $x^2(t) = \varphi(x^1(t))$ is a solution of Σ^2 for this same $u(t)$, yielding the same output $H_j^1(x(t)) = H_j^2(x(t))$, $j \in \underline{m}$, $\forall t$. Formally

$$\varphi_* X_{H_j^1} = X_{H_j^2}, \quad j = 0,1,\ldots,m, \qquad H_j^2 \circ \varphi = H_j^1, \quad j \in \underline{m}.$$

(a) Prove that φ is a Poisson bracket isomorphism, i.e. $\{F\circ\varphi, G\circ\varphi\}_1 = \{F,G\}_1\circ\varphi$, for any $F,G: M^2 \to \mathbb{R}$.

(b) Prove that $H_0^2\circ\varphi = H_0^1 + \text{constant}$, i.e. the internal energy (modulo constants) is the same.

12.6 **(a)** Show that a linear system $\dot{x} = Ax + Bu$, $y = Cx$, $x \in \mathbb{R}^{2n}$, $u,y \in \mathbb{R}^m$, is a Hamiltonian system if there exists a skew-symmetric invertible matrix W such that

(1) $$AW + WA^T = 0, \qquad B = WC^T$$

(or equivalently, if we let $J := -W^{-1}$, $A^TJ + JA = 0$, $B^TJ = C$). Then the system is called a linear Hamiltonian system.

(b) ([vdS6], compare also with Theorem 5.9) Consider a Hamiltonian system (12.54) with $X_{H_0}(x_0) = 0$ and $H_j(x_0) = 0$, $j \in \underline{m}$, satisfying $\dim d\mathcal{O}(x_0) = \dim M = 2n$. By Proposition 12.19 the observation space $\mathcal{O}$ is spanned by all functions

(2) $$F = \{F_1, \{F_2, \{\ldots\{F_k, H_j\}\}\ldots\}\}, \qquad j \in k = 0,1,2,\ldots$$

(for $k = 0$ we let $F = H_j$). Define the grade of F (denoted $gr(F)$) as the number of times that one of the functions H_j, $j \in \underline{m}$, appears in F. Show that we can choose coordinates z about x_0, with $z(x_0) = 0$, such that in these coordinates (12.54) is a linear Hamiltonian system $\dot{z} = Az + Bu$, $y = Cz$, if and only if all elements F as in (2) with $gr(F) = 3$ are zero in a neighborhood of x_0. Furthermore show that W in (1) equals $W(0)$, with $W(z)$ the structure matrix of the Poisson bracket in the coordinates z, and that $W(z) = W(0)$ for all z in a neighborhood of 0. (Hint: Prove that all elements F as in (2) with $gr(F) \geq 3$ are zero in a neighborhood of x_0 if and only if all elements F as in (2) with $gr(F) = 2$ are *constant* in a neighborhood of x_0.)

12.7 **(a)** Let $\mathcal{F}$ be a function group on the symplectic manifold M. Define the distribution $D_{\mathcal{F}}$ as $D_{\mathcal{F}}(x) = \text{span}\{X_F(x) \mid F \in \mathcal{F}\}$. Prove that $D_{\mathcal{F}}$ is involutive. A function group $\mathcal{F}$ is called invariant for the Hamiltonian system (12.54) if $\{H_j, \mathcal{F}\} \subset \mathcal{F}$, $j = 0,1,\ldots,m$. Prove that then $D_{\mathcal{F}}$ is an invariant distribution for (12.54).

(b) Let $\mathcal{F}$ be an invariant function group with dim $d\mathcal{F}(x)$ = constant. Suppose $\mathcal{F} \cap \mathcal{F}^{\perp}$ are the constant functions. Prove that H_j, $j = 0,1,\ldots,m$, are in suitable canonical coordinates of the form $H_j(q_1,\ldots,q_n,p_1,\ldots,p_n) = \sum_{i=1}^{n} p_i^T f_i(q)$. What can be said about the case that $\mathcal{F}$ satisfies $\mathcal{F}^{\perp} \subset \mathcal{F}$, or $\mathcal{F} \subset \mathcal{F}^{\perp}$?

(c) Let D be an involutive distribution of constant dimension on the symplectic manifold M. By Frobenius' theorem there exist locally independent functions $K_1,\ldots,K_k$ such that $D(x)$ = ker span$\{dK_1(x),\ldots$ $\ldots,dK_k(x)\}$. Assume that $K_1,\ldots,K_k$ are globally defined. Prove that there exists a function group $\mathcal{F}$ such that $D = D_{\mathcal{F}}$ if and only if span$\{K_1,\ldots,K_k\}$ is a function group ([vdS7]).

12.8 Consider the two-link robot manipulator from Example 12.3, see also Example 12.22. Consider only the first input $u = u_1$ and the corresponding output $y = q_1$. Compute the observation space $\mathcal{O}$, first for $g = 0$, and then for $g \neq 0$. Decide about local strong accessibility and observability. Do the same for input $u = u_2$ and output $y = q_2$.

12.9 Consider a Hamiltonian system (12.54) with $X_{H_0}(x_0) = 0$ and $H_j(x_0) = 0$, $j \in \underline{m}$, satisfying dim $d\mathcal{O}(x_0)$ = dim $M = 2n$. Define the function spaces

$$\mathcal{F}_k = \text{span}\{ad^r_{H_0}H_1,\ldots,ad^r_{H_0}H_m \mid r = 0,1,\ldots,k-1\}, \quad k = 1,2,\ldots$$

(a) Prove: the system is feedback linearizable around x_0 if and only if $\mathcal{F}_1,\ldots,\mathcal{F}_{2n}$ are function groups satisfying dim $d\mathcal{F}_k(x)$ = constant, x around x_0, $k \in \underline{2n}$ (compare with Chapter 6).

(b) Consider the simple Hamiltonian system of Example 12.38. Show that a necessary condition for feedback linearizability is that $G_{11}(q)$ only depends on $q_1,\ldots,q_m$.

12.10 Let $X(x) = \big(X_1(x),\ldots,X_n(x)\big)^T$ be a vectorfield on $\mathbb{R}^n$ with natural coordinates $x = (x_1,\ldots,x_n)$. The *divergence* of X, div(X), is defined as div$(X)(x) = \sum_{i=1}^{n} \frac{\partial X_i}{\partial x_i}(x)$.

(a) Show that the maps X^t: $\mathbb{R}^n \to \mathbb{R}^n$ are *volume-preserving*, i.e. $\det\big(\frac{\partial X^t}{\partial x}(x)\big) = 1$ for all $x \in \mathbb{R}^n$, if and only if div$(X)(x) = 0$ for all $x \in \mathbb{R}^n$.

(b) Prove that div$(X_H)(x) = 0$ for any Hamiltonian vectorfield X_H. Conclude that a Hamiltonian vectorfield cannot have locally asymptotically stable equilibria.

12.11 Consider a Hamiltonian system (12.54). Apply feedback $u = \alpha(x) + v$. Show that the feedback transformed system is again Hamiltonian (with respect to the same Poisson structure) if and only if there exists a function $P(y_1, \ldots, y_m)$ such that

$$\alpha_i(x) = -\frac{\partial P}{\partial y_i}\big(H_1(x), \ldots, H_m(x)\big), \qquad i \in \underline{m}.$$

Show that the internal energy of the resulting Hamiltonian system is given as $H_0(x) + P\big(H_1(x), \ldots, H_m(x)\big)$.

12.12 (a) Consider a Hamiltonian system (12.54) with non-singular decoupling matrix. Suppose that H_0 has a strict local minimum in x_0 (implying that x_0 is a locally stable equilibrium for X_{H_0}). Suppose $H_j(x_0) = 0$, $j \in \underline{m}$. Show that x_0 is also a locally stable equilibrium for the zero dynamics.

(b) Suppose that the conditions of Lemma 12.32 are satisfied. Prove that $x_0 = (q_0, 0)$ is a locally stable equilibrium for the zero dynamics.

12.13 (a) Consider a Hamiltonian system (12.54) with non-singular decoupling matrix. Define new outputs $\bar{y}_j = R_j(y_1, \ldots, y_m)$, $j \in \underline{m}$, where

$$\operatorname{rank}\left(\frac{\partial R_i}{\partial y_j}\right)(y) = m \text{ everywhere},$$

and $y_1 = \ldots = y_m = 0$ if and only if $\bar{y}_1 = \ldots = \bar{y}_m = 0$. Show that the decoupling matrix of the resulting system is still non-singular, and that the zero-dynamics remain the same.

(b) Apply (a) to the case of the two-link rigid robot-manipulator of Example 1.1 with $(\bar{y}_1, \bar{y}_2)$ the Cartesian coordinates of the endpoint. Are there any singularities?

12.14 ([CvdS]) Consider a nonlinear system

$$\Sigma: \dot{x} = f(x) + \sum_{j=1}^{m} g_j(x)u_j, \quad y_j = h_j(x), \qquad j \in \underline{m}, \ x \in M.$$

With any vectorfield X on M we can associate a function H^X from T^*M to $\mathbb{R}$ by setting $H^X(x,\alpha) = \alpha\big(X(x)\big)$, $\alpha \in T_x^*M$. In natural coordinates $(x,p) = (x_1, \ldots, x_n, p_1, \ldots, p_n)$ the function $H^X(x,p)$ is given as $p^T X(x)$. Furthermore for any $h: X \to \mathbb{R}$ define $h^\ell: T^*M \to \mathbb{R}$ by $h^\ell := h \circ \pi$, $\pi: T^*M \to M$ projection. Define now the following system on T^*M

$$\Sigma_e: \begin{cases} \dot{x}_e = X_{H^f}(x_e) + \sum_{j=1}^{m} X_{H^{g_j}}(x_e) - \sum_{j=1}^{m} X_{h_j^\ell}(x_e)u_j^a, & x_e \in T^*M, \\ y_j^a = -H^{g_j}(x_e), \quad y_j = h_j^\ell(x_e), & j \in \underline{m}, \end{cases}$$

and show that Σ_e is a Hamiltonian system on T^*M with inputs u, u^a

and outputs y^a, y, called the Hamiltonian extension.

Show that the observation space of Σ_e is spanned by all functions H^ℓ, with H in the observation space $\mathcal{O}$ of Σ, and all functions H^X, with X in the strong accessibility algebra C_0 of Σ (cf. Definition 3.19). Prove that the Hamiltonian extension is locally observable and locally strongly accessible if and only if Σ is locally observable and locally strongly accessible. Show that the observation space of Σ_e is not changed if we set $u_j^a = 0$, $j \in \underline{m}$.

12.15 Consider a particle in $\mathbb{R}^3$ with mass m in a potential field with potential V, subject to an external force u. The system is given by

$$\dot{q}_i = \frac{1}{m} p_i, \quad \dot{p}_i = -\frac{\partial V}{\partial q_i}(q) + u_i, \quad y_i = q_i, \quad i = 1,2,3.$$

Suppose that $V(q)$ is invariant under rotations around the e_1-axis. Show that (F, F^e) with $F(q,p) = q_2 p_3 - p_2 q_3$ and $F^e(y,u) = y_2 u_3 - u_2 y_3$ is a conservation law. (F equals the angular momentum). Apply Proposition 12.46 and subsequently Theorem 12.48.

12.16 Prove Remark 12.37 (see [DLT]).

12.17 Consider a simple Hamiltonian system (Definition 12.26). Show that the map $\varphi\colon T^*Q \to T^*Q$, in local natural coordinates defined as $\varphi(q,p) = (q,-p)$, satisfies

$$\varphi_* X_{H_j} = -X_{H_j}, \quad j = 0,1,\ldots,m, \qquad H_j \circ \varphi = H_j, \quad j \in \underline{m}.$$

Show that this implies that if $\big(u(t), x(t), y(t)\big)$ is a solution curve of the system, then also $\big((Ru)(t), \varphi\big((Rx)(t)\big), Ry(t)\big)$ is a solution curve, with $(Rf)(t) := f(-t)$. (The system is called *time-reversible*, [vdS1,5].)

13
Controlled Invariance and Decoupling for General Nonlinear Systems

In Chapters 7-11 we have confined ourselves to *affine* nonlinear control systems. The aim of the present chapter is to generalize the main results obtained to general smooth nonlinear dynamics

$$\dot{x} = f(x,u), \qquad u \in U, \tag{13.1}$$

where $x = (x_1,\ldots,x_n)$ are local coordinates for the (state space) manifold M. Throughout this chapter we will assume that the input space U is an *open subset* of $\mathbb{R}^m$, with local coordinates $u = (u_1,\ldots,u_m)$.

From Chapter 2 we recall that (13.1) can be regarded as the local coordinate expression of a smooth map (the *system map*)

$$F\colon M \times U \to TM, \tag{13.2}$$

satisfying the commutative diagram

$$\begin{array}{ccc} M \times U & \xrightarrow{\quad F \quad} & TM \\ & {\scriptstyle \pi}\searrow \quad \swarrow{\scriptstyle \pi_M} & \\ & M & \end{array} \tag{13.3}$$

with $\pi\colon M \times U \to M$ and $\pi_M\colon TM \to M$ being the natural projections. Indeed, let $x = (x_1,\ldots,x_n)$ be local coordinates for M, and let $(x,v) = (x_1,\ldots,x_n, v_1,\ldots,v_n)$ be corresponding natural coordinates for the tangent bundle TM, cf. (2.61). Furthermore let $u = (u_1,\ldots,u_m)$ be local coordinates for U. Then by the commutativity of (13.3) the map F is locally represented as

$$F(x,u) = (x,f(x,u)), \tag{13.4}$$

thereby recovering (13.1).

13.1 Locally Controlled Invariant Distributions

Analogously to Chapter 3 (Definition 3.44) and Chapter 7 (Definition 7.5) we state

Definition 13.1 *A distribution D on M is invariant for (13.1) if*

$$[f(\cdot,u),\ D] \subset D \qquad \text{for every } u \in U, \tag{13.5}$$

and locally controlled invariant for (13.1) if there exists locally a regular static state feedback (cf. (5.102)), briefly feedback,

$$u = \alpha(x,\tilde{u}), \qquad \operatorname{rank} \frac{\partial \alpha}{\partial \tilde{u}}(x,\tilde{u}) = m, \qquad \text{for all } x,\tilde{u}, \tag{13.6}$$

such that the closed-loop dynamics

$$\dot{x} = \tilde{f}(x,\tilde{u}) := f(x,\alpha(x,\tilde{u})) \tag{13.7}$$

satisfies

$$[\tilde{f}(\cdot,\tilde{u}),D] \subset D, \qquad \text{for every } \tilde{u}. \tag{13.8}$$

Remark For *global* controlled invariance we additionally have to require that $u = \alpha(x,\tilde{u})$ is globally defined with the property that for every x the mapping $\tilde{u} \to \alpha(x,\tilde{u})$ is a *diffeomorphism*.

If D is *involutive* and *constant-dimensional*, then by Frobenius' Theorem (Theorem 2.42) we can find local coordinates $x = (x^1,x^2)$ for M, with $x^1 = (x_1,\ldots,x_k)$, $x^2 = (x_{k+1},\ldots,x_n)$, $k = \dim D$, such that $D = \operatorname{span}\{\frac{\partial}{\partial x^1}\} := \operatorname{span}\{\frac{\partial}{\partial x_1},\ldots,\frac{\partial}{\partial x_k}\}$. It follows from (13.5) that if D is *invariant* for (13.1), then (13.1) locally decomposes as, cf. (3.109),

$$\begin{aligned} \dot{x}^1 &= f^1(x^1,x^2,u) \\ \dot{x}^2 &= f^2(x^2,u), \end{aligned} \tag{13.9}$$

and similarly for the closed-loop dynamics (13.7) if D is locally controlled invariant.

Our first goal is to obtain a generalization of Theorem 7.5, characterizing locally controlled invariant distributions for affine nonlinear systems. In order to do so we first set up some additional mathematical apparatus.

First, let $h\colon M \to \mathbb{R}$ be a *function* (everything in the sequel is assumed to be *smooth*). The *complete lift* (or *prolongation*) of h is defined as the function $\dot{h}\colon TM \to \mathbb{R}$ given by

$$\dot{h}(p,X_p) := dh(p)(X_p) = X_p(h), \quad X_p \in T_pM,\ p \in M. \tag{13.10}$$

Notice that for any set of local coordinates $(x_1,\ldots,x_n)$ for M, the resulting natural coordinates $(x_1,\ldots,x_n,v_1,\ldots,v_n)$ for TM, cf. (2.61), are also given as $(x_1,\ldots,x_n,\dot{x}_1,\ldots,\dot{x}_n)$. In the sequel we will therefore *throughout* denote the natural coordinates for TM by $(x,\dot{x}) := (x_1,\ldots,x_n,\dot{x}_1,\ldots,\dot{x}_n)$. It is easily seen that the prolongation $\dot{h}$ in natural coordinates $(x,\dot{x})$ is locally represented as

$$\dot{h}(x,\dot{x}) = \sum_{i=1}^{n} \frac{\partial h}{\partial x_i}(x)\ \dot{x}_i . \tag{13.11}$$

The *vertical lift* $h^{\ell}: TM \to \mathbb{R}$ of h is defined as

$$h^{\ell} := h \circ \pi_M, \qquad \pi_M: TM \to M \text{ projection}, \tag{13.12}$$

and thus is locally represented as

$$h^{\ell}(x,\dot{x}) = h(x) \tag{13.13}$$

Secondly, let X be a *vectorfield* on M. We define the *complete lift* (or *prolongation*) of X as the vectorfield $\dot{X}$ on TM satisfying

$$\dot{X}(\dot{h}) = \big(X(h)\big)^{\cdot} \tag{13.14}$$

for all $h: M \to \mathbb{R}$. This uniquely defines $\dot{X}$, as follows from

Lemma 13.2 *Let $\widetilde{X}$ be a vectorfield on TM satisfying $\widetilde{X}(\dot{h}) = 0$ for all $h: M \to \mathbb{R}$. Then $\widetilde{X} = 0$.*

Proof In natural coordinates $(x,\dot{x})$ we can write

$$\widetilde{X}(x,\dot{x}) = \sum_{i=1}^{n} X_i(x,\dot{x})\ \frac{\partial}{\partial x_i} + \sum_{i=1}^{n} Z_i(x,\dot{x})\ \frac{\partial}{\partial \dot{x}_i} \tag{13.15}$$

First take $h = x_i$, then it follows from $\widetilde{X}(\dot{x}_i) = 0$ that $Z_i = 0$, $i \in \underline{n}$. Hence, using (13.11), $0 = \widetilde{X}(\dot{h}) = \sum_{i,j=1}^{n} X_i(x,\dot{x})\dot{x}_j\ \frac{\partial^2 h}{\partial x_i \partial x_j}(x)$, or

$$\big(X_1(x,\dot{x}),\ldots,X_n(x,\dot{x})\big)\ D^2h(x) \begin{pmatrix} \dot{x}_1 \\ \vdots \\ \dot{x}_n \end{pmatrix} = 0 \tag{13.16}$$

with $D^2h(x)$ being the Hessian matrix of h. Since h is arbitrary (and thus $D^2h(x)$ is an arbitrary symmetric matrix) this implies

$$X_i(x,\dot{x})\dot{x}_j + X_j(x,\dot{x})\dot{x}_i = 0, \qquad i,j \in \underline{n}. \tag{13.17}$$

Let now $\dot{x}_1 \neq 0$. Putting $i = 1$ in (13.17) we obtain $X_j(x,\dot{x})\dot{x}_1 = -X_1(x,\dot{x})\dot{x}_j$, $j \in \underline{n}$, and thus

$$X_j(x,\dot{x}) = \alpha(x,\dot{x})\dot{x}_j\,, \quad j \in \underline{n}, \tag{13.18}$$

for some function $\alpha(x,\dot{x})$ (not depending on j). Substituting (13.18) into (13.17) we find $2\alpha(x,\dot{x})\dot{x}_i\dot{x}_j = 0$, $i,j \in \underline{n}$, and thus if $\dot{x}_i \neq 0$ and $\dot{x}_j \neq 0$ we obtain $\alpha(x,\dot{x}) = 0$, or $X_j(x,\dot{x}) = 0$, $j \in \underline{n}$. By continuity this yields $X_j(x,\dot{x}) = 0$ for all $(x,\dot{x})$, $j \in \underline{n}$, and thus $\tilde{X} = 0$. □

If X is locally represented as $\sum_{i=1}^{n} X_i(x)\frac{\partial}{\partial x_i}$ then it readily follows from (13.14) that $\dot{X}$ is locally given as

$$\dot{X}(x,\dot{x}) = \sum_{i=1}^{n} X_i(x)\frac{\partial}{\partial x_i} + \sum_{i,j=1}^{n} \frac{\partial X_i}{\partial x_j}(x)\dot{x}_j \frac{\partial}{\partial \dot{x}_i}\,. \tag{13.19}$$

The *vertical lift* of X is defined as the vectorfield X^ℓ on TM satisfying

$$X^\ell(\dot{h}) = \left(X(h)\right)^\ell \tag{13.20}$$

for all $h\colon M \to \mathbb{R}$. This uniquely defines X^ℓ as again follows from Lemma 13.2. Furthermore in local coordinates

$$X^\ell(x,\dot{x}) = \sum_{i=1}^{n} X_i(x)\frac{\partial}{\partial \dot{x}_i}. \tag{13.21}$$

Thirdly, let σ be a *differential one-form* on M. The *complete lift* (or *prolongation*) of σ is the differential one-form on TM defined by setting

$$\dot{\sigma}(q)(Z_q) := Z_q(\hat{\sigma}), \qquad Z_q \in T_qTM,\ q \in TM, \tag{13.22}$$

where $\hat{\sigma}\colon TM \to \mathbb{R}$ is the function given as

$$\hat{\sigma}(X_p) = \sigma(p)(X_p),\ X_p \in T_pM, \quad p \in M. \tag{13.23}$$

If σ is given in local coordinates as $\sigma = \sum_{i=1}^{n}\sigma_i(x)dx_i$, then it readily follows that

$$\dot{\sigma}(x,\dot{x}) = \sum_{i=1}^{n} \frac{\partial \sigma_i}{\partial x_j}(x)\dot{x}_j\,dx_i + \sum_{i=1}^{n} \sigma_i(x)d\dot{x}_i\,. \tag{13.24}$$

Finally the *vertical lift* of σ is the differential one-form σ^ℓ on TM defined as

$$\sigma^\ell := \pi_M^*\,\sigma, \qquad \pi_M\colon TM \to M \ \text{ projection}, \tag{13.25}$$

i.e. in local coordinates

$$\sigma^\ell(x,\dot{x}) = \sum_{i=1}^{n} \sigma_i(x)dx_i \,. \tag{13.26}$$

Using the local coordinate expressions the following identities are easily verified.

Proposition 13.3 *For any function $h\colon M \to \mathbb{R}$, any vectorfields X, X_1 and X_2 on M, and any differential one-form σ on M, we have*

$$\dot{X}(h^\ell) = (X(h))^\ell = X^\ell(\dot{h}), \qquad \dot{X}(\dot{h}) = (X(h))^{\cdot}, \qquad X^\ell(h^\ell) = 0\,, \tag{13.27}$$

$$\dot{\sigma}(X^\ell) = (\sigma(X))^\ell = \sigma^\ell(\dot{X}), \qquad \dot{\sigma}(\dot{X}) = (\sigma(X))^{\cdot}, \qquad \sigma^\ell(X^\ell) = 0\,, \tag{13.28}$$

$$[\dot{X}_1,\dot{X}_2] = ([X_1,X_2])^{\cdot}, \quad [\dot{X}_1,X_2^\ell] = [X_1,X_2]^\ell, \qquad [X_1^\ell,X_2^\ell] = 0\,, \tag{13.29}$$

$$d\dot{h} = (dh)^{\cdot}, \; dh^\ell = (dh)^\ell. \tag{13.30}$$

Prolongations of distributions and co-distributions are now defined as follows.

Definition 13.4 *Let the distribution D on M be given as $D(p) = \mathrm{span}\{X_i(p) \mid i \in I\}$, $p \in M$, with X_i vectorfields on M, then the prolongation of D is the distribution $\dot{D}$ on TM defined as*

$$\dot{D}(q) = \mathrm{span}\,\{\dot{X}_i(q),\, X_i^\ell(q) \mid i \in I\}, \qquad q \in TM. \tag{13.31}$$

Analogously, let the co-distribution P on M be given as $P(p) = \mathrm{span}\{\sigma_i(p) \mid p \in I\}$, $p \in M$, with σ_i differential one-forms on M, then the prolongation of P is the co-distribution $\dot{P}$ on TM defined as

$$\dot{P}(q) = \mathrm{span}\{\dot{\sigma}_i(q),\, \sigma_i^\ell(q) \mid i \in I\}, \qquad q \in TM. \tag{13.32}$$

In case D and P are involutive and constant-dimensional we obtain the following simple local representations of $\dot{D}$ and $\dot{P}$. By Frobenius' Theorem (Theorem 2.42) we can find local coordinates $x = (x_1,\ldots,x_n)$ such that $D = \mathrm{span}\,\{\frac{\partial}{\partial x_1},\ldots,\frac{\partial}{\partial x_k}\}$, $k = \dim D$. Then in the natural coordinates $(x,\dot{x})$ for TM we have

$$\dot{D} = \mathrm{span}\,\{\frac{\partial}{\partial x_1},\ldots,\frac{\partial}{\partial x_k},\, \frac{\partial}{\partial \dot{x}_1},\ldots,\frac{\partial}{\partial \dot{x}_k}\}. \tag{13.33}$$

Similarly if $P = \mathrm{span}\,\{dx_{\ell+1},\ldots,dx_n\}$, $n-\ell = \dim P$, then

$$\dot{P} = \mathrm{span}\,\{dx_{\ell+1},\ldots,dx_n, d\dot{x}_{\ell+1},\ldots,d\dot{x}_n\}. \tag{13.34}$$

For general (co-)distributions D and P we obtain

Proposition 13.5 *Let D be a distribution, and P be a co-distribution on M. Then*

(a) If $D = \ker P$ (see 2.173), then $\dot{D} = \ker \dot{P}$.

(b) If $P = \operatorname{ann} D$ (see 2.174), then $\dot{P} = \operatorname{ann} \dot{D}$.

(c) If D (resp. P) has constant dimension then $\dot{D}$ (resp. $\dot{P}$) has constant dimension.

(d) If D (resp. P) is involutive then $\dot{D}$ (resp. $\dot{P}$) is involutive.

(e) If $X^\ell \in \dot{D}$ for a vectorfield X on M, then $X \in D$.

Proof (a) and (b) follow from (13.28). Part (c) is trivial, while (d) follows from (13.29). Part (e) is proved as follows. Let $X^\ell \in \dot{D}$, where $D(p) = \operatorname{span}\{X_i(p) \mid i \in I\}$, $p \in M$. Then $X^\ell = \sum_{i\in I} (\alpha_i \dot{X}_i + \beta_i X_i^\ell)$ for certain functions α_i, β_i on TM. By (13.27) we obtain

$$0 = X^\ell(h^\ell) = \sum_{i\in I} \left((\alpha_i \dot{X}_i(h^\ell) + \beta_i X_i^\ell(h^\ell))\right) = \sum_{i\in I} \alpha_i \dot{X}_i(h^\ell)$$

for all $h: M \to \mathbb{R}$ and thus by Lemma 13.2, $\sum_{i\in I} \alpha_i \dot{X}_i = 0$. Hence $X^\ell = \sum_{i\in I} \beta_i X_i{}^\ell$, and from the local expression (13.20) it is easily seen that β_i, $i \in I$, can be chosen independent of $\dot{x}$, implying that $X = \sum_{i\in I} \beta_i X_i \in D$. □

A key observation with regard to the use of the prolonged distribution $\dot{D}$ for purposes of invariance is contained in

Proposition 13.6 *Consider the nonlinear dynamics (13.1) given by the system map $F: M \times U \to TM$ as in (13.2). Let D be a distribution on M. Define D_e as the unique distribution on $M \times U$ such that*

$$\pi_* D_e = D, \qquad \bar{\pi}_* D_e = 0 \,, \tag{13.35}$$

with π, resp. $\bar{\pi}$, being the natural projection of $M \times U$ on M, resp. U (i.e. for any $Z_q \in T_q(M \times U)$, $q \in M \times U$, with $Z_q \in D(q)$, we have $\pi_{*q} Z_q \in D(\pi(q))$ and $\bar{\pi}_{*q} Z_q = 0$, while moreover every element of $D(\pi(q))$ can be written as $\pi_{*q} Z_q$ for some $Z_q \in D(q)$). Then D is invariant for (13.1) (cf. Definition 13.1) if and only if

$$F_* D_e \subset \dot{D}, \tag{13.36}$$

i.e. for any $Z_q \in T_q(M{\times}U)$ satisfying $Z_q \in D_e(q)$, we have $F_{*q} Z_q \in \dot{D}(F(q))$,

$q \in M \times U$.

Proof The proof is based on the following formula. Let X be a vectorfield on M, and let X_e be the unique vectorfield on $M \times U$ with $\pi_* X_e = X$, $\bar{\pi}_* X_e = 0$. Then

$$F_{*(x,u)} X_e(x,u) = \begin{pmatrix} I_n & 0 \\ \frac{\partial f}{\partial x}(x,u) & \frac{\partial f}{\partial u}(x,u) \end{pmatrix} \begin{pmatrix} X(x) \\ 0 \end{pmatrix} = \begin{pmatrix} X(x) \\ \frac{\partial f}{\partial x}(x,u)X(x) \end{pmatrix},$$

where $X(x) = (X_1(x), \ldots, X_n(x))$ is the local representative of the vectorfield X on M, and where the last vector is taken at the point $(x, f(x,u)) \in TM$. This last vector can be trivially rewritten as

$$\begin{pmatrix} X(x) \\ \frac{\partial X}{\partial x}(x)f(x,u) \end{pmatrix} - \begin{pmatrix} 0 \\ \frac{\partial X}{\partial x}(x)f(x,u) - \frac{\partial f}{\partial x}(x,u)X(x) \end{pmatrix}.$$

In view of (13.19) and (13.21) we thus obtain

$$F_* X_e = \dot{X} - [f,X]^{\ell}, \tag{13.37}$$

with $[f,X] = [f(\cdot,u), X(\cdot)](x)$ depending on $(x,u) \in M \times U$.

Now let $X \in D$. Suppose that D is invariant for (13.1), i.e. for all u $[f(\cdot,u), X(\cdot)] \in D$. Then clearly the right-hand side of (13.37) is in $\dot{D}$, proving that $F_* D_e \subset \dot{D}$. Conversely, let $F_* X_e \in \dot{D}$ for any $X \in D$. Since $\dot{X} \in \dot{D}$ this implies by (13.37) that $[f(\cdot,u), X(\cdot)]^{\ell} \in \dot{D}$, and therefore by Proposition 13.5(e) $[f(\cdot(\cdot,u), X(\cdot)] \in D$. □

Remark If D is involutive and constant dimensional then (13.36) takes the following simple form. By Frobenius' Theorem we can find local coordinates $x = (x^1, x^2)$ such that $D = \text{span}\,\{\frac{\partial}{\partial x^1}\}$, and thus (cf. (13.34)) $\dot{D} = \text{span}\,\{\frac{\partial}{\partial x^1}, \frac{\partial}{\partial \dot{x}^1}\}$. Writing accordingly $f = (f^1, f^2)$, then (13.36) amounts to the equality $\frac{\partial f^2}{\partial x^1}(x^1, x^2, u) = 0$, which implies the local decomposition (13.9).

Now let us proceed to local *controlled invariance*. First let us associate with (13.1) the extended system (Definition 6.11)

$$\begin{aligned} \dot{x} &= f(x,u), \\ \dot{u} &= w, \end{aligned} \tag{13.38}$$

which is an *affine* system with state space $M \times U$ and inputs $w \in \mathbb{R}^m$. (Recall that we assumed U to be an open subset of $\mathbb{R}^m$.) The drift vectorfield $f(x,u)\frac{\partial}{\partial x}$ on $M \times U$ will be denoted by f_e, and the distribution of input vectorfields $\{\frac{\partial}{\partial u_1},\ldots,\frac{\partial}{\partial u_m}\}$ for (13.38) by G_e.

Theorem 13.7 *Consider the nonlinear system (13.1) with system map* F: $M \times U \to TM$. *Let* D *be an involutive distribution of constant dimension on* M. *Assume that the distribution*

$$\tilde{G}_e := \{Z \in G_e \mid F_* Z \in \dot{D}\} \tag{13.39}$$

on $M \times U$ *has constant dimension. Then* D *is locally controlled invariant if and only if*

$$F_* D_e \subset \dot{D} + F_* G_e \tag{13.40}$$

Remark Notice that (13.40) may be equivalently replaced by the requirement (with π: $M \times U \to M$ being the projection)

$$F_*(\pi_*^{-1}(D)) \subset \dot{D} + F_* G_e \tag{13.40'}$$

Proof By Frobenius' Theorem we can find local coordinates $x = (x_1,\ldots,x_n)$ such that $D = \text{span}\,\{\frac{\partial}{\partial x_1},\ldots,\frac{\partial}{\partial x_k}\}$. Write $x^1 = (x_1,\ldots,x_k)$, $x^2 = (x_{k+1},\ldots,x_n)$, and correspondingly $f = (f_1,\ldots,f_n)^T$, $f^1 = (f_1,\ldots,f_k)^T$, $f^2 = (f_{k+1},\ldots,f_n)^T$. Then (13.40) is equivalent to

$$\operatorname{Im} \frac{\partial f^2}{\partial x^1}(x,u) \subset \operatorname{Im} \frac{\partial f^2}{\partial u}(x,u), \quad \text{for every } (x,u). \tag{13.41}$$

Now suppose D is locally controlled invariant. Then there exists locally a feedback $u = \alpha(x,v)$ such that $[\tilde{f}(\cdot,v),D] \subset D$, with $\tilde{f}(x,v) := f(x,\alpha(x,v))$. Equivalently

$$\frac{\partial \tilde{f}^2}{\partial x^1}(x,v) = 0, \text{ or, } \quad \frac{\partial f^2}{\partial x^1}(x,u) + \frac{\partial f^2}{\partial u}(x,u)\,\frac{\partial \alpha}{\partial x^1}(x,v) = 0, \tag{13.42}$$

which implies (13.41). Conversely let (13.41) be satisfied. Denote $P = \text{ann}\, D$. Then $\dot{P}$ is an involutive codistribution on TM, in the above coordinates given as (cf. Exercise 2.8 and (13.34)) $\dot{P} = \text{span}\,\{dx^2, d\dot{x}^2\}$. This implies (see Exercise 2.14) that $F^*\dot{P}$ is an involutive codistribution on $M \times U$, locally given as

$$F^*\dot{P} = \text{span}\,\{dx^2, df^2\} = \text{span}\,\{dx^2, \frac{\partial f^2}{\partial x^1}dx^1 + \frac{\partial f^2}{\partial u}\,du\}. \tag{13.43}$$

By (13.41) there exist m-vectors $b_i(x,u)$, $i \in \underline{k}$, satisfying

$$\frac{\partial f^2}{\partial x_i}(x,u) + \frac{\partial f^2}{\partial u}(x,u)\, b_i(x,u) = 0, \qquad i \in \underline{k}, \tag{13.44}$$

and thus in view of (13.43), writing $b_i = (b_{1i},\ldots,b_{mi})^T$,

$$\ker F^*\dot{P} = \operatorname{span}\{\frac{\partial}{\partial x_i} + \sum_{s=1}^{m} b_{si}(x,u)\frac{\partial}{\partial u_s},\ i \in \underline{k}\} + \ker \frac{\partial f^2}{\partial u}\, du. \tag{13.45}$$

Notice furthermore that $\ker \frac{\partial f^2}{\partial u} du = \tilde{G}_e$ (cf. (13.39)). Let us first assume that $\tilde{G}_e = 0$. Since $\ker F^*\dot{P}$ is an involutive distribution the Lie brackets

$$\left[\frac{\partial}{\partial x_i} + \sum_{s=1}^{m} b_{si}(x,u)\frac{\partial}{\partial u_s},\ \frac{\partial}{\partial x_j} + \sum_{s=1}^{m} b_{sj}(x,u)\frac{\partial}{\partial u_s}\right] =$$

$$\sum_{s=1}^{m}\left[\frac{\partial b_{sj}}{\partial x_i} - \frac{\partial b_{si}}{\partial x_j} + \sum_{r=1}^{m}\left(\frac{\partial b_{sj}}{\partial u_r} b_{ri} - \frac{\partial b_{si}}{\partial u_r} b_{rj}\right)\right](x,u)\frac{\partial}{\partial u_s}$$

are contained in $\ker F^*\dot{P}$, and thus

$$\frac{\partial b_{sj}}{\partial x_i} - \frac{\partial b_{si}}{\partial x_j} + \sum_{r=1}^{m}\left(\frac{\partial b_{sj}}{\partial u_r} b_{ri} - \frac{\partial b_{si}}{\partial u_r} b_{rj}\right) = 0,\quad i,j \in \underline{k},\ s \in \underline{m}\ . \tag{13.46}$$

These partial differential equations are exactly the *integrability conditions* of the classical Frobenius' Theorem (Theorem 2.45), with the only difference that in (13.46) there are additional *parameters* $x_{k+1},\ldots,x_n$. Thus by Theorem 2.45 there exists locally an m-vector $\alpha(x_1,\ldots,x_n,v_1,\ldots,v_m)$ (regarded as a function of $(x_1,\ldots,x_k)$ and $(v_1,\ldots,v_m)$ parametrized by $(x_{k+1},\ldots,x_n)$) such that

$$\begin{aligned} &\frac{\partial \alpha}{\partial x_i}(x,v) = b_i(x,\alpha(x,v)), \qquad i \in \underline{k}\ , \\ &\operatorname{rank}\frac{\partial \alpha}{\partial v}(0,v) = m,\quad v \in \mathbb{R}^m . \end{aligned} \tag{13.47}$$

Moreover for any x and v the matrix $\frac{\partial \alpha}{\partial v}(x,v)$ has rank m, and thus $u = \alpha(x,v)$ defines a regular static state feedback. Denoting $\tilde{f}(x,v) := f(x,\alpha(x,v))$ it immediately follows from (13.47) that

$$\begin{aligned} &\frac{\partial \tilde{f}^2}{\partial x_i}(x,v) = \frac{\partial f^2}{\partial x_i}(x,\alpha(x,v)) + \frac{\partial f^2}{\partial u}(x,\alpha(x,v))\,\frac{\partial \alpha}{\partial x_i}(x,v) = \\ &\frac{\partial f^2}{\partial x_i}(x,\alpha(x,v)) + \frac{\partial f^2}{\partial u}(x,\alpha(x,v))b_i(x,\alpha(x,v)) = 0\quad ,\quad i \in \underline{k}\ , \end{aligned} \tag{13.48}$$

and thus D is invariant for the closed-loop system $\dot{x} = \tilde{f}(x,v)$.

Now let $\dim \ker \frac{\partial f^2}{\partial u}(x,u) = m - \bar{m} > 0$. Since for each x the distribution $\tilde{G}_e = \ker \frac{\partial f^2}{\partial u}(x,u)du$ is an involutive distribution of constant dimension on U we can find for each x local coordinates $\bar{u}_1(x),\ldots,\bar{u}_m(x)$ for U such that

$$\ker \frac{\partial f^2}{\partial u}(x,u)du = \text{span}\,\{\frac{\partial}{\partial \bar{u}_{\bar{m}+1}},\ldots,\frac{\partial}{\partial \bar{u}_m}\}. \tag{13.49}$$

Equivalently, there locally exists a mapping $u = \bar{\alpha}(x,\bar{u})$, with rank $\frac{\partial \bar{\alpha}}{\partial \bar{u}} = m$, such that (13.49) holds. The mapping $u = \bar{\alpha}(x,\bar{u})$ defines a (preliminary) feedback, transforming the system into $\dot{x} = \bar{f}(x,\bar{u}) := f(x,\bar{\alpha}(x,\bar{u}))$. It follows from (13.41) that

$$\text{Im}\,\frac{\partial \bar{f}^2}{\partial x^1}(x,\bar{u}) \subset \text{Im}\,\frac{\partial \bar{f}^2}{\partial \bar{u}^1}(x,\bar{u}), \tag{13.50}$$

where $\bar{u}^1 = (\bar{u}_1,\ldots,\bar{u}_{\bar{m}})$, and thus there exist $\bar{m}$-vectors $\bar{b}_i(x,\bar{u})$ such that

$$\frac{\partial \bar{f}^2}{\partial x_i}(x,\bar{u}) + \frac{\partial \bar{f}^2}{\partial \bar{u}^1}\,\bar{b}_i(x,\bar{u}) = 0, \quad i \in \underline{k}. \tag{13.51}$$

By considering the distribution (see (13.45))

$$\text{span}\,\{\frac{\partial}{\partial x_i} + \sum_{s=1}^{\bar{m}} \bar{b}_{si}(x,\bar{u})\,\frac{\partial}{\partial \bar{u}_s},\ i \in \underline{k}\}, \tag{13.52}$$

we now have reduced the problem to the case $\tilde{G}_e = 0$. Thus there exists locally a feedback $\bar{u}^1 = \alpha^1(x,v^1)$, $\bar{u}^1, v^1 \in \mathbb{R}^{\bar{m}}$ such that D is invariant for the system $\dot{x} = \bar{f}(x,\alpha^1(x,v^1),\bar{u}^2)$, with $\bar{u}^2 = (\bar{u}_{\bar{m}+1},\ldots,\bar{u}_m)$. The total feedback which renders D invariant is therefore given as

$$u = \alpha(x,v) := \bar{\alpha}(x,\alpha^1(x,v^1),v^2), \tag{13.53}$$

with $v = (v^1,v^2) = (v_1,\ldots,v_{\bar{m}},\ v_{\bar{m}+1},\ldots,v_m)$. □

From a geometric point of view the above theorem can be interpreted in the following manner. Let D satisfy the assumptions of Theorem 13.7, and denote $P = \text{ann}\,D$. Then $E := \ker F^*\dot{P}$ is an involutive distribution on $M \times U$. Moreover if (13.40) holds, then E is constant dimensional and satisfies

$$\pi_* E = D \quad (\text{i.e. } \pi_{*(x,u)}E(x,u) = D(x),\ \forall (x,u)). \tag{13.54}$$

Furthermore by definition of E and by Proposition 13.5

$$F_* E = F_*(\ker F^* \dot{P}) \subset \ker \dot{P} = \dot{D}. \tag{13.55}$$

Hence from Theorem 13.7 we obtain (compare with Proposition 13.6)

Corollary 13.8 *Let D be a distribution on M as in Theorem 13.7, i.e. involutive and constant dimensional and such that $\tilde{G}_e$ (see (13.39)) has constant dimension. Then D is locally controlled invariant if and only if there exists an involutive constant dimensional distribution E on $M \times U$, with $E \cap G_e$ constant dimensional, satisfying ($\pi: M \times U \to M$ being the natural projection)*

$$\begin{aligned} \pi_* E &= D, \\ F_* E &\subset \dot{D}. \end{aligned} \tag{13.56}$$

Notice that if $\tilde{G}_e = 0$ then $E = \ker F^*\dot{P}$ has dimension equal to $\dim D$, and is the *unique* distribution satisfying (13.56). If $\tilde{G}_e \neq 0$ then the proof of Theorem 13.7 shows that at least locally we can define a (non-unique) distribution E, *contained* in $\ker F^*\dot{P}$, satisfying (13.56) and $\dim E = \dim D$. Indeed we may take E as the distribution defined in (13.52).

Moreover, a distribution E satisfying (13.56) and $\dim E = \dim D$ is directly related to a feedback $u = \alpha(x,v)$ which renders D invariant. In fact, E is necessarily of the form

$$E = \operatorname{span}\left\{ \frac{\partial}{\partial x_i} + \sum_{s=1}^{m} b_{si}(x,u) \frac{\partial}{\partial u_s},\ i \in \underline{k} \right\}, \tag{13.57}$$

and the functions $b_{si}(x,u)$ determine $\alpha(x,v)$ by (13.47). Conversely if $u = \alpha(x,v)$ renders D invariant then denoting

$$b_{si}(x,u) := \left. \frac{\partial \alpha_s}{\partial x_i}(x,v) \right|_{v = \alpha^{-1}(x,u)}, \tag{13.58}$$

the distribution E defined by (13.57) satisfies (13.56). From a more geometrical viewpoint E is determined by $\alpha(x,v)$ as the distribution on $M \times U$ whose integral manifolds are of the form $\{(x,u) = \alpha(x,v)) \mid v \text{ is constant}\}$.

Let us now see how Theorem 13.7 specializes to affine systems

$$\dot{x} = f(x) + \sum_{j=1}^{m} g_j(x) u_j, \tag{13.59}$$

and how we recover the results of Chapter 7 (e.g. Theorem 7.5). First we note that condition (13.40), or equivalently (13.41), reduces to

$$\operatorname{Im}\left(\frac{\partial f^2}{\partial x^1}(x,u) + \sum_{j=1}^{m} \frac{\partial g_j^2}{\partial x^1} u_j \right) \subset G^2(x), \text{ for every } (x,u), \tag{13.60}$$

where $G^2(x)$ denotes the matrix composed of the last $n-k$ rows of the matrix $G(x)$ with columns $g_1(x),\ldots,g_m(x)$. It is easily seen that (13.60) is equivalent to

$$\begin{aligned} &[f,D](x) \subset D(x) + G(x), \\ &[g_j,D](x) \subset D(x) + G(x), \qquad j \in \underline{m}, \end{aligned} \tag{13.61}$$

as in Theorem 7.5. Furthermore, condition (13.60) is equivalent to the existence of m-vectors $b_i(x,u)$ such that (cf.(13.44))

$$\frac{\partial f^2}{\partial x_i}(x) + \sum_{j=1}^{m} \frac{\partial g_j^2}{\partial x_i} u_j + G^2(x)b_i(x,u) = 0, \qquad i \in \underline{k}. \tag{13.62}$$

It follows that in the affine case the vectors b_i are of the form

$$b_i(x,u) = \ell_i(x) + K_i(x)u \tag{13.63}$$

for certain m-vectors $\ell_1(x),\ldots,\ell_m(x)$, and $m \times m$ matrices $K_1(x),\ldots,K_m(x)$. Therefore the set of p.d.e.'s (13.47) takes the form

$$\frac{\partial \alpha}{\partial x_i}(x,v) = \ell_i(x) + K_i(x)\alpha(x,v), \qquad i \in \underline{k}. \tag{13.64}$$

It follows that $\gamma(x,v) := \alpha(x,v) - \alpha(x,0)$ satisfies

$$\frac{\partial \gamma}{\partial x_i}(x,v) = K_i(x)\gamma(x,v), \qquad i \in \underline{k}, \tag{13.65}$$

which implies that $\gamma(x,v)$ can be taken to be *linear* in v, i.e. $\gamma(x,v) = \bar{\beta}(x)v$, for some $m \times m$ matrix $\bar{\beta}(x)$ satisfying

$$\frac{\partial \bar{\beta}}{\partial x_i}(x) = K_i(x)\bar{\beta}(x), \qquad i \in \underline{k}. \tag{13.66}$$

Since $\frac{\partial \alpha}{\partial v}(x,v)$ has rank m it follows that rank $\bar{\beta}(x) = m$ everywhere. Denoting $\bar{\alpha}(x) = \alpha(x,0)$ it is concluded that the feedback which renders D invariant can be taken of the affine form $u = \bar{\alpha}(x) + \bar{\beta}(x)v$, in accordance with Theorem 7.5. Furthermore we see that $\bar{\alpha}(x)$ satisfies

$$\frac{\partial \bar{\alpha}}{\partial x_i}(x) = \ell_i(x) + K_i(x)\bar{\alpha}(x), \qquad i \in \underline{k}, \tag{13.67}$$

Finally, the integrability conditions (13.46) reduce to

$$\begin{aligned} &\frac{\partial K_i}{\partial x_j} - \frac{\partial K_j}{\partial x_i} + K_iK_j - K_jK_i = 0, \quad i,j \in \underline{k}, \\ &\frac{\partial \ell_i}{\partial x_j} - \frac{\partial \ell_j}{\partial x_i} + K_i\ell_j - K_j\ell_i = 0, \quad i,j \in \underline{k}, \end{aligned} \tag{13.68}$$

which are exactly the integrability conditions for the partial differential equations (13.66) and (13.67); compare with (7.40) and (7.37).

Motivated by Corollary 13.8 we will now relate controlled invariance for a general nonlinear system (13.1) to controlled invariance of its *extended system* (13.38).

Proposition 13.9 *Consider the nonlinear system (13.1) with its extended system (13.38). Denote as before* $f_e = f(x,u)\frac{\partial}{\partial x}$ *and* $G_e = \text{span}\{\frac{\partial}{\partial u_1},\dots,\frac{\partial}{\partial u_m}\}$.
(a) Let D be a distribution satisfying the assumptions of Theorem 13.7. Then D is locally controlled invariant for (13.1) if and only if there exists an involutive constant dimensional distibution E on $M \times U$, with $\pi_* E = D$ and $E \cap G_e$ constant dimensional, which is locally controlled invariant for (13.38), i.e.

$$[f_e, E] \subset E + G_e, \tag{13.69}$$
$$[\frac{\partial}{\partial u_j}, E] \subset E + G_e, \qquad j \in \underline{m}.$$

(b) Conversely, let E be an involutive constant dimensional distribution on $M \times U$ such that E satisfies (13.69), and $E \cap G_e$ is constant dimensional. Then $D := \pi_* E$ is a well-defined distribution on M, which is involutive and constant dimensional. Moreover assume that the distribution $\tilde{G}_e$ for D (cf. (13.39)) has constant dimension, then D is locally controlled invariant for (13.1).

Proof Part (a). In view of Corollary 13.8 we only have to show that (13.69) is equivalent to (13.56). By Frobenius' Theorem we can find local coordinates $x = (x^1, x^2)$ for M such that $D = \text{span}\{\frac{\partial}{\partial x^1}\}$. Since $\pi_* E = D$ and $E \cap G_e$ is constant dimensional it follows that we can find coordinate functions $v = (v^1, v^2)$ for $M \times U$ such that (x^1, x^2, v^1, v^2) is a coordinate system for $M \times U$, and $E = \text{span}\{\frac{\partial}{\partial x^1}, \frac{\partial}{\partial v^1}\}$. Denote correspondingly $f = (f^1, f^2)$, where $F(x,v) = (x, f(x,v))$ is the local representation of F: $M \times U \to TM$. In these coordinates $F_* E \subset \dot{D}$ is equivalent to

$$\frac{\partial f^2}{\partial x^1}(x,v) = 0, \quad \frac{\partial f^2}{\partial v^1}(x,v) = 0. \tag{13.70}$$

On the other hand from (13.69) the same equations are obtained.

For part (b) we observe that by (13.69) the distribution $E + G_e$ is involutive and has constant dimension. Hence by an application of Proposition 3.50 we can find local coordinates $z = (z^1, z^2, z^3, z^4)$ for $M \times U$ such that $E_e = \text{span}\{\frac{\partial}{\partial z^1}, \frac{\partial}{\partial z^3}\}$ and $G_e = \text{span}\{\frac{\partial}{\partial z^3}, \frac{\partial}{\partial z^4}\}$. Since $G_e =$

span $\{\frac{\partial}{\partial u_1},\ldots,\frac{\partial}{\partial u_m}\}$ it immediately follows that (z^1,z^2) is a coordinate system for M. Denote $x^1:= z^1$, $x^2:= z^2$, $v^1:= z^3$, $v^2:= z^4$. Then E = span $\{\frac{\partial}{\partial x^1},\frac{\partial}{\partial v^1}\}$, and $D = \pi_* E$ = span $\{\frac{\partial}{\partial x^1}\}$, and the result follows from Corollary 13.8. □

We conclude from Proposition 13.9 that there is a one-to-one correspondence between locally controlled invariant distributions for (13.1) and for its extended system (13.38). Also the *feedbacks* required to make these distributions invariant for (13.1), respectively for (13.38), are intimately related. Indeed, let D be a locally controlled invariant distribution for (13.1) and choose coordinates $x = (x^1,x^2)$ such that D = span $\{\frac{\partial}{\partial x^1}\}$. Let now $u = \alpha(x,v)$ be a regular feedback which locally renders D invariant, i.e. $[\tilde{f}(\cdot,v),D] \subset D$, for all v, where $\tilde{f}(x,v):= f(x,\alpha(x,v))$. Then (x,v) is a new coordinate system for $M \times U$. Defining in these new coordinates the distribution E as span$\{\frac{\partial}{\partial x^1}\}$, it immediately follows that (with $\tilde{f}_e := \tilde{f}(x,v)\frac{\partial}{\partial x}$)

$$
\begin{aligned}
&[\tilde{f}_e\,,\ E] \subset E, && \\
&[\frac{\partial}{\partial v_j},E] \subset E, && j \in \underline{m}.
\end{aligned}
\tag{13.71}
$$

Hence, E is invariant with respect to the dynamics

$$
\begin{aligned}
\dot{x} &= \tilde{f}(x,v)\ , \\
\dot{v} &= \tilde{w}\ ,
\end{aligned}
\tag{13.72}
$$

with state (x,v) and inputs $\tilde{w}$. The extended system (13.38) of (13.1) is related to (13.72) by the (extended) state space transformation $x = x$, $u = \alpha(x,v)$, and in the old coordinates (x,u) the system (13.72) takes the form

$$
\begin{aligned}
\dot{x} &= \tilde{f}(x,v)\ , \\
\dot{u} &= \frac{\partial\alpha}{\partial x}(x,v)\tilde{f}(x,v) + \frac{\partial\alpha}{\partial v}(x,v)\tilde{w}\ ,
\end{aligned}
\tag{13.73}
$$

where v is such that $\alpha(x,v) = u$. Comparing (13.73) with (13.38) we see that $\tilde{w}$ is related to w via the feedback transformation

$$
w = \frac{\partial\alpha}{\partial x}(x,v)\tilde{f}(x,v) + \frac{\partial\alpha}{\partial v}(x,v)\tilde{w}\ ,
\tag{13.74}
$$

with v satisfying $\alpha(x,v) = u$. Denoting $v = \alpha^{-1}(x,u)$ (abuse of notation) it follows that the affine feedback which renders E invariant for the extended system (13.38) is given as

$$w = \frac{\partial\alpha}{\partial x}(x,\alpha^{-1}(x,u))f(x,u) + \frac{\partial\alpha}{\partial v}(x,\alpha^{-1}(x,u))\tilde{w}. \tag{13.75}$$

13.2 Disturbance Decoupling

The results on local controlled invariance obtained in the previous section will now be used for solving the local disturbance decoupling problem for general nonlinear systems. Consider a general nonlinear system with disturbances q

$$\dot{x} = f(x,u,q), \qquad u \in U, \qquad q \in Q, \tag{13.76a}$$

$$y = h(x,u), \qquad y \in \mathbb{R}^p. \tag{13.76b}$$

Here, as before, $x = (x_1,\ldots,x_n)$ are local coordinates for an n-dimensional manifold M, u are coordinates for U (the input space), which is an open subset of $\mathbb{R}^m$, and q are coordinates for Q (the space of disturbances), which is assumed to be an open subset of $\mathbb{R}^\ell$. Everything is assumed to be smooth. Alternatively (13.76a) is given by a system map $F\colon M \times U \times Q \longrightarrow TM$, locally represented as $F(x,u,q) = (x,f(x,u,q))$. First we state the following generalization of Proposition 4.23 regarding output invariance with respect to the disturbance q.

Proposition 13.20 *Consider the system (13.76) with system map F. The output y is invariant under q if there exists an involutive and constant dimensional distribution D on M such that*

(i) $[f(\cdot,u,q),D] \subset D$, *for all* $(u,q) \in U \times Q$,

(ii) $F_*TQ \subset \dot{D}$,

(iii) $D \subset \ker d_x h(\cdot,u)$, *for all* u.

(Here TQ denotes the ℓ-dimensional distribution on $M \times U \times Q$ given in local coordinates (x,u,q) as span $\{\frac{\partial}{\partial q}\}$; and $d_x h(x,u) := \frac{\partial h}{\partial x}(x,u)dx$.)

Proof By Frobenius' Theorem we can find local coordinates $x = (x^1,x^2)$ for M such that $D = \text{span}\,\{\frac{\partial}{\partial x^1}\}$. Write accordingly $f = (f^1,f^2)$, then (i) implies that (13.76a) is of the form

$$\begin{aligned} \dot{x}^1 &= f^1(x^1,x^2,u,q) \;, \\ \dot{x}^2 &= f^2(x^2,u,q) \;. \end{aligned} \tag{13.77}$$

Furthermore by (ii) f^2 does not depend on q, and by (iii) $h(x,u)$ does not depend on x^1, implying that $y = h(x,u)$ is invariant under q. □

The local disturbance decoupling problem (cf. Problem 7.13) consists of finding a locally defined regular static state feedback, briefly feedback, $u = \alpha(x,v)$ for (13.76), such that in the feedback transformed system the output y is invariant under q. Following the same approach as in Chapter 7 this will be done by looking for a distribution D on M which satisfies the conditions of Proposition 13.20 with respect to a *feedback transformed* system. As in Chapter 7 the notion of local controlled invariance will be crucial in doing this.

Motivated by the one-to-one correspondence of locally controlled invariant distributions with respect to (13.1) and its extended system (13.38) (cf. Proposition 13.9) we consider the extended system of (13.76a)

$$\begin{aligned} \dot{x} &= f(x,u,q) \\ \dot{u} &= w \\ \dot{q} &= d \end{aligned} \tag{13.78}$$

with state (x,u,q), and inputs (w,d). (Notice that for defining (13.78) the disturbances q (as well as the inputs u) are assumed to be differentiable. However the final conditions for disturbance decoupling (Proposition 13.21) will be valid for *arbitrary* disturbance functions.) Since (13.78) is an affine system, we can compute under constant rank assumptions, using the algorithms treated in Chapter 7 (e.g. Algorithm 7.19), the maximal controlled invariant distribution contained in ker dh for (13.78), i.e. the maximal distribution E on $M \times U \times Q$, satisfying (with $f_e = f(x,u,q)\frac{\partial}{\partial x}$)

$$[f_e, E] \subset E + \text{span}\,\{\frac{\partial}{\partial u}\} + \text{span}\,\{\frac{\partial}{\partial q}\}, \tag{13.79a}$$

$$[\frac{\partial}{\partial u_j}, E] \subset E + \text{span}\,\{\frac{\partial}{\partial u}\} + \text{span}\,\{\frac{\partial}{\partial q}\}, \quad j \in \underline{m}, \tag{13.79b}$$

$$[\frac{\partial}{\partial q_j}, E] \subset E + \text{span}\,\{\frac{\partial}{\partial u}\} + \text{span}\,\{\frac{\partial}{\partial q}\}, \quad j \in \underline{\ell}, \tag{13.79c}$$

$$E \subset \ker d\bar{h} \quad (\text{with } \bar{h}(x,u,q) := h(x,u)). \tag{13.79d}$$

Denote this maximal distribution as E^*, and assume that E^* and the intersection $E^* \cap \left(\text{span}\,\{\frac{\partial}{\partial u}\} + \text{span}\,\{\frac{\partial}{\partial q}\}\right)$ have constant dimension. Then it follows as in the proof of Proposition 13.9, part (b), that $\pi_* E^* =: D^*$, with $\pi\colon M \times U \times Q \to M$ the natural projection, is a well-defined distribution on M, which is involutive and constant dimensional. We obtain

Proposition 13.21 *Let E^* and D^* be as above. Then the local disturbance decoupling problem is solvable if $F_* TQ \subset \dot{D}^*$.*

Proof By Frobenius' Theorem we can find local coordinates $x = (x^1,x^2)$ for M such that $D^* = \text{span}\,\{\frac{\partial}{\partial x^1}\}$. Write accordingly $f = (f^1,f^2)$. Regard now (13.76a) as a system with inputs u *and* q. Then by Proposition 13.9 D^* is locally controlled invariant with respect to (13.76a). Thus (cf. (13.44)) there exist m-vectors $b_i(x,u,q)$, $i \in \underline{k}$, and ℓ-vectors $c_i(x,u,q)$, $i \in \underline{k}$, such that for $i \in \underline{k}$

$$\frac{\partial f^2}{\partial x_i}(x,u,q) + \frac{\partial f^2}{\partial u}(x,u,q)b_i(x,u,q) + \frac{\partial f^2}{\partial q}(x,u,q)c_i(x,u,q) = 0. \tag{13.80}$$

Now $F_*TQ \subset \dot{D}^*$ implies that f^2 does not depend on q. Therefore the m-vectors b_i in (13.80) can be taken to be independent of q (and c_i can be taken arbitrarily, say equal to zero), thus reducing (13.80) to

$$\frac{\partial f^2}{\partial x_i}(x,u) + \frac{\partial f^2}{\partial u}(x,u)\; b_i(x,u) = 0, \quad i \in \underline{k}. \tag{13.81}$$

As in the proof of Theorem 13.7 the vectors b_i determine the feedback $u = \alpha(x,v)$ (cf. (13.49)) which renders D^* (or E^*) invariant. Indeed, if $\ker \frac{\partial f^2}{\partial u} = 0$ then $\alpha(x,v)$ is determined as the solution of (13.47), and in case $\ker \frac{\partial f^2}{\partial u} \neq 0$ we proceed as in the proof of Theorem 13.7. Finally since $E^* \subset \ker d\bar{h}$ it follows that D^* satisfies the conditions of Proposition 13.20 for the feedback transformed system $\dot{x} = \tilde{f}(x,v,q) := f(x,\alpha(x,v),q)$. □

13.3 Input-Output Decoupling

We will briefly show how the approach to the input-output decoupling of square affine systems as dealt with in Chapter 8 can be readily generalized to square general nonlinear systems

$$\begin{aligned} \dot{x} &= f(x,u), \qquad u \in U, \text{ open subset of } \mathbb{R}^m, \\ y &= h(x,u), \qquad y \in \mathbb{R}^m. \end{aligned} \tag{13.82}$$

As in Chapter 8 we will throughout assume that (13.82) is an *analytic* system, although the results will partly also hold for smooth systems as well (see Chapter 8).

First we give (compare with Definition 8.7)

Definition 13.22 *Consider the system (13.82). The characteristic numbers ρ_j are the smallest integers ≥ -1 such that for $j \in \underline{p}$*

$$\frac{\partial}{\partial u} L_f^k h_j(x,u) := \left(\frac{\partial}{\partial u_1} L_f^k h_j, \ldots, \frac{\partial}{\partial u_m} L_f^k h_j\right)(x,u) = 0,$$
$$k = 0,1,\ldots,\rho_j, \ \forall (x,u) \in M, \qquad (13.83)$$
$$\frac{\partial}{\partial u} L_f^{(\rho_j+1)} h_j(x,u) \neq 0, \qquad \textit{for some } (x,u) \in M \times U.$$

If $\frac{\partial}{\partial u} L_f^k h_j(x,u) = 0$ *for all* $k \geq 0$ *and* $x \in M$, *then we set* $\rho_j = \infty$.

Remark Notice that for an affine system Definition 13.22 reduces to Definition 8.7.

Analogously to Definition 8.3 we state

Definition 13.23 *Let* $(x_0, u_0) \in M \times U$. *The system (13.82) is said to be locally strongly input-output decoupled around* (x_0, u_0) *if there exists a neighborhood* V *of* (x_0, u_0) *such that*

$$\frac{\partial}{\partial u_j} L_f^k h_i(x,u) = 0, \quad k \geq 0, \ (x,u) \in V, \ i \neq j, \ i,j \in \underline{m}, \qquad (13.84)$$

and finite integers $\rho_1, \ldots, \rho_m$ *satisfying*

$$\frac{\partial}{\partial u_i} L_f^k h_i(x,u) = 0, \quad k = 0,1,\ldots,\rho_i, \ (x,u) \in V, \ i \in \underline{m},$$
$$\frac{\partial}{\partial u_i} L_f^{(\rho_i+1)} h_i(x,u) \neq 0, \ \textit{for some } (x,u) \in V, \ i \in \underline{m}, \qquad (13.85)$$

such that moreover the set

$$\{(x,u) \in M \times U \mid \frac{\partial}{\partial u_i} L_f^{(\rho_i+1)} h_i(x,u) \neq 0, \ i \in \underline{m}\} \qquad (13.86)$$

contains V.

From now on we will drop for simplicity *throughout* the adjective strongly. The *local regular static state feedback input-output decoupling problem* (cf. Problem 8.6) is to find a locally defined regular static state feedback $u = \alpha(x,v)$ for (13.82) such that the feedback transformed system is locally input-output decoupled. Analogously to Theorem 8.13 this problem is approached as follows. *Assume* throughout that $\rho_1, \ldots, \rho_m$ for (13.82) are *finite*. Then define the decoupling matrix of (13.82) as (compare with (8.25))

$$A(x,u) = \begin{pmatrix} \frac{\partial}{\partial u_1} L_f^{(\rho_1+1)} h_1(x,u) & \ldots & \frac{\partial}{\partial u_m} L_f^{(\rho_1+1)} h_1(x,u) \\ \vdots & & \vdots \\ \frac{\partial}{\partial u_1} L_f^{(\rho_m+1)} h_m(x,u) & \ldots & \frac{\partial}{\partial u_m} L_f^{(\rho_m+1)} h_m(x,u) \end{pmatrix} \qquad (13.87)$$

Theorem 13.24 *Consider the square system (13.82) with finite characteristic numbers $\rho_1, \ldots, \rho_m$. Let $(x_0, u_0) \in M \times U$. Then the local regular static state feedback input-output decoupling problem around (x_0, u_0) is locally solvable if and only if*

$$\operatorname{rank} A(x_0, u_0) = m \ . \tag{13.88}$$

Proof Let (13.88) be satisfied. Consider the equations

$$\begin{bmatrix} L_f^{(\rho_1+1)} h_1(x,u) \\ \vdots \\ L_f^{(\rho_m+1)} h_m(x,u) \end{bmatrix} = \begin{bmatrix} v_1 \\ \vdots \\ v_m \end{bmatrix} . \tag{13.89}$$

By the Implicit Function Theorem there exists locally about (x_0, u_0) a solution $u = \alpha(x,v)$ of (13.89) with $\frac{\partial \alpha}{\partial v}(x,v)$ invertible. Since $y_i^{(\rho_i+1)} = L_f^{(\rho_i+1)} h_i(x,u)$, $i \in \underline{m}$, this feedback solves the problem. For the converse direction we refer to the proof of the "only if" part of Theorem 8.9. □

Similarly to Chapter 8 (see Exercise 8.1), it readily follows that (13.88) is also for *smooth* systems (13.82) a sufficient condition for local input-output decoupling (see Exercise 13.4). For the generalization of the *geometric* theory of input-output decoupling (see Chapter 9) to general smooth systems (13.82) we refer to Exercise 13.5.

Now suppose $\operatorname{rank} A(x_0, u_0) = m$. Consider the functions

$$z^i = (h_i(x,u), \ldots, L_f^{\rho_i} h_i(x,u)), \quad i \in \underline{m}. \tag{13.90}$$

By definition of ρ_i, $i \in \underline{m}$, all these functions do not depend on u. Furthermore, an immediate extension of Proposition 8.11 shows that they are *independent* functions of x. Hence we can choose additional functions $\bar{z}$ about x_0 such that $(z^1, \ldots, z^m, \bar{z})$ are local coordinates for M about x_0 with $\bar{z}(x_0) = 0$. After applying the feedback resulting from solving (13.89) for u we obtain in these coordinates, analogously to (8.45), the following local "normal form" (valid for analytic as well as for smooth systems)

$$\begin{aligned} \dot{z}^i &= A_i z^i + b_i v_i , \qquad\qquad i \in m, \\ \dot{\bar{z}} &= \bar{f}(\bar{z}, z^1, \ldots, z^m, v_1, \ldots, v_m), \end{aligned} \tag{13.91}$$

where the pairs (A_i, b_i), $i \in \underline{m}$, are in Brunovsky canonical form (6.50).

Example 13.25 Consider the simplified model of a voltage fed induction motor, as dealt with in Example 8.21. As in (8.98) we let the supply voltage vector be expressed as

$$\begin{bmatrix} \bar{u}_1 \\ \bar{u}_2 \end{bmatrix} = \begin{bmatrix} V \cos \theta \\ V \sin \theta \end{bmatrix} \tag{13.92}$$

where V is the amplitude and θ the angular position. Now let us consider $V =: u_1$ and $\theta =: u_2$ as *inputs* to the system (compare with Example 8.21). Clearly for this choice of inputs the system is *not* affine in the inputs. As in Example 8.21 we consider the stator flux and stator torque as outputs, i.e.

$$\begin{aligned} y_1 &= x_3^2 + x_4^2 , \\ y_2 &= x_2 x_3 - x_1 x_4 . \end{aligned} \tag{13.93}$$

Then $\rho_1 = \rho_2 = 0$, and the equations (13.89) are given as (compare with (8.101))

$$-2\alpha\sigma L_s x_1 x_3 - 2\alpha\sigma L_s x_2 x_4 + 2x_3 u_1 \cos u_2 + 2x_4 u_1 \sin u_2 = v_1 \tag{13.94}$$

$$\omega x_1 x_3 + (\alpha + \beta) x_1 x_4 - (\alpha + \beta) x_2 x_3 + \omega x_2 x_4 - \omega\sigma^{-1} L_s^{-1} x_3^2 - \omega\sigma^{-1} L_s^{-1} x_4^2 + (-x_1 + \sigma^{-1} L_s^{-1} x_3) u_1 \sin u_2 + (x_2 - \sigma^{-1} L_s^{-1} x_4) u_1 \cos u_2 = v_2$$

Hence the decoupling matrix $A(x,u)$ equals

$$\begin{bmatrix} 2x_3 \cos u_2 + 2x_4 \sin u_2 & -2x_3 u_1 \sin u + 2x_4 u_1 \cos u_2 \\ (-x_1 + \sigma^{-1} L_s^{-1} x_3) \sin u_2 + (x_2 - \sigma^{-1} L_s^{-1} x_4) \cos u_2 & (-x_1 + \sigma^{-1} L_s^{-1} x_3) u_1 \cos u_2 \\ & -(x_2 - \sigma^{-1} L_s^{-1} x_4) u_1 \sin u_2 \end{bmatrix} \tag{13.95}$$

which is a non-singular matrix in all points (x,u) where

$$\det A(x,u) = 2u_1 \left(\frac{x_3^2 + x_4^2}{\sigma L_s} - (x_1 x_3 + x_2 x_4) \right) \neq 0 \tag{13.96}$$

Clearly u_1 is always different from zero, being the amplitude of the voltage. Furthermore the term inside the large brackets is proportional to the scalar product of the stator and rotor flux, and thus is non-zero during normal mode operation. Hence in view of Theorem 13.24, during normal mode operation the system is locally input-output decoupable by a static state feedback and this feedback is locally given as the solution of (13.94). □

Now, as in Chapter 8, we proceed to the problem of input-output decoupling of the square system (13.82) by *dynamic state feedback*

$$\begin{cases} \dot{z} = \gamma(z,x,v), & z \in \mathbb{R}^q, \quad v \in \mathbb{R}^m, \\ u = \alpha(z,x,v), \end{cases} \tag{13.97}$$

where α and γ are smooth mappings. As in Chapter 8 we assume throughout that (13.82) is an *analytic* system, and we only solve a *local* version of the problem. The key tool is the following generalization of Algorithm 8.18.

Algorithm 13.26 (Dynamic extension algorithm) Consider the analytic square system (13.82).

Step 1 Denote the characteristic numbers by $\rho_1^1,\ldots,\rho_m^1$, and the decoupling matrix (cf.(13.87)) by $D^1(x,u)$. Let $r_1(x,u) := \operatorname{rank} D^1(x,u)$. By analyticity $r_1(x,u)$ is constant, say r_1, on an open and dense subset B_1 of $M \times \mathbb{R}^m$. Assume we work on a neighborhood contained in B_1. Reorder the output functions $h_1,\ldots,h_m$ in such a way that the first rows of D^1 are independent. Write $h^0 = (h_1,\ldots,h_{r_1})$, $\bar{h}^0 = (h_{r_1+1},\ldots,h_m)$ and correspondingly $y = (y^1,\bar{y}^1)$. Consider the equations

$$\begin{bmatrix} y_1^{(\rho_1^1+1)} \\ \vdots \\ y_{r_1}^{(\rho_{r_1}^1+1)} \end{bmatrix} = \begin{bmatrix} L_f^{(\rho_1^1+1)} h_1(x,u) \\ \vdots \\ L_f^{(\rho_{r_1}^1+1)} h_{r_1}(x,u) \end{bmatrix} = \begin{bmatrix} v_1 \\ \vdots \\ v_{r_1} \end{bmatrix} . \tag{13.98}$$

By the Implicit Function Theorem we can locally solve for r_1 components of u as a function $\alpha^1(x,v)$ of x and v, so as to obtain

$$y_i^{(\rho_i^1+1)} = v_i , \qquad i = 1,\ldots,r_1 . \tag{13.99}$$

Thus after a possible relabeling of the input components $(u_1,\ldots,u_m)$ we obtain a "partial" state feedback $u^1 = \alpha^1(x,v^1)$, where $u^1 = (u_1,\ldots,u_{r_1})$ and $v^1 = (v_1,\ldots,v_{r_1})$. Leaving the remaining inputs $\bar{u}^1$ unchanged, the system transforms into $f^1(x,v^1,\bar{u}^1) := f(x,\alpha^1(x,v^1),\bar{u}^1)$ and $\bar{h}^1(x,v^1) := \bar{h}^0(x,\alpha^1(x,v^1))$. Renaming v^1 as u^1 we obtain the system

$$\begin{aligned} \dot{x} &= f^1(x,u^1,\bar{u}^1) , \\ \bar{y}^1 &= \bar{h}^1(x,u^1) , \end{aligned} \tag{13.100}$$

with inputs $\bar{u}^1$, outputs $\bar{y}^1$, and where u^1 is regarded as a set of *parameters*.

Step ℓ+1 Assume we have defined a sequence of integers $r_1,\ldots,r_\ell$, with $q_\ell := \sum_{i=1}^{\ell} r_i$, and we have a block partioning of h as $(h^1,\ldots,h^\ell,\bar{h}^\ell)$, with $\dim h^i = r_i$, $i \in \underline{\ell}$, and correspondingly $y = (y^1,\ldots,y^\ell,\bar{y}^\ell)$. Furthermore, assume we have obtained a system

$$\begin{aligned} \dot{x} &= f^{\ell}(x,\tilde{U}^{\ell-1},u^{\ell},\bar{u}^{\ell}) \,, \\ \bar{y}^{\ell} &= \bar{h}^{\ell}(x,\tilde{U}^{\ell-1},u^{\ell}) \,, \end{aligned} \tag{13.101}$$

Here the new controls u are correspondingly split as $u = (u^1,\ldots,u^{\ell},\bar{u}^{\ell})$, and $\tilde{U}^{\ell-1}$ denotes $U^{\ell-1} := (u^1,\ldots,u^{\ell-1})$ and suitable time-derivatives. We regard (13.101) as a system with inputs $\bar{u}^{\ell}$, parametrized by $\tilde{U}^{\ell-1}$ and u^{ℓ}. Denote the characteristic numbers of (13.101) by $\rho^{\ell+1}_{q_{\ell}+1},\ldots,\rho^{\ell+1}_{m}$, and the $(m-q_{\ell}) \times (m-q_{\ell})$ decoupling matrix of (13.101) by $D^{\ell+1}(x,\tilde{U}^{\ell},\bar{u}^{\ell})$. By analyticity the rank of this matrix is constant, say $r_{\ell+1}$, on an open and dense subset $B_{\ell+1}$ of points $(x,\tilde{U}^{\ell},\bar{u}^{\ell})$. Reorder the output functions $\bar{h}^{\ell}$ such that the first $r_{\ell+1}$ rows of $D^{\ell+1}$ are independent, and write with $q_{\ell+1} := q_{\ell} + r_{\ell+1}$, $h^{\ell+1} = (h_{q_{\ell}+1},\ldots,h_{q_{\ell+1}})$, $\bar{h}^{\ell+1} = (h_{q_{\ell+1}+1},\ldots,h_m)$, and denote accordingly $\bar{y}^{\ell} = (y^{\ell+1},\bar{y}^{\ell+1})$, $\bar{u}^{\ell} = (u^{\ell+1},\bar{u}^{\ell+1})$. Consider the equations

$$y_i^{(\rho_i^{\ell+1})} = L_f^{\rho_i^{\ell+1}} h_i(x,\tilde{U}^{\ell},\bar{u}^{\ell}) = v_i \,, \qquad i = q_{\ell}+1,\ldots,q_{\ell} \,. \tag{13.102}$$

By the Implicit Function Theorem we can locally solve for $r_{\ell+1} = q_{\ell+1} - q_{\ell}$ components of $\bar{u}^{\ell}$ as functions of x, $\tilde{U}^{\ell}$ and $(v_{q_{\ell}+1},\ldots,v_{q_{\ell+1}}) =: v^{\ell+1}$, so as to obtain

$$y_i^{(\rho_i^{\ell+1})} = v_i \,, \qquad i = q_{\ell}+1,\ldots,q_{\ell+1} \,. \tag{13.103}$$

Thus after a possible relabeling of the input components of $\bar{u}^{\ell}$ we obtain a "partial" state feedback $u^{\ell+1} = \alpha^{\ell+1}(x,\tilde{U}^{\ell},v^{\ell+1})$. Leaving the inputs $\bar{u}^{\ell+1}$ unaltered the system transforms into $f^{\ell+1}(x,\tilde{U}^{\ell},v^{\ell+1},\bar{u}^{\ell+1}) := f^{\ell}(x,\tilde{U}^{\ell-1},\alpha^{\ell+1}(x,\tilde{U}^{\ell},v^{\ell+1}),\bar{u}^{\ell+1})$ with output function $\bar{h}^{\ell+1}(x,\tilde{U}^{\ell},v^{\ell+1}) := \bar{h}^{\ell}(x,\tilde{U}^{\ell-1},\alpha^{\ell+1}(x,\tilde{u}^{\ell},v^{\ell+1}))$. Renaming again $u^{\ell+1} := v^{\ell+1}$ we obtain the system

$$\begin{aligned} \dot{x} &= f^{\ell+1}(x,\tilde{U}^{\ell},u^{\ell+1},\bar{u}^{\ell+1}) \,, \\ \bar{y}^{\ell+1} &= \bar{h}^{\ell+1}(x,\tilde{U}^{\ell},u^{\ell+1}) \,, \end{aligned} \tag{13.104}$$

□

As in Chapter 8 it follows that there exists some finite integer k such that

$$0 < q_1 < q_2 < \ldots < q_k = q_{k+1} = \ldots \leq m, \tag{13.105}$$

and the integer $q^* = q_k$ will be called the *rank* of the system (13.82).

Analogously to Theorem 8.19 we obtain

Theorem 13.27 *Consider the square analytic system (13.82). The following two conditions are equivalent:*

(i) The dynamic state feedback input-output decoupling problem is locally solvable on an open and dense subset of M.

(ii) The rank q^* of the system equals m.

For the proof of Theorem 13.27 we refer to the proof of Theorem 8.19; details will be left to the reader.

13.4 Locally Controlled Invariant Submanifolds

Analogously to Definition 11.1 we will call a submanifold $N \subset M$ *locally controlled invariant* for the nonlinear system (13.1) if there exists locally around any $x_0 \in N$ a strict static state feedback, briefly feedback, $u = \alpha(x)$, $x \in N$, such that

$$f(x,\alpha(x)) \in T_x N \quad , \quad x \in N \ . \tag{13.106}$$

Locally we may choose coordinates $x = (x^1,x^2)$ for M such that $N = \{x | x^1 = 0\}$. If we write accordingly

$$f(x,u) = \begin{bmatrix} f^1(x^1,x^2,u) \\ f^2(x^1,x^2,u) \end{bmatrix} \tag{13.107}$$

then we obtain the following analog of Proposition 11.2.

Proposition 13.27 *Consider the nonlinear system (13.1) with $N \subset M$, locally given as $\{x | x^1 = 0\}$. Assume that rank $\frac{\partial f^1}{\partial u}(0,x^2,u)$ does not depend on x^2,u. Then N is locally controlled invariant if and only if for any $x \in N$ there exists $u \in U$ such that $f(x,u) \in T_x N$.*

Proof Let rank $\frac{\partial f^1}{\partial u}(0,x^2,u) = s$. Then by the Implicit Function Theorem the equation $f^1(0,x^2,u) = 0$ can be solved for s of the u-variables as smooth functions of x^2. □

Remark 13.28 Moreover it follows that we can locally define a *degenerate* feedback $u = \alpha(x,v)$, $v \in \mathbb{R}^{\tilde{m}}$, $\tilde{m} := m - s$, rank $\frac{\partial x}{\partial v} = \tilde{m}$, such that

$$f(x,\alpha(x,v)) \in T_x N \quad , \qquad x \in N, \quad \text{for all } v \in \mathbb{R}^{\tilde{m}} \ . \tag{13.108}$$

Now let us consider a nonlinear system with *outputs*

$$\dot{x} = f(x,u) \quad , \quad x \in M \;,\; u \in U \;,$$
$$y = h(x,u) \quad , \quad y \in \mathbb{R}^p \;. \tag{13.109}$$

A submanifold $N \subset M$ will be called locally controlled invariant *output-nulling* if there exists locally a feedback $u = \alpha(x)$, $x \in N$, such that

$$f(x,\alpha(x)) \in T_xN \quad , \quad h(x,\alpha(x)) = 0 \quad , \quad x \in N \;. \tag{13.110}$$

In order to compute the *maximal* locally controlled invariant output-nulling submanifold for (13.109) we again take recourse to the *extended system*

$$\begin{cases} \dot{x} = f(x,u) & , \quad (x,u) \in M \times U \;, \\ \dot{u} = w & , \quad w \in \mathbb{R}^m \;, \end{cases}$$
$$y = h(x,u) \quad , \quad y \in \mathbb{R}^p \;, \tag{13.111}$$

which is an *affine* system with state (x,u) and input w. Now let $(x_0,u_0) \in M \times U$, satisfying $f(x_0,u_0) = 0$ and $h(x_0,u_0) = 0$, be a regular point for Algorithm 11.6 applied to the extended system (13.11). Then locally through (x_0,u_0) we obtain the maximal locally controlled invariant output-nulling submanifold $N_e^* := N_{k^*}^e \subset M \times U$, given in the form

$$N_e^* = \{(x,u) \text{ near } (x_0,u_0) \,|\, h(x,u) = \varphi_1(x,u) = \ldots = \varphi_{k^*-1}(x,u) = 0\} \tag{13.112}$$

By the first assumption of regularity in Algorithm 11.6 the matrix

$$\frac{\partial}{\partial u}\begin{bmatrix} h(x,u) \\ \varphi_1(x,u) \\ \vdots \\ \varphi_{k^*-1}(x,u) \end{bmatrix} \tag{13.113}$$

has constant rank $r^* := r_{k^*}$ on N_e^* (due to the particular form of the input vectorfields in (13.111)). An application of the implicit function theorem yields the existence of a regular feedback $u = \alpha(x,u',u")$, with dim $u"$ $= r^*$, such that locally around (x_0,u_0)

$$\text{rank}\,\frac{\partial}{\partial u"}\begin{bmatrix} h(x,\alpha(x,u',u")) \\ \vdots \\ \varphi_{k^*-1}(x,\alpha(x,u',u")) \end{bmatrix} = r^* \;,\; \frac{\partial}{\partial u'}\begin{bmatrix} h(x,\alpha(x,u',u")) \\ \vdots \\ \varphi_{k^*-1}(x,\alpha(x,u',u")) \end{bmatrix} = 0 \,, \tag{13.114}$$

and thus we can eliminate the variables $u"$ as a function $u"(x)$ of x. Furthermore, by the second regularity assumption in Algorithm 11.6 the matrix

$$\frac{\partial}{\partial x}\begin{bmatrix} h(x,\alpha(x,u',u''(x))) \\ \vdots \\ \varphi_{k^*-1}(x,\alpha(x,u',u''(x)) \end{bmatrix} \tag{13.115}$$

has constant rank on N_e^*. It follows that we may solve for a part of the state-variables, say x'', as a function $x'' = \psi(x')$ of the remaining state-variables, denoted as x'. Thus finally we obtain a system of the form

$$\dot{x}' = f'(x',u') \tag{13.116}$$

where $f'(x',u') := \tilde{f}(x',x''(x'),u')$, with $\tilde{f}(x,u') := f(x,\alpha(x,u',u''(x)))$. The set $\{x = (x',x'') \mid x'' = \psi(x')\}$ defines locally around x_0 a submanifold, denoted as N^*, of M. By construction N^* is a locally controlled invariant output-nulling submanifold for (13.109), which is *maximal* in the following sense. Let N be another locally controlled invariant output-nulling submanifold through x_0. Locally around x_0 we may represent N as $N = \{x \mid x^1 = 0\}$. *Assume* that the rank of the matrix (cf.(13.107))

$$\begin{bmatrix} \frac{\partial f^1}{\partial u}(0,x^2,u) \\ \frac{\partial h}{\partial u}(0,x^2,u) \end{bmatrix} \tag{13.117}$$

is constant, say r, around (x_0,u_0). Then (see Remark (13.28)), there exists locally a degenerate feedback $u = \alpha(x,v)$, $\alpha(x_0,0) = u_0$, $v \in \mathbb{R}^{\tilde{m}}$, $\tilde{m} := m - r$, rank $\frac{\partial \alpha}{\partial v} = \tilde{m}$, such that

$$\left.\begin{aligned} &f(x,\alpha(x,v)) \in T_xN\ , \\ &h(x,\alpha(x,v)) = 0\ , \end{aligned}\right\} \quad , \; x \in N, \text{ for all } v \in \mathbb{R}^{\tilde{m}}\ . \tag{13.118}$$

Now one easily checks that the following submanifold of $M \times U$

$$N_e := \{(x,\alpha(x,v)) \mid x \in N\ ,\ v \in \mathbb{R}^{\tilde{m}}\}$$

is a locally controlled invariant output-nulling submanifold for the extended system (13.111), and thus, since N_e^* is maximal, necessarily $N_e \subset N_e^*$.

Since x' serve as coordinates for N^*, (13.116) describes the dynamics on N^*, which will be called, in analogy with (11.31), the *constrained dynamics* for (13.109). (For a definition of *zero dynamics*, analogously to Definition 11.14, we refer to Exercise 13.8.)

As in the affine case (cf. Proposition 11.13) we have the following important special case where the computation of N^* becomes more easy.

Proposition 13.29 *Consider the system (13.109). Assume that the characte-*

ristic numbers $\rho_1,\dots,\rho_p$ as defined in Definition 13.22 are all finite. Furthermore assume that the $p \times m$ matrix

$$A(x,u) = \left[\frac{\partial}{\partial u_j} L_f^{(\rho_i+1)} h_i(x,u) \right]_{\substack{i = 1,\dots,p \\ j = 1,\dots,m}} \tag{13.119}$$

has rank p on the set

$$N^* = \{x \mid h_i(x,u) = \dots = L_f^{\rho_i} h_i(x,u) = 0 \ , \ i \in \underline{p}\} \subset M \ . \tag{13.120}$$

(Note that by definition of ρ_i the functions $h_i,\dots,L_f^{\rho_i}h_i$, $i \in \underline{p}$, do not depend on u!). If N^* is non-empty (for example if there exists (x_0,u_0) with $h(x_0,u_0) = 0$ and $f(x_0,u_0) = 0$), then N^* is a smooth submanifold of M of dimension $n - \sum_{i=1}^{p} (\rho_i+1)$, and is equal to the maximal (globally) controlled invariant output-nulling submanifold for (13.109). The feedbacks $u = \alpha^*(x)$ which render N^* invariant are given as the solutions of

$$L_f^{(\rho_i+1)} h_i(x,\alpha^*(x)) = 0 \ , \ x \in N^* \ , \ i \in \underline{p} \ . \tag{13.121}$$

In fact, there exists locally around any point in N^* a degenerate feedback $u = \alpha(x,v)$, $v \in \mathbb{R}^{\tilde{m}}$, $\tilde{m} := m - p$, $\operatorname{rank} \frac{\partial \alpha}{\partial v} = \tilde{m}$, such that

$$L_f^{(\rho_i+1)} h_i(x,\alpha(x,v)) = 0 \quad , \ x \in N^*, \text{ for all } v \in \mathbb{R}^{\tilde{m}}, \ i \in \underline{p} \ , \tag{13.122}$$

and the constrained dynamics are given as the dynamics $\dot{x} = f(x,\alpha(x,v))$ restricted to N^*.

Example 13.30 Consider the model of a rocket outside the atmosphere as treated in Example 6.13. Consider as output y the angle θ , i.e. $y = x_2$. Then clearly $\rho = 1$, and the decoupling matrix is given as $(T/mx_1)\cos u$, which is non-zero for u around 0. Thus, $N^* = \{(x_1,x_2,x_3,x_4) \in T(\mathbb{R}^+ \times S^1);\ x_2 = x_4 = 0\}$, and $\alpha^*(x)$ solving (13.121) equals 0, resulting in a constrained (and zero) dynamics

$$\begin{aligned} \dot{x}_1 &= x_3 \ , \\ \dot{x}_3 &= - gR^2/x_1^{\,2} + T/m \ . \end{aligned} \tag{13.123}$$

□

The theory of *interconnections* of systems, as dealt with in Chapter 11 for affine systems and for interconnections only involving the states of the systems, can now be immediately extended to general nonlinear systems

$$\dot{x}_i = f_i(x_i,u_i) \quad , \quad x \in M^i \ , \ u_i \in \mathbb{R}^{m_i} \quad , \quad i \in \underline{k} \ , \tag{13.124}$$

interconnected by a interconnection constraint of the form

$$\varphi(x_1,\ldots,x_k,u_1,\ldots,u_k) = 0 \,, \tag{13.125}$$

where φ is a smooth mapping. Indeed we have to compute the *constrained dynamics* for the general nonlinear system defined by the *product system* (cf. (11.52a)) of the systems (13.124), with output mapping φ.

Example 13.31 Consider two nonlinear systems $\dot{x}_i = f_i(x_i,u_i)$, $y_i = h_i(x_i,u_i)$, $i = 1,2$. Suppose the second system is placed in a feedback loop for the first system, resulting in the interconnection equations

$$\begin{aligned} u_2 &= y_1 = h_1(x_1,u_1), \\ u_1 &= y_2 = h_2(x_2,u_2). \end{aligned} \tag{13.126}$$

Clearly, all the characteristic numbers are equal to -1, while the decoupling matrix

$$\begin{bmatrix} -\dfrac{\partial h_1}{\partial u_1} & I \\ I & -\dfrac{\partial h_2}{\partial u_2} \end{bmatrix} \tag{13.127}$$

has full rank if and only if the matrix $\left(I - \dfrac{\partial h_1}{\partial u_1}\dfrac{\partial h_2}{\partial u_2}\right)$ is invertible. This condition is usually imposed as a prerequisite for "well-posedness" of the interconnection.

The extension of the theory of *inverse systems*, as dealt with affine systems in Chapter 11, to general nonlinear systems (13.109) will be left to the reader (see also Exercise 13.6).

13.5 Control Systems Defined on Fiber Bundles

As already argued in the Introduction (e.g. Example 1.7), the global definition of a nonlinear control system $\dot{x} = f(x,u)$ as a system map $F : M \times U \to TM$ has some serious drawbacks. The problem is that in some cases the input space U is not independent of the state, thus implying that the use of the product $M \times U$ of the state and the input is too restrictive. The mathematical generalization of a product space is a *fiber bundle*:

Definition 13.32 *$(B,M,\pi,U,\{O_i\}_{i\in I})$ is called a* fiber bundle *if the following holds. B,M and U are smooth manifolds, respectively called the* *total space, the base space and the standard fiber.* The map $\pi: B \to M$ *is a surjective submersion, and* $\{O_i\}_{i \in I}$ *is a covering family of open subsets for* M *such that for every* $i \in I$ *there is a diffeomorphism* $\Phi_i : \pi^{-1}(O_i) \to O_i \times U$, *satisfying the commutative diagram*

$$\begin{array}{ccc} \pi^{-1}(O_i) & \xrightarrow{\Phi_i} & O_i \times U \\ & \pi \searrow \quad \swarrow \text{projection} & \\ & O_i & \end{array} \tag{13.128}$$

The submanifolds $\pi^{-1}(x)$, $x \in M$, *are called the fibers at* x. *If no confusion can arise we denote the fiber bundle simply by* (B,M).

Notice that a fiber bundle is only *locally* isomorphic to a product space $O_i \times U$. Nonlinear systems on fiber bundles are now defined as follows.

Definition 13.33 *A smooth nonlinear control system on a fiber bundle* $(B,M,\pi,U,\{O_i\}_{i\in I})$, *is defined by a map (the* system map*)* $F : B \to TM$ *satisfying* ($\pi_M: TM \to M$ *being the projection)*

$$\begin{array}{ccc} B & \xrightarrow{F} & TM \\ & \pi \searrow \quad \swarrow \pi_M & \\ & M & \end{array} \tag{13.129}$$

The system will be briefly denoted by (B,M,F).

Let $x_0 \in O_i$ and $b_0 \in \pi^{-1}(x_0)$. Choose local coordinates $x = (x_1,\ldots,x_n)$ around x_0, and $u = (u_1,\ldots,u_m)$ around $u_0 := \Phi_i(b_0)$. Then

$$(x_1 \circ \Phi_i, \ldots, x_n \circ \Phi_i, u_1 \circ \Phi_i, \ldots, u_m \circ \Phi_i) \tag{13.130}$$

are local coordinates for B around b_0, which we will simply denote as $(x_1,\ldots,x_n,u_1,\ldots,u_m) = (x,u)$. In such coordinates the map F, in view of (13.129), is locally represented as $F(x,u) = (x,f(x,u))$, thus recovering the local coordinate expression $\dot{x} = f(x,u)$.

Observe that on any non-empty intersection $O_i \cap O_j$ we can take local coordinates (x,u) as above, either using the diffeomorphism $\Phi_i : \pi^{-1}(O_i) \to O_i \times U$, or the diffeomorphism $\Phi_j : \pi^{-1}(O_j) \to O_j \times U$. The relation is as follows. There exists a diffeomorphism Φ_{ij} rendering the following diagram commutative

$$\begin{array}{ccc} & \pi^{-1}(O_i \cap O_j) & \\ \swarrow & & \searrow \\ (O_i \cap O_j) \times U & \xrightarrow{\Phi_{ij}} & (O_i \cap O_j) \times U \end{array} \tag{13.131}$$

Let now $x_0 \in O_i \cap O_j$ and take local coordinates x around x_0. Furthermore let $b_0 \in \pi^{-1}(x_0)$, and take local coordinates $u^i = (u_1^i,\dots,u_m^i)$ around $\Phi_i(b_0)$, and $u^j = (u_1^j,\dots,u_m^j)$ around $\Phi_j(b_0)$. In such coordinates Φ_{ij} is locally represented as

$$\Phi_{ij}(x,u^i) = (x,\varphi_{ij}(x,u^i) = u^j) \tag{13.132}$$

Alternatively , as in (13.130) we obtain two coordinate systems (x,u^i) and (x,u^j) around $b_0 \in B$, which are related as $u^j = \varphi_{ij}(x,u^i)$. Furthermore since Φ_{ij} is a diffeomorphism it follows that rank $\frac{\partial}{\partial u^i}\varphi_{ij} = m$ everywhere, and thus this change in input coordinates can be interpreted as locally defined *feedback*. More generally, we give

Definition 13.34 *Let (B,M,F) be a nonlinear system. A (global)* feedback *is a diffeomorphism $A : B \to B$ satisfying*

$$\begin{array}{ccc} B & \xrightarrow{A} & B \\ & \pi\searrow \quad \swarrow\pi & \\ & M & \end{array} \tag{13.133}$$

(A is called a bundle isomorphism.) Let $x_0 \in O_i$ and $b_0 \in \pi^{-1}(x_0)$. Let (x,v) and (x,u) be two sets of local coordinates as in (13.130), respectively around $b_0 \in B$ and around $A(b_0) \in B$. Then A is locally represented as

$$A(x,v) = (x,\alpha(x,v) = u) \ . \tag{13.134}$$

The feedback transformed system is denoted as $(B,M,F\circ A)$.

The above definitions specialize to *affine* nonlinear control systems $\dot{x} = f(x) + \sum_{j=1}^{m} g_j(x)u_j$ as follows. A fiber bundle $(B,M,\pi,U,\{O_i\}_{i\in I})$ is called a *vector bundle* if U is a linear space, i.e. $U = \mathbb{R}^m$, and the diffeomorphisms $\Phi_i : \pi^{-1}(O_i) \to O_i \times U$, $i \in I$, are such that all the maps φ_{ij} as in (13.132), $i,j \in I$, are *linear* in u^i, if we take u^i and u^j to be linear coordinates for U. An affine control system (B,M,F) is now given by a *vector bundle* (B,M) and a system map $F\colon B \to TM$ which is *affine* as a map from $\pi^{-1}(x) \subset B$ to $\pi_M^{-1}(x) \subset TM$. Taking local coordinates x for M and linear coordinates for $U = \mathbb{R}^m$ we immediately obtain the local coordinate expression $F(x,u) = (x,\dot{x} = f(x) + \sum_{j=1}^{m} g_j(x)u_j)$. Finally a (global) feedback A will be defined as in Definition 13.34, with the restriction that A as a map from $\pi^{-1}(x)$ to itself is affine, resulting in the local coordinate expression (cf. (13.134)) $A(x,v) = (x,\alpha(x) + \beta(x)v = u)$, with $\beta(x)$ an invertible $m \times m$ matrix.

Let us now indicate how the material of the previous sections generalizes to nonlinear control systems (B,M,F). First let us consider Proposition 13.6. We immediately encounter the problem of defining D_e as a distribution on B instead of on $M \times U$. Clearly on any subset $\pi^{-1}(O_i) \subset B$ we can define, similarly to Proportion 13.6, the unique distribution D_e^i such that $\pi_* D_e^i = D$ on O_i and $\bar{\pi}_{i*} D_e^i = 0$, where $\bar{\pi}_i := pr_2 \circ \Phi_i$ and $pr_2 : 0_i \times U \to U$ is the projection onto the second factor. However this does not necessarily define a distribution D_e on the *whole* fiber bundle B; in fact we need that $D_e^i = D_e^j$ on $\pi^{-1}(O_i \cap O_j)$. The global definition of D_e is ensured by the following additional requirement on the bundle (B,M). Suppose that all the maps φ_{ij}, $i,j \in I$, as in (13.132) do not depend on $x \in O_i \cap O_j$. Defining on every $\pi^{-1}(O_i)$ the distribution H^i as $\bar{\pi}_{i*}^{-1}(0)$, it easily follows that $H^i = H^j$ on $\pi^{-1}(O_i \cap O_j)$, and thus there exists a *globally* defined distribution H on B with $H = H^i$ on $\pi^{-1}(O_i)$, $i \in I$. It follows that $\pi_* H = TM$ and that $\dim H = \dim M$. Now we define D_e on B by setting

$$D_e = \pi_*^{-1} D \cap H \ . \tag{13.135}$$

Contrary to Proposition 13.6, the generalization of Theorem 13.7 (characterizing locally controlled invariant distributions) to nonlinear control systems (B,M,F) does not pose any problems. Indeed, let D be an involutive distribution of constant dimension on M. Define the vertical distribution G_e on B as $G_e = \pi_*^{-1}(0)$, and denote $\tilde{G}_e = \{Z \in G_e \mid F_* Z \in \dot{D}\}$. Assuming that G_e has constant dimension we obtain that D is locally controlled invariant for (B,M,F) if and only if (cf. (13.40'))

$$F_*(\pi_*^{-1}(D)) \subset \dot{D} + F_* G_e \ . \tag{13.136}$$

Hence if the system (M,B,F) is locally represented as $\dot{x} = f(x,u)$, with (x,u) as in (13.130), then (13.136) implies the local existence of a regular static feedback $u = \alpha(x,v)$ such that D is invariant for $\dot{x} = \tilde{f}(x,v) := f(x,\alpha(x,v))$.

Let us now briefly elaborate on such local feedback transformations. Take local coordinates $x = (x_1,\ldots,x_n)$ for M. Identifying $x_i : M \to \mathbb{R}$ with $x_i \circ \pi : B \to \mathbb{R}$ we can regard $x_1,\ldots,x_n$ as coordinate functions on B. Thus we can take additional coordinate functions $u = (u_1,\ldots,u_m)$ on B such that (x,u) is a coordinate system for B. Such coordinates for B are called *fiber respecting*; indeed the fibers of B are given as the sets

x = constant. (Note that the coordinates (x,u) as in (13.130) are automatically fiber respecting.) Now let (x,v) be another set of fiber respecting coordinates on the same neighbourhood of B. It immediately follows that (x,v) is related to (x,u) by a mapping

$$u = \alpha(x,v) \quad , \quad \text{rank } \frac{\partial \alpha}{\partial v}(x,v) = m , \tag{13.137}$$

and conversely, every mapping $u = \alpha(x,v)$ as in (13.137) defines a new set of fiber respecting coordinates (x,v). We conclude that local regular static state feedback (13.37) can be regarded as the *transition* from one set of fiber respecting coordinates to another. Furthermore, in case (B,M) is a *vector bundle* it is natural to restrict to fiber respecting coordinates (x,u) such that $u = (u_1,\ldots,u_m)$ are *affine* functions on the fibers of B, and a regular static state feedback $u = \alpha(x) + \beta(x)v$, $\det \beta(x) \neq 0$, corresponds to the transition from one set of fiber respecting coordinates (in the above restricted sense) to another.

Let us now pass on to *disturbance decoupling*. First of all, what is the appropriate global description of a system with disturbances (13.76)? We will choose to model the input space U as state-dependent (as before), and the space Q of disturbances as state- *and* input-dependent. Formally we let $\pi : B \to M$ be a bundle with standard fiber U, and $\tilde{\pi} : \tilde{B} \to B$ be a fiber bundle (with base space B!) with standard fiber Q.

Definition 13.35 *A nonlinear control system with disturbances $(\tilde{B},B,M,F)$ is given by fiber bundles (B,M) and $(\tilde{B},B)$ as above, and a system map $F : \tilde{B} \to TM$ satisfying*

Let (x,u), respectively (x,u,q), be fiber respecting coordinates for B, respectively for $\tilde{B}$, then we immediately recover from (13.138) the local coordinate expression (13.76a).

Consider now a distribution D on M. We will say that D is *locally controlled invariant for* $(\tilde{B},B,M,F)$, if there exist fiber respecting coordinates (x,u) for B such that for *all* fiber respecting coordinates (x,u,q) for $\tilde{B}$ we have that $[f(.,u,q),D] \subset D$, for every u,q.

Proposition 13.36 *Let $(\tilde{B},B,M,F)$ be a nonlinear control system with distur-*

bances. Let D be an involutive distribution of constant dimension on M. Assume that the distribution $\{Z \in \pi_*'^{-1}(0) \mid F_* Z \in \dot{D}\}$ on $\tilde{B}$ has constant dimension. Then D is locally controlled invariant for $(\tilde{B},B,M,F)$ if and only if

$$\begin{aligned} &(i)\ F_*(\pi_*'^{-1}(D)) \subset \dot{D} + F_*(\pi_*'^{-1}(0)), \\ &(ii) F_*(\tilde{\pi}_*^{-1}(0)) \subset \dot{D}. \end{aligned} \tag{13.139}$$

Proof Use Theorem 13.7 together with the observations made in the proof of Proposition 13.21. □

Finally consider the system with disturbances $(\tilde{B},B,M,F)$, together with an output map

$$h : B \to \mathbb{R}^p . \tag{13.140}$$

It is easily seen that Proposition 13.21 immediately generalizes to this case. Furthermore, all the results obtained in Section 13.3 about input-output decoupling, being local in nature, immediately carry over to the case of a nonlinear control system (B,M,F) with output map (13.140). The same holds for the material on controlled invariant submanifolds and constrained dynamics as covered in Section 13.4.

Notes and References

The generalization of the notion of controlled invariant distributions to general nonlinear systems, including the basic Theorem 13.7, is due to [NvdS1]. The present proof is also partly based on [vdS2], while the definitions of prolongations are largely taken from [YI]. For a treatment of controlled invariance by *output* feedback for general nonlinear systems we refer to [NvdS2]. The relation between feedbacks for the system and its extended system were noticed in [vdS3]; see also [NS]. The treatment of the disturbance decoupling problem for general nonlinear systems (Section 13.2) is due to [NvdS3]. The input-block-output decoupling problem for general nonlinear systems was studied in [vdS3]; the present analytic approach (Theorem 13.24) and the extension to dynamic feedback (Theorem 13.27) seem not to have been stated explicitly before. Example 13.25 is based on [dLU]. The generalization of the notion of controlled invariant submanifold and of constrained dynamics to general nonlinear systems has been dealt with in [vdS4].

The definition of control systems on fiber bundles is due to [Br1,2], see also [Wi], and was further developed in [vdS1,2], [NvdS1,2,3]; in particular Definition 13.35 and Proposition 13.36 can be found in [NvdS3]. Relationships between controlled invariant distributions and integrable connections on fiber bundles were studied in [NvdS1].

[Br1] R.W. Brockett, "Control theory and analytical mechanics", in **Geometric Control Theory** (eds. C. Martin, R. Hermann), Vol VII of Lie Groups: History, Frontiers and Applications, Math Sci Press, Brookline, pp. 1-46, 1977.

[Br2] R.W. Brockett, "Global descriptions of nonlinear control problems; vector bundles and nonlinear control theory", manuscript 1980.

[CvdS] P.E. Crouch, A.J. van der Schaft, **Variational Hamiltonian Control Systems**, Lect. Notes Contr. Inf. Sci. 101, Springer, Berlin, 1987.

[dLU] A. de Luca, G. Ulivi, "Dynamical decoupling of voltage frequency controlled induction motors", in **Analysis and Optimization of Systems** (eds. A. Bensoussan, J.L. Lions), Lect. Notes Contr. Inf. Sci. 111, pp. 127-137, 1988.

[NS] H. Nijmeijer, J.M. Schumacher, "Input-output decoupling of nonlinear systems with an application to robotics", in **Analysis and Optimization of Systems**, Part II, (eds. A. Bensoussan, J.L Lions), Lect. Notes Contr. Inf. Sci. 63, Springer, Berlin, pp. 391-411, 1984.

[NvdS1] H. Nijmeijer, A.J. van der Schaft, "Controlled invariance for nonlinear systems", IEEE Trans. Aut. Contr., AC-27, pp. 904-914, 1982.

[NvdS2] H. Nijmeijer, A.J. van der Schaft, "Controlled invariance by static output feedback for nonlinear systems", System Control Lett., 2, pp. 39-47, 1982.

[NvdS3] H. Nijmeijer, A.J. van der Schaft, "The disturbance decoupling problem for nonlinear control systems", IEEE Trans. Aut. Contr., AC-28, pp. 621-623, 1983.

[NvdS4] H. Nijmeijer, A.J. van der Schaft, "Partial symmetries for non-linear systems", Math. Syst. Th. 18, pp. 79-96, 1985.

[vdS1] A.J. van der Schaft, "Observability and controllability for smooth nonlinear systems", SIAM J. Contr. Optimiz., 20, pp. 338-354, 1982.

[vdS2] A.J. van der Schaft, **System theoretic descriptions of physical systems**, CWI Tracts 3, CWI Amsterdam, 1984.

[vdS3] A.J. van der Schaft, "Linearization and input-output decoupling for general nonlinear systems", System Control Lett., 5, pp. 27-33, 1984.

[vdS4] A.J. van der Schaft, "On clamped dynamics of nonlinear systems", in **Analysis and Control of Nonlinear Systems** (eds. C.I. Byrnes, C.F. Martin, R.E. Saeks), North Holland, Amsterdam, pp. 499-506, 1988.

[Wi] J.C. Willems, "System theoretic models for the analysis of physical systems", Ricerche di Automatica, 10, pp. 71-106, 1979.

[YI] K. Yano, S. Ishihara, **Tangent and cotangent bundles**, Dekker, New York, 1973.

Exercises

13.1 ([CvdS]) Consider an affine nonlinear system

$$\Sigma : \quad \begin{array}{l} \dot{x} = f(x) + \sum_{j=1}^{m} g_j(x)u_j \qquad , \quad x \in M, \ u \in \mathbb{R}^m, \ y \in \mathbb{R}^p , \\ y_i = h_i(x) \quad , \quad i \in \underline{p} . \end{array}$$

Along every solution $(u(t), x(t), y(t))$ of Σ we can define the *linearized* system, which is a linear time-varying system. If we take Σ together with all its linearized systems we obtain the system

$$\Sigma^p : \quad \begin{cases} \dot{x} = f(x) + \sum_{j=1}^{m} g_j(x)u_j \\ \dot{v} = \frac{\partial f}{\partial x}(x)v + \sum_{j=1}^{m} \frac{\partial g_j}{\partial x}(x)v\, u_j + \sum_{j=1}^{m} g_j(x)u_j^v \end{cases} \qquad \begin{cases} y_i = h_i(x) \quad , \quad i \in \underline{p} , \\ y_i^v = \frac{\partial h_i}{\partial x}(x)v \ , \quad i \in \underline{p} , \end{cases}$$

with v the variational state, u_j^v the variational inputs and y_i^v the variational outputs. (Σ^p is called the *prolongation* of Σ.)

(a) Show that in a coordinate-free way Σ^p is given as

$$\begin{aligned} \dot{x}_p &= \dot{f}(x_p) + \sum_{j=1}^{m} \dot{g}_j(x_p)u_j + \sum_{j=1}^{m} g_j^{\ell}(x_p)u_j^v \ , \ x_p \in TM , \\ y_i &= h_i^{\ell}(x_p) \quad , \quad i \in \underline{p} , \\ y_i^v &= \dot{h}_i(x_p) \quad , \quad i \in \underline{p} , \end{aligned}$$

with state space TM (natural coordinates (x,v)), input space $T\mathbb{R}^m$ (natural coordinates (u,u^v)) and output space $T\mathbb{R}^p$ (natural coordinates (y,y^v)).

(b) Let $\mathcal{C}_0$ be the strong accessibility algebra and $\mathcal{O}$ the observation space of Σ. Show that the strong accessibility algebra $\mathcal{C}_0^p$ and the observation space $\mathcal{O}^p$ of Σ^p are given as

$$\mathcal{C}^p = \{ \dot{X} \mid X \in \mathcal{C}_0 \} + \{ X^{\ell} \mid X \in \mathcal{C}_0 \} ,$$

$$\mathcal{O}^p = \{ \dot{H} \mid H \in \mathcal{O} \} + \{ H^{\ell} \mid H \in \mathcal{O} \} .$$

Furthermore prove that $\dim C_0^p = \dim TM$ if and only if $\dim C_0 = \dim M$, and that $\dim d\mathcal{O}^p = \dim TM$ if and only if $\dim d\mathcal{O} = \dim M$.

13.2 Consider E^* and D^* as in Proposition 13.21. Suppose that $f(x,u,q)$ happens to be affine, i.e.

$$f(x,u,q) = f(x) + \sum_{j=1}^{m} g_j(x)u_j + \sum_{i=1}^{\ell} e_i(x)q_i \ ,$$

while $h(x,u)$ only depends on x. Show that D^* is equal to the maximal distribution D satisfying

$$[f,D] \subset D + \mathrm{span}\{g_1,\ldots,g_m\} + \mathrm{span}\{e_1,\ldots,e_\ell\} ,$$

$$[g_j,D] \subset D + \mathrm{span}\{g_1,\ldots,g_m\} + \mathrm{span}\{e_1,\ldots,e_\ell\} , \ j \in \underline{m} ,$$

$$[e_i,D] \subset D + \mathrm{span}\{g_1,\ldots,g_m\} + \mathrm{span}\{e_1,\ldots,e_\ell\} , \ i \in \underline{\ell} .$$

Furthermore show that the condition $F_*TQ \subset \dot{D}^*$ reduces to $\mathrm{span}\{e_1,\ldots,e_\ell\} \subset D^*$. Finally show how the conditions of Proposition 13.21 reduce to the conditions of Theorem 7.14.

13.3 Consider a general nonlinear system $\dot{x} = f(x,u)$, $y = h(x,u)$, with finite characteristic numbers ρ_i, $i \in \underline{p}$. Show that the characteristic numbers of the extended system $\dot{x} = f(x,u)$, $\dot{u} = w$, $y = h(x,u)$ are equal to $\rho_i + 1$, $i \in \underline{p}$. Show that the decoupling matrix of the extended system *equals* the decoupling matrix $A(x,u)$ defined in (13.87). Relate the decoupling feedbacks for the original system to the decoupling feedbacks of the extended system.

13.4 (a) Consider a *smooth* square system (13.82). Show that the rank condition (13.88) is a *sufficient* condition for the local solvability of the regular static state feedback input-output decoupling problem (using the obvious generalization of Definition 8.1 for an input-output decoupled system), see also Exercise 8.1.

(b) Consider the analytic system $\dot{x}_1 = u_1$, $\dot{x}_2 = u_1 + u_2^3$, $y_1 = x_1$, $y_2 = x_2$. Verify that the system cannot be decoupled using regular static state feedback. On the other hand, show that the C^0-feedback $u_1 = v_1$, $u_2 = (v_2 - v_1)^{1/3}$, does decouple the system.

13.5 ([vdS3]) Consider a general smooth nonlinear system $\dot{x} = f(x,u)$, $y_i = h_i(x,u)$, where $h_i : M \to \mathbb{R}^{p_i}$, $i \in \underline{m}$, $u \in \mathbb{R}^m$. Show that the *block* input-output decoupling problem (as treated in Chapter 9 for affine smooth nonlinear systems) can be solved as follows. Consider the extended system, and define for this system the distributions $D^*_{i\,e}$, $i \in \underline{m}$, as in Chapter 9. Let the same constant dimension assumptions hold as in Chapter 9. Prove that the *block* input-output decoupling problem is solvable if and only if the block input-output decoupling problem is solvable for the extended system, if and only if (with $G_e = \mathrm{span}\{\frac{\partial}{\partial u_m}, \ldots, \frac{\partial}{\partial u_m}\}$)

$$G_e = D^*_{1\,e} \cap G_e + \ldots\ldots + D^*_{m\,e} \cap G_e .$$

Show that the distributions $D^*_{i\,e}$ immediately determine a decoupling feedback in the following way. Construct involutive constant dimensional distributions E_i, $i \in \underline{m}$, on $M \times \mathbb{R}^m$, all of dimension $n + m - 1$, such that

$$\sum_{j\neq i} D^*_{je} \subset E_i \quad , \text{ and } \quad \pi_{M*}E_i = TM \ .$$

Show that $E := \bigcap_{i=1}^{m} E_i$ is again an involutive constant dimensional distribution on $M \times \mathbb{R}^m$, of dimension n and satisfying $\pi_{M*}E = TM$. The required decoupling feedback $u = \alpha(x,v)$ has the property that the sets $\{(x,u = \alpha(x,v)) \mid v \text{ is constant}\}$ are integral manifolds of E.

13.6 Consider the normal form (13.91). Prove that we can choose the local coordinates $\overline{z}$ in such a way that $\overline{f}$ in (13.91) does not depend on $v_1,\ldots,v_m$, if and only if the distribution (with $f_e := f(x,u)\frac{\partial}{\partial x}$)

$$\text{span}\{\frac{\partial}{\partial u_1},\ldots,\frac{\partial}{\partial u_m} \ , \ [f_e,\frac{\partial}{\partial u_1}],\ldots,[f_e,\frac{\partial}{\partial u_1}]\}$$

on $M \times U$ is involutive.

13.7 Show that the *feedback linearization* problem for a system $\dot{x} = f(x,u)$, $f(x_0,u_0) = 0$, $x \in \mathbb{R}^n$, $u \in \mathbb{R}^m$ (cf. Definition 6.10) can be equivalently rephrased as follows: Find m functions $h_i(x,u)$, with $h_i(x_0,u_0) = 0$, such that the decoupling matrix $A(x_0,u_0)$ has rank m and $\sum_{i=1}^{m} (\rho_i + 1) = n$.

13.8 Consider the square system (13.82), and assume that rank $A(x,u) = m$ for all (x,u). Show that the *inverse system* is locally given by the equations

$$\dot{\overline{z}} = \overline{f}(\overline{z},\ z^1,\ldots,z^m,\ y_1^{(\rho_1+1)},\ldots,\underline{y}_m^{(\rho_m+1)}) \ ,$$

for suitable initial conditions $z(0)$, where $\overline{z}$ and $\overline{f}$ are as in (13.91) and $z^i := (y_i,\ldots,y_i^{(\rho_i)})$, $i \in \underline{m}$, together with the equation

$$u = \alpha(\overline{z},\ z^1,\ldots,z^m,\ y_1^{(\rho_1+1)},\ldots,y_m^{(\rho_m+1)}) \ ,$$

where $\alpha(\overline{z},z^1,\ldots,z^m,\ y_1^{(\rho_1+1)},\ldots,y_m^{(\rho_m+1)})$ is the unique solution of (cf. (13.89))

$$\begin{pmatrix} L_f^{(\rho_1+1)}h_1(\overline{z}^1,z,\ldots,z^m,u) \\ \vdots \\ L_f^{(\rho_m+1)}h_m(\overline{z},z^1,\ldots,z^m,u) \end{pmatrix} = \begin{pmatrix} y_1^{(\rho_1+1)} \\ \vdots \\ y_m^{(\rho_m+1)} \end{pmatrix}$$

(Here $h_i(\overline{z},z^1,\ldots,z^m,u)$ denotes $h_i(x,u)$ expressed in the new coordinates $\overline{z},z^1,\ldots,z^m,u$.) Generalize Exercise 11.8 to the system (13.82).

13.9 Generalize Exercises 7.10 and 11.9 (*Model Matching*) to a general nonlinear square system $\dot{x} = f(x,u)$, $y = h(x,u)$, satisfying rank $A(x,u) = m$ for every (x,u).

13.10 (*Zero-dynamics for general nonlinear systems*, [vdS4]). Consider the constrained dynamics (13.116), where x' are coordinates for N^*.

Compute for its extended system $\dot{x}' = f'(x',u')$, $\dot{u}' = w'$, the strong accessibility distribution, and show that under suitable constant rank conditions this distribution projects to a distribution C_0' on N^*, which is invariant for the constrained dynamics. Assuming that C_0' has constant dimension, show, using Theorem 3.35, that the constrained dynamics projects to the lower dimensional dynamics (the *zero-dynamics*)

$$\dot{x}'^2 = f'^2(x'^2)$$

Show that this definition is consistent with Definition 11.14.

13.11 A vectorfield X on M is called a *symmetry* for the vectorfield f on M if $[X,f] = 0$. (Indeed, in view of Lemma 2.34 this implies that $X^t \circ f^s = f^s \circ X^t$, and thus X^t maps solutions of $\dot{x} = f(x)$ onto solutions of f.) Show that $[X,f] = 0$ if and only if $F_* X = \dot{X}$, where $F : M \to TM$ is the map given in natural coordinates as $F(x) = (x,f(x))$.

Analogously, X is called a symmetry for $\dot{x} = f(x,u)$ if $F_* X_e = \dot{X}$, with $F : M \times U \to TM$ given as $F(x,u) = (x,f(x,u))$, and with X_e as in the proof of Proposition 13.6. Let D be a distribution spanned by symmetries $X_1,\ldots,X_k$ for $\dot{x} = f(x,u)$. Show that D is an invariant distribution. Conversely, can every invariant distribution be written as the span of symmetries? (see [NvdS4]).

13.12 Prove Proposition 13.29, and show that N^* as given by (13.120) can be also obtained by the general algorithm given above Proposition 13.29 (cf. (13.111) - (13.118)).

14
Discrete-Time Nonlinear Control Systems

In the preceding chapters we have restricted ourselves to continuous-time nonlinear control systems, and their discrete-time counterparts have been ignored so far. Although most engineering applications are concerned with (physical) continuous-time systems, discrete-time systems naturally occur in various situations. Most commonly discrete-time nonlinear systems appear as the discretization of continuous-time nonlinear systems. For a continuous-time nonlinear system, locally described as

$$\dot{\bar{x}} = \bar{f}(\bar{x},\bar{u}) \ , \tag{14.1}$$

$$\bar{y} = \bar{h}(\bar{x},\bar{u}) \ , \tag{14.2}$$

the *discretization* or *sampled-data representation* of (14.1,2) is formed in the following manner. Suppose the controls in (14.1,2) are applied in a piecewise constant fashion so that $\bar{u}$ is constant over the intervals $[kT,(k+1)T)$, $k = 1,2,\ldots$, where $T > 0$ is the *sampling time*. Then, letting $x(k)$, $u(k)$ and $y(k)$ denote $\bar{x}(kT)$, $\bar{u}(kT)$ and $\bar{y}(kT)$ respectively, one obtains a discrete-time system

$$x(k+1) = f(x(k),u(k)) \ , \tag{14.3}$$

$$y(k) \quad = h(x(k),u(k)) \ . \tag{14.4}$$

The relation between (14.1,2) and (14.3,4) is then obviously given by the fact that $x(k+1) = f(x(k),u(k))$ equals the solution at time $(k+1)T$ of the differential equation (14.1) starting at time kT in $\bar{x}(kT) = x(k)$ and with a constant control $\bar{u} = u$ applied, and similarly, $y(k) = h(x(k),u(k))$ equals the output y evaluated at time kT. So we have

$$f(x,u) = \bar{f}_u^T(x) \ , \tag{14.5}$$

$$h(x,u) = \bar{h}(x,u) \ , \tag{14.6}$$

where $\bar{f}_u^T$ is the time T-integral of the vectorfield $\bar{f}_u(x) = f(x,u)$. Clearly the sampled-data representation (14.3,4) depends on the sampling time T.

Discretizations of continuous-time systems are especially important because in the control of continuous-time systems present-day technology often asks for digitally implemented controllers, and hence is operated in discrete time. This leads, of course, to several interesting questions regarding the sampled-data representation (14.3,4) of (14.1,2). For example, what can be said about controllability and observability of (14.3,4) in relation to the same properties of (14.1,2)? Or, is the discrete-time system (14.3,4) input-output decouplable whenever (14.1,2) is? Here we will not pursue problems of this type, but instead refer to the relevant literature cited at the end of this chapter.

The purpose of this chapter is to study the discrete-time nonlinear system (14.3,4) *per se*, and not necessarily as the sampled-data representation of a continuous-time system. In this regard we emphasize that various models in biology, economics and econometrics are naturally formulated in discrete time, see for instance Example 1.3 as an illustration.

Several questions we have studied so far for continuous-time nonlinear systems can also be stated for the discrete-time system (14.3,4). However, their solutions will not always directly parallel the continuous-time results, since typical operations associated with a system in continuous-time do not immediately pass over to a discrete-time system. We will discuss here only some of the continuous-time problems of the foregoing chapters. In Section 14.1 we deal with the feedback linearization problem for the system (14.3), which forms the discrete-time counterpart of Chapter 6. Next, in Section 14.2, we study controlled invariance and the Disturbance Decoupling Problem for the system (14.3,4), compare with Chapters 7 and 13. Finally we discuss in Section 14.3 the input-output decoupling problem for the system (14.3,4), like we have done in Chapters 8, 9 and 13.

14.1 Feedback Linearization of Discrete-Time Nonlinear Systems

Consider the smooth discrete-time nonlinear dynamics

$$x(k+1) = f(x(k),u(k)) \,, \tag{14.3}$$

where $x = (x_1,\dots,x_n)$ and $u = (u_1,\dots,u_m)$ are smooth local coordinates for the state space M and input space U respectively. Before defining feedback linearizability of (14.3) we first introduce the notion of a regular static state feedback for (14.3). Similary as for a continuous-time nonlinear system, see Chapter 5, we call a relation

$$u = \alpha(x,v) \tag{14.6}$$

a *regular static state feedback*, whenever $\frac{\partial\alpha}{\partial v}(x,v)$ is nonsingular at every point (x,v). Notice that this implies locally a one-to-one relation between the old inputs u and the new controls v. Analogously to Definition 6.10, we formulate the feedback linearizability of (14.3).

Definition 14.1 *Let* (x_0,u_0) *be and equilibrium point for (14.3), i.e.* $f(x_0,u_0) = x_0$. *The system (14.3) is* feedback linearizable *around* (x_0,u_0) *if there exist*

(i) a coordinate transformation $S: V \subset \mathbb{R}^n \to S(V) \subset \mathbb{R}^n$ *defined on a neighborhood* V *of* x_0 *with* $S(x_0) = 0$,

(ii) a regular static state feedback $u = \alpha(x,v)$ *satisfying* $\alpha(x_0,0) = u_0$ *and defined on a neighborhood* $V \times O$ *of* $(x_0,0)$ *with* $\frac{\partial\alpha}{\partial v}(x,v)$ *non-singular on* $V \times O$,

such that in the new coordinates $z = S(x)$ *the closed loop dynamics are linear*

$$z(k+1) = Az(k) + Bv(k) \ , \tag{14.7}$$

for some matrices A *and* B.

At this point it is useful to note that a coordinate change $z = S(x)$ transforms (14.3) in a different manner than for a continuous-time system. In the z-coordinates we obtain from (14.3) the discrete-time dynamics

$$z(k+1) = S(f(S^{-1}(z(k)),u(k))) \ , \tag{14.8}$$

and so the feedback linearizability of Definition 14.1 amounts to the equation

$$S\big(f(S^{-1}(z(k)),\ \alpha(S^{-1}(z(k)),v(k)))\big) = Az(k) + Bv(k), \tag{14.9}$$

(compare with equation (6.67)).

The following sequence of distributions will be instrumental in the solution of the feedback linearization problem for (14.3). Let $\pi : M \times U \to M$ be the canonical projection and K the distribution defined by

$$K = \ker f_* \ . \tag{14.10}$$

Algorithm 14.2 Assume f_* has full rank around (x_0,u_0).
Step 0 Define in a neighborhood of (x_0,u_0) in $M \times U$ the distribution

$$D_0 = \pi_*^{-1}(0) \ , \tag{14.11}$$

Step i + 1 Suppose that around (x_0,u_0) $D_i + K$ is an involutive constant dimensional distribution on $T(M \times U)$. Then define in a neighborhood of (x_0,u_0)

$$D_{i+1} = \pi_*^{-1} f_*(D_i), \tag{14.12}$$

and stop if $D_i + K$ is not involutive or constant dimensional.

The effectiveness of the above algorithm rests upon the following observation.

Lemma 14.3 *Let (x_0,u_0) be an equilibrium point of (14.3) and assume that f_* has full rank around (x_0,u_0). Let D be an involutive constant dimensional distribution on $M \times U$ such that $D + K$ is also involutive and constant dimensional. Then there exists a neighborhood O of (x_0,u_0) such that $f_*(D|_O)$ is an involutive constant dimensional distribution around x_0.*

Proof Choose local coordinates on M such that $x_0 = 0$. From the fact that f_* has full rank around (x_0,u_0) it follows that K is a constant dimensional involutive distribution around (x_0,u_0). Therefore, see Theorem 2.42, there exist local coordinates $z = (z^1,z^2)$ around (x_0,u_0) in $M \times U$ such that

$$K = \operatorname{span}\{\frac{\partial}{\partial z^2}\} , \tag{14.13}$$

where z^2 is an m-dimensional vector. This implies that in these coordinates $f(z) = f(z^1)$. Moreover $\frac{\partial f}{\partial z^1}(z^1)$ is a nonsingular $(n \times n)$-matrix around z_0. So, using the Inverse Function Theorem, we may introduce new local coordinates $(f(z^1),z^2)$ around (x_0,u_0) in $M \times U$. With respect to these coordinates the function f takes the form $f(\bar{z}^1,\bar{z}^2) = \bar{z}^1$ (see also Exercise 2.5), and thus locally f is a projection. In the rest of the proof we will use these coordinates and drop the bar notation. Obviously, we will have that (14.13) holds true in these coordinates. Next, let $X_1,\dots,X_\ell$ be a basis for D. The involutivity of $K + D$ implies that

$$[X_i,K] \subset D + K \;\; , \;\; i \in \underline{\ell} \; , \tag{14.14}$$

and the constant dimensionality of $K + D$, and so of $K \cap D$, implies that we may apply Theorem 7.5, yielding a basis $\{\tilde{X}_1,\dots,\tilde{X}_\ell\}$ for D with

$$[\tilde{X}_i,K] \subset K \;\;\; , \;\; i \in \underline{\ell} \; . \tag{14.15}$$

In the above coordinates this implies that the vectorfields $\tilde{X}_i$, $i \in \underline{\ell}$, have the form

$$\begin{pmatrix} \tilde{X}_i^1(z^1,z^2) \\ \tilde{X}_i^2(z^1) \end{pmatrix}, \quad i \in \underline{\ell} . \tag{14.16}$$

Clearly, $f_*(D)$ is then a distribution spanned by the vectorfields $\tilde{X}_i^2(z^1)$, $i \in \underline{\ell}$, which by the constant dimensionality of $K \cap D$ has constant dimension. Also, the involutivity of $\text{span}\{\tilde{X}_1^2,\ldots,\tilde{X}_\ell^2\}$ immediately follows from the involutivity of D. □

Using inductively Lemma 14.3 we obtain the following result.

Corollary 14.4 *Locally around* (x_0,u_0) *(14.12) defines an involutive constant dimensional distribution* D_{i+1}.

Theorem 14.5 *Consider the discrete-time nonlinear system (14.3) about the equilibrium point* (x_0,u_0). *The system (14.3) is linearizable around* (x_0,u_0) *to a controllable linear system if and only if Algorithm 14.2 applied to the system (14.3) gives distributions* $D_0,\ldots,D_n$ *such that* $\dim D_n = n + m$.

Proof First, suppose (14.3) is feedback linearizable about (x_0,u_0) into a controllable linear system (14.7). In these coordinates we find that

$$f_*(x_0,u_0) = \left(A \;\vdots\; B\right) , \tag{14.17}$$

which by the controllability of (14.7) must have full rank. One immediately calculates that

$$D_0 = \text{span}\{\frac{\partial}{\partial v}\} , \tag{14.18a}$$

$$D_i = \mathcal{B} + \ldots + A^{i-1}\mathcal{B} + \text{span}\{\frac{\partial}{\partial v}\} , \quad i = 1,2,\ldots , \tag{14.18b}$$

where $\mathcal{B} = \text{Im}\, B$. Hence the involutivity and constant dimensions conditions of Algorithm 14.2 hold. Moreover, the controllability of (14.7) implies that $\dim D_n = n + m$.

In order to prove the converse, we proceed as follows. Let locally around (x_0,u_0)

$$\Delta_i = \pi_* D_i , \quad i \in \underline{n} . \tag{14.19}$$

By Lemma 14.3 and Corollary 14.4, Δ_i is a constant dimensional involutive distribution on a neighborhood of x_0 in M, $i \in \underline{n}$. Clearly we have that $\Delta_i \subset \Delta_{i+1}$. Let $m_i = \dim \Delta_i$, $i \in \underline{n}$ and set $\rho_i = m_i - m_{i-1}$, $i \in \underline{n}$, with $m_0 = 0$. From the fact that $\dim D_n = n + m$ one obtains the existence of a minimal number $\kappa \le n$ such that $\dim \Delta_\kappa = n$. Lemma 6.4 applied to the sequence of distributions $\Delta_1 \subset \Delta_2 \subset \ldots \subset \Delta_\kappa$ yields local coordinates x around x_0 such that

$$\Delta_i = \text{span}\{\frac{\partial}{\partial x^1}, \ldots, \frac{\partial}{\partial x^i}\}, \quad i \in \underline{\kappa}, \tag{14.20}$$

where $\dim x^i = \rho_i$, $i \in \underline{\kappa}$. With respect to the above coordinates x we write $f(x,u) = (f^1(x,u), \ldots, f^\kappa(x,u))^T$ accordingly. We investigate next the particular structure of f with respect to the distributions Δ_i, $i \in \underline{\kappa}$. We have, see (14.19) and (14.12), that $f_* D_0 = \Delta_1$, which implies that in a neighborhood of (x_0, u_0)

$$\text{span}\{\sum_{j=1}^{m} \frac{\partial f^j}{\partial u_i}(x,u) \frac{\partial}{\partial x^j}, \; i \in \underline{m}\} = \text{span}\{\frac{\partial}{\partial x^1}\}, \tag{14.21}$$

yielding $\frac{\partial f^j}{\partial u}(x,u) = 0$ for $j = 2, \ldots, \kappa$, and also rank $\frac{\partial f^1}{\partial u} = \rho_1$. Similarly, $f_* D_1 = \Delta_2$ gives

$$\text{span}\{\sum_{j=1}^{k} \frac{\partial f^j}{\partial x^1}(x,u) \frac{\partial}{\partial x^j}, \frac{\partial f^1}{\partial u}(x,u) \frac{\partial}{\partial x^1}\} = \text{span}\{\frac{\partial}{\partial x^1}, \frac{\partial}{\partial x^2}\}, \tag{14.22}$$

from which we obtain $\frac{\partial f^j}{\partial x^1}(x,u) = 0$ for $j = 3, \ldots, \kappa$ and rank $\frac{\partial f^2}{\partial x^1}(x,u) = \rho_2$. A repetition of the above argument, using $f_* D_{i-1} = \Delta_i$, $i \in \underline{\kappa}$, yields the following form for f:

$$f(x,u) = \begin{pmatrix} f^1(x^1, x^2, \ldots, x^\kappa, u) \\ f^2(x^1, x^2, \ldots, x^\kappa) \\ f^3(x^2, \ldots, x^\kappa) \\ \vdots \\ f^\kappa(x^{\kappa-1}, x^\kappa) \end{pmatrix}. \tag{14.23}$$

Note that this is exactly the form as obtained in (6.35) or (6.77). Next we exploit the fact that rank $\frac{\partial f^i}{\partial x^{i-1}}(x,u) = \rho_i$, $i \in \underline{\kappa}$, in order to successively change the coordinates $(x^1, \ldots, x^\kappa)$ (analogously to the proof of Theorem 6.12). Observe first that $\rho_i \ge \rho_{i+1}$ for $i \in \underline{\kappa}$. In the first step we introduce new coordinates $(z^1, \ldots, z^\kappa)$ via $z^j = x^j$, for $j \ne \kappa-1$ and $z^{\kappa-1} = (f^\kappa(x^{\kappa-1}, x^\kappa), \tilde{x}^{\kappa-1})$ where $\tilde{x}^{\kappa-1}$ are $\rho_{k-1} - \rho_k$ components of $x^{\kappa-1}$

chosen in such a way that rank $(\frac{\partial z^{\kappa-1}}{\partial x^{\kappa-1}}) = \rho_{\kappa-1}$. Denoting $(z^1,\dots,z^\kappa)$ once again as $(x^1,\dots,x^\kappa)$ f can be written as $(f^1,\dots,f^{\kappa-1}, \bar{x}^{\kappa-1})^T$, where $\bar{x}^{\kappa-1}$ are the first ρ_κ components of $x^{\kappa-1}$ and $f^1,\dots,f^{\kappa-1}$ are now the component functions expressed in the new coordinates. Repeating this procedure successively for $f^{\kappa-2},\dots,f^1$ we arrive at the following expression for f

$$f(z,u) = \begin{bmatrix} f^1(z^1,z^2\dots,z^\kappa,u) \\ z^1 \\ \vdots \\ z^{\kappa-1} \end{bmatrix}, \tag{14.24}$$

where $(z^1,\dots,z^\kappa)$ are the local coordinates obtained in the last step. Finally we perform a state feedback transformation as follows. Define v as $v = (f^1(z,u),\tilde{u})$ where $\tilde{u}$ are $m - \rho_1$ components of u selected in such a way that rank $\frac{\partial v}{\partial u} = m$. In the (x,v) coordinates we obtain the system

$$z(k+1) = \begin{bmatrix} \bar{v}\ (k) \\ z^1(k) \\ \vdots \\ z^{\kappa-1}(k) \end{bmatrix}, \tag{14.25}$$

with $\bar{v}$ denoting the first ρ_1 components of v. Clearly (14.25) is a controllable linear system. □

Remark 14.6 Like in the continuous-time case, the coordinate transformations used in the above proof change at each step the component functions $f^1,\dots,f^\kappa$ of f (while leaving them of the form (14.23)). However, the modifications on f in discrete-time, respectively in continuous-time, are in general different, see (14.8), respectively (6.67).

Example 14.7 Consider as in Example 1.3 the dynamics of a controlled closed economy, which are described by equations of the form (see (1.33))

$$\begin{cases} Y(k+1) = f_1(Y(k),R(k),K(k),W(k),G(k)) , \\ R(k+1) = f_2(Y(k),R(k),K(k),W(k),M(k)) , \\ K(k+1) = f_3(Y(k),R(k),K(k)) . \end{cases} \tag{14.26}$$

In (14.26) $G(k)$ and $M(k)$ denote the controls on the system and $W(k)$ is an exogenous variable. Assume we are working around an equilibrium point $(\bar{Y},\bar{R},\bar{K},\bar{G},\bar{M},\bar{W})$ and suppose the exogenous variable $W(k)$ equals $\bar{W}$ for all k. In order to see if the model (14.26) is feedback linearizable around

$(\bar{Y},\bar{R},\bar{K},\bar{G},\bar{M},\bar{W})$ we have to verify the conditions of Theorem 14.5. We make the following assumptions

$$\frac{\partial f_1}{\partial G}(\bar{Y},\bar{R},\bar{K},\bar{W},\bar{G}) \neq 0 \ , \tag{14.27a}$$

$$\frac{\partial f_2}{\partial M}(\bar{Y},\bar{R},\bar{K},\bar{W},\bar{M}) \neq 0 \ , \tag{14.27b}$$

$$\left(\frac{\partial f_3}{\partial Y}(\bar{Y},\bar{R},\bar{K}), \ \frac{\partial f_3}{\partial K}(\bar{Y},\bar{R},\bar{K})\right) \neq (0,0) \ . \tag{14.27c}$$

Note that for the specific dynamics of Example 1.3 the condition (14.27a) is automatically satisfied. Define a preliminary feedback so that

$$\begin{aligned} u_1 &= f_1(Y,R,K,\bar{W},G) + \bar{Y} \ , \\ u_2 &= f_2(Y,R,K,\bar{W},M) + \bar{R} \ . \end{aligned} \tag{14.28}$$

Provided (14.27a,b) holds, the Inverse Function Theorem assures us that we can achieve (14.28) by means of a regular state feedback $G = \alpha_1(Y,R,K,\bar{W},u_1)$, $M = \alpha_2(Y,R,K,\bar{W},u_2)$. The feedback modified system then has the form

$$\begin{cases} Y(k+1) - \bar{Y} = u_1(k) \ , \\ R(k+1) - \bar{R} = u_2(k) \ , \\ K(k+1) - \bar{K} = f_3(Y(k),R(k),K(k)) - \bar{K} \ . \end{cases} \tag{14.29}$$

The right-hand side of (14.29), seen as a mapping from the (Y,R,K,u_1,u_2)-space into the (Y,R,K)-space, has full rank, cf. (14.27c). We now apply Algorithm 14.2. Denoting the state variables $(Y-\bar{Y},R-\bar{R},K-\bar{K})$ as x, we compute the distribution $\hat{K}$ as in (14.10) as the 2-dimensional distribution on the (x,u) space of the form

$$\hat{K}(x,u) = \text{span}\{X_1(x)\frac{\partial}{\partial x} \ , \ X_2(x)\frac{\partial}{\partial x}\} \ , \tag{14.30}$$

with X_1 and X_2 satisfying

$$\frac{\partial f_3}{\partial x}(x)X_i(x) = 0 \ , i = 1,2 \ . \tag{14.31}$$

Since, see (14.11), $D_0 = \text{span}\{\partial/\partial u\}$, we immediately have that $K_1 + D_0$ is a constant dimensional involutive distribution. Using again (14.27c) we obtain from (14.12) that D_1 is an involutive distribution of dimension 5 on the (x,u)-space. Therefore, the conditions (14.27a,b,c) guarantee that the dynamics (14.26) are feedback linearizable about the equilibrium point. □

14.2 Controlled Invariant Distributions and the Disturbance Decoupling Problem in Discrete-Time

We now discuss the notion of local controlled invariance for the discrete-time system (14.3). Afterwards we show how controlled invariant distributions are instrumental in the solution of the disturbance decoupling problem for discrete-time systems. The theory we develop here very much resembles the corresponding continuous-time theory of Chapters 7 and 13, see in particular Sections 13.1 and 13.2, and therefore some of the proofs and results will only briefly be sketched. The following definition is the discrete-time version of Definition 13.1.

Definition 14.8 *A distribution D on M is invariant for the smooth dynamics (14.3) if*

$$f_*(\cdot,u)\,D \subset D, \qquad \text{for every } u \in U\,, \tag{14.32}$$

and locally controlled invariant for (14.3) if there locally exists a regular state feedback, briefly feedback, $u = \alpha(x,v)$ such that the closed loop dynamics

$$x(k+1) = \tilde{f}(x(k),v(k)) := f(x(k),\alpha(x(k),v(k))) \tag{14.33}$$

satisfies

$$\tilde{f}_*(\cdot,v)\,D \subset D\,, \qquad \textit{for every } v. \tag{14.34}$$

An involutive constant dimensional distribution D which is invariant under (14.3), induces a local decomposition for the dynamics, which slightly differs from the continuous-time case, see e.g. (13.9). This difference is due to the fact that if we apply $f(\cdot,u)$ to a point x in a local coordinate chart, then $f(x,u)$ may leave the chart, no matter how the control was chosen. We therefore introduce a *pair* of coordinate charts. Let $(x_0,u_0) \in M \times U$ and choose local coordinates on a neighborhood V of $f(x_0,u_0)$. Consider the open set $f^{-1}(V)$ containing (x_0,u_0) and choose a local coordinate chart $\tilde{V}$ in $M \times U$ about (x_0,u_0) such that $\tilde{V} \subset f^{-1}(V)$. Denote the coordinates for $\tilde{V}$ as (x,u) and for V as x. This will be referred to as the coordinate chart pair $(V,\tilde{V})$. Such a coordinate chart pair permits us to do local computations. Note that for an equilibrium point (x_0,u_0) we may always select $\tilde{V}$ such that $\pi(\tilde{V}) \subset V$, so that in this case the local coordinates on $\tilde{V}$ may be chosen in such way that the coordinates for $\pi(\tilde{V})$ and V coincide. The invariance condition (14.32) for an involutive constant dimensional distribution D now translates as follows. Using Frobenius' Theorem (Corollary 2.43) we select a coordinate chart pair $(V,\tilde{V})$ such that D is described by $D = \operatorname{span}\{\frac{\partial}{\partial x^1}\}$, where $x = (x^1,x^2)$, $x^1 = (x_1,\ldots,x_k)$ and $x^2 = (x_{k+1},\ldots,x_n)$. Then, writing accordingly $f = (f^1,f^2)$, (14.32) implies the local decomposition

$$\begin{aligned} x^1(k+1) &= f^1(x^1(k),x^2(k),u(k)) , \\ x^2(k+1) &= f^2(x^2(k),u(k)) , \end{aligned} \tag{14.35}$$

and similarly for the closed loop dynamics (14.33) if D is locally controlled invariant. The next theorem provides a coordinate-free test for checking the local controlled invariance of a given distribution D. Its proof is very much inspired by the corresponding continuous-time result, cf. Theorem 13.7 and Corollary 13.8.

Theorem 14.9 *Consider the smooth discrete-time system (14.3) and let D be an involutive constant dimensional distribution on M. Assume $f(x,\cdot): U \to M$ has constant rank for every x, and suppose that $f_*^{-1}(D) \cap \pi_*^{-1}(0)$ is constant dimensional. Then D is locally controlled invariant if and only if*

$$f_*(\pi_*^{-1}(D)) \subset D + f_*(\pi_*^{-1}(0)) . \tag{14.36}$$

Proof Let $(x_0,u_0) \in M \times U$ and choose a coordinate chart pair so that $D = \operatorname{span}\{\frac{\partial}{\partial x^1}\}$, where $x = (x^1,x^2)$, $x^1 = (x_1,\ldots,x_k)$ and $x^2 = (x_{k+1},\ldots,x_n)$. In these coordinates the distribution $\pi_*^{-1}(D)$ is then given as $\operatorname{span}\{\frac{\partial}{\partial x^1},\frac{\partial}{\partial u}\}$ and the condition (14.36) is equivalent to

$$\operatorname{Im} \frac{\partial f^2}{\partial x^1}(x,u) \subset \operatorname{Im} \frac{\partial f^2}{\partial u}(x,u) \text{ for all } (x,u) . \tag{14.37}$$

Now suppose D is locally controlled invariant around (x_0,u_0). Then there is locally a feedback $u = \alpha(x,v)$ such that $\tilde{f}_*D \subset D$, where $\tilde{f}(x,v) = f(x,\alpha(x,v))$. Equivalently $\frac{\partial \tilde{f}^2}{\partial x^1}(x,v) = 0$, or

$$\frac{\partial f^2}{\partial x^1}(x,u)\Big|_{u=\alpha(x,v)} + \frac{\partial f^2}{\partial u}(x,u)\Big|_{u=\alpha(x,v)} \cdot \frac{\partial \alpha}{\partial x^1}(x,v) = 0 \tag{14.38}$$

for all (x,u) around (x_0,u_0). Clearly (14.38) implies (14.37).

Conversely let (14.37) be satisfied. Then there exist smooth m-vectors $b_i(x,u)$, $i \in \underline{k}$, satisfying

$$\frac{\partial f^2}{\partial x_i}(x,u) + \frac{\partial f^2}{\partial u}(x,u)\; b_i(x,u) = 0,\; i \in \underline{\kappa}\;. \tag{14.39}$$

Let $P = \text{ann}\, D$, then f^*P and π^*P are involutive codistributions on $M \times U$ (see Exercise 2.14), and so also $f^*P + \pi^*P$ is an involutive codistribution on $M \times U$. In the given coordinates

$$f^*P + \pi^*P = \text{span}\{dx^2,\; \frac{\partial f^2}{\partial x^1}\, dx^1 + \frac{\partial f^2}{\partial u}\, du\}. \tag{14.40}$$

Writing $b_i = (b_{1i},\ldots,b_{mi})^T$, we have in view of (14.39) that

$$\ker(f^*P + \pi^*P) = \text{span}\,\{\frac{\partial}{\partial x_i} + \sum_{s=1}^{m} b_{si}(x,u)\frac{\partial}{\partial u_s}\;,\; i \in \underline{\kappa}\} + \ker\{\frac{\partial f^2}{\partial u} du\}. \tag{14.41}$$

Note that the last term in the right hand side of (14.41) equals the constant dimensional distribution $f_*^{-1}(D) \cap \pi_*^{-1}(0)$, and thus $\ker(f^*P + \pi^*P)$ is a constant dimensional distribution. Now define

$$E = \ker(f^*P + \pi^*P), \tag{14.42}$$

then it clearly follows that

$$\begin{aligned} \pi_* E &\subset D\;, \\ f_* E &\subset D\;, \end{aligned} \tag{14.43}$$

which is the discrete-time analog of (13.56). In order to construct locally the feedback $u = \alpha(x,v)$ which makes D invariant we proceed as follows (compare with the proof of Theorem 13.7). First assume that $E \cap \pi_*^{-1}(0) = 0$, or equivalently $f_*^{-1}(D) \cap \pi_*^{-1}(0) = 0$. Then $\dim E = \dim D = k$ and we may take a new coordinate chart pair such that $D = \text{span}\{\frac{\partial}{\partial x^1}\}$, and moreover by the fact that $\pi_*: E \to D$ is an isomorphism, there exists a state-dependent change of coordinates for U, resulting in coordinates (x,v) such that $E = \text{span}\{\frac{\partial}{\partial x^1}\}$. The input coordinate change $v = \tilde{\alpha}(x,u)$ is precisely the inverse of the desired feedback $u = \alpha(x,v)$. In case $E \cap \pi_*^{-1}(0) \neq 0$, a slight modification as has been given in the proof of Theorem 13.7 again yields a feedback $u = \alpha(x,v)$ which renders D invariant. □

As in the continuous-time case an important role is played by the maximal (locally) controlled invariant distribution for the dynamics (14.3) contained in some given involutive distribution $\mathcal{K}$ on M. That such a maximal locally controlled invariant distribution in $\mathcal{K}$ exists, follows in a similar way as in Chapter 7, see in particular Propositions 7.10 and 7.11 and Corollary 7.12. Without proof we give the following result.

Proposition 14.10 *Let $\mathcal{K}$ be an involutive distribution on M. Then we have*

(i) *If D_1 and D_2 are distributions on M contained in $\mathcal{K}$, satisfying (14.36) on $M \times U$, then also $D_1 + D_2$ satisfies (14.36) on $M \times U$.*

(ii) *If D is a distribution on M contained in $\mathcal{K}$, satisfying (14.36) on $M \times U$, then the same holds true for the involutive closure $\overline{D}$ of D.*

(iii) *There exists a largest involutive distribution D^* on M which satisfies (14.36) on $M \times U$.*

Clearly, the distribution D^*, whose existence is guaranteed by Proposition 14.10 (iii) is locally controlled invariant on the open and dense subsets of $M \times U$ where the constant dimension hypothesis of Theorem 14.9 are met. The distribution D^* may be computed analogously to the continuous-time algorithm (7.53) in the following way. Define recursively

$$
\begin{aligned}
E^0 &= T(M \times U) \\
E^{\mu+1} &= \pi_*^{-1}(\mathcal{K}) \cap \{X \in T(M \times U) \mid f_* X \in \pi_* E^{\mu} + f_*(\pi_*^{-1}(0)) \\
&\qquad \text{on an open and dense subset of } M \times U\}
\end{aligned}
\tag{14.44}
$$

The following result is immediate.

Corollary 14.11 *Let $E^* = \lim\limits_{\mu \to \infty} E^{\mu}$, then*

$$
D^* = \pi_* E^* . \tag{14.45}
$$

We will now briefly discuss the Disturbance Decoupling Problem in discrete-time. Consider the system

$$
\begin{cases}
x(k+1) = f(x(k), u(k), q(k)) \\
y(k) \quad\; = h(x(k), u(k))
\end{cases}
\tag{14.46}
$$

where the system map $f: M \times U \times Q \to M$ now also depends upon the disturbances $q \in Q$, the ℓ-dimensional disturbance manifold. Regarding the output invariance with respect to the disturbances q in (14.46) we have the following result which parallels Proposition 13.20.

Proposition 14.12 *The output y of the system (14.46) is invariant under q if there exists an involutive constant dimensional distribution D on M such that*

$$f_*(\cdot,u,q)\, D \subset D \quad , \quad \textit{for all } (u,q) \in U \times Q \ , \tag{14.47a}$$

$$f_* TQ \subset D \ , \tag{14.47b}$$

$$D \subset \ker d\, h(\cdot,u) \ , \textit{ for all } u \in U \tag{14.47c}$$

(Here TQ denotes the ℓ-dimensional distribution on $M \times U \times Q$ given in local coordinates (x,u,q) by $\mathrm{span}\{\frac{\partial}{\partial q}\}$*, and $d_x h(x,u) := \frac{\partial h}{\partial x}(x,u)dx$.)*

Proof Using Frobenius' Theorem we select a coordinate chart pair $(V,\tilde{V})$ such that D is described by $D = \mathrm{span}\{\frac{\partial}{\partial x^1}\}$ where $x = (x^1,x^2)$, $x^1 = (x_1,\ldots,x_k)$ and $x^2 = (x_{k+1},\ldots,x_n)$. Then writing accordingly $f = (f^1,f^2)$ we obtain from (14.47a) the local decomposition

$$\begin{aligned} x^1(k+1) &= f^1(x^1(k),x^2(k),u(k),q(k)) \ , \\ x^2(k+1) &= f^2(x^2(k),u(k),q(k)) \ . \end{aligned} \tag{14.48}$$

Furthermore (14.47b) implies that f^2 does not depend upon q and from (14.47c) it follows that $h(x,u)$ does not depend on x^1, thereby implying that $y(k) = h(x(k),u(k))$ is invariant under q. □

The Disturbance Decoupling Problem consists of finding a regular static state feedback $u = \alpha(x,v)$ for the system (14.46) such that in the feedback transformed system the output y is invariant under q, cf. Problem 7.7. Analogously to Chapters 7 and 13 we search for a local solution of the Disturbance Decoupling Problem in the following manner. Let (x_0,u_0) be an arbitrary point in $M \times U$, then the Disturbance Decoupling Problem is locally solvable around (x_0,u_0) if there exists a feedback $u = \alpha(x,v)$ defined around (x_0,u_0) such that in the feedback modified system

$$\begin{cases} x(k+1) = f(x(k),\alpha(x(k),v(k)),q(k)) =: \tilde{f}(x(k),v(k),q(k)) \\ y(k) \quad\ \ = h(x(k),\alpha(x(k),v(k))) \qquad =: \tilde{h}(x(k),v(k)) \end{cases} \tag{14.49}$$

the output $y(k)$ is invariant under $q(k)$. Let $\bar{\pi} : M \times U \times Q \to M$ be the canonical projection and define the distribution K as

$$K = \bigcap_{u \in U} \ker d_x h(\cdot,u) \ . \tag{14.50}$$

Now, viewing the discrete-time dynamics of (14.46) as a system with inputs u *and* q, we know that according to Proposition 14.10 there exists a maximal involutive distribution $\bar{D}^*$ in K on M which satisfies

$$f_*(\bar{\pi}_*^{-1}(\bar{D}^*)) \subset \bar{D}^* + f_*(\bar{\pi}_*^{-1}(0)) \ . \tag{14.51}$$

With the aid of Proposition 14.12 and using the results about local controlled invariance we arrive at the next result.

Theorem 14.13 *Consider the system (14.46) and assume* $f(x,\cdot,\cdot): U \times Q \to M$ *has constant rank, and* $\overline{D}^*$ *and* $\overline{D}^* \cap f_*(\overline{\pi}_*^{-1}(0))$ *are constant dimensional distributions. Then locally about each point* (x_0, u_0) *there exists a regular static state feedback* $u = \alpha(x,v)$ *such that the feedback modified system (14.49) satisfies the conditions (14.47a,b,c) around* (x_0, u_0), *if and only if*

$$f_*(TQ) \subset \overline{D}^* . \tag{14.52}$$

Proof The necessity of (14.52) follows directly from Proposition 14.12 and the fact that the feedback $u = \alpha(x,v)$ does not depend upon the disturbances. The sufficiency of (14.52) for the local existence of a feedback $u = \alpha(x,v)$ for which the modified system (14.49) satisfies (14.47a,b,c), is a consequence of the following facts. As in the proof of Theorem 14.9 the distribution $\overline{D}^*$ induces a distribution $\overline{E}$ on $M \times U \times Q$ such that, cf. (14.42,43),

$$\begin{aligned} \overline{\pi}_* \overline{E} &\subset \overline{D}^* , \\ f_* \overline{E} &\subset \overline{D}^* . \end{aligned} \tag{14.53}$$

This already implies the local existence of a feedback also depending upon the disturbances, which makes the distribution $\overline{D}^*$ invariant. Since $f_*(TQ) \subset \overline{D}^*$ it follows that $TQ \subset \overline{E}$. Completely analogous to the proof of Proposition 13.21 it follows that there locally exists a feedback $u = \alpha(x,v)$ which makes $\overline{D}^*$ invariant. □

Theorem 14.13 is a local result, since the feedback $u = \alpha(x,v)$ is only defined in a neighborhood of an initial point $(x(0),u(0))$. Of course one may apply the result again about the point $(x(1),u(1))$, where $x(1) = f(x(0),u(0),q(0))$ and so on (provided the assumptions of Theorem 14.13 are met), but in general these local feedbacks need not patch together into a globally defined feedback. Of course, the output invariance with respect to q remains as long as the state and input evolve in such a way that we do not leave the neighborhood of $(x(0),u(0))$ on which the feedback $u = \alpha(x,v)$ was constructed. This happens in particular if the system evolves for all time on a sufficiently small neighborhood of an equilibrium point (x_0, u_0).

14.3 Input-Output Decoupling in Discrete-Time

We next discuss, in analogy with Chapters 8 and 13, the input-output decoupling problem for a *square analytic* discrete-time nonlinear system

$$x(k+1) = f(x(k),u(k)) \ , \tag{14.3}$$

$$y(k) = h(x(k),u(k)) \ , \tag{14.4}$$

where u and y are m-dimensional control and output vectors respectively. (The smooth case can be treated more or less similarly, see Chapters 8 and 13.) With each component of the output, y_i, we can associate a *characteristic number* ρ_i in the following way. Let $(x,u) \in M \times U$, then we compute for $i \in \underline{m}$ the derivative

$$\frac{\partial h_i}{\partial u}(x,u) = \left(\frac{\partial h_i}{\partial u_1}(x,u),\dots,\frac{\partial h_i}{\partial u_m}(x,u)\right) \ . \tag{14.54}$$

From the analyticity of the discrete-time system it follows that this vector is either nonzero for all (x,u) in an open and dense subspace O_i of $M \times U$, or the vector (14.54) vanishes at all points (x,u). In the first case we define $\rho_i = -1$, whereas in the latter case we continue by observing that the function $h_i(x,u)$ does *not* depend upon u; so we may write $h_i(x,u) = h_i^0(x)$. Next we compute the vector $\frac{\partial}{\partial u} h_i^0(f(x,u))$. If this vector is nonzero at an open and dense subset O_i of $M \times U$ we set $\rho_i = 0$, otherwise we continue with the function $h_i^1(x) = h_i^0(f(x,u))$. In this way the number ρ_i - if it exists - determines the inherent delay between the inputs and the i-th output; that is given a point $(x,u) \in O_i \subset M \times U$, then we can "see" the input $u(0) = u$ after $\rho_i + 1$ steps in the i-th output. In case none of the iterated functions $h_i^{k+1}(x,u) = h_i^k(f(x,u))$ depends on u we define $\rho_i = \infty$. Note that the above definition of a characteristic number is consistent with the continuous-time analogues, Definitions 8.7 and 13.23. Next, assume the analytic discrete-time system (14.3,4) has finite characteristic numbers $\rho_1,\dots,\rho_m$. Then define the decoupling matrix of (14.3,4) as (see also (8.25) and (13.87))

$$A(x,u) = \begin{bmatrix} \frac{\partial}{\partial u} h_1^{\rho_1}(f(x,u)) \\ \vdots \\ \frac{\partial}{\partial u} h_m^{\rho_m}(f(x,u)) \end{bmatrix} \tag{14.55}$$

where, as before $h_i^0(x) = h_i(x,u)$, $h_i^1(x) = h_i^0(f(x,u))$ etc., $i \in \underline{m}$.

In analogy with Definitions 8.3 and 13.23 we say that the system (14.3,4) is locally strongly input-output decoupled about (x_0, u_0) if the decoupling matrix $A(x,u)$ is a diagonal nonsingular matrix for all (x,u) on a neighborhood of (x_0, u_0). Since we only deal with *strong* input-output decoupling in this chapter we will drop for simplicity the adjective "strongly" *throughout*. The *regular static state feedback input-output decoupling problem* now consists of finding a regular static state feedback $u = \alpha(x,v)$ such that the feedback modified system

$$x(k+1) = f(x(k),\alpha(x(k),v(k))) =: \tilde{f}(x(k),v(k)) , \tag{14.56}$$

$$y(k) = h(x(k),\alpha(x(k),v(k))) =: \tilde{h}(x(k),v(k)) , \tag{14.57}$$

is input-output decoupled. As in Section 14.1 we will give a local solution of this problem in a neighborhood of an equilibrium point (x_0, u_0), i.e. $f(x_0, u_0) = x_0$. This will be referred to as the local regular static state feedback input-output decoupling problem around (x_0, u_0). We have the following result, which parallels Theorems 8.13 and 13.24.

Theorem 14.14 *Consider the analytic system (14.3,4) with finite characteristic numbers* $\rho_1, \ldots, \rho_m$ *defined on open and dense subsets* $O_1, \ldots, O_m$ *of* $M \times U$. *Let* (x_0, u_0) *be an equilibrium point of (14.3) belonging to* $O_1 \cap \ldots . \cap O_m$. *Then the local regular static state feedback input-output decoupling problem is solvable around* (x_0, u_0) *if and only if*

$$\operatorname{rank} A(x_0, u_0) = m . \tag{14.58}$$

Proof First assume (14.58) holds and let $h_i(x_0, u_0) = y_{0i}$, $i \in \underline{m}$. Consider the set of equations

$$\begin{bmatrix} h_1^{\rho_1}(f(x,u)) \\ \vdots \\ h_m^{\rho_m}(f(x,u)) \end{bmatrix} - \begin{bmatrix} y_{01} \\ \vdots \\ y_{0m} \end{bmatrix} = \begin{bmatrix} v_1 \\ \vdots \\ v_m \end{bmatrix} . \tag{14.59}$$

By the Implicit Function Theorem there exists locally about (x_0, u_0) a solution $u = \alpha(x,v)$ of (14.59), with $u_0 = \alpha(x_0, 0)$ and $\frac{\partial \alpha}{\partial v}(x,v)$ invertible. Since $y_i(k+\rho_i+1) = h_i^{\rho_i}(f(x(k),u(k)))$, $i \in \underline{m}$, the above defined feedback solves the input-output decoupling problem around (x_0, u_0). Next, suppose $u = \alpha(x,v)$ solves the input-output decoupling problem around (x_0, u_0). Consider the feedback modified system (14.56,57) for which the i-th input

v_i only influences the i-th output, implying that (14.56,57) has a nonsingular diagonal decoupling matrix. Let us write $\tilde{A}(x,v)$ as the decoupling matrix of (14.56,57). Then the decoupling matrix (14.55) and $\tilde{A}(x,v)$ are related via

$$\tilde{A}(x,v) = A(x,u)\Big|_{u=\alpha(x,v)} \cdot \frac{\partial\alpha}{\partial v}(x,v) \tag{14.60}$$

in a neighborhood of (x_0,u_0). To see that (14.60) is true, we first note that the characteristic numbers $\rho_1,\ldots,\rho_m$ of the system (14.3,4) are invariant under the application of a regular static state feedback. Namely, let $i \in m$, and suppose $\rho_i = -1$. Then

$$\frac{\partial\tilde{h}_i}{\partial v}(x,v) = \frac{\partial}{\partial v}(h_i(x,\alpha(x,v))) = \frac{\partial h_i}{\partial u}(x,u)\Big|_{u=\alpha(x,v)} \cdot \frac{\partial\alpha}{\partial v}(x,v) \tag{14.61}$$

which already establishes (14.60) in case $\rho_1=\ldots=\rho_m = -1$. Now suppose $\rho_i \geq 0$. Then the functions $h_i^0(x) = h_i(x,u)$, $h_i^1(x) = h_i(f(x,u)),\ldots,h_i^{\rho_i}(x) = h_i^{\rho_i-1}(f(x,u))$ are only depending upon x and not on u. Obviously, applying a state feedback $u = \alpha(x,v)$ does not alter these functions. Now the i-th row of the matrix $A(x,u)$ is given as

$$\frac{\partial}{\partial u} h_i^{\rho_i}(f(x,u)) = \frac{\partial h_i^{\rho_i}}{\partial x}(f(x,u)) \cdot \frac{\partial f}{\partial u}(x,u), \tag{14.62}$$

whereas the i-th row of $\tilde{A}(x,v)$ equals

$$\frac{\partial}{\partial v} h_i^{\rho_i}(\tilde{f}(x,v)) = \frac{\partial h_i^{\rho_i}}{\partial x}(f(x,u))\Big|_{u=\alpha(x,v)} \cdot \frac{\partial f}{\partial u}(x,u)\Big|_{u=\alpha(x,v)} \cdot \frac{\partial\alpha}{\partial v}(x,v). \tag{14.63}$$

Combining the expressions (14.62) and (14.63) exactly yields (14.60). Since in (14.60) the matrix $\tilde{A}(x,v)$ is nonsingular, the result (14.58) is immediate. □

Next we discuss as in Chapters 8 and 13 the *dynamic state feedback input-output decoupling problem*. For the analytic discrete-time nonlinear system (14.3,4) a dynamic state feedback or dynamic precompensator is given as

$$\begin{cases} z(k+1) = \varphi(z(k),x(k),v(k)) \\ u(k) = \psi(z(k),x(k),v(k)) \end{cases} \tag{14.64}$$

with $z \in \mathbb{R}^q$ and $v \in U$, denoting the new inputs. The dynamic state feedback input-output decoupling problem consists of finding - if possible - a precompensator (14.63) such that the closed loop system

$$\begin{cases} x(k+1) = f(x(k),\psi(z(k),x(k),v(k))) \\ z(k+1) = \varphi(z(k),x(k),v(k)) \\ y(k) = h(x(k),\psi(z(k),x(k),v(k))) \end{cases} \tag{14.65}$$

is input-output decoupled. As before we will produce a local solution of this problem in a neighborhood of an equilibrium point (x_0,u_0). The essential tool in obtaining a local solution is a discrete-time version of the dynamic extension algorithm as was discussed in Chapters 8 and 13.

Algorithm 14.15 (Discrete-time Dynamic Extension Algorithm) Consider the square analytic system (14.3,4).

Step 1 Assume the characteristic numbers of (14.3,4) are defined in a neighborhood containing (x_0,u_0) and denote them as $\rho_1^1,\ldots,\rho_m^1$. Write the decoupling matrix (cf. 14.55)) as $D^1(x,u)$. Let $r_1(x,u) := \operatorname{rank} D^1(x,u)$. By the analyticity of (14.3,4), $r_1(x,u)$ is constant, say r_1, on an open and dense subset B_1 of $M \times U$. Assume $(x_0,u_0) \in B_1$. Reorder the output functions $h_1,\ldots,h_m$ in such a way that the first r_1 rows of D^1 are linearly independent. Write $h^0 = (h_1,\ldots,h_{r_1})$, $\bar{h}^0 = (h_{r_1+1},\ldots,h_m)$ and correspondingly $y = (y^1,\bar{y}^1)$. Let $h(x_0,u_0) = y_0$ and consider the equations

$$\begin{bmatrix} h_1^{\rho_1}(f(x,u)) \\ \vdots \\ h_{r_1}^{\rho_{r_1}}(f(x,u)) \end{bmatrix} - \begin{bmatrix} y_{01} \\ \vdots \\ y_{0r_1} \end{bmatrix} = \begin{bmatrix} v_1 \\ \vdots \\ v_{r_1} \end{bmatrix} . \tag{14.66}$$

By the Implicit Function Theorem we can locally solve around (x_0,u_0) r_1 components of u as a function of x and v. So after a possible relabeling of the input components $(u_1,\ldots,u_m)$ we obtain a "partial" state feedback $u^1 = \alpha^1(x,v^1)$ with $u_0^1 = \alpha^1(x_0,0)$, where $u^1 = (u_1,\ldots,u_{r_1})$ and $v^1 = (v_1,\ldots,v_{r_1})$. Leaving the remaining inputs $\bar{u}^1$ unchanged the system map (14.34) transforms into $f^1(x,v^1,\bar{u}^1) := f(x,\alpha^1(x,v^1),\bar{u}^1)$ and $\bar{h}^1(x,v^1) := \bar{h}^0(x,\alpha^1(x,v^1))$. Renaming v^1 as u^1 we obtain the system

$$\begin{aligned} x(k+1) &= f^1(x(k)),u^1(k),\bar{u}^1(k)) , \\ \bar{y}^1(k) &= \bar{h}^1(x(k),u^1(k)) , \end{aligned} \tag{14.67}$$

with inputs $\bar{u}^1$, outputs $\bar{y}^1$ and where u^1 is regarded as a set of parameters. By the above chosen "partial" feedback the first r_1 outputs are already input-output decoupled, see Theorem 14.14.

Step $\ell + 1$ Assume we have defined a sequence of integers $r_1,\dots,r_\ell$ with $q_\ell := \sum_{i=1}^{\ell} r_i$, and we have a block partitioning of h as $(h^1,h^2,\dots,h^\ell,\bar{h}^\ell)$ with $\dim h^i = r_i$, $i \in \underline{\ell}$, and correspondingly $y = (y^1,\dots,y^\ell,\bar{y}^\ell)$. Furthermore we have obtained a system

$$\begin{aligned} x(k+1) &= f^\ell(x(k),\, \tilde{U}^{\ell-1}(k),\, u^\ell(k),\, \bar{u}^\ell(k)) \,, \\ \bar{y}^\ell(k) &= \bar{h}^\ell(x(k),\, \tilde{U}^{\ell-1}(k),\, u^\ell(k)) \,. \end{aligned} \tag{14.67}$$

Here the (new) controls u are correspondingly split into $u = (u^1,\dots,u^\ell,\bar{u}^\ell)$ and $\tilde{U}^{\ell-1}$ denotes $U^{\ell-1}(k) = (u^1(k),\dots,u^{\ell-1}(k))$ together with suitable future inputs $u^i(k+s)$, $s > 0$, $i = 1,\dots,\ell-1$. The system (14.67) is regarded as a system with inputs $\bar{u}^{\ell-1}$, parametrized by $\tilde{U}^{\ell-1}(k)$ and $u^\ell(k)$. Denote the characteristic numbers of (14.67) by $\rho^{\ell+1}_{q_\ell+1},\dots,\rho^{\ell+1}_m$ and the decoupling matrix of (14.67) by $D^{\ell+1}(x,\tilde{U}^\ell,\bar{u}^\ell)$. From the analyticity of the system (14.67) the rank of this matrix is constant, say $r_{\ell+1}$ on an open dense subset $B_{\ell+1}$ of points $(x,\tilde{U}^\ell,\bar{u}^\ell)$. Assume (x_0,u_0) belongs to the projection of $B_{\ell+1}$ on to $M \times U$. Reorder the output functions $\bar{h}^\ell$ such that the first $r_{\ell+1}$ rows of $D^{\ell+1}$ are linearly independent, and write $q_{\ell+1} := q_\ell + r_{\ell+1}$, $h^\ell = (h_{q_\ell+1},\dots,h_{q_{\ell+1}})$, $\bar{h}^\ell = (h_{q_{\ell+1}+1},\dots,h_m)$ and similarly $y = (y^{\ell+1},\bar{y}^{\ell+1})$, $\bar{u}^\ell = (u^{\ell+1},\bar{u}^{\ell+1})$. Consider the equations

$$h_i^{\rho_i^{\ell+1}}(f(x,\tilde{U}^\ell,\bar{u}^\ell)) - y_{0i} = v_i \quad , \quad i = q_\ell + 1,\dots,q_{\ell+1} \,. \tag{14.68}$$

By the Implicit Function Theorem we can locally solve $r_{\ell+1} := q_{\ell+1} - q_\ell$ components of $\bar{u}^\ell$ as a function of x, $\tilde{U}^\ell$ and $(v_{q_\ell+1},\dots,v_{q_{\ell+1}})$. So after a possible relabeling of the input components of $\bar{u}^\ell$ we obtain a parametrized "partial" feedback $u^{\ell+1} = \alpha^{\ell+1}(x,\tilde{U}^\ell,v^{\ell+1})$. Leaving the other inputs $\bar{u}^{\ell+1}$ unaltered and renaming again $u^{\ell+1} := (v_{q_\ell+1},\dots,v_{q_{\ell+1}})$, we obtain a system of the form

$$\begin{aligned} x(k+1) &= f^{\ell+1}(x(k),\tilde{U}^\ell(k),u^{\ell+1}(k),\bar{u}^{\ell+1}(k)) \,, \\ \bar{y}^{\ell+1}(k) &= \bar{h}^{\ell+1}(x(k),\tilde{U}^\ell(k),u^{\ell+1}(k)) \,, \end{aligned} \tag{14.69}$$

where we set $f^{\ell+1}(x,\tilde{U}^\ell,v^{\ell+1},\bar{u}^{\ell+1}) := f^\ell(x,\tilde{U}^{\ell-1},\alpha^{\ell+1}(x,\tilde{U}^\ell,v^{\ell+1}),\, \bar{u}^{\ell+1})$ and $\bar{h}^{\ell+1}(x,\tilde{U}^\ell,v^{\ell+1}) := \bar{h}^\ell(x,\tilde{U}^{\ell-1},\alpha^{\ell+1}(x,\tilde{U}^\ell,v^{\ell+1}))$. Notice that the "partial" feedback defined above achieves input-output decoupling between $y^{\ell+1}$ and $v^{\ell+1}$.

Assuming that the open and dense subsets of the $(x,\tilde{U}^{\ell},\bar{u}^{\ell})$-spaces, $\ell = 1,2,\ldots$ project onto a neighborhood of (x_0,u_0) we obtain as in Algorithm 8.18 a finite list of integers $q_1,q_2,\ldots,q_k$ such that

$$0 < q_1 <\ldots\ldots< q_k = q_{k+1} \ , \tag{14.70}$$

and the integer $q^* := q_k$ will be called the *rank* of the system (14.3,4).

Remark 14.16 Although the discrete-time Dynamic Extension Algorithm as given here only works on a neighborhood of an equilibrium point (x_0,u_0), it can also be used at an arbitrary point $(x,u) \in M \times U$, provided the constant rank assumptions are met. The analyticity of the system guarantees that the rank q^*, detemined on a neighborhood of an equilibrium point (x_0,u_0), equals the rank of the system (14.3,4) on an open and dense subset of $M \times U$. Therefore we may refer to q^* as being the rank of the system (14.3,4), compare with Chapter 8, cf. (8.88).

Theorem 14.17 *Consider the square analytic system (14.3,4) around the equilibrium point (x_0,u_0). Suppose all assumptions made in Algorithm 14.15 are satisfied. Then the following two conditions are equivalent:*

(i) The dynamic state feedback input-output decoupling problem is locally solvable around (x_0,u_0).

(ii) The rank q^* of the system equals m.

Proof The proof of this result completely parallels the continuous-time result, cf. proof of Theorem 8.19, once we have observed that at each step in Algorithm 14.15 the vector $\tilde{U}^r(k)$ can be replaced by a suitable s-fold integrator defined for each of the inputs $u_1,\ldots,u_{q_r}$ in $\tilde{U}^r$. Namely let for $i = 1,\ldots,q_r$,

$$z_{ij}(k+1) = z_{i\,j+1}(k) \quad , \quad 1 \le j \le s_r \ , \tag{14.71a}$$

$$z_{is_r}(k+1) = w_i(k) \quad , \tag{14.71b}$$

where s_r stands for the largest value s for which some $u_i(k+s)$ appears in $\tilde{U}^r(k)$. Using (14.71a,b) the vector $\tilde{U}^r(k)$ can be replaced by the vector $(z_{11}(k),\ldots,z_{1s_r}(k),\ldots,z_{q_r1}(k),\ldots,z_{q_rs_r}(k),\ w_1(k),\ldots,w_{q_r}(k))$ and the result follows similarly as in Theorem 8.19. □

We terminate with an illustrative example.

Example 14.18 (see Example 14.7) The dynamic equations for a controlled closed economy have the form

$$\begin{cases} Y(k+1) = f_1(Y(k),R(k),K(k),W(k),G(k)) \\ R(k+1) = f_2(Y(k),R(k),K(k),M(k)) \\ K(k+1) = f_3(Y(k),R(k),K(k)). \end{cases} \tag{14.26}$$

As in Example 14.7 we assume we are working around an equilibrium point $(\bar{Y},\bar{R},\bar{K},\bar{G},\bar{M},\bar{W})$ and furthermore the exogenous variable $W(k)$ in (14.26) equals $\bar{W}$ for all k. Similarly as in Example 1.3, see (1.34) we consider outputs of the form

$$\begin{aligned} Q_1(k) &= Y(k) \ , \\ Q_2(k) &= h(\bar{W},Y(k),K(k)) \ . \end{aligned} \tag{14.72}$$

Note that both target variables Q_1 and Q_2 are not directly influenced by the control variables G and M, and so $\rho_i \geq 0$, $i = 1,2$. In order to determine ρ_1 and ρ_2 we make the following assumptions

$$\frac{\partial f_1}{\partial G}(\bar{Y},\bar{R},\bar{K},\bar{W},\bar{G}) \neq 0 \ , \tag{14.73a}$$

$$\frac{\partial h}{\partial y}(\bar{W},\bar{Y},\bar{K}) \neq 0 \ . \tag{14.73b}$$

Then $\rho_1 = \rho_2 = 0$ and the decoupling matrix of the system (14.26,72) is given as

$$A(Y,R,K,G,M,\bar{W}) = \begin{bmatrix} \frac{\partial f_1}{\partial G}(Y,R,K,\bar{W},G) & 0 \\ \frac{\partial h}{\partial Y}(\bar{W},Y,K)\cdot\frac{\partial f_1}{\partial G}(Y,R,K,\bar{W},G) & 0 \end{bmatrix} , \tag{14.74}$$

which clearly has rank 1 about the equilibrium point. According to Theorem 14.14 the system (14.26,72) is therefore not input-output decoupable via static state feedback. So we have to use the Dynamic Extension Algorithm to see whether or not the system is input-output decoupable by dynamic state feedback. As in step 1 of Algorithm 14.15 we introduce a "partial" state feedback as the inverse relation of

$$f_1(Y,R,K,\bar{W},G) - \bar{Y} = \tilde{G} \ , \tag{14.75}$$

yielding the input G as a state-dependent function of the new input $\tilde{G}$. As in the proof of Theorem 14.17 we introduce an s-fold delay on the input $\tilde{G}$, where s depends on the characteristic numbers in each step of the algorithm. In this specific example we will see that $s = 1$ suffices. The precompensated system (14.26) then reads as

$$\begin{cases} Y(k+1) = Z(k) \\ Z(k+1) = \hat{G}(k) \\ R(k+1) = f_2(Y(k),R(k),K(k),\overline{W}(k)M(k)) \\ K(k+1) = f_3(Y(k),R(k),K(k)) \end{cases} \tag{14.76}$$

where $\hat{G}$ and M are the new inputs of the system. Note that $\tilde{G}(k+1) = \hat{G}(k)$. Clearly, in the precompensated system (14.76,72) we already have input-output decoupling between the control $\hat{G}$ and the output Q_1. Next we determine the characteristic number of the second output with respect to the input M. We compute

$$Q_2(k+1) = h(\overline{W},Y(k+1),K(k+1) = h(\overline{W},Z(k),f_3(Y(k),R(k),K(k))) \ , \tag{14.77}$$

which implies $\rho_2^2 \geq 0$, and

$$Q_2(k+1) = h(\overline{W},Z(k+1),f_3(Y(k+1),R(k+1),K(k+1))) \ . \tag{14.78}$$

From (14.78) and (14.76) it follows that $Q_2(k+2)$ explicitly depends upon $M(k)$ provided the following assumption holds

$$\frac{\partial h}{\partial K}(\overline{W},\overline{Y},\overline{K}) \cdot \frac{\partial f_3}{\partial R}(\overline{Y},\overline{R},\overline{K}) \cdot \frac{\partial f_2}{\partial M}(\overline{Y},\overline{R},\overline{K},\overline{W},\overline{M}) \neq 0 \ , \tag{14.79}$$

or, equivalently,

$$\frac{\partial h}{\partial K}(\overline{W},\overline{Y},\overline{K}) \neq 0 \ , \ \frac{\partial f_3}{\partial R}(\overline{Y},\overline{R},\overline{K}) \neq 0, \ \frac{\partial f_2}{\partial M}(\overline{Y},\overline{R},\overline{K},\overline{W},\overline{M}) \neq 0 \ . \tag{14.80}$$

In case (14.73a,b), and (14.79) or (14.80) hold, the system (14.26,72) is locally dynamic input-output decouplable about the equilibrium point $(\overline{Y},\overline{R},\overline{K},\overline{G},\overline{M},\overline{W})$. Notice that the forementioned conditions for dynamic input-output decoupling imply that the discrete-time dynamics (14.26) is feedback linearizable about the equilibrium point, cf. the conditions (14.27a,b,c) of Example 14.7. □

Finally we note that, completely analogous to the continuous-time theory of Chapters 8 and 13, a discrete-time nonlinear system which can be locally input-output decoupled via either static state feedback or dynamic state feedback possesses, after applying such a decoupling feedback, a local "normal form". We leave the details for the reader (see Exercise 14.6).

Notes and References

This chapter has been devoted to a discussion of some problems for discrete-time nonlinear systems which already have been treated in the previous chapters in the continuous-time case. We have limited ourselves to some results which more or less parallel the continuous-time theory.

A different approach for studying discrete-time nonlinear systems is based upon discrete-time generating series, see e.g. [No],[MN1],[MN2],[MN3],[MNI],[Ko]. Polynomial discrete-time systems have been treated in [So1],[SR]. Invertible discrete-time nonlinear systems have been studied by associating with the system a family of vectorfields, see [FN],[JN],[JS]. In [Fl] discrete-time systems have been studied using tools from difference algebra. As already noticed, discrete-time nonlinear systems may appear as the discretization of a system in continuous-time. Various questions regarding the sampled-data representation of a continuous-time system have been studied, see e.g. [AHS],[Ch],[GS] for results on sampling of linear systems, and [AJLMS],[MN3],[So2] for their nonlinear versions.

The feedback linearization problem of Section 14.1 has been studied in [Gr4]. Other results about this problem may be found in [JA], [LAM],[LM1],[LM2]. Controlled invariant distributions for discrete-time nonlinear systems have been introduced in [Gr1],[Gr2], see also [MN2] and [Nij1]. A solution of the disturbance decoupling problem using controlled invariant distributions has been given in [Gr1], see [Ko] for another approach. A geometric approach to the input-output decoupling problem has been given in [Gr3],[GN]. The solution of Section 14.3 has been taken from [Nij2], see [MNI] for a similar treatment. The Examples 4.7 and 4.18 are based upon [WK] and have appeared in [Nij3],[Nij4].

[AHS] K.J. Aström, P. Hagander, J. Sternby, "Zeros of sampled systems", Automatica 20, pp. 31-38, 1984.

[AJLMS] A. Arapostathis, B. Jakubczyk, H.G. Lee, S.I. Marcus, E.D. Sontag, "The effect of sampling on linear equivalence and feedback linearization", preprint, 1989.

[Ch] C.T. Chen, **Linear System Theory and Design**, Holt, Rinehart and Winston, New York, 1984.

[FN] M. Fliess, D. Normand-Cyrot, "A group theoretic approach to discrete-time nonlinear controllability", Proc. 20th. IEEE Conf. Decision Control, San Diego, pp. 551-557, 1981.

[Fl] M. Fliess, "Esquisses pour une théorie des systèmes nonlinéaires en temps discret", Rend. Sem. Mat. Univers. Politechnico Torino, pp. 55-67, 1986.

[Gr1] J.W. Grizzle, "Controlled invariance for discrete-time nonlinear systems with an application to the disturbance decoupling problem", IEEE Trans. Aut. Contr. AC-30, pp. 868-874, 1985.

[Gr2] J.W. Grizzle, "Distributions invariantes commandées pour les systèmes nonlinéaires en temps discret", C.R. Acad. Sci. Paris, t.300, Série I, pp. 447-450, 1985.

[Gr3] J.W. Grizzle, "Local input-output decoupling of discrete-time nonlinear systems", Int. J. Contr. 43, pp. 1517-1530, 1986.

[Gr4] J.W. Grizzle, "Feedback linearization of discrete-time systems", in **Analysis and Optimization of Systems** (eds. A. Bensoussan, J.L Lions) Lect. Notes Contr. Inf. Sci. 83, Springer, Berlin, pp. 273-281, 1986.

[GN] J.W. Grizzle, H. Nijmeijer, "Zeros at infinity for nonlinear discrete-time systems", Math. Syst. Th. 19, pp. 79-93, 1986.

[GS] J.W. Grizzle, M.H. Shor, "Sampling, infinite zeros and decoupling of linear systems", Automatica 24, pp. 387-396, 1988.

[JN] B. Jakubczyk, D. Normand-Cyrot, "Orbites de pseudo-groupes de difféomorphismes et commandabilité des systèmes nonlinéaires en temps discret", C.R. Acad. Sci. Paris, t.298, Série I, pp. 257-260, 1984.

[Ja] B. Jakubczyk, "Feedback linearization of discrete-time systems", Systems Control Lett. 9, pp. 411-416, 1987.

[JS] B. Jakubczyk, E.D. Sontag, "Controllability of nonlinear discrete-time systems: a Lie-algebraic approach", Report SYCON-88-09, Rutgers Center for Systems and Control, 1988.

[Ko] U. Kotta, "The disturbance decoupling problem in nonlinear discrete time systems", Preprints IFAC-Symposium Nonlinear Control Systems Design, Capri, Italy, pp. 59-63, 1989.

[LAM] H.G. Lee, A. Arapostathis, S.I. Marcus, "On the linearization of discrete-time systems", Int. J. Contr., 45, pp. 1803-1822, 1987.

[LM1] H.G. Lee, S.I. Marcus, "Approximate and local linearizability of nonlinear discrete-time systems", Int. J. Contr., 44, pp. 1103-1124, 1986.

[LM2] H.G. Lee, S.I. Marcus, "On input-output linearization of discrete-time nonlinear systems", Syst. Contr. Lett. 8, pp. 249-260, 1987.

[MN1] S. Monaco, D. Normand-Cyrot, "Sur la commande non interactive des systèmes nonlinéaires en temps discret", in **Analysis and Optimization of Systems** (eds. A. Bensoussan, J.L. Lions) Lect. Notes Contr. Inf. Sci. 63, Springer, Berlin, pp. 364-377, 1984.

[MN2] S. Monaco, D. Normand-Cyrot, "Invariant distributions for discrete time nonlinear systems", Systems Control Lett. 5, pp. 191-196, 1984.

[MN3] S. Monaco, D. Normand-Cyrot, "Zero dynamics of sampled nonlinear systems", Systems Control Lett. 11, pp. 229-234, 1988.

[MN1] S. Monaco, D. Normand-Cyrot, I. Isola, "Nonlinear decoupling in discrete time", Preprints IFAC-Symposium Nonlinear Control Systems Design, Capri, Italy, pp. 48-55, 1989.

[Nij1] H. Nijmeijer, "Observability of autonomous discrete-time nonlinear systems, a geometric approach", Int. J. Contr. 36, pp. 867-874, 1982.

[Nij2] H. Nijmeijer, "Local (dynamic) input-output decoupling of discrete-time nonlinear systems", IMA J. Math. Contr. Inf. 4, pp. 237-250, 1987.

[Nij3] H. Nijmeijer, "On dynamic decoupling and dynamic path controllability in economic systems", J. Econ. Dyn. Contr. 13, pp. 21-39, 1989.

[Nij4] H. Nijmeijer, "Remarks on the control of discrete-time nonlinear systems", in **Perspectives in control theory** (eds. B. Jakubczyk, K. Malanowski, W. Respondek), Birkhäuser, Boston, 1989.

[No] D. Normand-Cyrot, "Théorie et pratique des systèmes nonlinéaires en temps discret", Thèse de Docteur d'Etat, Université de Paris-Sud, Centre d'Orsay, 1983.

[SR] E.D. Sontag, Y. Rouchaleau, "On discrete-time polynomial systems", J. Nonl. Anal. 1, pp. 55-64, 1976.

[So1] E.D. Sontag, **Polynomial response maps**, Springer, Berlin, 1979.

[So2] E.D. Sontag, "An eigenvalue condition for sampled weak controllability of bilinear systems", Systems Control Lett. 7, pp. 313-316, 1986.

[WK] H.W. Wohltmann, W. Krömer, "Sufficient conditions for dynamic path controllability of economic systems", J. Econ. Dyn. Contr. 7, pp. 315-330, 1984.

Exercises

14.1 Prove that Algorithm 14.2 is feedback invariant.

14.2 Consider the discrete-time nonlinear system $x(k+1) = f(x(k), u(k))$ with $(x,u) \in \mathbb{R}^n \times \mathbb{R}^m$. Assume that the dynamics are affine with respect to the given coordinates (x,u), i.e. $f(x,u) = \tilde{f}(x) + \sum_{i=1}^{m} \tilde{g}_i(x)u_i$. Let $z = S(x)$ be a coordinate transformation. Show that in general the system in the z-coordinates is not affine.

14.3 Consider the system (14.1) together with the sampled-data representation (14.3) with sampling time T. Prove that in general feedback linearization is not preserved under sampling, i.e. feedback linearizability of (14.1) does not imply feedback linearizability of (14.3).

14.4 Consider an analytic single input single output nonlinear system $\Sigma : \dot{\bar{x}} = f(\bar{x}) + g(\bar{x})u$, $\bar{y} = h(\bar{x})$. Let $\Sigma_d : x(k+1) = f(x(k),u(k))$, $y(k) = h(x(k))$ be the sampled-data representation of this system. Let ρ and ρ_d denote the characteristic numbers of Σ respectively Σ_d.

(a) Prove $\rho = \infty$ implies $\rho_d = \infty$.

(b) Prove that whenever ρ is finite then $\rho_d = 0$.

14.5 Consider the discrete-time nonlinear system (14.3,4) with finite characteristic numbers ρ_i. Prove that $\rho_i \leq n - 1$.

14.6 Consider a smooth discrete-time nonlinear system meeting the requirements of Theorem 14.14. Derive, in analogy with the continuous-time result (cf. Chapter 13, (13.91)), a local normal form for the input-output decoupled system.

14.7 Consider an analytic discrete-time nonlinear system (14.3,4) about an equilibrium point (x_0, u_0). Assume $m = p$. Let Σ_ℓ: $x_{k+1} = A\bar{x}_k + B\bar{u}_k$, $\bar{y}_k = C\bar{x}_k + D\bar{u}_k$, be the linearization of (14.3,4) around (x_0, u_0). Prove that under generic conditions the local regular static feedback input-output decoupling problem for the system (14.3,4) is solvable around (x_0, u_0), if and only if the static state feedback input-output decoupling problem for Σ_ℓ is solvable.

14.8 Consider the system (14.26,72) of Example 14.18. Interpret the exogenous variable $W(k)$ as a, not necessarily constant, disturbance for the system. Assume that $\frac{\partial h}{\partial W} = 0$ in equation (14.72). Show that the local Disturbance Decoupling Problem is not solvable for this system if $\frac{\partial f_3}{\partial Y} \cdot \frac{\partial f_1}{\partial W} + \frac{\partial f_3}{\partial R} \cdot \frac{\partial f_2}{\partial W} \neq 0$.

14.9 Prove Proposition 14.10.

14.10 Consider an analytic single-input single-output discrete-time nonlinear system $x(k+1) = f(x(k),\ u(k))$, $y(k) = h(x(k))$. Let ρ be the characteristic number of this system. Prove that $D^* = \ker dh^0 \cap \ker dh^1 \cap \ldots \ker dh^\rho$.

14.11 ([Ja]) Consider the discrete-time system $x(k+1) = f(x(k),u(k))$ with $(x,u) \in \mathbb{R}^n \times \mathbb{R}^m$ about an equilibrium point (x_0, u_0). Let $A(x,u) = \frac{\partial f}{\partial x}(x,u)$ and $B(x,u) = \frac{\partial f}{\partial u}(x,u)$. Define on $\mathbb{R}^n$ the u-dependent distributions $\Delta_i(x,u)$, $i \in \underline{n}$, via $\Delta_1(x,u) = A^{-1}(x,u)\operatorname{Im} B(x,u)$, $\Delta_{i+1}(x,u) = A^{-1}(x,u)\big(\Delta_i(f(x,u),u) + \operatorname{Im} B(x,u)\big)$, where $W = A^{-1}V$ is the pre-image of A, i.e. $AW = V$. Assume that rank $f_* = n$. Prove that the system is locally feedback linearizable about (x_0, u_0) if and only if the distributions $\Delta_1, \ldots, \Delta_n$ are of constant dimension and independent of u and $\dim \Delta_n = n$.

Subject Index